Mathematical Quantum Physics for Engineers and Technologists

The ACES Series on Computational and Numerical Modelling in Electrical Engineering

Andrew F. Peterson, PhD – Series Editor

The volumes in this series will encompass the development and application of numerical techniques to electrical and electronic systems, including the modelling of electromagnetic phenomena over all frequency ranges and closely related techniques for acoustic and optical analysis. The scope includes the use of computation for engineering design and optimization, as well as the application of commercial modelling tools to practical problems. The series will include titles for senior undergraduate and postgraduate education, research monographs for reference, and practitioner guides and handbooks.

Titles in the Series

K. Warnick, **"Numerical Methods for Engineering,"** 2010.

W. Yu, X. Yang and W. Li, **"VALU, AVX and GPU Acceleration Techniques for Parallel FDTD Methods,"** 2014.

A.Z. Elsherbeni, P. Nayeri and C.J. Reddy, **"Antenna Analysis and Design Using FEKO Electromagnetic Simulation Software,"** 2014.

A.Z. Elsherbeni and V. Demir, **"The Finite-Difference Time-Domain Method in Electromagnetics with MATLAB® Simulations 2nd Edition,"** 2015.

M. Bakr, A.Z. Elsherbeni and V. Demir, **"Adjoint Sensitivity Analysis of High Frequency Structures with MATLAB®,"** 2017.

O. Ergul, **"New Trends in Computational Electromagnetics,"** 2019.

D. Werner, **"Nanoantennas and Plasmonics: Modelling, design and fabrication,"** 2020.

K. Kobayashi and P.D. Smith, **"Advances in Mathematical Methods for Electromagnetics,"** 2020.

V. Lancellotti, **"Advanced Theoretical and Numerical Electromagnetics, Volume 1: Static, stationary and time-varying fields,"** 2021.

V. Lancellotti, **"Advanced Theoretical and Numerical Electromagnetics, Volume 2: Field representations and the method of moments,"** 2021.

S. Roy, **"Uncertainty Quantification of Electromagnetic Devices, Circuits, and Systems,"** 2021.

R. Araneo, **"Modeling for Electrical Engineering,"** 2022.

C. Gennarelli, F. Ferrara, R. Guerriero, and F. D'Agostino, **"Non-Redundant Near-Field to Far-Field Transformation Techniques,"** 2022.

M. Li and M. Salucci, **"Applications of Deep Learning in Electromagnetics: Teaching Maxwell's equations to machines,"** 2022.

E.K. Miller, **"Charge Acceleration and the Spatial Distribution of Radiation Emitted by Antennas and Scatterers,"** 2022.

Mathematical Quantum Physics for Engineers and Technologists

Volume 1: Fundamentals

Alireza Baghai-Wadji

ACES

The Institution of Engineering and Technology

Published by SciTech Publishing, an imprint of The Institution of Engineering and Technology, London, United Kingdom

The Institution of Engineering and Technology is registered as a Charity in England & Wales (no. 211014) and Scotland (no. SC038698).

The Institution of Engineering and Technology
Futures Place
Kings Way, Stevenage
Hertfordshire SG1 2UA, United Kingdom

www.theiet.org

British Library Cataloguing in Publication Data
A catalogue record for this product is available from the British Library

ISBN 978-1-83953-866-7 (Volume 1 hardback)
ISBN 978-1-83953-867-4 (Volume 1 PDF)
ISBN 978-1-83953-868-1 (Volume 2 hardback)
ISBN 978-1-83953-869-8 (Volume 2 PDF)

Typeset in India by MPS Limited

Cover Image: virtualphoto via Getty Images

To my wife
Elisabeth

Contents

About the author

Alireza Baghai-Wadji is a professor emeritus of electronics and computational engineering at the University of Cape Town, South Africa. He has occupied academic, executive and principal engineering consulting positions on five continents. Currently, he is promoting meso- and nanoscopic device modelling and simulation across a consortium of universities and high-tech industries. His contributions in mathematical physics include the diagonalization of PDEs, Green's function-induced generalized Dirac delta functions, algebraic-, exponential regularization techniques for taming infinities in near-fields, and the discovery of the Discrete Taylor Transform and Inverse Transform.

Preface

Quantum Physics, Engineering & Technology: The fabrication of miniaturized devices in nano-electronics, nano-photonics, and nano-plasmonics, the synthesis of application-tailored materials, the design of sophisticated quantum computing algorithms, the realization of quantum computers, all require that engineers and technologists have a working knowledge of quantum physics and, more importantly, feel confident to maneuver through the intricate calculations involved.

The vast majority of engineers and technologists who have not enjoyed formal training in quantum physics perceive the theory shrouded in a mist of inaccessible collection of mystical sounding notions, ideas and concepts, and consequently, feel discouraged from pursing it to their own satisfaction. Filled with the desire to utilize the Entanglement Phenomenon and understand the consequences of the Heisenberg Uncertainty Principle, some engineers end up getting entangled in the uncertainty itself, thus reaffirming their discouragement.

Studying the history of the development of new ideas in science and engineering, i.e., of rare occasions when breakthroughs take place, of eras during which new paradigms manifest themselves, of circumstances in which apparently something emerges out of nothing, of circumstances when synchronism arises out of disorder, is reassuring. By definition, creative ideas draw on elements of thought and abstract structures in the mind, which have little resemblance to what have existed before. Thereby, the degree of dissimilarity with existing concepts can serve as an adequate measure of their novelty and of the ingenuity of their creators.

Quantum theory deals with Objects and Phenomena to which observers are denied direct access, despite the fact that every bit of their physical existence is governed by quantum phenomena. No doubt this inability adds to the difficulty.

Quantum physics is not the only theory which presents itself as elusive. Examples are abound. Despite the fact that the Special Theory of Relativity, which only requires elementary mathematics and can be taught at high school, it is perceived as difficult, primarily due to the fact that its manifestation takes place at velocities which are utterly uncommon in daily lives, and thus inconceivable. Similarly, the General Theory of Relativity deals with such massive objects which are unthinkable to ordinary humans bound to the earth or even to those heroes who have been in regions in the proximity of earth. The mathematics involved in the General Theory of Relativity is admittedly more difficult than the mathematics used in the Special Theory of Relativity. However, the complexity of the required mathematics, the Differential Geometry, is presumably not the reason why the theory is perceived as difficult. Engineers active in geosciences, and the designers of satellites for maintaining GPS services learn to master the geometry of curved surfaces and the dynamics of objects moving on or around them.

Two more examples should help illuminating the main point here. The first one being the attributes Irrational, Imaginary, and Complex, all referring to archaic and misplaced denotations in today's undergraduate mathematics education, which cause unnecessary obscurity and resentment amongst contemporary learners. Viewed from the right perspective, and through the lenses of an experienced practitioner, these number concepts are instead exemplarily rational, real, and simple, and above all admirably beautiful, and immensely powerful in their utility. Nonetheless, these antiquated names are still unnecessarily in use to denote mathematical objects, despite the fact that they can be understood reasonably well, and with considerably less effort. The second example concerns the notions of Covariant and Contravariant Transforms, which contain the prefixes "Co-" and "Contra" of "variant" to something which is not explicated in the naming. This important concept will be necessarily and indirectly touched upon in this text. Naming concepts is contextual and extraordinarily crucial. This is particularly the case, if names carry along negative connotations, or even more, if they cause wrong outdated associations. A possible alternative approach would be to start with the current state-of-the-art understanding of the subject matter, and then describe, or even explain the historical development of the ideas leading to present days' perception.

Coming back to the main object of the inquiry in this text, i.e., the Quantum Physics, a few more ideas are in place to ensure that the reader obtains a proper appreciation of this powerful theory, which is undoubtedly the pinnacle of humankind's collective achievement. (i) There are still several competing interpretations of Quantum Physics in existence today. (ii) These historically highly charged intellectual debates should not be confused with the fact that the theory itself is astonishingly accurate and possesses unparalleled predictive powers. Irrespective of whatever technological application it has been used for, or whichever experimental test it has been exposed to, quantum theory has survived all stringent scrutinies. Its incompatibility with General Theory of Relativity should not be of any concern relevant to the present discussion. And this positive outlook, with highest likelihood, is expected to persist and support engineers and technologist in pursing their design objectives, in foreseeable future. (iii) Consequently, the interpretation of Quantum Physics must be separated from its powerful predictive powers. Engineers and technologists cannot effort to ignore the availability of a jewel in their toolbox and wait until a more satisfactory answer has been found to the ontological nature of the Entanglement Phenomenon, Superposition Principle, or the Collapse of Wavefunctions, and the Measurement Problem. The theory of Quantum Physics exists and it can be utilized. Engineers and technologists need to master an adequate knowledge of quantum theory as a tool to more confidently unleash their design capabilities and manufacturing abilities. (iv) Quantum Physics and the Special Theory of Relativity must be demystified and become an integral part of undergraduate engineering curricula, in particular and foremost, in electrical engineering departments. Thereby, the required mathematics is no more difficult than the mathematics which is ordinarily taught at contemporary universities. All what is needed is a slight change in storytelling. The current text is just an initial experiment to this purpose. (v) This text does not include the Special Theory of Relativity. Nonetheless, an insight gained from author's innumerable classroom experimentations deserves mentioning. Many if not exclusively all attempts to explain the Special Theory of Relativity involve trains, lifts, space capsules, with caricature depictions of a person being carried inside those vehicles. Objects from daily routine are employed to explain phenomenae which defy themselves from daily experiences. The tale of a few featureless dimensionless colored points, individually, pairwise, or assembled in small groups, serving as actors and speeding around in an otherwise empty space might be more appealing and perhaps more naturally conducive to the mind.

The many strategies employed in conceptualizing the current text require additional explanation. Thoughts in the remaining part of the Preface as well as throughout the text have been itemized, they can be regarded as Questions and Answer of two friends walking in a mesmerizingly beautiful natural surrounding and describing what they observe (the overwhelming majesty of the nature's wonders in the Botanical Garden on Table Mountain in Cape Town comes to mind). The questions and answers are communicated by sentences, which necessarily leave out certain important nuances. The human language seems to be inadequate of conceptualizing, accounting for, and communicating all experienced contents at once. The missing contents are subsequently formulated in terms of Remarks. In some rare occasions, significant contents have been framed as Lemmas, which are less pretentious than Theorems. In a few places in the text, the reader will experience repetitions. On these occasions the reader is strongly encouraged to continue the reading and reinforce what they have learned. It should be emphasized that repeated passages are never exact copies of the preceding instances. The careful and patient reader will discover important subtleties which imperfect repetitions carry with themselves.

Celebrating Engineering and Technology: This text is to celebrate the awe-inspiring achievements of fellow engineers and technologists by assisting them to more easily acquire and employ the works of fellow applied mathematicians and theoretically inclined scientists who have the desire to test their ideas. The very identification of Engineers and Technologists designates groups of professionals who are concerned with accuracy, precision and rigor in designing, manufacturing, and testing of the products they create. A mere reference to their breathtaking achievements in recent decades should suffice to realize that there is no end in sight to their stunningly creative accomplishments. The group of Applied Mathematicians and Theoretically Minded Scientists, on the other hand, refers to leagues of professionals who are often interested in the implementation of their ideas to solve real-world problems. Here too, the impact of Mathematicians' and Computer Scientists' engagements in proving deep theorems and designing powerful algorithms command earnest admiration and deep appreciation. The prime objective in this text is to contribute to building bridges over existing or perceived gaps between these two groups of professionals. Along the way opportunities arise to scrutinize the process itself. It is up to the reader's judgment whether or not this pretentious goal has been achieved. If the author has been successful, it is only because of his meticulous use of several powerful techniques in the common toolkits of philosophers, mathematicians, engineers, and artists. These techniques, as the author interprets and utilizes them, are briefly sketched further below.

The Envisioned Readership: The reader is an engineer, or a technologist, a senior undergraduate, a junior post-graduate, a researcher in engineering, computer science, or in applied mathematics with an interest in engineering. The reader is an active learner, who has set out to acquire a solid understanding of the mathematical intricacies in Quantum Physics. The reader is eager to use pencil and paper and tries to repeat and invent alternative solutions by themselves, enjoys adding comments and remarks into the margins of the book. The reader has previously found the mathematics and the formalism in Quantum Physics daunting or difficult, and thus genuinely appreciates going though detailed explanations, to understand every facet of the story satisfactorily.

Author's World View: The author firmly believes that every sophisticated thought that has been formed by human beings can in principle be communicated to everyone who has the urge to grasp the content of the realized thought. He believes that complex thoughts can be broken down into digestible pieces, while he is a holistic thinker. He is convinced that practice and repetition lead to mastery and that there is no mastery without devotion. He is an engineer by training and a self-taught applied mathematician. He has been a devote learner of philosophy, mathematical logic, and has a weakness for foundational questions in physics and mathematics. Throughout his career he has been working with engineers and technologists and has tried to formulate their technical problems mathematically and contribute to their solutions. He highly values the significance of verbalization and articulation of problems and their solutions, as effective means for weeding out inconsistencies, and capturing what is left out.

What Does the Book Contain: The title of the Book reveals that the mathematics in Quantum Physics is expected to be prepared and presented to engineers and technologists. Engineers learn functions, transforms, linear algebra, and favorably work with matrices. Some specialize in solving partial differential equations arising in boundary value problems, and others like to think in terms of iterative feedback, control theory, and states. There is increasing desire amongst engineers to know more about Quantum Physics, and understand what is going on in small scales. There is a sense among engineers and technologists that the Schrödinger cat, entanglement, spooky actions at a distance, the notion of abstract Hilbert Space, and Dirac Bra-Ket notation, all allude to the fact that the mathematics underlying Quantum Physics is different than what they have learned. The ambition of this book is to show that this view is not entirely correct. The book utilizes what engineers and technologists already know and guides them step-by-step to master new mathematical tools which they require. The book is the first volume in a series of works that develop mathematical techniques as they are required to solve next generations of nano-electronic, nano-photonic, nano-plasmonic devices, and synthetic materials. The book challenges several deeply-rooted believes and proposes a new School-of-Thought based on obviousness, clarity, recognizing recurring patterns, avoiding out-dated notions and concepts, which have hampered students' understanding and possibly discouraged them to pursue rigorous applied and engineering mathematics. There is no need to introduce designations, e.g., imaginary unit, complex numbers, covariant and contra-variant coordinate transformations. On the other hand, there are the notions of planewave and the Dirac delta function, which are pervasive in electrical engineering and technical physics, and yet their meanings have not always been communicated deservedly clearly. The concepts of planewave and the Dirac delta function will be dissected meticulously, examined critically, and presented exhaustively. The book develops a handful of techniques to translate abstract mathematical ideas into the language of engineers and technologists. This process extends the existing vocabulary and refines the grammar in use.

No Interpretation of Quantum Physics: The book does not take any position in interpreting the foundational questions. It refrains from reviewing the current competing interpretations. These topics can be read in nearly every popular work on Quantum Theory. This author is convinced that the basic obstacles engineers and technologists encounter are of mathematical nature, and due to their unfamiliarity with overly abstract and encrypted symbols. It is immaterial to which School-of-Interpretation they subscribe, if they can confidently solve the Schrödinger Equation and apply the Density Functional Theory to their nano-scopic devices and design materials with specified properties, their objectives have been achieved. This book promises to prepare the reader exactly to those ends. It is just an enabling start. Along the way the reader sharpens their understanding of the mathematical tools and gets encouraged to design and develop customized tools in mathematics, as they might be needed. The latter creative scenario is the ideal objective of the book.

Exploiting Reductive and Deductive Reasoning: The attempt to reduce statements to a few independent Axioms and identify a small number of consistent Rules of Inference are the hallmarks of Reductive and Deductive Thinking, which are indispensable devices in mathematicians' as well as engineers' tool boxes. Starting from statements and breaking them down into axioms (analysis), and conversely, beginning from the axioms and constructing theorems (synthesis) allow mathematicians to gain access and create deep insights into the abstract mathematical machinery they create. Engineers do the same thing when they talk about various transforms and the corresponding inverse transforms. Mathematicians' world views undergo a metamorphosis in the course of collaborating with engineers and technologists. They start seeing the world through the lenses of engineers. The Mathematicians' abstract universe in which they operate takes on new shapes and accommodates new unexpected objects with rich and versatile features previously unknown to them, neither to their fellow engineers. Engineers and technologists proceed and experience similarly, even though their tools and approaches are substantially different. Engineers think in terms of elementary building blocks and set up rules for creating complex structures with prescribed functionality. Often, if not as a rule, they must operate under stringent conditions, time-, material-, energy-, and human resources being usually amongst the most pressing ones. While many of the constraints are real, some are due to inaccessibility and unfamiliarity with existing theoretical methods. The mere ability to formulate, articulate, express, and relay the experienced constraints to fellow mathematicians often leads to new insights and discoveries. The ability to verbalize and articulate technical problems becomes as important as honing the skills of Reductive and Deductive Thinking.

The Omnipresence of Inductive Reasoning: Another technique that mathematicians employ is Inductive Reasoning. By demonstrating that a statement is valid for a few Initial Induction Steps and thereby refining their intuition, mathematicians progress to formulating an Induction Hypothesis, and then they attempt to proof it. Even though not explicitly, surely however implicitly, engineers work inductively too. Observing that certain things behave in certain ways, they form an intuition for the existence of certain regularities, and then draw conclusions. In contrast to mathematicians' idealized world, engineers operate in rapidly changing environments which prompt them to frequently modify or even redesign their products. Nonetheless, there is no scape from Inductive Thinking, which seems to be ingrained in the fabric of human brains and thus, inescapably determines how the world is perceived. In mathematics, this is known as the Proof by Mathematical Induction. The Formal Taylor Series Expansion and the Proof by Mathematical Induction are two of the pillars this text is built upon. The fact that these universal tools can be easily grasped and employed catapults them to stardom. They are ubiquitous throughout the present exposition.

Trial and Removal of Error: Finally, the method of Trial and Removal of Error, with a strong emphasis on conscious Removal, should be hailed as the archetypical technique of learning, sharpening one's skills, and generally cultivating thinking. Every child knows it naturally, and some visionaries in education and industry have realized the significance of the process. Many optimization, learning, and iterative algorithms tacitly or explicitly implement this concept or variations thereof. In developing many of the ideas in this text countless experiments were carried out and important lessons learned. At times, new ideas and relationships emerged, sometimes due to concerted directed uncompromising efforts, and in rare cases serendipitously. It is highly probable that a trained mind is tuned to more easily See, Capture, and Benefit from unexpected virtuous events, in the course of experimentation. Ordinarily, the "Failed" experiments are discarded and the "Successful" ones are polished and presented. This obscures the line of reasoning and distorts the reality of the Process leading to the resulting progress. The reader should be reminded that the seamless continuity of arguments is based upon countless experimentations. Whenever, unexpectedly insightful passages are encountered in the text, the reader can confidently assume that they are the results of numerous experimentations. While carrying out the process of Trial and Removal of Error one should resist that the steps are done automatically. Consciously being aware of minute changes in trial steps and diligent diagnosis of outcomes are key to the process. Without any doubt Trial and Removal of Error, which means experimentation combined with the habit of critically scrutinizing the results, and weeding out the undesirable outcomes, should be embraced as one of the most empowering techniques in creativity and problem solving.

The Roles of Epistemology and Ontology: It is not the purpose of this text to promote or demote any positions in Epistemology (philosophy of knowledge) or Ontology (metaphysics of being). Even though a healthy dose of exposure to them early on in the intellectual and academic upbringing of engineers and technologists would probably be beneficial. The primary intention is here to sensitize the reader to these concepts, as far as they are relevant to the pursuit of knowledge. Two names deserve to be mentioned representatively, Karl Popper and Thomas Kuhn. On the one hand, Popper's writings on building conjectures, refining one's skills to refute and falsify statements, maybe thought of sources for sharpening engineers' and technologists' thinking abilities. Popper's view fine tunes one's argumentative and reasoning powers, and promotes scientific thinking. On the other hand, Kuhn's writings on the continual accumulation of knowledge (collectively or individually), leading to the emergence of new paradigms, describe an enabling dynamic process: one becomes aware of one's own progress in understanding and conceptualizing the world, and assumes an active role in becoming the architect of one's own intellectual capabilities. Very often, Popper and Kuhn are contrasted in their views. A better approach might be to view Popper's work as a method for reasoning and understanding how to work critically and scientifically. And to consider Kuhn's work as an explanation of the dynamics of scientific progress, the way how new paradigms force the emergence of new theories. Following this interpretation, Popper's and Kuhn's views are complementary. Consequently, both theories can be pursued simultaneously and incessantly.

Interest in the Foundations and Fundamental Questions: Many research supervisors in engineering and technology schools tend to discourage their students (perhaps wisely) from pursuing fundamental research questions. It is generally perceived that the pursuance of fundamentals is the privilege of tenured academics or researchers in dedicated research laboratories. While this view might be correct, it is certainly not advisable to take Foundational Research Questions out of the vocabulary of aspiring researchers. The complete opposite is true. The desire for answering foundational questions must be nurtured early on in engineers and technologists education and recognized as a deriving force for creativity.

Creative Powers Unleashed Through Bi- and Multilingualism: This author has been working in the intersection of a few related fields including mathematics, physics, and engineering, across several cultures, philosophies, and schools of thought. Perhaps circumstantially and inevitably, emphasizing Clarity of Thought, nurturing the notion of Obviousness, and understanding Simply-by-Inspection have become his uncompromising guiding principles in pursuing and communicating knowledge. In due course, several observations have been made: engineers wish to know more mathematics and acquire the necessary skills to develop well-structured computational recipes, understand important algorithms designed by computer scientists, and follow the chain of arguments of mathematicians in proving theorems. On the other hand, mathematicians, applied and computational mathematicians anyway, wish to get inspiration from the engineering and technology laboratories; they find deep satisfaction in their theories and algorithms being successfully tested in real-life conditions. Often, establishing artificial, facilitative links between the two Camps do not work effectively. More importantly, in many instances just interpreting objects between the languages $\mathcal{L}_{math}$ and $\mathcal{L}_{eng}$ does not suffice.

Better results can be achieved if members of each community sense the urge to reach out to members of other groups and learn to speak their language. Being able to formulate thoughts in other languages and appreciate the wisdom of other learned cultures, expressed in their native languages, open doors and boost the creative powers immeasurably.

An even more powerful approach is the promotion of a third category of professionals who aspire to be conversant in both $\mathcal{L}_{math}$ and $\mathcal{L}_{eng}$. The need for Multilingual professionals is imminent, as advances in nano-bio-technology, learning-and-thinking machines, system-of-systems, and other technological developments herald. The academia and the high-tech industries are called upon to respond to this need, by rethinking the way how to convey the accumulated knowledge to the next generation of Universal Thinking Specialists. There is a sense that Atomistic, Black-Box reductive thinking should give way to a more holistic-style of conceptualization which is leaned toward the Emergent Phenomenae and Complexity. That is, a School-of-Thought fit for the 21st century.

The Communication Style in This Book: The style in the text is axiomatic throughout the discourse. Starting with an ultimate minimum of givens, the ideas are deduced and concepts developed step-by-step. The lengthy derivations and the many Remarks dispersed in the text are meant to nurture the reader's perception in scrutinizing the chain of arguments, and understanding the relationships simply-by-inspection. Despite proclaiming earlier the need for holistic thinking in the philosophy of problem solving, the Idea of Black Boxes and Modules cannot or should not be entirely abandoned in advanced engineering. The ideas and concepts are introduced in the form of Solved Problems. In each case, the statement of the Problem represents a box, and the proposed Solution corresponds to the opening of the box and peering into the box for the purpose of illuminating its inner functioning. Detailed Solutions of Problems should guide the intuition of the reader and make the contents of Problems obvious and render them self-explanatory. Actively going through the Solution is reassuring and prompts the reader to come up with their own way of solving the problem. Problems and their Solutions are meant to emulate an active conversation with the reader. The many Remarks anticipate possible questions by the reader.

A New School-of-Thought: Every meaningful, well-formed thought, irrespective of its depth and scope, must be communicable. The uncompromising attempt to clarify thoughts, ideas, and notions, and render them obvious is itself a powerful source of creativity which leads to discovery and unexpected points of views. Engineers and technologists will be enabled to do more with less effort, and more easily to connect with mathematicians and theoretical physicists.

Anything New in This Book: The book is the first outcome of an ambitious project which sets out to build from scratch a minimum number of tools necessary and sufficient for a critical understanding of the Mathematical Quantum Physics. It resorts to and substantially refines the tools that engineers and technologists are already familiar with and routinely employ. The starting points are the infinite set of monomials $\{1, x, x^2, \cdots\}$ which can arguably be regarded as the archetypical basis for synthesizing functions. By choosing the monomials as the preferred basis, several goals have been achieved in one stroke. The preeminent significance of the Position Variable x, and thereof the monomials x^n, in Quantum Physics have been emphasized right from the beginning. In the course of discussing the synthesis of functions from monomials the notion of derivative – the differential operator d/dx emerges automatically. The Formal Taylor Series Expansion has been introduced and praised as a universal tool in understanding the Mathematical Quantum Physics. The wide-ranging implications of the Formal- as well as Practical Taylor Series Expansion leading to the Taylor Transform and Inverse Transform (TTIT) have been addressed in two book chapters elsewhere. An independent book wholly dedicated to TTIT is in completion. The trigonometric sine-, cosine-, and exponential functions are introduced as solutions to specific Ordinary Differential Equations, subject to judiciously chosen boundary conditions, following the tradition of introducing Special Functions in Mathematical Physics. The results dispel the uneasy connotations of imaginary, and complex numbers, and replaces them with real and simple mathematical objects. This insight will also enable the reader to enter the world of Geometric Algebra if they need to do so. The Geometric Algebra is not treated in this text. Along with Wannier functions, spinors, and the Chern number, they constitute the contents of an independent Volume. After demystifying the i (the imaginary unit), it is used in the text with the conviction that the readers associate them with the process of rotation in the plane rather than with something imaginary and obscure. These considerations lead to the notion of the planewave. Working with planewaves is second nature to engineers and technologists, in particular the electrical engineers, who most likely constitute the main group of readers. Planewaves involve the position-, and time variables, along with the corresponding dual variables the wavenumber and the frequency. Planewaves are periodic in space as well as in time. The periodicities are characterized, respectively, by the wavelength λ and the wave duration (period) T. The wavenumber specifies the number of wave undulations (the number of λs) in a chosen unit interval of length. The frequency is a measure of how often the wave oscillates in a chosen unit interval of time. Engineers and technologists are also familiar with the facts that the wavenumber is associated with the derivative with respect to the position variable, and that the frequency is associated with the derivative with respect to the time variable, of the corresponding planewave. De Broglie and Einstein proposed a linear relationship between the wavenumber and the linear momentum, respectively, frequency and the total energy. Being cognizant of these two simple, yet ingenious and groundbreaking hypotheses, fully suffices to understand the Annihilation- and Creation Operators, the Commutator Algebra, and the rest of the present Volume 1, in addition to the Hamiltonian function (total energy of a system), the Hamiltonian Operator, Quantum Harmonic Oscillator, the Perturbed Quantum Oscillator, the Squeezed States, the Schrödinger Picture, the Heisenberg Picture, the Interaction Picture, the Quantum Electrodynamics, and the Feynman Path Integral in the Volume 2.

The Contents of the Book: The book is divided into four comparatively large chapters, and each chapter into several sections. The ideas are communicated in terms of Problems and their Solutions, along with countless Remarks which have been disseminated throughout the book. The idea of communicating the contents in terms of Problems and their Solutions embedded into a continuous storytelling process is to identify and demarcate contents which can stand alone. Remarks additionally comment on certain features in the Solutions which needed to be excluded. The additional features are not necessarily parts of the Solutions. Remarks are statements concerning the method that has been employed in the Solution, or aim at clarifying noteworthy consequences of the Solution. Each chapter starts with a brief guide through the chapter and completes the discussion with concluding remarks. Each chapter has its own References. The References are exclusively from the author's private library. They are representatives of a considerably larger number of books which are available to the reader. The brief guides through the chapters are collated into the Introduction to provide an overview of the contents.

Acknowledgments: Parts of this book, in various states of its development, have been instructed as short courses at various IEEE (Institute of Electrical and Electronics Engineers) and ACES (Applied Computational Electromagnetics Society) Conferences. I also had the privilege to carry out research and teach at universities around the globe on five continents: Vienna University of Technology (Austria), California University Irvine (USA), Arizona State University (USA), Aalto University (formerly, Helsinki University of Technology) (Finland), National University of Singapore, Nanyang Technological University (Singapore), Institute for High Performance Computing (Singapore), RMIT University (Australia), Beijing Institute of Technology (China), Xian University of Science and Technology (China), Amity University (India), Vellore Institute of Technology (India), Taylor's University (Malaysia), University of Cape Town (South Africa). By writing this book, I express my gratitude to the participating course attendees and students in my classes for their interest in the methodology promoted in this writing.

I thank Ms Nicki Dennis, Consultant Senior Books Commissioning Editor, Radar, Manufacturing and Electromagnetics, IET, The Institution of Engineering and Technology, for inviting me to write this book. I also thank Ms Olivia Wilkins, Assistant Editor, IET, The Institution of Engineering and Technology, for her continuous encouragements. It has been an utmost privilege and a delight working with Ms Dennis and Ms Wilkins and experiencing their splendid professionalism, and constant unwavering support. I am grateful to Mr N. Srinivasan and his team from MPS Limited for their copyediting and typesetting work. I thank the anonymous reviewers for their time and encouraging reviews.

I thank professor Andrew F. Peterson, Georgia Institute of Technology, USA, Series Editor, for inviting me to contribute to the ACES Series on Computational and Numerical Modelling in Electrical Engineering. The Mathematical Quantum Physics for Engineers and Technologists (MQPET) project is expected to comprise several volumes beyond the current Volume 1: Fundamentals and the upcoming Volume 2: Governing Equations.

With profound gratitude I dedicate this book to my wife Elisabeth for her love and friendship.

Introduction

This text presents the first of a Series of Volumes dedicated to the Mathematical Quantum Physics for Engineers and Technologists (MQPET). The primary goal of the MQPET Series is to empower the engineers and technologists to read relevant technical books and research papers with confidence and develop the skills required for judging for themselves any of the prevailing interpretations in quantum physics. The second objective is to convince the reader of the significance of the basics and the creative powers hidden in the fundamentals. To hone this critical ability, several techniques have been employed while composing the texts of the volumes in the MQPET Series. The third aim is to consciously condition the mind to see one and the same idea or concept from different perspectives. The current volume epitomizes the style of presentation of the developed and envisioned engineering mathematics in the MQPET Series. Unlike the standard texts on the subject, this book develops every single mathematical notion and idea from scratch, by exclusively employing tools which are most likely familiar to the reader or have been rendered self-explanatory by the author. The chapters in this book exemplify this significant idea on various occasions. No doubt most readers will notice and appreciate the distinctive mode of reasoning-and-training pursued in the book. Delving into the details of any subject matter of interest is essential to the profound understanding of the intricacies hidden in seemingly obvious relationships. It is deeply satisfying to experience that starting with utterly humble and elementary notions and ideas, a multitude of interesting relationships establish themselves systematically, emerge surprisingly, or appear serendipitously. Geometrical and algebraic tools and methodologies underly the story telling. The powerful methods in the Geometric Algebra will be explicated in later volumes after sharpening and readying the intuition of the readers. This book consists of four comparatively long chapters. Each chapter opens with a brief guide through the chapter and closes with concluding remarks. Every chapter is furnished with its own References. The four guides through the chapters collated below provide an overview.

Chapter 1: Starting with 1 and x, the set of monomials has been constructed and their properties are discussed. From the outset, an intentional emphasis has been put on the position variable x. The monomials (x^n) simplify the discussion of the derivatives in a useful manner. The paramount role of the differential operator d/dx alongside the position variable x has been deliberately stressed, since they feature in quantum physics prominently. Next it is shown how general functions can be synthesized from the set of monomials. The discussion leads to the all-embracing powerful concept of the formal Taylor series expansion. The realization that the formal Taylor series expansion can be seen as a *bona fide* transform and inverse transform has been merely touched upon in this text. This finding constitutes the subject matter of an independent upcoming book. Employing the monomials as a basis (complete collection of building blocks), several elementary functions have been constructed. Chief among them are $\cos \alpha$ and $\sin \alpha$ and their combination of a sort. Other important elementary functions are $\cosh \alpha$ and $\sinh \alpha$ and their combination of a different sort. These considerations lead to three 2×2 matrices, i.e., the identity matrix $\mathbf{I}$, the $\pi/2-$rotation matrix $\mathbf{J}$, and the matrix $\mathbf{K}$ for performing reflection over the line $y = x$. The corresponding higher dimensional matrices have not been included in this text. Their inclusion would have been a convincing testimony to the power of the techniques developed in this text. However, these omissions do not obscure the chain of arguments in any ways. Until much later in the chapter, there is no mention of two unfortunate designations in mathematics, i.e., the imaginary numbers and the complex numbers. The aim has been to demonstrate that everything is real and simple, rather than imaginary and complex, provided a suitable framework has been created. Following a thorough preparatory discussion, the standard symbol i (or j as electrical engineers refer to it) has been resort to. In this chapter, as in all other chapters, the story-telling follows a principle which has been tested by the author for several decades: First, establish the desired relationships in the light of contemporary (state-of-the-art) understanding of ideas and concepts. Then refer to the historical facts and anecdotes. This strategy prevents installing in the minds of the readers ideas which subsequently must be corrected or replaced by more clear and simpler ones. Starting with unfortunate notations, concepts, and traditions not only tends to mystify the underlying ideas but also hampers creativity. It is often more expensive to remove (or, replace) them, if possible at all, than to install them. Proceeding along the proposed path, one might serendipitously or actively find the one or the other gem, which otherwise would have remained hidden. The last part of chapter briefly discusses the genesis of π, the equation of a plane, and it culminates in the introduction of the concept of the planewave in one-, two, and three spatial dimensions. Considerable effort has been undertaken to breakdown the ideas into simpler ones. Planewaves are crucially important and they, on their own behalf, constitute alternative building blocks for synthesizing and analyzing functions. More crucially, planewaves are joint eigenfunctions of the differential operators $\partial/\partial x$ and $\partial/\partial t$ with x and t referring to space and time coordinates, respectively. It turns out that the corresponding eigenvalues are inverse (reciprocal) space and time variables. These findings are core concepts which are used in Quantum Physics, and luckily second nature to engineers and technologists, as will be shown in the following sections.

Chapter 2: While the Kronecker delta symbol is a simple and useful bookkeeping vehicle, the Dirac delta function is not even a function despite what its name suggests. It is rather a functional: its application onto an arbitrarily smooth Test Function followed by an integration establishes the sifting property of the Dirac delta function. In virtue of its sifting property the Dirac delta function is defined. In engineering community, the Dirac delta function is associated with the sampling of a function at an intended value of the independent variable. The Dirac delta function and its allied topics play prominent roles in this chapter. The discussion builds primarily upon planewaves and generalizes the underlying ideas in many ways. It is shown that customized Dirac delta functions can be constructed. Original problem-specific Dirac delta functions in one and two spatial dimensions, designed by the author, are briefly touched upon. While Dirac delta functions in real domain are generally not factorizable, their symbolic integral representations in spectral domain permit factorization. These ideas motivate connecting to exciting developments in signal processing (more generally, functional analysis) encompassing topics such Multiresolution Analysis, the theory of Wavelets and Dual Wavelets, and the theory of Frames and Dual Frames. The discussion in this chapter does not delve into these areas. It however, prepares the reader to do so if they wanted to. The fact that considerations in engineering and Quantum Physics contributed significantly to the development of the latter theories, and the fact that these theories in turn promise to contribute to Quantum Physics can be viewed as a triumph of applied mathematics. Epistemologically, the formation and further development of the Applied Mathematics is largely inspired by applications. Its ontology is however independent of physics, be Quantum Physics, or for this matter, Theory of General Relativity, Quantum Electrodynamics, or String Theory. This insight is revealing. The Applied Mathematics can be acquired as a collection of tools. The tools, however, must be distinctively separated from the interpretations of the theory they are helping to deliver. The tools and techniques in Applied Mathematics can be developed based on Toy Models. This idea has been taken seriously in this chapter. Based on simplest possible Toy Models, the notions of orthonormal bases and their dual bases, non-normal and dual non-normal bases, non-orthogonal and dual non-orthogonal bases, frames and dual frames, and the notion of generalized transform and inverse transform have been introduced and made plausible. The Dirac bracket notation, the inner product, the exterior product, the resolution of identity, and abstract Hilbert space have been employed to represent Fourier and inverse Fourier transform, the transformation of complete or over-complete coordinates, the Nyquist–Shannon sampling theorem, and the interpolation of functions. A brief discussion of the author's recently developed Discrete Taylor Transform and Inverse Transform has also been included. The proper design of elementary Toy Models promises to convey a host of intriguing ideas in simplest possible ways.

Chapter 3: This chapter promises to be particularly appealing to engineers and technologists, and at the same time extremely empowering. The procedures are algebraic and thus easy to follow. Not unlike the processes involved in Iterative Techniques, Self-Similarity, and Fractal Analysis, simple rules upon repeated application lead to the emergence of miraculously powerful and aesthetic results. Given the humble beginning of each category of examples, it is astonishing to observe the variety of powerful outcomes. In spite of the many Superlatives listed, the entire discussion is built upon three major pillars. (i) formal Taylor series expansion, (ii) Canonical Commutator Relationship, and (iii) Proof by Mathematical Induction. The reader is expected to be, by now, fully familiar with (i) and (iii) as these ideas were practiced dozens of times in the previous two chapters. The formal Taylor series expansion in (i) helps to get rid of the Complexity of Functions of operators $f(\mathbf{A})$ and temporarily replaces $f(\mathbf{A})$ by powers of their operators $\mathbf{A}^n$ (Opening the Box Procedure). After the required manipulations have been carried out, and $\mathbf{A}$ has been transformed to, say, $\mathbf{Q}$, the resulting $\mathbf{Q}$ replaces $\mathbf{A}$ leading to $f(\mathbf{Q})$. The method of Proof by Mathematical Induction in (iii) takes advantage of humans' ability to build intuition and generalize. Merely by experimenting with a few initial steps, the mind recognizes the underlying pattern, which in turn enables one to state the Induction Hypothesis. Ordinarily, the skills acquired during the initial steps suffice to perform the Induction Step, and complete the Proof. While (i) and (iii) are impressively powerful tools, the impact of the property in (ii) is astonishingly aesthetic. The Canonical Commutator Relationship of two operators $\mathbf{A}$ and $\mathbf{B}$, i.e., $[\mathbf{A}, \mathbf{B}] = \mathbf{I}$, with $\mathbf{I}$ standing for the Identity Operator, is the genesis of far reaching implications. Several types of composite operators and functions of operators will be considered. In a few instances problems have been solved in two ways, contrasting short, insightful, and elegant solutions against long, atomistic and algorithmic solutions. Both skills must be honed, as an interplay between the two leads to mastery. Furthermore, first and higher derivatives of Operators with respect to a parameter $\mathbf{A}(\lambda)$, and first and higher derivatives of Composite Operators with respect to a parameter are investigated. Also, derivatives of functions $f(\mathbf{A})$ with respect to operators are scrutinized. Several relationships involving commutators of composite operators are analyzed. The chapter concludes with studying the properties of an important category of commutators which involves the position and differential operators. Being truthful to the title, each section starts with simple exercises to warm up, proceeds to rigorous training, and concludes with the satisfaction of what has been achieved.

Chapter 4: Standard introduction of the differential operator d/dx requires the "vague" idea of Evaluating a Function at a Given Point. The discussion in this chapter starts with rendering this crucial notion precise. The concepts of the Dirac delta function $\delta(\cdot)$, and in particular, its γ-parametrized representation $\delta_\gamma(\cdot)$, loom large in interpreting what the Evaluation of a Function at a Certain Point means. Varying γ, the γ-parametrized representations $\delta_\gamma(\cdot)$ constitute a continuous sequence of well-behaved arbitrarily differentiable functions. The presented analysis promises to provide a fresh perspective of the ideas underlying the first-, and consequently, the higher-order differential operators. The n-order differential operators d^n/dx^n are defined in terms of integrals involving n-order derivatives of $\delta_\gamma(\cdot)$. The developed formulations are algorithmic, and provide easy-to-implement recipes for numerical calculations. Subsequently, new symbols have been introduced to simplify and unify the calculation of derivatives of the products of functions, e.g., $f(x)g(x)$ and $f(x)g(x)h(x)$. As elsewhere in this book, the introduced procedures are based on Inductive Reasoning: the ideas propel their own further developments. The discussion continues with thoroughly examining several important commutator relationships involving the differential operator $\hat{D}$ (standing for d/dx) and the position operator $\hat{x}$, in one-, two-, and three spatial dimensions. The presentation in this chapter culminates in the introduction of an original versatile scheme for the construction of a class of novel generalized Annihilation Operators ($\hat{b}$) and Creation Operators ($\hat{b}^\dagger$). The generalized formula includes the Canonical Commutation Relationship $[\hat{D}, \hat{x}] = \hat{\mathbb{I}}$, and the widely-used fundamental commutation relationship $[\hat{b}, \hat{b}^\dagger] = \hat{\mathbb{I}}$, as simplest possible realizations. Consecutively, two further major generalizations are presented. These findings are summarized and presented as a Theorem alongside its Proof. The entirety of the development in this chapter has its genesis in the position operator $\hat{x}$, the differential operator $\hat{D}$, and the Canonical Commutation Relationship, $[\hat{A}, \hat{B}] = \hat{A}\hat{B} - \hat{B}\hat{A} = \hat{\mathbb{I}}$. The fact that the formal Taylor series expansion plays an important role in these developments should not be surprising by now. The formal Taylor series expansion breaks arbitrary functions down into ingredients which can readily be interpreted, augmented, and incorporated into Commutators. One last insight should be brought to the attention of the reader. The notions of the Differentiation and the Commutation both require building the Difference of two terms. The existence of the Minus Sign leads to the "astonishing proliferation" of results and relationships. The reader would benefit from the discussion in this chapter greatly by keeping in their minds the following fact: the presence of the Minus Sign in the definitions of the Differentiation and the Commutation brings along inherent self-regularization and self-renormalization effects. The large number of Solved Problems are aimed to make this idea clear. Even though not covered in this chapter for completeness it should be mentioned that the Lagrangian (as the difference between the kinetic energy and the potential energy) also plays in significant role in Quantum Physics.

Chapter 1
The tale of "1" and "*x*"

1.1 A brief guide through the chapter

Starting with 1 and x, the set of monomials have been constructed and their properties discussed. From the outset, an intentional emphasis has been put on the Position Variable x. The monomials (x^n) simplify the discussion of the derivatives in a useful manner. The paramount role of the Differential Operator d/dx alongside the position variable x has been deliberately stressed, since they feature in Quantum Physics prominently. Next it is shown how general functions can be synthesized from the set of monomials. The discussion leads to the all-embracing powerful concept of the Formal Taylor Series Expansion. The realization that the Formal Taylor Series Expansion can be seen as a *bona fide* transform and inverse transform has been merely touched upon in this text. This finding constitutes the subject matter of an independent upcoming book. Employing the monomials as a basis (complete collection of building blocks), several elementary functions have been constructed. Chief among them are $\cos \alpha$ and $\sin \alpha$ and their combination of a sort. Other important elementary functions are $\cosh \alpha$ and $\sinh \alpha$ and their combination of a different sort. These considerations lead to three 2×2 matrices, i.e., the identity matrix $\mathbf{I}$, the $\pi/2-$rotation matrix $\mathbf{J}$, and the matrix $\mathbf{K}$ for performing reflection over the line $y = x$. The corresponding higher dimensional matrices have not been included in this text. Their inclusion would have been a convincing testimony to the power of the techniques developed in this text. However, these omissions do not obscure the chain of arguments in any ways. Until much later in the chapter, there is no mention of two unfortunate designations in mathematics, i.e., the imaginary numbers and the complex numbers. The aim has been to demonstrate that everything is real and simple, rather than imaginary and complex, provided a suitable framework has been created. Following a thorough preparatory discussion, the standard symbol i (or j as electrical engineers refer to it) has been resort to. In this chapter, as in all other chapters, the story-telling follows a principle which has been tested by the author for several decades: first, establish the desired relationships in the light of contemporary (state-of-the-art) understanding of ideas and concepts. Then refer to the historical facts and anecdotes. This strategy prevents installing in the minds of the readers ideas which subsequently must be corrected or replaced by more clear and simpler ones. Starting with unfortunate notations, concepts, and traditions not only tends to mystify the underlying ideas but also hampers creativity. It is often more expensive to remove (or, replace) them, if possible at all, than to install them. Proceeding along the proposed path, one might serendipitously or actively find the one or the other gem, which otherwise would have remained hidden. The last part of chapter briefly discusses the genesis of π, the equation of a plane, and it culminates in the introduction of the concept of the planewave in one, two, and three spatial dimensions. Considerable effort has been undertaken to breakdown the ideas into simpler ones. Planewaves are crucially important and they, on their own behalf, constitute alternative building blocks for synthesizing and analyzing functions. More crucially, planewaves are joint eigenfunctions of the differential operators $\partial/\partial x$ and $\partial/\partial t$ with x and t referring to space and time coordinates. It turns out that the corresponding eigenvalues are inverse (reciprocal) space and time variables. These findings are core concepts which are used in quantum physics, and luckily second nature to engineers and technologists, as will be shown in the following sections.

1.2 Monomials

The Tale of 1 and x: Consider the infinite sequence of the monomials x^0, x^1, x^2, x^3, ..., x^{n-1}, x^n, x^{n+1}, ..., constituting the set $\mathbb{X}$. The terms with odd- and even exponents are, respectively, odd- and even functions of x. The odd- and even functions follow (precede) each other alternately. By definition $x^0 \overset{\text{def.}}{=} 1$ and $x^n \overset{\text{def.}}{=} x \cdots x$ (x recurring n times)*. Starting with the second term, each term in the sequence is built from the preceding term by being multiplied by x: $x^{n+1} = x \cdot x^n$ ($n \in \mathbb{N}_0$). The sequence is constructed from $\mathbb{X}_G$, a generic set consisting of the elements 1 and x: $\mathbb{X}_G = \{1, x\}$. $\mathbb{X}_G$ equipped with the multiplication operator, "$\cdot$", is viewed as the genesis of $\mathbb{X}$. The multiplication of any two elements in $\mathbb{X}$ is another element of $\mathbb{X}$ ($\mathbb{X}$ is called to be closed, complete):

$$x^m \cdot x^n = \underbrace{x \cdot \ldots \cdot x}_{\text{m-times}} \cdot \underbrace{x \cdot \ldots \cdot x}_{\text{n-times}} \tag{1.1a}$$

$$= \underbrace{x \cdot \ldots \cdot x \cdot x \cdot \ldots \cdot x}_{\text{(m+n)-times}} \tag{1.1b}$$

$$= x^{m+n}. \tag{1.1c}$$

In virtue of,

$$x^m \cdot x^n = x^{m+n} \tag{1.2a}$$

$$= x^{n+m} \tag{1.2b}$$

$$= x^n \cdot x^m, \tag{1.2c}$$

the order of the multiplicative terms is immaterial. The derivative of each term with respect to x is proportional to the preceding term, except the first term, which by definition possesses no precedent. The integral of each term with respect to x is proportional to the following term: there is no last term in the sequence, as there is no last element in $\mathbb{N}$. Defining $x \overset{\text{def.}}{=} x^1$, the set of monomials $\mathbb{X} = \{1, x, x^2, x^3, \ldots, x^{n-1}, x^n, x^{n+1}, \ldots\}$ splits into two subsets $\mathbb{X}_{\text{even}} = \{1, x^2, \ldots, x^{2n}, \ldots\}$, and $\mathbb{X}_{\text{odd}} = \{x, x^3, \ldots, x^{2n+1}, \ldots\}$, the set of even- and odd monomials, respectively. The even monomials in $\mathbb{X}_{\text{even}}$ assume the value of 1 for $x = \pm 1$, are zero at $x = 0$ (with multiplicity $2n$), and approach ∞ for $x \to \pm\infty$ (with multiplicity $2n$). The first term in the set $\mathbb{X}_{\text{even}}$, "1," shares these properties as well.[†] The odd monomials in $\mathbb{X}_{\text{odd}}$ assume the value of -1 for $x = -1$ and 1 for $x = 1$, respectively. All terms in $\mathbb{X}_{\text{odd}}$ are zero at $x = 0$ (with multiplicity n), and approach, respectively, $-\infty$ for $x \to -\infty$, and ∞ for $x \to \infty$ (with multiplicity n). The set $\mathbb{X}$ is an archetypal complete set of elementary functions for "formally" synthesizing arbitrary functions.[‡,§]

1.3 Derivative of monomials—extracting of a linear term

The standard notion of derivative of a function $f(x)$ with respect to its independent variable x in the analysis requires an expression for $f(x + \Delta x)$ for vanishingly small values of the increment Δx, which is tantamount to using the Taylor Series Expansion of $f(x + \Delta x)$, at x. In the case of monomials, the notion of derivative can be accomplished merely by extracting a linear term from the monomials and thus avoiding the circularity which is otherwise required for

---

*These definitions permit building a "consistent" framework for the current discussion. The uncompromising notion of "consistency," the holly grail of sound and valid reasoning, is going to become increasingly more clear as the discussion unfolds.

†The constant term "1" in $\mathbb{X}$ (and, consequently, $\mathbb{X}_{\text{even}}$) fully respects the consistency criterion: the term x^n is n-fold zero at $x = 0$. (Crossing the x-axis $n(= 2k + 1)$-times, or touching the x-axis $n(= 2k)$-times, as the case maybe.) Thus, in view of $1 = x^n$ ($n = 0$), the constant term 1 is zero-fold (none-fold, empty-fold) zero at $x = 0$, i.e., it is a non-zero constant ($c \neq 0$). Furthermore, the monomials in $\mathbb{X}$ are alternately even- or odd functions, according to the parity (even-, oddness) of their respective exponents. Finally, the values of x^n are exclusively unity at $x = 1$. Since x^1 is odd and is zero at $x = 0$ with multiplicity one, x^0, preceding x^1, must consequently be even and zero at $x = 0$ with multiplicity zero, i.e., possessing no touching point with the x-axis. Since x^0 is even it must be 1 at $x = -1$ as well. The latter properties suggest defining x^0 as a constant with its value being one: $x^0 \overset{\text{def.}}{=} 1$. Similarly, the constant term 1 is zero-fold (none-fold, empty fold) infinity in the limit x → ... . In this sense, the constant term 1 satisfies all the conditions for belonging to the $\mathbb{X}$ (and thus $\mathbb{X}_{\text{even}}$) club, while being truthful to its very nature, as a constant.

‡The attribute "archetypal" alludes to the fact that $\mathbb{X}$ is viewed here as the mother of all elementary functions (ur-building block functions) for constructing other functions, including other basis-, primitive building block functions. The connotation "complete" refers to the fact that there are sufficient building functions available to accomplish what is aimed to be done.

§At this stage, there is no need to obscure the discussion by considering the set $\{x^{-1}, x^{-2}, \ldots, x^{-n}, \ldots\}$ with negative exponents.

determining the derivative of functions. As will be shown momentarily, building the derivative of x^n essentially requires writing $x^n = x^{n-1}x$ or $x^n = xx^{n-1}$, and following a simple recipe. In the following, the idea will be first illustrated in terms of a few elementary examples and then generalized by the method of Mathematical Induction.

Problem: The derivative of x.

Determine the derivative of x.

Solution:

$$\frac{d}{dx}x \overset{\text{def.}}{=} \lim_{\Delta x \to 0} \frac{(x + \Delta x) - x}{\Delta x} \tag{1.3a}$$

$$\overset{x \text{ cancels out}}{=} \lim_{\Delta x \to 0} \frac{\Delta x}{\Delta x} \tag{1.3b}$$

$$\overset{\Delta x \text{ cancels}}{=} \lim_{\Delta x \to 0} 1 \tag{1.3c}$$

$$\overset{1 \text{ indep. of } \Delta x}{=} 1. \tag{1.3d}$$

∎

Problem: The derivative of x^2.

Determine the derivative of x^2.

Solution:

$$\frac{d}{dx}x^2 \overset{\text{def.}}{=} \lim_{\Delta x \to 0} \frac{(x + \Delta x)^2 - x^2}{\Delta x}. \tag{1.4}$$

Consider the numerator in isolation and extract the term $(x + \Delta x)$,

$$(x + \Delta x)^2 \overset{\text{extraction}}{=} (x + \Delta x)(x + \Delta x) \tag{1.5a}$$

$$= x^2 + x(\Delta x) + (\Delta x)x + (\Delta x)^2. \tag{1.5b}$$

The order of multiplication is irrelevant in the current context, $(\Delta x)x = x(\Delta x)$. Consequently,

$$(x + \Delta x)^2 = x^2 + 2x(\Delta x) + (\Delta x)^2. \tag{1.6}$$

Subtracting x^2 from both sides, factoring out Δx from the remaining two terms at the R.H.S., and dividing through by Δx,

$$\frac{(x + \Delta x)^2 - x^2}{\Delta x} = 2x + \Delta x. \tag{1.7}$$

In the limit $\Delta x \to 0$,

$$\lim_{\Delta x \to 0} \frac{(x + \Delta x)^2 - x^2}{\Delta x} = \lim_{\Delta x \to 0} (2x + \Delta x) \tag{1.8a}$$

$$= 2x. \tag{1.8b}$$

Comparing (1.4) with (1.8),

$$\frac{d}{dx}x^2 = 2x.$$

(1.9)

■

Problem: The derivative of x^n proof by the method of mathematical induction.

Determine the derivative of x^n.

Solution: Inspired by (1.9), the induction hypothesis can be formulated,

$$\frac{d}{dx}x^n = nx^{n-1}.$$

(1.10)

In carrying out the induction step, the following self-explanatory manipulations offer themselves,

$$\frac{d}{dx}x^{n+1} \overset{\text{def.}}{=} \lim_{\Delta x \to 0} \frac{(x+\Delta x)^{n+1} - x^{n+1}}{\Delta x}$$

(1.11a)

$$\overset{\text{extraction}}{=} \lim_{\Delta x \to 0} \frac{(x+\Delta x)(x+\Delta x)^n - xx^n}{\Delta x}$$

(1.11b)

$$\overset{\text{distributive law}}{=} \lim_{\Delta x \to 0} \frac{x(x+\Delta x)^n + \Delta x(x+\Delta x)^n - xx^n}{\Delta x}.$$

(1.11c)

Focus on the first and the third terms in the numerator at the R.H.S. Factoring out x, and splitting terms,

$$\frac{d}{dx}x^{n+1} = \lim_{\Delta x \to 0} \frac{x\left[(x+\Delta x)^n - x^n\right] + \Delta x(x+\Delta x)^n}{\Delta x}$$

(1.12a)

$$= \lim_{\Delta x \to 0} \left\{ \frac{x\left[(x+\Delta x)^n - x^n\right]}{\Delta x} + \frac{\Delta x(x+\Delta x)^n}{\Delta x} \right\}.$$

(1.12b)

The variable x in the first term is independent of Δx, and Δx in the second term cancels from the numerator and denominator, thus leading to,

$$\frac{d}{dx}x^{n+1} = x \lim_{\Delta x \to 0} \frac{(x+\Delta x)^n - x^n}{\Delta x} + \lim_{\Delta x \to 0}(x+\Delta x)^n.$$

(1.13)

The first limit at the R.H.S. expresses the derivative of x^n. According to the induction hypothesis, (1.10), it is equal to nx^{n-1}. The second limit is just x^n. Consequently, considering the latter two facts,

$$\frac{d}{dx}x^{n+1} = xnx^{n-1} + x^n$$

(1.14)

$$= nx^n + x^n$$

(1.15)

$$= (n+1)x^n.$$

(1.16)

It should be reiterated that taking the derivative of monomials, as illustrated above, can be accomplished with mere reference to $x^{n+1} = x \cdot x^n$, i.e., by extracting a linear term, a property which was utilized to define the monomials recursively. The significance of this fact will become apparent when more general functions are synthesized using monomials. The monomials and their properties are exclusively all what is needed, in the theory proposed here.

■

1.4 Analysis and synthesis of functions in terms of monomials

Analysis: Given the function $f(x)$, determine the coefficients c_n ($n \in \mathbb{N}_0$) such that,

$$f(x) = c_0 + c_1 x + c_2 x^2 + c_3 x^3 + c_4 x^4 + c_5 x^5 + \cdots . \tag{1.17}$$

Given $f(x)$, the procedure of determining the coefficients,

$$f(x) \implies \{c_0, c_1, c_2, c_3, c_4, c_5, \dots \}, \tag{1.18}$$

is referred to as the analysis problem, the analysis step, and is treated in the next section. It turns out that $c_n = f^{(n)}(0)/n!$, with $f^{(n)}(0)$ referring to the nth derivative of $f(x)$ evaluated at $x = 0$.

Synthesis: Any given set of coefficients c_n ($n \in \mathbb{N}_0$) uniquely synthesizes a function $f(x)$,

$$f(x) \impliedby \{c_0, c_1, c_2, c_3, c_4, c_5, \dots \} , \tag{1.19}$$

with $f(x)$ as given in (1.17).

Remark: Formal Power Series Expansion

Viewing the Power Series Expansion in (1.17) as Formal is enabling and grants great utility. The notion of Formal Power Series Expansion unleashes creativity in a myriad of ways and leads to a wealth of useful procedures and relationships. The attribute Formal implicates the existence of the derivatives $f^{(n)}(x)$ to any order required. Furthermore, it implies that there is no concern about the sense in which the equality in (1.17) is valid, whether or not the series converges, in which sense it converges, and to which function. The concept of Formal Power Series Expansion is an all-purpose tool that engineers and scientists cannot afford to miss out from their manipulation repertoire and creative toolboxes. There are a plenty of occasions in this text to substantiate this claim.

$\square$

1.5 Determining the expansion coefficients c_n (the analysis step)

Given $f(x)$ $(= f^{(0)}(x))$, in terms of a Formal Power Series Expansion, successively differentiate $f(x)$ and its corresponding synthesis formula, to develop an intuition about the resulting relationships,

$$f^{(0)}(x) = c_0 + c_1 x + c_2 x^2 + c_3 x^3 + c_4 x^4 + c_5 x^5 + \cdots \tag{1.20a}$$

$$f^{(1)}(x) = c_1 + 2c_2 x + 3c_3 x^2 + 4c_4 x^3 + 5c_5 x^4 + \cdots \tag{1.20b}$$

$$f^{(2)}(x) = 2c_2 + 2 \cdot 3c_3 x + 3 \cdot 4c_4 x^2 + 4 \cdot 5c_5 x^3 + \cdots \tag{1.20c}$$

$$f^{(3)}(x) = 2 \cdot 3c_3 + 2 \cdot 3 \cdot 4c_4 x + 3 \cdot 4 \cdot 5c_5 x^2 + \cdots \tag{1.20d}$$

$$f^{(4)}(x) = 2 \cdot 3 \cdot 4c_4 + 2 \cdot 3 \cdot 4 \cdot 5c_5 x + \cdots \tag{1.20e}$$

$$\vdots$$

Evaluating the derivatives $f^{(n)}(x)$ at $x = 0$,

$$f^{(0)}(0) = c_0 \tag{1.21a}$$

$$f^{(1)}(0) = c_1 \tag{1.21b}$$

$$f^{(2)}(0) = 2c_2 \tag{1.21c}$$

$$f^{(3)}(0) = 2 \cdot 3 c_3 \tag{1.21d}$$

$$f^{(4)}(0) = 2 \cdot 3 \cdot 4 c_4 \tag{1.21e}$$

$$\vdots$$

Recognizing the factorials $0!, 1!, 2!, 3!, 4!$ at the R.H.S., and rearranging,

$$c_0 = \frac{1}{0!} f^{(0)}(0) \tag{1.22a}$$

$$c_1 = \frac{1}{1!} f^{(1)}(0) \tag{1.22b}$$

$$c_2 = \frac{1}{2!} f^{(2)}(0) \tag{1.22c}$$

$$c_3 = \frac{1}{3!} f^{(3)}(0) \tag{1.22d}$$

$$c_4 = \frac{1}{4!} f^{(4)}(0) \tag{1.22e}$$

$$\vdots$$

These relationships suggest the induction hypothesis,

$$c_n = \frac{1}{n!} f^{(n)}(0), \quad n \in \mathbb{N}_0. \tag{1.23}$$

1.6   Formal Taylor Series Expansion (the synthesis step)

Substituting (1.23) into (1.20a), results in the Formal Taylor Series Expansion,

$$f(x) = \sum_{n=0}^{\infty} \frac{1}{n!} f^{(n)}(0) x^n. \tag{1.24}$$

Exercise: Shifted elementary monomials.

Let x_0 be an arbitrary real number ($x_0 \in \mathbb{R}$). Shifting the monomials x^n ($n \in \mathbb{N}_0$) by x_0 results in the set $\mathbb{X}_{x_0}$,

$$\mathbb{X}_{x_0} = \left\{ 1, (x - x_0), (x - x_0)^2, (x - x_0)^3, \ldots, (x - x_0)^{n-1}, (x - x_0)^n, (x - x_0)^{n+1}, \ldots \right\}. \tag{1.25}$$

Let a given function $f(x)$ be synthesized by the linear combination (superposition) of the shifted monomials in (1.25). Denoting the coefficient of the nth shifted term by κ_n,

$$f(x) = \kappa_0 + \kappa_1 (x - x_0) + \kappa_2 (x - x_0)^2 + \kappa_3 (x - x_0)^3 + \kappa_4 (x - x_0)^4 + \kappa_5 (x - x_0)^5 + \cdots . \tag{1.26}$$

Successively differentiating the terms on both sides of (1.26),

$$f^{(0)}(x) = \kappa_0 + \kappa_1(x - x_0) + \kappa_2(x - x_0)^2 + \kappa_3(x - x_0)^3 + \kappa_4(x - x_0)^4 + \kappa_5(x - x_0)^5 + \cdots \qquad (1.27a)$$

$$f^{(1)}(x) = \kappa_1 + 2\kappa_2(x - x_0) + 3\kappa_3(x - x_0)^2 + 4\kappa_4(x - x_0)^3 + 5\kappa_5(x - x_0)^4 + \cdots \qquad (1.27b)$$

$$f^{(2)}(x) = 2\kappa_2 + 2 \cdot 3\kappa_3(x - x_0) + 3 \cdot 4\kappa_4(x - x_0)^2 + 4 \cdot 5\kappa_5(x - x_0)^3 + \cdots \qquad (1.27c)$$

$$f^{(3)}(x) = 2 \cdot 3\kappa_3 + 2 \cdot 3 \cdot 4\kappa_4(x - x_0) + 3 \cdot 4 \cdot 5\kappa_5(x - x_0)^2 + \cdots \qquad (1.27d)$$

$$f^{(4)}(x) = 2 \cdot 3 \cdot 4\kappa_4 + 2 \cdot 3 \cdot 4 \cdot 5\kappa_5(x - x_0) + \cdots \qquad (1.27e)$$

$$\vdots$$

Setting $x = x_0$,

$$f^{(0)}(x_0) = \kappa_0 \qquad (1.28a)$$

$$f^{(1)}(x_0) = \kappa_1 \qquad (1.28b)$$

$$f^{(2)}(x_0) = 2\kappa_2 \qquad (1.28c)$$

$$f^{(3)}(x_0) = 2 \cdot 3\kappa_3 \qquad (1.28d)$$

$$f^{(4)}(x_0) = 2 \cdot 3 \cdot 4\kappa_4 \qquad (1.28e)$$

$$\vdots$$

Recognizing the factorials at the R.H.S., and rearranging,

$$\kappa_0 = \frac{1}{0!}f^{(0)}(x_0) \qquad (1.29a)$$

$$\kappa_1 = \frac{1}{1!}f^{(1)}(x_0) \qquad (1.29b)$$

$$\kappa_2 = \frac{1}{2!}f^{(2)}(x_0) \qquad (1.29c)$$

$$\kappa_3 = \frac{1}{3!}f^{(3)}(x_0) \qquad (1.29d)$$

$$\kappa_4 = \frac{1}{4!}f^{(4)}(x_0) \qquad (1.29e)$$

$$\vdots$$

By induction,

$$\kappa_n = \frac{1}{n!}f^{(n)}(x_0), \quad n \in \mathbb{N}_0. \qquad (1.30)$$

Substituting (1.30) into (1.26),

$$f(x) = \sum_{n=0}^{\infty} \frac{1}{n!}f^{(n)}(x_0)(x - x_0)^n , \qquad (1.31)$$

which is the Formal Taylor Series Expansion of the function $f(x)$ at $x = x_0$.

∎

1.7   Introducing $f_{\text{even}}(x)$ and $f_{\text{odd}}(x)$

Given the arbitrary function $f(x)$, define the functions $f_{\text{even}}(x)$ and $f_{\text{odd}}(x)$, according to,

$$f_{\text{even}}(x) = \frac{1}{2}\big[f(x) + f(-x)\big] \tag{1.32a}$$

$$f_{\text{odd}}(x) = \frac{1}{2}\big[f(x) - f(-x)\big]. \tag{1.32b}$$

As the subscripts suggest, and it can readily be shown, $f_{\text{even}}(x)$ and $f_{\text{odd}}(x)$ are even- and odd functions, respectively,

$$f_{\text{even}}(-x) \overset{(1.32a)}{=} \frac{1}{2}\big[f(-x) + f(x)\big] = \frac{1}{2}\big[f(x) + f(-x)\big] = f_{\text{even}}(x) \tag{1.33a}$$

$$f_{\text{odd}}(-x) \overset{(1.32b)}{=} \frac{1}{2}\big[f(-x) - f(x)\big] = -\frac{1}{2}\big[f(x) - f(-x)\big] = -f_{\text{odd}}(x). \tag{1.33b}$$

Problem: Formal Power Series Expansion for $f_{\text{even}}(x)$ and $f_{\text{odd}}(x)$.

Employ the Formal Power Series Expansion of $f(x)$, (1.17), to obtain Formal Power Series Expansions of $f_{\text{even}}(x)$ and $f_{\text{odd}}(x)$.

Solution: Substituting (1.17) into (1.32), quite unsurprisingly, the results

$$f_{\text{even}}(x) = c_0 + c_2 x^2 + c_4 x^4 + c_6 x^6 + c_8 x^8 + \cdots \tag{1.34a}$$

$$f_{\text{odd}}(x) = c_1 x + c_3 x^3 + c_5 x^5 + c_7 x^7 + c_9 x^9 + \cdots \tag{1.34b}$$

are obtained. The functions $f_{\text{even}}(x)$ and $f_{\text{odd}}(x)$ comprise the even- and odd polynomials, respectively, which constitute $f(x)$,

$$f(x) = f_{\text{even}}(x) + f_{\text{odd}}(x). \tag{1.35}$$

∎

1.8   Elementary functions

This section presents the Formal Power Series Expansion as a universal tool for investigating the existence of a collection of elementary functions with prescribed properties.

Problem: Does any function $f(x)$ with $f^{(1)}(x) = f(x)$ and $f(0) = 1$ exist?

Solution: The formulation in the statement of the problem alludes to the possibility that the desired function might not exist. If a finite number or all the coefficients c_n ($n \in \mathbb{N}_0$) are non-zero, and they can be (uniquely) determined, it is said that the function $f(x)$ does exist. Given the Formal Power Series Expansion,

$$f(x) = c_0 + c_1 x + c_2 x^2 + c_3 x^3 + c_4 x^4 + c_5 x^5 + \cdots, \tag{1.36}$$

differentiate both sides of (1.36) with respect to x,

$$f^{(1)}(x) = c_1 + 2c_2 x + 3c_3 x^2 + 4c_4 x^3 + 5c_5 x^4 + \cdots. \tag{1.37}$$

Set $f^{(1)}(x) = f(x)$,

$$\underbrace{c_1} + \underbrace{2c_2x} + \underbrace{3c_3x^2} + \underbrace{4c_4x^3} + 5c_5x^4 + \cdots$$

$$= \underbrace{c_0} + \underbrace{c_1x} + \underbrace{c_2x^2} + \underbrace{c_3x^3} + c_4x^4 + c_5x^5 + \cdots. \tag{1.38}$$

Compare the series on both sides of the equality sign term-by-term, as indicated by horizontal curly brackets,

$$c_1 = c_0 \tag{1.39a}$$

$$2c_2 = c_1 \implies c_2 = \frac{1}{2}c_1 = \frac{1}{2}c_0 = \frac{1}{2!}c_0 \tag{1.39b}$$

$$3c_3 = c_2 \implies c_3 = \frac{1}{3}c_2 = \frac{1}{3}\frac{1}{2!}c_0 = \frac{1}{3!}c_0 \tag{1.39c}$$

$$4c_4 = c_3 \implies c_4 = \frac{1}{4}c_3 = \frac{1}{4}\frac{1}{3!}c_0 = \frac{1}{4!}c_0 \tag{1.39d}$$

$$\vdots$$

$$nc_n = c_{n-1} \implies c_n = \frac{1}{n!}c_0 \tag{1.39e}$$

$$\vdots$$

The equation for the general term in (1.39e) has been hypothesized (judiciously guessed) by induction. The formulations of the induction hypothesis for the elementary examples considered in this chapter suggest themselves Simply-by-Inspection. No further justification or proof seem necessary. In view of (1.39), the coefficients c_n ($n \in \mathbb{N}$) can be expressed in terms of c_0, leading to,

$$f(x) = c_0 + c_0x + \frac{1}{2!}c_0x^2 + \frac{1}{3!}c_0x^3 + \frac{1}{4!}c_0x^4 + \frac{1}{5!}c_0x^5 + \cdots. \tag{1.40}$$

Setting $f(0) = 1$ results in $c_0 = 1$. Denoting the resulting function by e^x or $\exp(x)$,

$$e^x = \exp(x) = 1 + x + \frac{1}{2!}x^2 + \frac{1}{3!}x^3 + \frac{1}{4!}x^4 + \frac{1}{5!}x^5 + \cdots. \tag{1.41}$$

Thus, the function e^x, or equivalently $\exp(x)$, represented by the Formal Power Series Expansion (1.41), satisfies the defining equation $de^x/dx = e^x$ subject to the condition $e^x|_{x=0} = 1$. Stated differently, given the set of monomials $\{1, x, x^2, x^2, x^3, x^4, x^5 \cdots\}$, the coefficients $\{1, \frac{1}{2!}, \frac{1}{3!}, \frac{1}{4!}, \frac{1}{5!}, ...\}$ define the function e^x. Since the set of monomials is universal, the set of coefficients suffices to encode the information necessary to synthesize e^x. ∎

Remark: Considering the fact that the coefficients c_n in the Formal Power Series Expansion are equal to $\frac{1}{n!}f^{(n)}(0)$, (1.39e) implies $f^{(n)}(0) = f^{(0)}(0) = f(0)$. This property is manifestly expressed in (1.40). The "obviously simplest" choice $c_0 = f(0) = 1$ resulted in (1.41) with $f^{(n)}(0) = 1$ for $n \in \mathbb{N}_0$. In view of the Formal Power Series Expansion representation in (1.41), characterized by $f^{(n)}(0) = 1$ for $n \in \mathbb{N}_0$, the function e^x is the simplest and most elegant Formal Power Series Expansion of any function conceivable. This "queen" of all functions reproduces itself no matter how often it is differentiated with respect to x. As demonstrated in the sequel, e^x passes some of her noble properties onto her successor functions, albeit in a slightly weaker form. The argument of e^x, i.e., x does not need to be a scalar quantity either. In fact this property is all about what the Formal Power Series Expansion is supposed to accomplish. The role that e^x plays among other functions, can be likened to the role of x in generating all the monomials, and of 1 in being the genesis of producing all its successor integers. The author has been putting forward the conjecture that the governing equations of natural phenomena must be diagonalizable. It turns out that solutions of the diagonalized set of equations are exponential functions of certain mathematical structures. It might be a worthwhile research undertaking to find out whether there is a more fundamental relationship underlying these properties.

□

Problem: Does any function $f_{\text{even}}(x)$ with $f_{\text{even}}^{(1)}(x) = f_{\text{even}}(x)$ and $f_{\text{even}}(0) = 0$ exist?

Solution: Consider the Formal Power Series Expansion of $f_{\text{even}}(x)$ alongside its first derivative $f_{\text{even}}^{(1)}(x)$,

$$f_{\text{even}}(x) = c_0 + c_2 x^2 + c_4 x^4 + c_6 x^6 + c_8 x^8 + \cdots \tag{1.42a}$$

$$f_{\text{even}}^{(1)}(x) = 2c_2 x + 4c_4 x^3 + 6c_6 x^5 + 8c_8 x^7 + \cdots . \tag{1.42b}$$

In virtue of the fact that $f_{\text{even}}(x)$ is an even-, and $f_{\text{even}}^{(1)}(x)$ an odd function, the equation $f_{\text{even}}(x) = f_{\text{even}}^{(1)}(x)$ can only be satisfied for the trivial choice of the coefficients $c_{2n} = 0$ ($n \in \mathbb{N}_0$). Thus, this case possesses the trivial solution $f_{\text{even}}(x) \equiv 0$, which also satisfies the condition $f_{\text{even}}(0) = 0$.

■

Problem: Does any function $f_{\text{even}}(x)$ with $f_{\text{even}}^{(2)}(x) = f_{\text{even}}(x)$ and $f_{\text{even}}(0) = 1$ exist?

Solution: The Formal Power Series Expansion of the second derivative of $f_{\text{even}}(x)$ can be obtained by differentiating the expressions on both sides of (1.42b),

$$f_{\text{even}}^{(2)}(x) = 2c_2 + 3 \cdot 4c_4 x^2 + 5 \cdot 6c_6 x^4 + 7 \cdot 8c_8 x^6 + \cdots . \tag{1.43}$$

Since $f_{\text{even}}(x)$ and $f_{\text{even}}^{(2)}(x)$ are both even functions, it is legitimate to inquire whether the requirement $f_{\text{even}}^{(2)}(x) = f_{\text{even}}(x)$ is realizable. Equating the Formal Power Series Expansions of $f_{\text{even}}^{(2)}(x)$ and $f_{\text{even}}(x)$,

$$2c_2 + 3 \cdot 4c_4 x^2 + 5 \cdot 6c_6 x^4 + 7 \cdot 8c_8 x^6 + \cdots$$
$$= c_0 + c_2 x^2 + c_4 x^4 + c_6 x^6 + c_8 x^8 + \cdots . \tag{1.44}$$

A term-by-term comparison of the coefficients associated with the monomials x^{2n} ($n \in \mathbb{N}_0$) results in,

$$2c_2 = c_0 \implies c_2 = \frac{1}{2}c_0 = \frac{1}{2!}c_0 \tag{1.45a}$$

$$3 \cdot 4c_4 = c_2 \implies c_4 = \frac{1}{3 \cdot 4}c_2 = \frac{1}{4!}c_0 \tag{1.45b}$$

$$5 \cdot 6c_6 = c_4 \implies c_6 = \frac{1}{5 \cdot 6}c_4 = \frac{1}{6!}c_0 \tag{1.45c}$$

$$7 \cdot 8c_8 = c_6 \implies c_8 = \frac{1}{7 \cdot 8}c_6 = \frac{1}{8!}c_0 \tag{1.45d}$$

$$\vdots$$

$$(2n-1) \cdot 2n c_{2n} = c_{2n-2} \implies c_{2n} = \frac{1}{(2n)!}c_0. \tag{1.45e}$$

Substituting c_{2n} ($n \in \mathbb{N}$) into (1.42a),

$$f_{\text{even}}(x) = c_0 + \frac{1}{2!}c_0 x^2 + \frac{1}{4!}c_0 x^4 + \frac{1}{6!}c_0 x^6 + \frac{1}{8!}c_0 x^8 + \cdots \tag{1.46}$$

Setting $f_{\text{even}}(0) = 1$ implies $c_0 = 1$. Denoting the resulting function by $\cosh(x)$,

$$\cosh(x) = 1 + \frac{1}{2!}x^2 + \frac{1}{4!}x^4 + \frac{1}{6!}x^6 + \frac{1}{8!}x^8 + \cdots . \tag{1.47}$$

Thus, $\cosh(x)$, defined by the Formal Power Series Expansion (1.47), satisfies the defining equation $d^2\cosh(x)/dx^2 = \cosh(x)$ subject to the constraint $\cosh(0) = 1$. The $\cosh(x)$ function consists of the even expansion terms of e^x. Thus, being an even function, $\cosh(x) = \left(e^x + e^{-x}\right)/2$. Stated differently, given the set of monomials $\{1, x^2, x^4, x^6, x^8, \dots\}$, the coefficients $\left\{1, \frac{1}{2!}, \frac{1}{4!}, \frac{1}{6!}, \frac{1}{8!}, \dots\right\}$ define the function $\cosh(x)$. The coefficient $\frac{1}{(2n)!}$ uniquely specifies the monomials x^{2n} ($n \in \mathbb{N}_0$). Since the set of monomials is universal, the set of coefficients suffices to encode the information necessary to synthesize $\cosh(x)$.

■

Problem: Does any function $f_{\text{odd}}(x)$ with $f_{\text{odd}}^{(1)}(x) = f_{\text{odd}}(x)$ and $f_{\text{odd}}(0) = 0$ exist?

Solution: Consider the Formal Power Series Expansion of $f_{\text{odd}}(x)$ alongside that of its first derivative $f_{\text{odd}}^{(1)}(x)$,

$$f_{\text{odd}}(x) = c_1 x + c_3 x^3 + c_5 x^5 + c_7 x^7 + \cdots \tag{1.48a}$$

$$f_{\text{odd}}^{(1)}(x) = c_1 + 3c_3 x^2 + 5c_5 x^4 + 7c_7 x^6 + \cdots . \tag{1.48b}$$

In virtue of the fact that $f_{\text{odd}}(x)$ is an odd- and $f_{\text{odd}}^{(1)}(x)$ an even function, the equation $f_{\text{odd}}(x) = f_{\text{odd}}^{(1)}(x)$ can only be satisfied for the trivial choice of the coefficients $c_{2n+1} = 0$ ($n \in \mathbb{N}_0$). Thus, this case possesses the trivial solution $f_{\text{odd}}(x) \equiv 0$, also satisfying the condition $f_{\text{odd}}(0) = 0$.

■

Problem: Does any function $f_{\text{odd}}(x)$ with $f_{\text{odd}}^{(2)}(x) = f_{\text{odd}}(x)$ and $f_{\text{odd}}^{(1)}(0) = 0$ exist?

Solution: Considering (1.48b), determine the Formal Power Series Expansion of the second derivative $f_{\text{odd}}^{(2)}(x)$,

$$f_{\text{odd}}^{(2)}(x) = 2 \cdot 3c_3 x + 4 \cdot 5c_5 x^3 + 6 \cdot 7c_7 x^5 + \cdots . \tag{1.49}$$

Since $f_{\text{odd}}(x)$ and $f_{\text{odd}}^{(2)}(x)$ are both odd functions, it is legitimate to ask the whether $f_{\text{odd}}(x) = f_{\text{odd}}^{(2)}(x)$ possesses a valid non-trivial solution. Setting $f_{\text{odd}}(x)$ equal to $f_{\text{odd}}^{(2)}(x)$, a term-by-term comparison results in,

$$2 \cdot 3c_3 = c_1 \implies c_3 = \frac{1}{2 \cdot 3} c_1 = \frac{1}{3!} c_1 \tag{1.50a}$$

$$4 \cdot 5c_5 = c_3 \implies c_5 = \frac{1}{4 \cdot 5} c_3 = \frac{1}{5!} c_1 \tag{1.50b}$$

$$6 \cdot 7c_7 = c_5 \implies c_7 = \frac{1}{6 \cdot 7} c_5 = \frac{1}{7!} c_1 \tag{1.50c}$$

$$\vdots$$

$$(2n) \cdot (2n + 1)c_{2n+1} = c_{2n-1} \implies c_{2n+1} = \frac{1}{(2n + 1)!} c_1. \tag{1.50d}$$

Substituting c_{2n+1} ($n \in \mathbb{N}$) into (1.48a),

$$f_{\text{odd}}(x) = c_1 x + \frac{1}{3!} c_1 x^3 + \frac{1}{5!} c_1 x^5 + \frac{1}{7!} c_1 x^7 + \cdots . \tag{1.51}$$

Setting $f_{\text{odd}}^{(1)}(0) = 1$ results in $c_1 = 1$. Denoting the resulting function by $\sinh(x)$,

$$\sinh(x) = x + \frac{1}{3!} x^3 + \frac{1}{5!} x^5 + \frac{1}{7!} x^7 + \cdots . \tag{1.52}$$

Thus, the function $\sinh(x)$, defined by the Formal Power Series Expansion (1.52), satisfies the defining conditions $d^2\sinh(x)/dx^2 = \sinh(x)$ subject to the constraint $d\sinh(x)/dx|_{x=0} = 1$. Stated differently, given the set of monomials $\{x, x^3, x^5, x^7, \dots\}$, the coefficients $\left\{\frac{1}{1!}, \frac{1}{3!}, \frac{1}{5!}, \frac{1}{7!}, \dots\right\}$ define the function $\sinh(x)$. The coefficient $\frac{1}{(2n+1)!}$ uniquely specifies the monomial x^{2n+1} ($n \in \mathbb{N}_0$). Since the set of monomials is universal, the set of coefficients suffices to encode the information necessary to synthesize $\sinh(x)$.

■

Problem: Does any function $f_{\text{even}}(x)$ with $f_{\text{even}}^{(2)}(x) = -f_{\text{even}}(x)$ and $f_{\text{even}}(0) = 1$ exist?

Solution: Consider $f_{\text{even}}(x)$ along with its first and second derivatives

$$f_{\text{even}}(x) = c_0 + c_2 x^2 + c_4 x^4 + c_6 x^6 + c_8 x^8 + \cdots \tag{1.53a}$$

$$f_{\text{even}}^{(1)}(x) = 2c_2 x + 4c_4 x^3 + 6c_6 x^5 + 8c_8 x^7 + \cdots \tag{1.53b}$$

$$f_{\text{even}}^{(2)}(x) = 2c_2 + 3 \cdot 4c_4 x^2 + 5 \cdot 6c_6 x^4 + 7 \cdot 8c_8 x^6 + \cdots . \tag{1.53c}$$

Since $f_{\text{even}}(x)$ and $f_{\text{even}}^{(2)}(x)$ are both even functions, it is legitimate to ask which coefficients c_{2n} ($n \in \mathbb{N}_0$) would render the equation $f_{\text{even}}^{(2)}(x) = -f_{\text{even}}(x)$ valid. Using (1.53a) and (1.53c),

$$2c_2 + 3 \cdot 4c_4 x^2 + 5 \cdot 6c_6 x^4 + 7 \cdot 8c_8 x^6 + \cdots$$
$$= -c_0 - c_2 x^2 - c_4 x^4 - c_6 x^6 - c_8 x^8 - \cdots . \tag{1.54}$$

A term-by-term comparison results in

$$2c_2 = -c_0 \implies c_2 = -\frac{1}{2}c_0 = -\frac{1}{2!}c_0 \tag{1.55a}$$

$$3 \cdot 4c_4 = -c_2 \implies c_4 = -\frac{1}{3 \cdot 4}c_2 = \frac{1}{4!}c_0 \tag{1.55b}$$

$$5 \cdot 6c_6 = -c_4 \implies c_6 = -\frac{1}{5 \cdot 6}c_4 = -\frac{1}{6!}c_0 \tag{1.55c}$$

$$7 \cdot 8c_8 = -c_6 \implies c_8 = -\frac{1}{7 \cdot 8}c_6 = \frac{1}{8!}c_0 \tag{1.55d}$$

$$\vdots$$

$$(2n-1) \cdot (2n)c_{2n} = -c_{2n-2} \implies c_{2n} = (-1)^n \frac{1}{(2n)!}c_0. \tag{1.55e}$$

It is satisfying to observe the source of alternating signs of the coefficients. Substituting c_{2n} ($n \in \mathbb{N}$) into (1.53a),

$$f_{\text{even}}(x) = c_0 - \frac{1}{2!}c_0 x^2 + \frac{1}{4!}c_0 x^4 - \frac{1}{6!}c_0 x^6 + \cdots . \tag{1.56}$$

Thus, for arbitrary nonzero c_0, the RHS of (1.56) represents an even function which is equal to the negative of its second derivative. The coefficient c_0 is arbitrary. Setting $f_{\text{even}}(0) = 1$ results in $c_0 = 1$. Denoting the resulting function by $\cos(x)$,

$$\cos(x) = 1 - \frac{1}{2!}x^2 + \frac{1}{4!}x^4 - \frac{1}{6!}x^6 + \cdots . \tag{1.57}$$

Given the set of monomials $\{1, x^2, x^4, x^6, \dots\}$, the coefficients $\{\frac{1}{0!}, -\frac{1}{2!}, \frac{1}{4!}, -\frac{1}{6!}, \dots\}$ define the function $\cos(x)$. The coefficient $(-1)^n \frac{1}{(2n)!}$ uniquely specifies the monomial x^{2n} ($n \in \mathbb{N}_0$). Since the set of monomials is universal, the set of coefficients suffices to encode the information necessary to synthesize $\cos(x)$.

■

Problem: Does any function $f_{\text{odd}}(x)$ with $f_{\text{odd}}^{(2)}(x) = -f_{\text{odd}}(x)$ and $f_{\text{odd}}^{(1)}(0) = 1$ exist?

Solution: Consider $f_{\text{odd}}(x)$ alongside its first and second derivatives,

$$f_{\text{odd}}(x) = c_1 x + c_3 x^3 + c_5 x^5 + c_7 x^7 + \cdots \tag{1.58a}$$

$$f_{\text{odd}}^{(1)}(x) = c_1 + 3c_3 x^2 + 5c_5 x^4 + 7c_7 x^6 + \cdots \tag{1.58b}$$

$$f_{\text{odd}}^{(2)}(x) = 2 \cdot 3c_3 x + 4 \cdot 5c_5 x^3 + 6 \cdot 7c_7 x^5 + \cdots . \tag{1.58c}$$

Since $f_{\text{odd}}(x)$ and $f_{\text{odd}}^{(2)}(x)$ are both odd functions, it is legitimate to expect that a certain set of coefficients c_{2n+1} ($n \in \mathbb{N}_0$) would ensure the equality $f_{\text{odd}}^{(2)}(x) = -f_{\text{odd}}(x)$. Employing (1.58a) and (1.58c),

$$2 \cdot 3c_3 x + 4 \cdot 5c_5 x^3 + 6 \cdot 7c_7 x^5 + \cdots$$
$$= -c_1 x - c_3 x^3 - c_5 x^5 - c_7 x^7 - \cdots . \tag{1.59}$$

A term-by-term comparison results in

$$2 \cdot 3c_3 = -c_1 \implies c_3 = -\frac{1}{2 \cdot 3} c_1 = -\frac{1}{3!} c_1 \tag{1.60a}$$

$$4 \cdot 5c_5 = -c_3 \implies c_5 = -\frac{1}{4 \cdot 5} c_3 = \frac{1}{5!} c_1 \tag{1.60b}$$

$$6 \cdot 7c_7 = -c_5 \implies c_7 = -\frac{1}{6 \cdot 7} c_5 = -\frac{1}{7!} c_1 \tag{1.60c}$$

$$\vdots$$

$$(2n) \cdot (2n+1) c_{2n+1} = -c_{2n-1} \implies c_{2n+1} = (-1)^n \frac{1}{(2n+1)!} c_1. \tag{1.60d}$$

This result reveals the source for alternating signs of the coefficients. Substituting c_{2n+1} ($n \in \mathbb{N}$) into (1.58a),

$$f_{\text{odd}}(x) = c_1 x - \frac{1}{3!} c_1 x^3 + \frac{1}{5!} c_1 x^5 - \frac{1}{7!} c_1 x^7 + \cdots . \tag{1.61}$$

Thus, for arbitrary $c_1 \neq 0$ the R.H.S. of (1.61) represents an odd function which is equal to the negative of its second derivative. The coefficient c_1 is arbitrary. Setting $f_{\text{odd}}^{(1)}(0) = 1$ results in $c_1 = 1$. Denoting the resulting function by $\sin(x)$,

$$\sin(x) = x - \frac{1}{3!} x^3 + \frac{1}{5!} x^5 - \frac{1}{7!} x^7 + \cdots . \tag{1.62}$$

Given the set of monomials $\{x, x^3, x^5, x^7, \dots\}$, the coefficients $\left\{ \frac{1}{1!}, -\frac{1}{3!}, \frac{1}{5!}, -\frac{1}{7!}, \dots \right\}$ define the function $\sin(x)$. The coefficient $(-1)^n \frac{1}{(2n+1)!}$ uniquely specifies the monomial x^{2n+1} ($n \in \mathbb{N}_0$). Since the set of monomials is universal, the set of coefficients suffices to encode the information necessary to synthesize $\sin(x)$. ∎

Problem: Do any coupled functions $g_{\text{odd}}(x)$ and $h_{\text{even}}(x)$, satisfying $g_{\text{odd}}^{(1)}(x) = h_{\text{even}}(x)$ and $h_{\text{even}}^{(1)}(x) = -g_{\text{odd}}(x)$ and subject to the constraints, $h_{\text{even}}(0) = 1$ and $g_{\text{odd}}^{(1)}(0) = 1$ exist?

Solution: Consider $g_{\text{odd}}(x)$ and $h_{\text{even}}(x)$ alongside their first derivatives, in terms of their Formal Power Series Expansions,

$$g_{\text{odd}}(x) = c_1 x + c_3 x^3 + c_5 x^5 + c_7 x^7 + \cdots \tag{1.63a}$$

$$g_{\text{odd}}^{(1)}(x) = c_1 + 3c_3 x^2 + 5c_5 x^4 + 7c_7 x^6 + \cdots \tag{1.63b}$$

$$h_{\text{even}}(x) = c_0 + c_2 x^2 + c_4 x^4 + c_6 x^6 + c_8 x^8 + \cdots \tag{1.63c}$$

$$h_{\text{even}}^{(1)}(x) = 2c_2 x + 4c_4 x^3 + 6c_6 x^5 + 8c_8 x^7 + \cdots . \tag{1.63d}$$

The functions $g_{\text{odd}}^{(1)}(x)$ and $h_{\text{even}}(x)$ are both even, and the functions $h_{\text{even}}^{(1)}(x)$ and $g_{\text{odd}}(x)$ are both odd functions, respectively. Thus, it is justified to ask which coefficients c_n ($n \in \mathbb{N}_0$) would simultaneously render the equations,

$$g_{\text{odd}}^{(1)}(x) = h_{\text{even}}(x) \tag{1.64a}$$

$$h_{\text{even}}^{(1)}(x) = -g_{\text{odd}}(x) , \tag{1.64b}$$

valid. Employing (1.63), the coupled equations in (1.64) yield,

$$\begin{aligned} c_1 + 3c_3x^2 + 5c_5x^4 + 7c_7x^6 + \cdots \\ = c_0 + c_2x^2 + c_4x^4 + c_6x^6 + c_8x^8 + \cdots \end{aligned} \tag{1.65a}$$

$$\begin{aligned} 2c_2x + 4c_4x^3 + 6c_6x^5 + 8c_8x^7 + \cdots \\ = -c_1x - c_3x^3 - c_5x^5 - c_7x^7 - \cdots \end{aligned} \tag{1.65b}$$

Exercise: Show that using (1.64a) alone, which implies (1.65a), would be insufficient for the determination of the coefficients.

■

Exercise: Show that using (1.64b) alone, which implies (1.65b), would be insufficient for the determination of the coefficients.

■

A term-by-term comparison in (1.65a) results in

$$c_1 = c_0 \tag{1.66a}$$
$$3c_3 = c_2 \tag{1.66b}$$
$$5c_5 = c_4 \tag{1.66c}$$
$$7c_7 = c_6 \tag{1.66d}$$
$$\vdots \tag{1.66e}$$

A term-by-term comparison in (1.65b) results in

$$2c_2 = -c_1 \tag{1.67a}$$
$$4c_4 = -c_3 \tag{1.67b}$$
$$6c_6 = -c_5 \tag{1.67c}$$
$$8c_8 = -c_7 \tag{1.67d}$$
$$\vdots \tag{1.67e}$$

Observe the "interwoven" manner in which the coefficients are interrelated. Adequately interlacing the equations in (1.66) and (1.67),

$$c_1 = c_0 \tag{1.68a}$$

$$2c_2 = -c_1 \tag{1.68b}$$

$$3c_3 = c_2 \tag{1.68c}$$

$$4c_4 = -c_3 \tag{1.68d}$$

$$5c_5 = c_4 \tag{1.68e}$$

$$6c_6 = -c_5 \tag{1.68f}$$

$$7c_7 = c_6 \tag{1.68g}$$

$$8c_8 = -c_7 \tag{1.68h}$$

$$\vdots$$

The coefficients c_n can be determined successively,

$$c_1 = c_0 \tag{1.69a}$$

$$2c_2 = -c_1 \implies c_2 = -\frac{1}{2}c_1 = -\frac{1}{2!}c_0 \tag{1.69b}$$

$$3c_3 = c_2 \implies c_3 = \frac{1}{3}c_2 = -\frac{1}{3!}c_0 \tag{1.69c}$$

$$4c_4 = -c_3 \implies c_4 = -\frac{1}{4}c_3 = \frac{1}{4!}c_0 \tag{1.69d}$$

$$5c_5 = c_4 \implies c_5 = \frac{1}{5}c_4 = \frac{1}{5!}c_0 \tag{1.69e}$$

$$6c_6 = -c_5 \implies c_6 = -\frac{1}{6}c_5 = -\frac{1}{6!}c_0 \tag{1.69f}$$

$$7c_7 = c_6 \implies c_7 = \frac{1}{7}c_6 = -\frac{1}{7!}c_0 \tag{1.69g}$$

$$8c_8 = -c_7 \implies c_8 = -\frac{1}{8}c_7 = \frac{1}{8!}c_0 \tag{1.69h}$$

$$\vdots$$

$$(2n+1)c_{2n+1} = c_{2n} \implies c_{2n+1} = \frac{1}{2n+1}c_{2n} = (-1)^n \frac{1}{(2n+1)!}c_0 \tag{1.69i}$$

$$(2n+2)c_{2n+2} = -c_{2n+1} \implies c_{2n+2} = -\frac{1}{2n+2}c_{2n+1} = (-1)^n \frac{1}{(2n+2)!}c_0 \tag{1.69j}$$

$$\vdots$$

Note that starting with c_2, any two consecutive coefficients of the same parity (sign) are followed by two coefficients of the opposite parity.

Even though not required for the solution of the present problem, it is illuminating to split these equations into two decks of equations involving, respectively, the odd,

$$c_1 = c_0 \tag{1.70a}$$

$$c_3 = -\frac{1}{3!}c_0 \tag{1.70b}$$

$$c_5 = \frac{1}{5!}c_0 \tag{1.70c}$$

$$c_7 = -\frac{1}{7!}c_0 \tag{1.70d}$$

$$c_{2n+1} = (-1)^n\frac{1}{(2n+1)!}c_0 \tag{1.70e}$$

$$\vdots \quad ,$$

and the even coefficients,

$$c_2 = -\frac{1}{2!}c_0 \tag{1.71a}$$

$$c_4 = = = \frac{1}{4!}c_0 \tag{1.71b}$$

$$c_6 = -\frac{1}{6!}c_0 \tag{1.71c}$$

$$c_8 = \frac{1}{8!}c_0 \tag{1.71d}$$

$$\vdots$$

$$c_{2n+2} = (-1)^n\frac{1}{(2n+2)!}c_0 \tag{1.71e}$$

$$\vdots \quad .$$

Using $h_{even}(0) = 1$ and $g_{odd}^{(1)}(0) = 1$ results (consistently) in $c_0 = c_1 = 1$. Identifying the functions $h_{even}(x) = \cos(x)$ and $g_{odd}(x) = \sin(x)$, respectively, in terms of the coefficients in (1.71) and (1.70),

$$\cos(x) = 1 - \frac{1}{2!}x^2 + \frac{1}{4!}x^4 - \frac{1}{6!}x^6 + \cdots \tag{1.72a}$$

$$\sin(x) = x - \frac{1}{3!}x^3 + \frac{1}{5!}x^5 - \frac{1}{7!}x^7 + \cdots . \tag{1.72b}$$

Further above, it was established that $h_{even}(x) = \cos(x)$ and $g_{odd}(x) = \sin(x)$. Considering these relationships in (1.64),

$$\cos^{(1)}(x) = -\sin(x) \tag{1.73a}$$

$$\sin^{(1)}(x) = \cos(x). \tag{1.73b}$$

Remark: It is revealing to observe the sequel of $\pm$ in the Formal Taylor Series Expansions in (1.72): consider the expansion of cos (x) talking the derivative of $\cos(x)$ the constant first-term vanishes and the expansion of $\cos^{(1)}(x)$ starts with $-x$. On the other hand, the expansion of $\sin(x)$ starts with x. This explains the minus sign in $\cos^{(1)}(x) = -\sin(x)$. In contrast, the derivative of the expansion of $\sin(x)$ starts with 1. Since the expansion of $\cos(x)$ also starts with 1, there is no explicit minus sign in the equation $\sin^{(1)}(x) = \cos(x)$.

□

Problem: Express (1.73) as one system of coupled equations and generalize the result by mathematical induction.

Solutions: Explicating the differential operators in (1.73),

$$\frac{d}{dx}\cos(x) = -\sin(x) \tag{1.74a}$$

$$\frac{d}{dx}\sin(x) = \cos(x), \tag{1.74b}$$

or, more compactly,

$$\frac{d}{dx}\begin{bmatrix} \cos(x) \\ \sin(x) \end{bmatrix} = \begin{bmatrix} -\sin(x) \\ \cos(x) \end{bmatrix}. \tag{1.75}$$

"Transforming" the vector at the R.H.S. into the vector at the L.H.S.,

$$\frac{d}{dx}\begin{bmatrix} \cos(x) \\ \sin(x) \end{bmatrix} = \begin{bmatrix} 0 & -1 \\ 1 & 0 \end{bmatrix}\begin{bmatrix} \cos(x) \\ \sin(x) \end{bmatrix}. \tag{1.76}$$

The innocent looking matrix at the R.H.S. is of paramount significance, as it will turn out momentarily. Differentiating the terms on both sides with respect to x,

$$\frac{d^2}{dx^2}\begin{bmatrix} \cos(x) \\ \sin(x) \end{bmatrix} = \begin{bmatrix} 0 & -1 \\ 1 & 0 \end{bmatrix}\frac{d}{dx}\begin{bmatrix} \cos(x) \\ \sin(x) \end{bmatrix}. \tag{1.77}$$

Substituting (1.76) into (1.77),

$$\frac{d^2}{dx^2}\begin{bmatrix} \cos(x) \\ \sin(x) \end{bmatrix} = \begin{bmatrix} 0 & -1 \\ 1 & 0 \end{bmatrix}\begin{bmatrix} 0 & -1 \\ 1 & 0 \end{bmatrix}\begin{bmatrix} \cos(x) \\ \sin(x) \end{bmatrix} \tag{1.78a}$$

$$= \begin{bmatrix} 0 & -1 \\ 1 & 0 \end{bmatrix}^2\begin{bmatrix} \cos(x) \\ \sin(x) \end{bmatrix}. \tag{1.78b}$$

Reasoning by induction,

$$\frac{d^n}{dx^n}\begin{bmatrix} \cos(x) \\ \sin(x) \end{bmatrix} = \begin{bmatrix} 0 & -1 \\ 1 & 0 \end{bmatrix}^n\begin{bmatrix} \cos(x) \\ \sin(x) \end{bmatrix}, \tag{1.79}$$

which is hypothesized to be valid for arbitrary $n \in \mathbb{N}_0$.

Remark: The proof by mathematical induction in general.

The proof by mathematical induction involves the initial step, (1.76), the induction hypothesis, (1.79), and the final step. The final step consists of inferring from the induction hypothesis which is a statement valid for $n \in \mathbb{N}$, to a statement, which is valid for $n + 1$. Thereby, ordinarily, the induction hypothesis, the initial step, and any other contextually valid statement can be employed. In elementary cases, guessing the induction hypothesis and carrying out the final step are self-evident and the process suggests itself. Solving more advanced and thus interesting cases, both the formulation of the induction hypothesis and the accomplishment of the final step require a certain level of sophistication, which necessarily reveals deeper insight into the hidden structure of the problem. What mathematicians do is that they incorporate these new insights into the statement of the problem. The resulting reformulated problem in turn, merely and straightforwardly requires an initial step, an induction hypothesis, and a final step. It can happen that the embedding of new insights into the statement of the problem is performed not system-characteristically, i.e., in an *ad hoc* manner. Very often this is the reason why the reader may find the formulation (statement) of the problem not intuitive. From the epistemological and problem solving perspectives, these considerations may become instrumental when the engineer ventures into new territories.

□

For establishing the final step, differentiate both sides of (1.79) with respect to x,

$$\frac{d^{n+1}}{dx^{n+1}} \begin{bmatrix} \cos(x) \\ \sin(x) \end{bmatrix} = \begin{bmatrix} 0 & -1 \\ 1 & 0 \end{bmatrix}^n \frac{d}{dx} \begin{bmatrix} \cos(x) \\ \sin(x) \end{bmatrix}. \tag{1.80}$$

Considering the induction initial step, (1.76), at the R.H.S.,

$$\frac{d^{n+1}}{dx^{n+1}} \begin{bmatrix} \cos(x) \\ \sin(x) \end{bmatrix} = \begin{bmatrix} 0 & -1 \\ 1 & 0 \end{bmatrix}^n \begin{bmatrix} 0 & -1 \\ 1 & 0 \end{bmatrix} \begin{bmatrix} \cos(x) \\ \sin(x) \end{bmatrix}, \tag{1.81}$$

results in,

$$\frac{d^{n+1}}{dx^{n+1}} \begin{bmatrix} \cos(x) \\ \sin(x) \end{bmatrix} = \begin{bmatrix} 0 & -1 \\ 1 & 0 \end{bmatrix}^{n+1} \begin{bmatrix} \cos(x) \\ \sin(x) \end{bmatrix}. \tag{1.82}$$

This completes the proof by mathematical induction.

∎

Remark: From an epistemological perspective, inductive reasoning (logic) which gives rise to the "Induction Problem" in philosophy, may not be regarded as watertight as the deductive reasoning. However, within the scope of arithmetic and analysis, the "Proof by Mathematical Induction" deserves to be viewed as a sound and valid way of proving hypotheses and establishing relationships.

□

Remark: For completeness, it should be mentioned that proofs by contradiction, even though logically valid and equivalent with direct proofs, are not highly valued by engineers, who prefer direct proofs: direct proofs are often constructive. To engineers and technologists, direct proofs are preferred techniques, whenever there is a choice to apply them. Resorting to proofs by contradiction and other proof strategies are mostly pursued when direct proofs are excessively difficult to conceive, or out of reach.

□

Remark: Consider,

$$\begin{bmatrix} 0 & -1 \\ 1 & 0 \end{bmatrix} \begin{bmatrix} 0 & -1 \\ 1 & 0 \end{bmatrix} = \begin{bmatrix} -1 & 0 \\ 0 & -1 \end{bmatrix}, \tag{1.83}$$

in (1.78h),

$$\frac{d^2}{dx^2} \begin{bmatrix} \cos(x) \\ \sin(x) \end{bmatrix} = \begin{bmatrix} -1 & 0 \\ 0 & -1 \end{bmatrix} \begin{bmatrix} \cos(x) \\ \sin(x) \end{bmatrix}. \tag{1.84}$$

The two equations decouple into the defining equations for $\cos(x)$ and $\sin(x)$,

$$\frac{d^2}{dx^2}\cos(x) = -\cos(x) \tag{1.85a}$$

$$\frac{d^2}{dx^2}\sin(x) = -\sin(x), \tag{1.85b}$$

in agreement with the results established earlier.

$\square$

1.9   The Formal Taylor Series Expansion as a powerful manipulatory tool

The following examples demonstrate the versatility and utility of the Formal Taylor Series Expansion in establishing trigonometric and other important relationships.

Problem: Apply the Formal Taylor Series Expansion to the function $\cos(\alpha + \Delta\alpha)$.

Solution: Consider, the Formal Taylor Series Expansion of $\cos(\alpha + \Delta\alpha)$ at α,

$$\cos(\alpha + \Delta\alpha) = \sum_{n=0}^{\infty}\left[\frac{d^n}{d\alpha^n}\cos(\alpha + \Delta\alpha)\right]_{\Delta\alpha=0}\frac{1}{n!}(\Delta\alpha)^n. \tag{1.86}$$

Expanding the series,

$$\begin{aligned}
\cos(\alpha + \Delta\alpha) = &\left[\frac{d^0}{d\alpha^0}\cos(\alpha + \Delta\alpha)\right]_{\Delta\alpha=0}\frac{1}{0!}(\Delta\alpha)^0\\
&+\left[\frac{d^1}{d\alpha^1}\cos(\alpha + \Delta\alpha)\right]_{\Delta\alpha=0}\frac{1}{1!}(\Delta\alpha)^1\\
&+\left[\frac{d^2}{d\alpha^2}\cos(\alpha + \Delta\alpha)\right]_{\Delta\alpha=0}\frac{1}{2!}(\Delta\alpha)^2\\
&+\left[\frac{d^3}{d\alpha^3}\cos(\alpha + \Delta\alpha)\right]_{\Delta\alpha=0}\frac{1}{3!}(\Delta\alpha)^3\\
&+\left[\frac{d^4}{d\alpha^4}\cos(\alpha + \Delta\alpha)\right]_{\Delta\alpha=0}\frac{1}{4!}(\Delta\alpha)^4\\
&+\left[\frac{d^5}{d\alpha^5}\cos(\alpha + \Delta\alpha)\right]_{\Delta\alpha=0}\frac{1}{5!}(\Delta\alpha)^5\\
&+\left[\frac{d^6}{d\alpha^6}\cos(\alpha + \Delta\alpha)\right]_{\Delta\alpha=0}\frac{1}{6!}(\Delta\alpha)^6\\
&+\left[\frac{d^7}{d\alpha^7}\cos(\alpha + \Delta\alpha)\right]_{\Delta\alpha=0}\frac{1}{7!}(\Delta\alpha)^7\\
&\vdots
\end{aligned} \tag{1.87}$$

Recall the conventions $0! = 1$, $(\Delta\alpha)^0 = 1$ and $(\Delta\alpha)^1 = (\Delta\alpha)$. In view of $\left(d^0/d\alpha^0\right)\cos(\alpha) = \cos(\alpha)$, and employing the coupled equations,

$$\frac{d}{d\alpha}\cos(\alpha) = -\sin(\alpha) \tag{1.88a}$$

$$\frac{d}{d\alpha}\sin(\alpha) = \cos(\alpha), \tag{1.88b}$$

explicit expressions for the higher-order derivatives of $\cos(\alpha)$ can be derived successively. The first few results are listed below for easy reference,

$$\frac{d^0}{d\alpha^0}\cos(\alpha) = \cos(\alpha) \tag{1.89a}$$

$$\frac{d^1}{d\alpha^1}\cos(\alpha) = -\sin(\alpha) \tag{1.89b}$$

$$\frac{d^2}{d\alpha^2}\cos(\alpha) = \frac{d}{d\alpha}\left[\frac{d}{d\alpha}\cos(\alpha)\right] = \frac{d}{d\alpha}[-\sin(\alpha)] = -\frac{d}{d\alpha}\sin(\alpha) = -\cos(\alpha) \tag{1.89c}$$

$$\frac{d^3}{d\alpha^3}\cos(\alpha) = \frac{d}{d\alpha}\left[\frac{d^2}{d\alpha^2}\cos(\alpha)\right] = \frac{d}{d\alpha}[-\cos(\alpha)] = -\frac{d}{d\alpha}\cos(\alpha) = \sin(\alpha) \tag{1.89d}$$

$$\frac{d^4}{d\alpha^4}\cos(\alpha) = \frac{d}{d\alpha}\left[\frac{d^3}{d\alpha^3}\cos(\alpha)\right] = \frac{d}{d\alpha}\sin(\alpha) = \cos(\alpha) \tag{1.89e}$$

$$\frac{d^5}{d\alpha^5}\cos(\alpha) = \frac{d}{d\alpha}\left[\frac{d^4}{d\alpha^4}\cos(\alpha)\right] = \frac{d}{d\alpha}\cos(\alpha) = -\sin(\alpha) \tag{1.89f}$$

$$\frac{d^6}{d\alpha^6}\cos(\alpha) = \frac{d}{d\alpha}\left[\frac{d^5}{d\alpha^5}\cos(\alpha)\right] = \frac{d}{d\alpha}[-\sin(\alpha)] = -\cos(\alpha) \tag{1.89g}$$

$$\frac{d^7}{d\alpha^7}\cos(\alpha) = \frac{d}{d\alpha}\left[\frac{d^6}{d\alpha^6}\cos(\alpha)\right] = \frac{d}{d\alpha}[-\cos(\alpha)] = \sin(\alpha) \tag{1.89h}$$

$$\vdots$$

A recurring pattern emerges: as a consequence of (1.88), the occurrence of the derivatives repeats itself with the "period length" of four. More specifically, the zeroth, first, second, and the third derivatives of $\cos(\alpha)$ are, respectively, $\cos(\alpha)$, $-\sin(\alpha)$, $-\cos(\alpha)$, and $\sin(\alpha)$. The fourth, fifth, sixth, and the seventh derivatives result in the same sequence of derivatives, and consequently, in $\cos(\alpha)$, $-\sin(\alpha)$, $-\cos(\alpha)$, and $\sin(\alpha)$, respectively. Let $(n \bmod 4)$, $(n \in \mathbb{N}_0)$ stand for the remainder of n after repeated division by 4. Then, $(n \bmod 4)$ generates the set $\{0, 1, 2, 3\}$, allowing to write,

$$\frac{d^n}{d\alpha^n}\cos(\alpha) = \frac{d^{(n \bmod 4)}}{d\alpha^{(n \bmod 4)}}\cos(\alpha) = \begin{cases} \cos(\alpha) & n \bmod 4 = 0 \\ -\sin(\alpha) & n \bmod 4 = 1 \\ -\cos(\alpha) & n \bmod 4 = 2 \\ \sin(\alpha) & n \bmod 4 = 3 \end{cases} \tag{1.90}$$

Thus, (1.87),

$$\cos(\alpha + \Delta\alpha) = \left[\cos(\alpha + \Delta\alpha)|_{\Delta\alpha=0}\right]$$

$$+ \left[-\sin(\alpha + \Delta\alpha)|_{\Delta\alpha=0}\right](\Delta\alpha)$$

$$+ \left[-\cos(\alpha + \Delta\alpha)|_{\Delta\alpha=0}\right]\frac{1}{2!}(\Delta\alpha)^2$$

$$+ \left[\sin(\alpha + \Delta\alpha)|_{\Delta\alpha=0}\right]\frac{1}{3!}(\Delta\alpha)^3$$

$$+ \left[\cos(\alpha + \Delta\alpha)|_{\Delta\alpha=0}\right]\frac{1}{4!}(\Delta\alpha)^4$$

$$+ \left[-\sin(\alpha + \Delta\alpha)|_{\Delta\alpha=0}\right]\frac{1}{5!}(\Delta\alpha)^5$$

$$+ \left[-\cos(\alpha + \Delta\alpha)|_{\Delta\alpha=0}\right]\frac{1}{6!}(\Delta\alpha)^6$$

$$+ \left[\sin(\alpha + \Delta\alpha)|_{\Delta\alpha=0}\right]\frac{1}{7!}(\Delta\alpha)^7$$

$$\vdots \qquad (1.91)$$

Simplifying the terms in the square brackets,

$$\cos(\alpha + \Delta\alpha) = \cos(\alpha)$$

$$- \sin(\alpha)(\Delta\alpha)$$

$$- \cos(\alpha)\frac{1}{2!}(\Delta\alpha)^2$$

$$+ \sin(\alpha)\frac{1}{3!}(\Delta\alpha)^3$$

$$+ \cos(\alpha)\frac{1}{4!}(\Delta\alpha)^4$$

$$- \sin(\alpha)\frac{1}{5!}(\Delta\alpha)^5$$

$$- \cos(\alpha)\frac{1}{6!}(\Delta\alpha)^6$$

$$+ \sin(\alpha)\frac{1}{7!}(\Delta\alpha)^7$$

$$\vdots \qquad (1.92)$$

Grouping the terms in accordance with their association with $\cos(\alpha)$ or $\sin(\alpha)$,

$$\cos(\alpha + \Delta\alpha) = \cos(\alpha)\left[1 - \frac{1}{2!}(\Delta\alpha)^2 + \frac{1}{4!}(\Delta\alpha)^4 - \frac{1}{6!}(\Delta\alpha)^6 + \cdots\right]$$

$$- \sin(\alpha)\left[\Delta\alpha - \frac{1}{3!}(\Delta\alpha)^3 + \frac{1}{5!}(\Delta\alpha)^5 - \frac{1}{7!}(\Delta\alpha)^7 + \cdots\right]. \qquad (1.93)$$

Identifying the series in the first- and the second square brackets, as the Formal Taylor Series Expansions of $\cos(\Delta\alpha)$ and $\sin(\Delta\alpha)$, respectively,

$$\cos(\alpha + \Delta\alpha) = \cos(\alpha)\cos(\Delta\alpha) - \sin(\alpha)\sin(\Delta\alpha).\tag{1.94}$$

Note that nowhere in the derivation it was required that $\Delta\alpha$ must be small. Consequently, $\Delta\alpha$ can be replaced by any angle θ of finite magnitude. The elementary trigonometric identity manifests itself,

$$\cos(\alpha + \theta) = \cos(\alpha)\cos(\theta) - \sin(\alpha)\sin(\theta).\tag{1.95}$$

$\square$

Problem: Apply the Formal Taylor Series Expansion to the function $\sin(\alpha + \Delta\alpha)$.

Solution:

$$\sin(\alpha + \Delta\alpha) = \left[\frac{d^0}{dx^0}\sin(\alpha + \Delta\alpha)\right]_{\Delta\alpha=0}\frac{1}{0!}(\Delta\alpha)^0$$

$$+ \left[\frac{d^1}{dx^1}\sin(\alpha + \Delta\alpha)|\right]_{\Delta\alpha=0}\frac{1}{1!}(\Delta\alpha)^1$$

$$+ \left[\frac{d^2}{dx^2}\sin(\alpha + \Delta\alpha)\right]_{\Delta\alpha=0}\frac{1}{2!}(\Delta\alpha)^2$$

$$+ \left[\frac{d^3}{dx^3}\sin(\alpha + \Delta\alpha)\right]_{\Delta\alpha=0}\frac{1}{3!}(\Delta\alpha)^3$$

$$+ \left[\frac{d^4}{dx^4}\sin(\alpha + \Delta\alpha)\right]_{\Delta\alpha=0}\frac{1}{4!}(\Delta\alpha)^4$$

$$+ \left[\frac{d^5}{dx^5}\sin(\alpha + \Delta\alpha)\right]_{\Delta\alpha=0}\frac{1}{5!}(\Delta\alpha)^5$$

$$+ \left[\frac{d^6}{dx^6}\sin(\alpha + \Delta\alpha)\right]_{\Delta\alpha=0}\frac{1}{6!}(\Delta\alpha)^6$$

$$+ \left[\frac{d^7}{dx^7}\sin(\alpha + \Delta\alpha)\right]_{\Delta\alpha=0}\frac{1}{7!}(\Delta\alpha)^7$$

$$\vdots\tag{1.96}$$

Employing (1.88),

$$\begin{aligned}
\sin(\alpha + \Delta\alpha) = & [\sin(\alpha + \Delta\alpha)]_{\Delta\alpha=0} \\
& + [\cos(\alpha + \Delta\alpha)]_{\Delta\alpha=0}(\Delta\alpha) \\
& + [-\sin(\alpha + \Delta\alpha)]_{\Delta\alpha=0}\frac{1}{2!}(\Delta\alpha)^2 \\
& + [-\cos(\alpha + \Delta\alpha)]_{\Delta\alpha=0}\frac{1}{3!}(\Delta\alpha)^3 \\
& + [\sin(\alpha + \Delta\alpha)]_{\Delta\alpha=0}\frac{1}{4!}(\Delta\alpha)^4 \\
& + [\cos(\alpha + \Delta\alpha)]_{\Delta\alpha=0}\frac{1}{5!}(\Delta\alpha)^5 \\
& + [-\sin(\alpha + \Delta\alpha)]_{\Delta\alpha=0}\frac{1}{6!}(\Delta\alpha)^6 \\
& + [-\cos(\alpha + \Delta\alpha)]_{\Delta\alpha=0}\frac{1}{7!}(\Delta\alpha)^7 \\
& \quad\vdots
\end{aligned} \tag{1.97}$$

Simplifying the terms in square brackets,

$$\begin{aligned}
\sin(\alpha + \Delta\alpha) = & \sin(\alpha) \\
& + \cos(\alpha)(\Delta\alpha) \\
& - \sin(\alpha)\frac{1}{2!}(\Delta\alpha)^2 \\
& - \cos(\alpha)\frac{1}{3!}(\Delta\alpha)^3 \\
& + \sin(\alpha)\frac{1}{4!}(\Delta\alpha)^4 \\
& + \cos(\alpha)\frac{1}{5!}(\Delta\alpha)^5 \\
& - \sin(\alpha)\frac{1}{6!}(\Delta\alpha)^6 \\
& - \cos(\alpha)\frac{1}{7!}(\Delta\alpha)^7 \\
& \quad\vdots
\end{aligned} \tag{1.98}$$

Grouping the terms in accordance with their association with $\sin(\alpha)$ or $\cos(\alpha)$,

$$\begin{aligned}
\sin(\alpha + \Delta\alpha) = & \sin(\alpha)\left[1 - \frac{1}{2!}(\Delta\alpha)^2 + \frac{1}{4!}(\Delta\alpha)^4 - \frac{1}{6!}(\Delta\alpha)^6 + \cdots\right] \\
& + \cos(\alpha)\left[\Delta\alpha - \frac{1}{3!}(\Delta\alpha)^3 + \frac{1}{5!}(\Delta\alpha)^5 - \frac{1}{7!}(\Delta\alpha)^7 + \cdots\right].
\end{aligned} \tag{1.99}$$

Identifying the terms in the first and the second square brackets as the Formal Taylor Series Expansion of $\cos(\Delta\alpha)$ and $\sin(\Delta\alpha)$, respectively,

$$\sin(\alpha + \Delta\alpha) = \sin(\alpha)\cos(\Delta\alpha) + \cos(\alpha)\sin(\Delta\alpha). \tag{1.100}$$

Replacing $\Delta\alpha$ with θ, the elementary trigonometric identity,

$$\sin(\alpha + \theta) = \sin(\alpha)\cos(\theta) + \cos(\alpha)\sin(\theta), \tag{1.101}$$

manifests itself.

□

Problem: Obtain an equivalent representation for the column vector $\begin{bmatrix} \cos(\alpha + \Delta\alpha) \\ \sin(\alpha + \Delta\alpha) \end{bmatrix}$.

Solution: Utilizing the results obtained in the preceding two problems,

$$\begin{bmatrix} \cos(\alpha + \Delta\alpha) \\ \sin(\alpha + \Delta\alpha) \end{bmatrix} = \begin{bmatrix} \cos(\alpha)\cos(\Delta\alpha) - \sin(\alpha)\sin(\Delta\alpha) \\ \sin(\alpha)\cos(\Delta\alpha) + \cos(\alpha)\sin(\Delta\alpha) \end{bmatrix}. \tag{1.102}$$

Substitute the Formal Taylor Series Expansion of $\cos(\Delta\alpha)$ and $\sin(\Delta\alpha)$,

$$\cos(\Delta\alpha) = 1 - \frac{1}{2!}(\Delta\alpha)^2 + \frac{1}{4!}(\Delta\alpha)^4 + \cdots \tag{1.103a}$$

$$\sin(\Delta\alpha) = \Delta\alpha - \frac{1}{3!}(\Delta\alpha)^3 + \frac{1}{5!}(\Delta\alpha)^5 + \cdots, \tag{1.103b}$$

at the R.H.S. of (1.102),

$$\begin{bmatrix} \cos(\alpha + \Delta\alpha) \\ \sin(\alpha + \Delta\alpha) \end{bmatrix} = \begin{bmatrix} \cos(\alpha)\left\{1 - \frac{1}{2!}(\Delta\alpha)^2 + \cdots\right\} - \sin(\alpha)\left\{\Delta\alpha - \frac{1}{3!}(\Delta\alpha)^3 + \cdots\right\} \\ \sin(\alpha)\left\{1 - \frac{1}{2!}(\Delta\alpha)^2 + \cdots\right\} + \cos(\alpha)\left\{\Delta\alpha - \frac{1}{3!}(\Delta\alpha)^3 + \cdots\right\} \end{bmatrix}. \tag{1.104}$$

Separating the terms associated with the Markers $(\Delta\alpha)^0 = 1$, $(\Delta\alpha)^1 = \Delta\alpha$, $(\Delta\alpha)^2$, ..., at the R.H.S.,

$$\begin{bmatrix} \cos(\alpha + \Delta\alpha) \\ \sin(\alpha + \Delta\alpha) \end{bmatrix} = \begin{bmatrix} \cos(\alpha) \\ \sin(\alpha) \end{bmatrix} + (\Delta\alpha)\begin{bmatrix} -\sin(\alpha) \\ \cos(\alpha) \end{bmatrix} + \frac{1}{2!}(\Delta\alpha)^2\begin{bmatrix} -\cos(\alpha) \\ -\sin(\alpha) \end{bmatrix}$$

$$+ \frac{1}{3!}(\Delta\alpha)^3\begin{bmatrix} \sin(\alpha) \\ -\cos(\alpha) \end{bmatrix} + \cdots. \tag{1.105}$$

Transforming the vectors at the R.H.S. into the canonical (standard) form,

$$
\begin{bmatrix} \cos(\alpha + \Delta\alpha) \\ \sin(\alpha + \Delta\alpha) \end{bmatrix} = \begin{bmatrix} \cos(\alpha) \\ \sin(\alpha) \end{bmatrix} + (\Delta\alpha) \begin{bmatrix} 0 & -1 \\ 1 & 0 \end{bmatrix} \begin{bmatrix} \cos(\alpha) \\ \sin(\alpha) \end{bmatrix}
$$
$$
+ \frac{1}{2!}(\Delta\alpha)^2 \begin{bmatrix} -1 & 0 \\ 0 & -1 \end{bmatrix} \begin{bmatrix} \cos(\alpha) \\ \sin(\alpha) \end{bmatrix}
$$
$$
+ \frac{1}{3!}(\Delta\alpha)^3 \begin{bmatrix} 0 & 1 \\ -1 & 0 \end{bmatrix} \begin{bmatrix} \cos(\alpha) \\ \sin(\alpha) \end{bmatrix} + \cdots . \tag{1.106}
$$

It is immediate that the second and the third matrices at the R.H.S. are the second and the third powers of the fundamental matrix $\begin{bmatrix} 0 & -1 \\ 1 & 0 \end{bmatrix}$. Upon extracting the fundamental matrix,

$$
\begin{bmatrix} -1 & 0 \\ 0 & -1 \end{bmatrix} = \begin{bmatrix} 0 & -1 \\ 1 & 0 \end{bmatrix} \begin{bmatrix} 0 & -1 \\ 1 & 0 \end{bmatrix}
$$
$$
= \begin{bmatrix} 0 & -1 \\ 1 & 0 \end{bmatrix}^2 . \tag{1.107}
$$

Furthermore, employing the result just obtained, and extracting the fundamental matrix,

$$
\begin{bmatrix} 0 & 1 \\ -1 & 0 \end{bmatrix} = \underbrace{\begin{bmatrix} -1 & 0 \\ 0 & -1 \end{bmatrix}} \begin{bmatrix} 0 & -1 \\ 1 & 0 \end{bmatrix}
$$
$$
\overset{(1.107)}{=} \begin{bmatrix} 0 & -1 \\ 1 & 0 \end{bmatrix}^2 \begin{bmatrix} 0 & -1 \\ 1 & 0 \end{bmatrix}
$$
$$
= \begin{bmatrix} 0 & -1 \\ 1 & 0 \end{bmatrix}^3 . \tag{1.108}
$$

Considering the above results along with the relationship,

$$
\begin{bmatrix} \cos(\alpha) \\ \sin(\alpha) \end{bmatrix} = \begin{bmatrix} 1 & 0 \\ 0 & 1 \end{bmatrix} \begin{bmatrix} \cos(\alpha) \\ \sin(\alpha) \end{bmatrix} , \tag{1.109}
$$

Equation (1.106) reads

$$
\begin{bmatrix} \cos(\alpha + \triangle\alpha) \\ \sin(\alpha + \triangle\alpha) \end{bmatrix} = \begin{bmatrix} 1 & 0 \\ 0 & 1 \end{bmatrix} \begin{bmatrix} \cos(\alpha) \\ \sin(\alpha) \end{bmatrix}
$$

$$
+ (\triangle\alpha) \begin{bmatrix} 0 & -1 \\ 1 & 0 \end{bmatrix} \begin{bmatrix} \cos(\alpha) \\ \sin(\alpha) \end{bmatrix}
$$

$$
+ \frac{1}{2!}(\triangle\alpha)^2 \begin{bmatrix} 0 & -1 \\ 1 & 0 \end{bmatrix}^2 \begin{bmatrix} \cos(\alpha) \\ \sin(\alpha) \end{bmatrix}
$$

$$
+ \frac{1}{3!}(\triangle\alpha)^3 \begin{bmatrix} 0 & -1 \\ 1 & 0 \end{bmatrix}^3 \begin{bmatrix} \cos(\alpha) \\ \sin(\alpha) \end{bmatrix} + \cdots . \tag{1.110}
$$

Factoring out the common vector $\begin{bmatrix} \cos(\alpha) \\ \sin(\alpha) \end{bmatrix}$ at the R.H.S.,

$$
\begin{bmatrix} \cos(\alpha + \triangle\alpha) \\ \sin(\alpha + \triangle\alpha) \end{bmatrix} = \left\{ \begin{bmatrix} 1 & 0 \\ 0 & 1 \end{bmatrix} \right.
$$

$$
+ (\triangle\alpha) \begin{bmatrix} 0 & -1 \\ 1 & 0 \end{bmatrix}
$$

$$
+ \frac{1}{2!}(\triangle\alpha)^2 \begin{bmatrix} 0 & -1 \\ 1 & 0 \end{bmatrix}^2
$$

$$
\left. + \frac{1}{3!}(\triangle\alpha)^3 \begin{bmatrix} 0 & -1 \\ 1 & 0 \end{bmatrix}^3 + \cdots \right\} \begin{bmatrix} \cos(\alpha) \\ \sin(\alpha) \end{bmatrix} . \tag{1.111}
$$

The expression in the curly brackets is the Formal Taylor Series Expansion of the exponential function $e^{\begin{bmatrix} 0 & -1 \\ 1 & 0 \end{bmatrix} \triangle\alpha}$:

$$
e^{\begin{bmatrix} 0 & -1 \\ 1 & 0 \end{bmatrix} \triangle\alpha} = \begin{bmatrix} 1 & 0 \\ 0 & 1 \end{bmatrix} + (\triangle\alpha) \begin{bmatrix} 0 & -1 \\ 1 & 0 \end{bmatrix} + \frac{1}{2!}(\triangle\alpha)^2 \begin{bmatrix} 0 & -1 \\ 1 & 0 \end{bmatrix}^2
$$

$$
+ \frac{1}{3!}(\triangle\alpha)^3 \begin{bmatrix} 0 & -1 \\ 1 & 0 \end{bmatrix}^3 + \cdots . \tag{1.112}
$$

Thus, (1.111) can be written in the form,

$$
\begin{bmatrix} \cos(\alpha + \triangle\alpha) \\ \sin(\alpha + \triangle\alpha) \end{bmatrix} = e^{\begin{bmatrix} 0 & -1 \\ 1 & 0 \end{bmatrix}\triangle\alpha} \begin{bmatrix} \cos(\alpha) \\ \sin(\alpha) \end{bmatrix}
\tag{1.113}
$$

Denoting the matrix at the R.H.S. by $\mathbf{J}$,

$$
\mathbf{J} = \begin{bmatrix} 0 & -1 \\ 1 & 0 \end{bmatrix},
\tag{1.114}
$$

establishes the transformation equation,

$$
\begin{bmatrix} \cos(\alpha + \triangle\alpha) \\ \sin(\alpha + \triangle\alpha) \end{bmatrix} = e^{\mathbf{J}\triangle\alpha} \begin{bmatrix} \cos(\alpha) \\ \sin(\alpha) \end{bmatrix}.
\tag{1.115}
$$

∎

Remark: Since $\triangle\alpha$ is arbitrary (not necessarily small), it can be replaced with any angle, say, θ,

$$
\begin{bmatrix} \cos(\alpha + \theta) \\ \sin(\alpha + \theta) \end{bmatrix} = e^{\mathbf{J}\theta} \begin{bmatrix} \cos(\alpha) \\ \sin(\alpha) \end{bmatrix}.
\tag{1.116}
$$

The motivation for carrying out the calculations up to this point in terms of $\triangle\alpha$ rather than θ, has been the tacit assumption that $\triangle\alpha$ is small compared with α. The aim has been not to distract the reader with possible convergence questions, even though the employment of Formal Taylor Series Expansion is exactly to liberate one from such concerns.

□

Problem: Establish an alternative (equivalent) representation of the column vector $\begin{bmatrix} \cos(\alpha + \theta) \\ \sin(\alpha + \theta) \end{bmatrix}$.

Solution: The starting point is

$$
\begin{bmatrix} \cos(\alpha + \theta) \\ \sin(\alpha + \theta) \end{bmatrix} = \begin{bmatrix} \cos(\alpha)\cos(\theta) - \sin(\alpha)\sin(\theta) \\ \sin(\alpha)\cos(\theta) + \cos(\alpha)\sin(\theta) \end{bmatrix}.
\tag{1.117}
$$

It is an easy exercise to write the vector at the R.H.S. as the product of a matrix (the entries of which are merely functions of θ) and a vector (the components of which exclusively involve α),

$$
\begin{bmatrix} \cos(\alpha)\cos(\theta) - \sin(\alpha)\sin(\theta) \\ \sin(\alpha)\cos(\theta) + \cos(\alpha)\sin(\theta) \end{bmatrix} = \begin{bmatrix} \cos(\theta) & -\sin(\theta) \\ \sin(\theta) & \cos(\theta) \end{bmatrix} \begin{bmatrix} \cos(\alpha) \\ \sin(\alpha) \end{bmatrix}.
\tag{1.118}
$$

Thus, (1.117) can be written as,

$$
\begin{bmatrix} \cos(\alpha + \theta) \\ \sin(\alpha + \theta) \end{bmatrix} = \begin{bmatrix} \cos(\theta) & -\sin(\theta) \\ \sin(\theta) & \cos(\theta) \end{bmatrix} \begin{bmatrix} \cos(\alpha) \\ \sin(\alpha) \end{bmatrix},
\tag{1.119}
$$

which is the desired representation alternative to (1.115).

∎

1.10   Euler identity in matrix form

Comparing the results from the previous two problems, (1.115) and (1.119),

$$e^{\mathbf{J}\theta} = \begin{bmatrix} \cos(\theta) & -\sin(\theta) \\ \sin(\theta) & \cos(\theta) \end{bmatrix}. \tag{1.120}$$

Writing the R.H.S. in the form

$$\begin{bmatrix} \cos(\theta) & -\sin(\theta) \\ \sin(\theta) & \cos(\theta) \end{bmatrix} = \begin{bmatrix} \cos(\theta) & 0 \\ 0 & \cos(\theta) \end{bmatrix} + \begin{bmatrix} 0 & -\sin(\theta) \\ \sin(\theta) & 0 \end{bmatrix}. \tag{1.121}$$

Or, equivalently,

$$\begin{bmatrix} \cos(\theta) & -\sin(\theta) \\ \sin(\theta) & \cos(\theta) \end{bmatrix} = \cos(\theta)\underbrace{\begin{bmatrix} 1 & 0 \\ 0 & 1 \end{bmatrix}}_{\mathbf{I}} + \sin(\theta)\underbrace{\begin{bmatrix} 0 & -1 \\ 1 & 0 \end{bmatrix}}_{\mathbf{J}} \tag{1.122a}$$

$$= \cos(\theta)\mathbf{I} + \sin(\theta)\mathbf{J}, \tag{1.122b}$$

where the 2×2 placeholder, (scaffolding) matrices $\mathbf{I}$ and $\mathbf{J}$ have compellingly offered themselves. Equations (1.120) and (1.122b) imply,

$$e^{\mathbf{J}\theta} = \cos(\theta)\mathbf{I} + \sin(\theta)\mathbf{J}. \tag{1.123}$$

This is a representation of the Euler Identity in the matrix form.

Problem: Consider, the Formal Taylor Series Expansions of $\cos(\alpha)$ and $\sin(\alpha)$,

$$\cos(\alpha) = 1 - \frac{1}{2!}\alpha^2 + \frac{1}{4!}\alpha^4 - \frac{1}{6!}\alpha^6 + \frac{1}{8!}\alpha^8 + \mathcal{O}\left(\alpha^{10}\right) \tag{1.124a}$$

$$\sin(\alpha) = \alpha - \frac{1}{3!}\alpha^3 + \frac{1}{5!}\alpha^5 - \frac{1}{7!}\alpha^7 + \frac{1}{9!}\alpha^9 + \mathcal{O}\left(\alpha^{11}\right). \tag{1.124b}$$

The symbol $\mathcal{O}\left(\alpha^n\right)$ denotes the collection of the terms of the order α^n or higher. Verify that,

$$\cos^2(\alpha) + \sin^2(\alpha) = 1. \tag{1.125}$$

Solution: A simple calculation reveals that,

$$\cos^2(\alpha) = 1 - \left(\frac{1}{2!} + \frac{1}{2!}\right)\alpha^2 + \left(\frac{1}{4!} + \frac{1}{2!\,2!} + \frac{1}{4!}\right)\alpha^4 - \left(\frac{1}{6!} + \frac{1}{2!\,4!} + \frac{1}{4!\,2!} + \frac{1}{6!}\right)\alpha^6$$

$$+ \left(\frac{1}{8!} + \frac{1}{2!\,6!} + \frac{1}{4!\,4!} + \frac{1}{6!\,2!} + \frac{1}{8!}\right)\alpha^8 + \mathcal{O}\left(\alpha^{10}\right) \tag{1.126a}$$

$$\sin^2(\alpha) = \alpha^2 - \left(\frac{1}{3!} + \frac{1}{3!}\right)\alpha^4 + \left(\frac{1}{5!} + \frac{1}{3!\,3!} + \frac{1}{5!}\right)\alpha^6$$

$$- \left(\frac{1}{7!} + \frac{1}{3!\,5!} + \frac{1}{5!\,3!} + \frac{1}{7!}\right)\alpha^8 + \mathcal{O}\left(\alpha^{10}\right). \tag{1.126b}$$

General compact formulae for the coefficients in these series will be provided after investigating a few initial cases to gain insight into how the corresponding terms in (1.126a) and (1.126b) cancel out when taken together. The general idea can be communicated in terms of the expressions in (1.126) though. It is easily seen that the coefficients associated with the coefficients α^2, α^4, α^6, and α^8 in the two series are equal in magnitude and opposite in sign. While this relationship is immediate in the case of α^2, in cases α^{2n} with large values of n, is quite tedious. However, there is a useful hidden pattern in the coefficients which renders the conclusions much easier. Consider the following case:

Add the coefficients of α^2 appearing in the power series expansions of $\cos^2(\alpha)$ and $\sin^2(\alpha)$:

$$-\frac{1}{2!} - \frac{1}{2!} + 1. \tag{1.127}$$

Remark: In the blink of an eye, it is seen that $-1/2 - 1/2 + 1 = 0$. Similarly, the coefficients associated with α^{2n} ($n \in \mathbb{N}$) can be directly calculated albeit increasingly more tedious and time consuming. The point here is actually the identification of an algorithmic and (hopefully) elegant way of calculation. This aim requires identifying patterns and recognizing structures which are otherwise hidden. There are occasions where finding gems can be demonstrated particularly well. The mathematical induction builds upon forming intuition by considering a few first steps. There is often an awe-inspiring experience when intricate details match together and one wonders why mathematics works in the way it does. □

Using $0! = 1$ and $1! = 1$, (1.127) can be replaced with judiciously augmented alternative,

$$-\frac{1}{0!2!} - \frac{1}{2!0!} + \frac{1}{1!1!}. \tag{1.128}$$

Ordering in terms of increasing first numbers in the denominators, $0!$, $1!$, and $2!$,

$$-\frac{1}{0!2!} + \frac{1}{1!1!} - \frac{1}{2!0!}. \tag{1.129}$$

If this expression turns out to be zero, multiplying it by $-2!$ would not affect this property. Therefore,

$$\frac{2!}{0!2!} - \frac{2!}{1!1!} + \frac{2!}{2!0!}. \tag{1.130}$$

Thus, if it can be shown that (1.130) is zero, (1.129) and all the preceding terms, in particular (1.127), are zero. Taking advantage of the powers of 1 and -1 prudently, the minus sign can be absorbed into the respective terms, generating a unified representation of all terms,

$$\frac{2!}{0!2!}(1)^0(-1)^2 + \frac{2!}{1!1!}(1)^1(-1)^1 + \frac{2!}{2!0!}(1)^2(-1)^0. \tag{1.131}$$

Making use of the notation,

$$\binom{n}{k} = \frac{n!}{k!(n-k)!}, \tag{1.132}$$

$$\binom{2}{0}(1)^0(-1)^2 + \binom{2}{1}(1)^1(-1)^1 + \binom{2}{2}(1)^2(-1)^0. \tag{1.133}$$

This expression is revealing—it is the binomial expansion of,

$$\left((1) + (-1)\right)^2 = 0. \tag{1.134}$$

Consequently, it can be concluded that (1.133) and all its preceding terms, in particular (1.127), are zero.

Add the coefficients of α^4 appearing in the power series expansions of $\cos^2(\alpha)$ and $\sin^2(\alpha)$:

$$\frac{1}{4!} + \frac{1}{2!2!} + \frac{1}{4!} - \frac{1}{3!} - \frac{1}{3!} \tag{1.135a}$$

$$\overset{\text{augmentation}}{\Longrightarrow} \quad \frac{1}{0!4!} + \frac{1}{2!2!} + \frac{1}{4!0!} - \frac{1}{1!3!} - \frac{1}{3!1!} \tag{1.135b}$$

$$\overset{\text{ordering}}{\Longrightarrow} \quad \frac{1}{0!4!} - \frac{1}{1!3!} + \frac{1}{2!2!} - \frac{1}{3!1!} + \frac{1}{4!0!} \tag{1.135c}$$

$$\overset{\text{mult. by } 4!}{\Longrightarrow} \quad \frac{4!}{0!4!} - \frac{4!}{1!3!} + \frac{4!}{2!2!} - \frac{4!}{3!1!} + \frac{4!}{4!0!} \tag{1.135d}$$

$$\overset{\text{unifying signs}}{\Longrightarrow} \quad \frac{4!}{0!4!}(1)^0(-1)^4 + \frac{4!}{1!3!}(1)^1(-1)^3 + \frac{4!}{2!2!}(1)^2(-1)^2$$

$$+ \frac{4!}{3!1!}(1)^3(-1)^1 + \frac{4!}{4!0!}(1)^4(-1)^0 \tag{1.135e}$$

$$\overset{\text{factorial rep.}}{\Longrightarrow} \quad \binom{4}{0}(1)^0(-1)^4 + \binom{4}{1}(1)^1(-1)^3 + \binom{4}{2}(1)^2(-1)^2$$

$$+ \binom{4}{3}(1)^3(-1)^1 + \binom{4}{4}(1)^4(-1)^0 \tag{1.135f}$$

$$\overset{\text{binomial form}}{\Longrightarrow} \quad \left(1 + (-1)\right)^4 = 0. \tag{1.135g}$$

Thus, it can be inferred that (1.135a) is zero.

Add the coefficients of α^6 appearing in the power series expansions of $\cos^2(\alpha)$ and $\sin^2(\alpha)$:

$$-\frac{1}{6!} - \frac{1}{2!4!} - \frac{1}{4!2!} - \frac{1}{6!} + \frac{1}{5!1!} + \frac{1}{3!3!} + \frac{1}{1!5!} \tag{1.136a}$$

$$\overset{\text{augmentation}}{\Longrightarrow} \quad -\frac{1}{0!6!} - \frac{1}{2!4!} - \frac{1}{4!2!} - \frac{1}{6!0!} + \frac{1}{5!1!} + \frac{1}{3!3!} + \frac{1}{1!5!} \tag{1.136b}$$

$$\overset{\text{ordering}}{\Longrightarrow} \quad -\frac{1}{0!6!} + \frac{1}{1!5!} - \frac{1}{2!4!} + \frac{1}{3!3!} - \frac{1}{4!2!} + \frac{1}{5!1!} - \frac{1}{6!0!} \tag{1.136c}$$

$$\overset{\text{mult. by } -6!}{\Longrightarrow} \quad \frac{6!}{0!6!} - \frac{6!}{1!5!} + \frac{6!}{2!4!} - \frac{6!}{3!3!} + \frac{6!}{4!2!} - \frac{6!}{5!1!} + \frac{6!}{6!0!} \tag{1.136d}$$

$$\overset{\text{unifying signs}}{\Longrightarrow} \quad \frac{6!}{0!6!}(1)^0(-1)^6 + \frac{6!}{1!5!}(1)^1(-1)^5 + \frac{6!}{2!4!}(1)^2(-1)^4 + \frac{6!}{3!3!}(1)^3(-1)^3$$

$$+ \frac{6!}{4!2!}(1)^4(-1)^2 + \frac{6!}{5!1!}(1)^5(-1)^1 + \frac{6!}{6!0!}(1)^6(-1)^0 \ldots \tag{1.136e}$$

$$\xRightarrow{\text{factorial rep.}} \binom{6}{0}(1)^0(-1)^6 + \binom{6}{1}(1)^1(-1)^5 + \binom{6}{2}(1)^2(-1)^4 + \binom{6}{3}(1)^3(-1)^3$$

$$+ \binom{6}{4}(1)^4(-1)^2 + \binom{6}{5}(1)^5(-1)^1 + \binom{6}{6}(1)^6(-1)^0 \tag{1.136f}$$

$$\xRightarrow{\text{binomial form}} \left(1 + (-1)\right)^6 = 0. \tag{1.136g}$$

Add the coefficients of α^8 appearing in the power series expansions of $\cos^2(\alpha)$ and $\sin^2(\alpha)$:

$$\frac{1}{8!} + \frac{1}{2!6!} + \frac{1}{4!4!} + \frac{1}{6!2!} + \frac{1}{8!} - \frac{1}{7!} - \frac{1}{3!5!} - \frac{1}{5!3!} - \frac{1}{7!} \tag{1.137a}$$

$$\xRightarrow{\text{augmentation}} \frac{1}{0!8!} + \frac{1}{2!6!} + \frac{1}{4!4!} + \frac{1}{6!2!} + \frac{1}{8!0!} - \frac{1}{1!7!} - \frac{1}{3!5!} - \frac{1}{5!3!} - \frac{1}{7!1!} \tag{1.137b}$$

$$\xRightarrow{\text{ordering}} \frac{1}{0!8!} - \frac{1}{1!7!} + \frac{1}{2!6!} - \frac{1}{3!5!} + \frac{1}{4!4!} - \frac{1}{5!3!} + \frac{1}{6!2!} - \frac{1}{7!1!} + \frac{1}{8!0!} \tag{1.137c}$$

$$\xRightarrow{\text{mult. by 8!}} \frac{8!}{0!8!} - \frac{8!}{1!7!} + \frac{8!}{2!6!} - \frac{8!}{3!5!} + \frac{8!}{4!4!} - \frac{8!}{5!3!} + \frac{8!}{6!2!} - \frac{8!}{7!1!} + \frac{8!}{8!0!} \tag{1.137d}$$

$$\xRightarrow{\text{unifying signs}} \frac{8!}{0!8!}(1)^0(-1)^8 + \frac{8!}{1!7!}(1)^1(-1)^7 + \frac{8!}{2!6!}(1)^2(-1)^6$$

$$+ \frac{8!}{3!5!}(1)^3(-1)^5 + \frac{8!}{4!4!}(1)^4(-1)^4 + \frac{8!}{5!3!}(1)^5(-1)^3$$

$$+ \frac{8!}{6!2!}(1)^6(-1)^2 + \frac{8!}{7!1!}(1)^7(-1)^1 + \frac{8!}{8!0!}(1)^8(-1)^0 \tag{1.137e}$$

$$\xRightarrow{\text{factorial rep.}} \binom{8}{0}(1)(1)^0(-1)^8 + \binom{8}{1}(1)^1(-1)^7 + \binom{8}{2}(1)^2(-1)^6$$

$$+ \binom{8}{3}(1)^3(-1)^5 + \binom{8}{4}(1)^4(-1)^4 + \binom{8}{5}(1)^5(-1)^3$$

$$+ \binom{8}{6}(6)^6(-1)^2 + \binom{8}{7}(1)^7(-1)^1 + \binom{8}{9}(1)^8(-1)^0 \tag{1.137f}$$

$$\xRightarrow{\text{binomial form}} \left(1 + (-1)\right)^8 = 0. \tag{1.137g}$$

Reasoning by mathematical induction implies that this statement is true for general terms α^{2n} ($n \in \mathbb{N}$). Consequently, adding the equations in (1.126), all terms cancel out, except the constant term 1, which lacks a Balancing (Compensating) counterpart. This sketches the derivation of (1.125) employing the Formal Taylor Series Expansion in (1.124). ∎

Problem: Consider the power series expansions of $\cos^2(\alpha)$ and $\sin^2(\alpha)$, which have been reproduced here for easy reference,

$$\cos^2(\alpha) = 1 - \left(\frac{1}{2!} + \frac{1}{2!}\right)\alpha^2 + \left(\frac{1}{4!} + \frac{1}{2!}\frac{1}{2!} + \frac{1}{4!}\right)\alpha^4 - \left(\frac{1}{6!} + \frac{1}{2!}\frac{1}{4!} + \frac{1}{4!}\frac{1}{2!} + \frac{1}{6!}\right)\alpha^6$$

$$+ \left(\frac{1}{8!} + \frac{1}{2!}\frac{1}{6!} + \frac{1}{4!}\frac{1}{4!} + \frac{1}{6!}\frac{1}{2!} + \frac{1}{8!}\right)\alpha^8 + \mathcal{O}\left(\alpha^{10}\right) \tag{1.138a}$$

$$\sin^2(\alpha) = \alpha^2 - \left(\frac{1}{3!} + \frac{1}{3!}\right)\alpha^4 + \left(\frac{1}{5!} + \frac{1}{3!}\frac{1}{3!} + \frac{1}{5!}\right)\alpha^6$$

$$- \left(\frac{1}{7!} + \frac{1}{3!}\frac{1}{5!} + \frac{1}{5!}\frac{1}{3!} + \frac{1}{7!}\right)\alpha^8 + \mathcal{O}\left(\alpha^{10}\right). \tag{1.138b}$$

Provide general compact expressions for the terms associated with α^2 in the above power series expansions of $\cos^2(\alpha)$ and $\sin^2(\alpha)$.

Solution: Consider (1.138a). The coefficients of α^6 and α^8 are sufficient to capture all relevant details. In particular, the sign of the coefficients of α^{2n} must be accounted for properly. A little thought reveals that the sign of α^{2n} is given by $(-1)^n$. Furthermore, the two constituent factorials are even numbers and they add up to $2n$. Finally, the number of the constituents terms of the coefficient of α^{2n} is $n + 1$. Accounting for these considerations, α^{2n} together with its coefficient is,

$$\left((-1)^n \sum_{k=0}^{n} \frac{1}{(2k)!(2n - 2k)!}\right)\alpha^{2n} \qquad n \in \mathbb{N}_0. \tag{1.139}$$

Note that this expression is valid in the case $n = 0$ as well. If $n = 0$ then k takes on only one value, i.e., $k = 0$. In this case the expression within the large round brackets gives unity.

Next consider (1.138b). Again the coefficients of α^6 and α^8 are sufficient to reveal all minute details. In particular, the sign of the coefficients of α^{2n} must be accounted for properly. The sign of α^{2n} is given by $(-1)^{n-1}$. The two constituent factorials are odd numbers and they add up to $2n$. Finally, the number of the constituents terms of the coefficient of α^{2n} is n. Accounting for these observations, α^{2n} together with its coefficient is,

$$\left((-1)^{n-1} \sum_{k=1}^{n} \frac{1}{(2k - 1)!(2n - (2k - 1))!}\right)\alpha^{2n} \qquad n \in \mathbb{N}. \tag{1.140}$$

This expression starts with $n = 1$.

■

Lemma: *For arbitrary $n \in \mathbb{N}$,*

$$\left((-1)^n \sum_{k=0}^{n} \frac{1}{(2k)!(2n - 2k)!}\right) + \left((-1)^{n-1} \sum_{k=1}^{n} \frac{1}{(2k - 1)!(2n - (2k - 1))!}\right) = 0. \tag{1.141}$$

Solution: Multiplying by $(-1)^{n+1}$, bringing the first term to the R.H.S., and multiplying through by $(2n)!$,

$$\sum_{k=1}^{n} \frac{(2n)!}{(2k - 1)!(2n - (2k - 1))!} = \sum_{k=0}^{n} \frac{(2n)!}{(2k)!(2n - 2k)!} \qquad n \in \mathbb{N}. \tag{1.142}$$

Using, the factorial notation, the following equation must be shown to be valid,

$$\sum_{k=1}^{n} \binom{2n}{2k-1} = \sum_{k=0}^{n} \binom{2n}{2k} \qquad n \in \mathbb{N}. \tag{1.143}$$

Consider the binomial expansion,

$$0 = \left(1 + (-1)\right)^{2n} = \sum_{k=0}^{n} (1)^k (-1)^{2n-k} \binom{2n}{k} \tag{1.144a}$$

$$0 = \left(1 + (-1)\right)^{2n} = \underbrace{\sum_{k=0}^{2n} (1)^{2k} (-1)^{2n-2k} \binom{2n}{2k}}_{\text{even terms}} + \underbrace{\sum_{k=1}^{n} (1)^{2k-1} (-1)^{2n-(2k-1)} \binom{2n}{2k-1}}_{\text{odd terms}}. \tag{1.144b}$$

Noting $(-1)^{2n-(2k-1)} = -1$,

$$0 = \sum_{k=0}^{n} \binom{2n}{2k} - \sum_{k=1}^{n} \binom{2n}{2k-1}, \tag{1.145}$$

which completes the proof.

∎

Problem: Show that

$$\cos^2(\alpha) + \sin^2(\alpha) = 1. \tag{1.146}$$

Solution:

$$\cos^2(\alpha) + \sin^2(\alpha) = \cos(\alpha)\cos(\alpha) + \sin(\alpha)\sin(\alpha) \tag{1.147}$$

$$\cos^2(\alpha) + \sin^2(\alpha) = \cos(\alpha - \alpha) \tag{1.148}$$

$$\cos^2(\alpha) + \sin^2(\alpha) = \cos(0) = 1. \tag{1.149}$$

∎

Remark:

- The logical structure of the axiomatic reasoning presented so far follows from simple rules of inference. Using a minimum number of assumptions and two manipulation techniques, several "elementary" functions are introduced. The employed tools consist of the Formal Taylor Series Expansion and the notion of the derivative of monomials using the limit-process $\lim_{x \to x_0}$ applied to fractions of the type $((x + x_0)^n - x^n)/x_0$. To this end, merely the multiplication properties of the monomials are required. This is substantial to the arguments put forward here: the linearization of the functions is not necessary.
- The set of "elementary" functions constructed includes e^x, $\cosh(x)$, $\sinh(x) \cos(\alpha)$, $\sin(\alpha)$, and $e^{\mathbf{J}\alpha}$. At this point a few words of caution might be in order: The function names, e.g., "cos" and "sin," are chosen since it turns out that they are the same widely-used functions $\cos(\alpha)$ and $\sin(\alpha)$. However, the "synthesized" functions $\cos(\alpha)$ and $\sin(\alpha)$ are not required to possess any properties beyond those stipulated in the text. This is the reason why the relationship $\cos^2(\alpha) + \sin^2(\alpha) = 1$ was established, rather being taken as known. Similarly, the result $\sin(0) = 0$ was concluded rather than obtained by resorting to known facts. In this sense, it would have been preferable to introduce the functions, say, $\mathcal{C}(\alpha)$ and $\mathcal{S}(\alpha)$, and then establish the facts that they are, respectively, identical with the functions $\cos(\alpha)$ and $\sin(\alpha)$. However, this detour might have obscured the discussion.
- The relationship $\cos^2(\alpha) + \sin^2(\alpha) = 1$, together with $\cos(\alpha) \in \mathbf{R}$ and $\sin(\alpha) \in \mathbf{R}$ imply that $-1 \le \cos(\alpha) \le 1$, $-1 \le \sin(\alpha) \le 1$, and $|\cos(\alpha)|^2 + |\sin(\alpha)|^2 = 1$ for arbitrary α.

- The triple values $|\cos\alpha|$, $|\sin\alpha|$, and 1, respectively, can be interpreted as the side lengths and the hypotenuse of a right triangle.
- The equation $\cos^2(\alpha) + \sin^2(\alpha) = 1$ describes a circle with radius unity and centered at the origin of the xy-Cartesian coordinate system (unit circle).
- Let $P(\alpha)$ denote the point on the perimeter of the unit circle associated with the angle α. $P(\alpha)$ traverses the unit circle counterclockwise with increasing α. Let the intersection of the unit circle with the x-axis mark the value $\alpha = 0$. Thus, the point $P(0)$ in the assumed xy-Cartesian coordinate system possesses the coordinates 1, and 0, and can therefore be assigned to the ordered pair $(1, 0)$. On the other hand, setting $\alpha = 0$ in (1.124),

$$P(0) \implies \begin{bmatrix} \cos(0) \\ \sin(0) \end{bmatrix} = \begin{bmatrix} 1 \\ 0 \end{bmatrix}. \tag{1.150}$$

- Consequently, for arbitrary α, the corresponding point $P(\alpha)$ has the coordinates $\cos(\alpha)$ and $\sin(\alpha)$, respectively, on the x- and y-axis,

$$P(\alpha) \implies \begin{bmatrix} \cos(\alpha) \\ \sin(\alpha) \end{bmatrix}. \tag{1.151}$$

- Connecting the origin of the xy-Cartesian coordinate system, $(0, 0)$, with the point $P(\alpha)$ on the perimeter of the unit circle, a unique "position-vector" $\mathbf{r}_0(\alpha)$ can be defined, which is characterized by the coordinates $\cos(\alpha)$ and $\sin(\alpha)$ along the x- and y-axis, respectively,

$$\mathbf{r}_0(\alpha) = \begin{bmatrix} \cos(\alpha) \\ \sin(\alpha) \end{bmatrix}. \tag{1.152}$$

The subindex 0 should allude to the fact that, r_0, the magnitude of the $\mathbf{r}_0(\alpha)$ is unity.
- The length of the circumference of the unit circle is, $2\pi r_0 = 2\pi r|_{r=1} = 2\pi$.
- The intersection of the ordinate (y-axis) with the unit circle is the point $P\left(\frac{\pi}{2}\right)$ with the Cartesian coordinates 0 and 1,

$$P\left(\frac{\pi}{2}\right) \implies \mathbf{r}_0\left(\frac{\pi}{2}\right) = \begin{bmatrix} \cos\left(\frac{\pi}{2}\right) \\ \sin\left(\frac{\pi}{2}\right) \end{bmatrix} = \begin{bmatrix} 0 \\ 1 \end{bmatrix}. \tag{1.153}$$

Consequently, the position vector $\mathbf{r}_0\left(\frac{\pi}{2}\right)$ is associated with the ordered pair $(0, 1)$.
- On the other hand, setting $\alpha = \pi/2$ into (1.124),

$$\cos\left(\frac{\pi}{2}\right) = 1 - \frac{1}{2!}\left(\frac{\pi}{2}\right)^2 + \frac{1}{4!}\left(\frac{\pi}{2}\right)^4 - \frac{1}{6!}\left(\frac{\pi}{2}\right)^6 + \frac{1}{8!}\left(\frac{\pi}{2}\right)^8 + \mathcal{O}\left[\left(\frac{\pi}{2}\right)^{10}\right] \tag{1.154a}$$

$$\sin\left(\frac{\pi}{2}\right) = \left(\frac{\pi}{2}\right) - \frac{1}{3!}\left(\frac{\pi}{2}\right)^3 + \frac{1}{5!}\left(\frac{\pi}{2}\right)^5 - \frac{1}{7!}\left(\frac{\pi}{2}\right)^7 + \frac{1}{9!}\left(\frac{\pi}{2}\right)^9 + \mathcal{O}\left[\left(\frac{\pi}{2}\right)^{11}\right]. \tag{1.154b}$$

- Comparing (1.153) with (1.154),

$$0 = 1 - \frac{1}{2!}\left(\frac{\pi}{2}\right)^2 + \frac{1}{4!}\left(\frac{\pi}{2}\right)^4 - \frac{1}{6!}\left(\frac{\pi}{2}\right)^6 + \frac{1}{8!}\left(\frac{\pi}{2}\right)^8 + \mathcal{O}\left[\left(\frac{\pi}{2}\right)^{10}\right] \tag{1.155a}$$

$$1 = \left(\frac{\pi}{2}\right) - \frac{1}{3!}\left(\frac{\pi}{2}\right)^3 + \frac{1}{5!}\left(\frac{\pi}{2}\right)^5 - \frac{1}{7!}\left(\frac{\pi}{2}\right)^7 + \frac{1}{9!}\left(\frac{\pi}{2}\right)^9 + \mathcal{O}\left[\left(\frac{\pi}{2}\right)^{11}\right]. \tag{1.155b}$$

- A myriad of nontrivial identities involving π can be established. Any attempt of proofing these and related relationships directly must be a daunting undertaking, if not impossible.

$\square$

Remark: From the above, $\mathbf{r}_0(\alpha) = \begin{bmatrix} \cos(\alpha) \\ \sin(\alpha) \end{bmatrix}$ represents the position vector with its endpoint being on the perimeter of the unit circle and building the angle α with the horizontal axis. Similarly, $\mathbf{r}_0(\alpha + \theta) = \begin{bmatrix} \cos(\alpha + \theta) \\ \sin(\alpha + \theta) \end{bmatrix}$ defines the position vector with its endpoint being on the perimeter of the unit circle and building the angle $\alpha + \theta$ with the horizontal axis. Thus, rotating the vector $\mathbf{r}_0(\alpha) = \begin{bmatrix} \cos(\alpha) \\ \sin(\alpha) \end{bmatrix}$ by the angle θ results in the vector $\mathbf{r}_0(\alpha + \theta) = \begin{bmatrix} \cos(\alpha+\theta) \\ \sin(\alpha+\theta) \end{bmatrix}$.

Consequently, in view of (1.116), $e^{\mathbf{J}\theta}$ can be interpreted as a matrix, which rotates position vectors by the angle θ. □

Consider $\mathbf{r}(\alpha) = r \begin{bmatrix} \cos(\alpha) \\ \sin(\alpha) \end{bmatrix} = r\mathbf{r}_0(\alpha)$ with r referring to the magnitude of $\mathbf{r}(\alpha)$. Letting r assume values from the interval $[0, \infty)$, the "ray" $\mathbf{r}(\alpha)$ in the direction of the position vector $\mathbf{r}_0(\alpha)$ can be defined which stretches from the origin $(0,0)$ to infinity: by properly choosing r in $\mathbf{r}(\alpha) = r\mathbf{r}_0(\alpha)$, any point on the ray $\mathbf{r}(\alpha)$ can be reached. Consequently, letting α to vary on the interval $[0, 2\pi)$ and r on the interval $[0, \infty)$, every point in the xy-plane can be addressed. Conversely, every point in the plane characterized by $\mathbf{r}(\alpha)$ can be transformed onto the unit circle, by adequately scaling the length of the corresponding position vector $\mathbf{r}_0(\alpha)$. These considerations suggest focusing the discussion on the points of the unit circle, an undertaking that leads to the construction of useful structures.

Given the point $P(\alpha)$ on the unit circle, fixes the values $\cos(\alpha)$ and $\sin(\alpha)$. Arranging $\cos(\alpha)$ and $\sin(\alpha)$ as the coordinates of a vector, the position vector

$$\mathbf{r}_0(\alpha) = \begin{bmatrix} \cos(\alpha) \\ \sin(\alpha) \end{bmatrix} \tag{1.156}$$

manifests itself.

However, viewing $\cos(\theta)$ and $\sin(\theta)$ as the entries of a 2×2 matrix, and arranging them in the form,

$$\mathbf{R}_0(\theta) = \begin{bmatrix} \cos(\theta) & -\sin(\theta) \\ \sin(\theta) & \cos(\theta) \end{bmatrix}, \tag{1.157}$$

a rotation matrix reveals itself, as it is the exact explication of $e^{\mathbf{J}\theta}$. The addition of the subindex 0 in (1.157) should be a reminder of the genesis of $\mathbf{R}_0(\theta)$; i.e., $\cos(\theta)$ and $\sin(\theta)$, the coordinates of $P_0(\theta)$ on the unit circle.

Every arbitrary point $P_0(\alpha)$ on the unit circle can be uniquely mapped onto the associated position vector $\mathbf{r}_0(\alpha)$,

$$P_0(\alpha) \iff \mathbf{r}_0(\alpha) = \begin{bmatrix} \cos(\alpha) \\ \sin(\alpha) \end{bmatrix}. \tag{1.158}$$

Every arbitrary point $P_0(\theta)$ on the unit circle can be uniquely mapped onto the associated rotation matrix $\mathbf{R}_0(\theta)$,

$$P_0(\theta) \iff \mathbf{R}_0(\theta) = \begin{bmatrix} \cos(\theta) & -\sin(\theta) \\ \sin(\theta) & \cos(\theta) \end{bmatrix}. \tag{1.159}$$

Both of the mathematical objects $\mathbf{r}_0(\alpha)$ and $\mathbf{R}_0(\theta)$ manifested themselves inevitably: (i) the effort to simultaneously relate the derivatives of the elementary functions $\cos(\alpha)$ and $\sin(\alpha)$ mutually to themselves, led to the emergence of matrix $\mathbf{J}$, and ultimately to the position vector $\mathbf{r}_0(\alpha)$. (ii) Consecutively, the effort to establish a relationship between $\mathbf{r}_0(\alpha + \theta)$ and $\mathbf{r}_0(\alpha)$ led to the rotation matrix $\mathbf{R}_0(\theta)$. The remarkable fact about both structures is that their building blocks are the elementary functions $\cos(\cdot)$ and $\sin(\cdot)$.

In the following subsection, relevant properties of $\mathbf{R}_0(\theta)$ are investigated. It turns out that $\mathbf{R}_0(\theta)$ rather than $\mathbf{r}_0(\alpha)$ possesses several utilitarian properties.

1.10.1   Properties of $\mathbf{R}_0(\theta)$

1.10.1.1   Multiplication of rotation matrices

Consider θ_1 and θ_2 on the unit circle and their associated rotation matrices $\mathbf{R}_0(\theta_1)$ and $\mathbf{R}_0(\theta_2)$, respectively. Then,

$$\mathbf{R}_0(\theta_1)\mathbf{R}_0(\theta_2) = \begin{bmatrix} \cos(\theta_1) & -\sin(\theta_1) \\ \sin(\theta_1) & \cos(\theta_1) \end{bmatrix}\begin{bmatrix} \cos(\theta_2) & -\sin(\theta_2) \\ \sin(\theta_2) & \cos(\theta_2) \end{bmatrix} \tag{1.160a}$$

$$= \begin{bmatrix} \cos(\theta_1)\cos(\theta_2) - \sin(\theta_1)\sin(\theta_2) & -\cos(\theta_1)\sin(\theta_2) - \sin(\theta_1)\cos(\theta_2) \\ \sin(\theta_1)\cos(\theta_2) + \cos(\theta_1)\sin(\theta_2) & -\sin(\theta_1)\sin(\theta_2) + \cos(\theta_1)\cos(\theta_2) \end{bmatrix} \tag{1.160b}$$

$$= \begin{bmatrix} \cos(\theta_1 + \theta_2) & -\sin(\theta_1 + \theta_2) \\ \sin(\theta_1 + \theta_2) & \cos(\theta_1 + \theta_2) \end{bmatrix} \tag{1.160c}$$

$$= \mathbf{R}_0(\theta_1 + \theta_2). \tag{1.160d}$$

Multiplication of the rotation matrices amounts to adding their defining rotation angles.

1.10.1.2   The inverse of rotation matrix

What is the inverse of $\mathbf{R}_0(\theta)$? Noting that $\mathbf{R}_0(\theta)$ is unitary (its determinant is unity),

$$\mathbf{R}_0^{-1}(\theta) = \begin{bmatrix} \cos(\theta) & -\sin(\theta) \\ \sin(\theta) & \cos(\theta) \end{bmatrix}^{-1} \tag{1.161a}$$

$$= \begin{bmatrix} \cos(\theta) & \sin(\theta) \\ -\sin(\theta) & \cos(\theta) \end{bmatrix} \tag{1.161b}$$

$$= \begin{bmatrix} \cos(-\theta) & -\sin(-\theta) \\ \sin(-\theta) & \cos(-\theta) \end{bmatrix} \tag{1.161c}$$

$$= \mathbf{R}_0(-\theta). \tag{1.161d}$$

In the transition from (1.161b) to (1.161c) the entry $\sin(\theta)$ is replaced with $-\sin(-\theta)$, and $-\sin(\theta)$ with $\sin(-\theta)$, because $\sin(\theta)$ is an odd function. Furthermore, $\cos(\theta)$ is replaced with $\cos(-\theta)$ since $\cos(\theta)$ is an even function. The result in (1.161) is quite intuitive: the inverse of a rotation can be achieved by undoing the rotation which amounts to a rotation in the reverse direction.

1.10.1.3   The square of the rotation matrix

Given $\mathbf{R}_0(\theta)$ determine $\mathbf{R}_0^2(\theta)$.

$$\mathbf{R}_0^2(\theta) = \mathbf{R}_0(\theta)\mathbf{R}_0(\theta) = \mathbf{R}_0(2\theta). \tag{1.162}$$

1.10.1.4   The square root of the rotation matrix

Taking the square root,

$$\{\mathbf{R}_0(2\theta)\}^{\frac{1}{2}} = \{\mathbf{R}_0^2(\theta)\}^{\frac{1}{2}} = \mathbf{R}_0(\theta). \tag{1.163}$$

Which setting $\eta = 2\theta$ amounts to,

$$\{\mathbf{R}_0(\eta)\}^{\frac{1}{2}} = \mathbf{R}_0\left(\frac{\eta}{2}\right). \tag{1.164}$$

And since η is arbitrary,

$$\{\mathbf{R}_0(\theta)\}^{\frac{1}{2}} = \mathbf{R}_0\left(\frac{\theta}{2}\right). \tag{1.165}$$

Thus, taking the square root of $\mathbf{R}_0(\theta)$ amounts to halving the rotation angle.

1.10.1.5   The rotation matrix to the power n

By induction,

$$\mathbf{R}_0^n(\theta) = \mathbf{R}_0(n\theta). \tag{1.166}$$

1.10.1.6   The nth root of the rotation matrix

Taking the nth root,

$$\{\mathbf{R}_0(n\theta)\}^{\frac{1}{n}} = \{\mathbf{R}_0^n(\theta)\}^{\frac{1}{n}} = \mathbf{R}_0(\theta). \tag{1.167}$$

Which setting $\eta = n\theta$ amounts to,

$$\{\mathbf{R}_0(\eta)\}^{\frac{1}{n}} = \mathbf{R}_0\left(\frac{\eta}{n}\right). \tag{1.168}$$

And since η is arbitrary,

$$\{\mathbf{R}_0(\theta)\}^{\frac{1}{n}} = \mathbf{R}_0\left(\frac{\theta}{n}\right). \tag{1.169}$$

Thus, taking the nth root of $\mathbf{R}_0(\theta)$ amounts to dividing the rotation angle by n.

Problem: Consider the rotation matrix for $\theta = \pi$,

$$\mathbf{R}_0(\pi) = \begin{bmatrix} \cos(\pi) & -\sin(\pi) \\ \sin(\pi) & \cos(\pi) \end{bmatrix}. \tag{1.170}$$

Determine the square root of $\mathbf{R}_0(\pi)$.

Solution: Nothing seems to be more straightforward than this task,

$$\{\mathbf{R}_0(\pi)\}^{\frac{1}{2}} = \mathbf{R}_0\left(\frac{\pi}{2}\right) = \begin{bmatrix} \cos\left(\frac{\pi}{2}\right) & -\sin\left(\frac{\pi}{2}\right) \\ \sin\left(\frac{\pi}{2}\right) & \cos\left(\frac{\pi}{2}\right) \end{bmatrix}. \tag{1.171}$$

∎

Remark: With $\cos(\pi) = -1$ and $\sin(\pi) = 0$ substituted into (1.170),

$$\mathbf{R}_0(\pi) = \begin{bmatrix} -1 & 0 \\ 0 & -1 \end{bmatrix} = -\begin{bmatrix} 1 & 0 \\ 0 & 1 \end{bmatrix} = -\mathbf{I}. \tag{1.172}$$

Similarly, taking into account $\cos(\pi/2) = 0$ and $\sin(\pi/2) = 1$ in (1.171),

$$\mathbf{R}_0\left(\frac{\pi}{2}\right) = \begin{bmatrix} 0 & -1 \\ 1 & 0 \end{bmatrix} = \mathbf{J}. \tag{1.173}$$

In view of the first equation in (1.171), along with (1.172) and (1.173), the following equivalent representations suggest themselves,

$$\{\mathbf{R}_0(\pi)\}^{\frac{1}{2}} = \mathbf{R}_0\left(\frac{\pi}{2}\right) \tag{1.174a}$$

$$\left\{-\begin{bmatrix} 1 & 0 \\ 0 & 1 \end{bmatrix}\right\}^{\frac{1}{2}} = \begin{bmatrix} 0 & -1 \\ 1 & 0 \end{bmatrix} \tag{1.174b}$$

$$\{-\mathbf{I}\}^{\frac{1}{2}} = \mathbf{J}. \tag{1.174c}$$

□

1.11 Real and simple rather than imaginary and complex

A history shrouded in the introduction of the mysterious notions of "imaginary" and "complex" along with the introduction of the "dreaded" symbol $\sqrt{-1}$ have misled, or, to say the least, confused generations of engineers and technologists, by confronting them with the minus sign under the square root sign! Terms such as $\sqrt{4}$ make perfect sense, since $\sqrt{4}$ is defined to be a value $\in \mathbb{R}$ which, when multiplied by itself (squared) should result in 4. Thus, $\sqrt{4} = \pm 2$, since $(\pm 2)^2 = 4$.

Certain problems in mathematics, applied sciences or engineering, however, lead to equations of the form $x^2 + 1 = 0$, to quote the simplest possible instance. While the constituent terms in this equation, i.e., x^2, 1, and 0 are all inhabitants of $\mathbb{R}$, there is no $x \in \mathbb{R}$ which would satisfy $x^2 + 1 = 0$. Associating the points on the horizontal x-axis with the elements of $\mathbb{R}$, no point on the x-axis solves the equation $x^2 + 1 = 0$. Expanding the computational universe from the x-axis to the xy-plane, and extending the rules of game is a strategy which is at the heart of many discoveries in mathematics and physics. Very often than not the augmented theory contains the less-developed theory as a special case.

Embed the x-axis into the xy-plane. Then, for moving from $A \in \mathbb{R}$ to $B \in \mathbb{R}$ there is now a plethora of options available beyond the path constraint to the x-axis. The embedding process extends the scope of the calculation beyond just enabling and facilitating computations which were not possible prior to the embedding. A few examples should clarify the ideas.

Problem: Determine the square root of 4.

Solution: Embed the x-axis into the (x,y)-plane. Endow the (x,y)-plane with its "natural" (x,y)-Cartesian coordinate system (horizontal x-axis and vertical y-axis). Superimpose an (r,θ)-polar coordinate system onto the (x,y)-Cartesian coordinate system. Working simultaneously in the (x,y)- and the (r,θ)-coordinate systems allows to simplify the discussion. Then, the point $P_1(x = 4)$, on the x-axis is at the same time the point $P_1(x = 4, y = 0)$ in the (x,y)-Cartesian coordinate system, and the point $P_1(r = 4, \theta = 0)$ in the (r,θ)-polar coordinate system. These equivalences are represented by the notation,

$$P_1(x = 4) \iff P_1(x = 4, y = 0) \iff P_1(r = 4, \theta = 0). \tag{1.175}$$

The point $P_1(r = 4, \theta = 0)$ is at the distance $r = 4$ from the origin $O(x = 0, y = 0)$ of the Cartesian (and poplar) coordinate system, and builds the angle $\theta = 0$ with the positive x-axis.

To determine the square root of 4, associated with the point $P_1(r = 4, \theta = 0)$ (in the (r,θ) plane rather than the point $P_1(x = 4)$ on the x-axis), take the square root of (the positive radius) $r = 4$ and halve the angle $(\theta = 0)$, to obtain the point $Q_1(r = 2, \theta = 0)$. The point $Q_1(r = 2, \theta = 0)$, as a point in the (x,y)-plane has the coordinates $Q_1(x = 2, y = 0)$, and is consequently a point on the x-axis: $Q_1(x = 2)$: the point $x = 2$ on the x-axis is the square root of 4.

Next add 2π to the $\theta = 0$ to obtain $P_2(r = 4, \theta = 2\pi)$ which coincides with $P_1(r = 4, \theta = 0)$. Proceed as before: take the square root of the positive radius $r = 4$ and halve the angle $(\theta = 2\pi)$, to obtain $Q_2(r = 2, \theta = \pi)$. In Cartesian coordinate system, this is the point $Q_2(x = -2, y = 0)$. Coincidentally, this point is also on the x-axis: $Q_2(x = -2)$: the point $x = -2$ on the x-axis is also the square root of 4.

Thus, adding 2π to the original $\theta = 0$ resulted in a second solution. What about adding another 2π to the θ? That means what about considering the point $P_3(r = 4, \theta = 4\pi)$ to obtain a third solution? To investigate this case, take the square root of $r = 4$ and halve the angle $(\phi = 4\pi)$, to obtain $Q_3(r = 2, \phi = 2\pi)$. This solution is, however, indistinguishable from the first solution $Q_1(r = 2, \theta = 0)$.

As is readily seen, adding an even or odd multiple of 2π to $\theta = 0$ results in $Q_1(r = 2, \theta = 0)$ or $Q_2(r = 2, \theta = \pi)$, respectively. Consequently, the square root of 4 has the solutions $x_1 = 2$ and $x_2 = -2$.

■

Problem: Determine the cube root of 27.

Solution: Embedding the x-axis into the (x,y)-plane, introducing the (x,y)-Cartesian coordinate system and superposing the (r,θ)-polar coordinate system, the starting points are the equivalent points $P_1(x = 27)$, on the x-axis,

$P_1(x = 27, y = 0)$ in the (x,y)-Cartesian coordinate system, and $P_1(r = 27, \theta = 0)$ in the (r, θ)-polar coordinate system, which are by construction equivalent,

$$P_1(x = 27) \iff P_1(x = 27, y = 0) \iff P_1(r = 27, \theta = 0). \tag{1.176}$$

The point $P_1(r = 27, \theta = 0)$ is at the distance $r = 27$ from the origin $O(x = 0, y = 0)$ of the Cartesian (and poplar) coordinate system, and builds the angle $\theta = 0$ with the positive x-axis.

To determine the cubic root of 27, associated with the point $P_1(r = 27, \theta = 0)$ (in the (r, θ) plane rather than the point $P_1(x = 27)$ on the x-axis), take the cubic root of (the positive radius) $r = 27$ and divide by three the angle ($\theta = 0$), to obtain the point $Q_1(r = 3, \theta = 0)$. The point $Q_1(r = 3, \theta = 0)$, as a point in the (x, y)-plane has the coordinates $Q_1(x = 3, y = 0)$, and is consequently a point on the x-axis: $Q_1(x = 3)$: the point $x = 3$ on the x-axis is the cubic root of 27.

Next add 2π to the $\theta = 0$ to obtain $P_2(r = 27, \theta = 2\pi)$ which coincides with $P_1(r = 27, \theta = 0)$. Proceed as before: take the cubic root of the positive radius $r = 27$ and divide by three the angle ($\theta = 2\pi$), to obtain $Q_2(r = 3, \theta = 2\pi/3)$. In Cartesian coordinate system, this is the point $Q_2(x = 3\cos(2\pi/3), y = 3\sin(2\pi/3)) = Q_2(x = -3/2, y = 3\sqrt{3}/2)$. This point is obviously no longer on the x-axis, and markedly different than $Q_1(x = 3)$.

This analysis shows that first, roots are not necessarily confined to the x-axis. And second, it still needs to be shown that $Q_2(x = 3\cos(2\pi/3), y = 3\sin(2\pi/3))$ is indeed the cubic root of 27 and in which sense? To answer this question associate with the point $Q_2(x = 3\cos(2\pi/3), y = 3\sin(2\pi/3))$ the (stretching by the factor 3 and counter-clockwise rotation by the angle $2\pi/3$) matrix

$$\mathbf{Q}_2 = 3\mathbf{R}_2(\cos(2\pi/3), \sin(2\pi/3)) = 3 \begin{bmatrix} \cos\left(\frac{2\pi}{3}\right) & -\sin\left(\frac{2\pi}{3}\right) \\ \sin\left(\frac{2\pi}{3}\right) & \cos\left(\frac{2\pi}{3}\right) \end{bmatrix}.$$

Multiplying $\mathbf{Q}_2$ three times by itself,

$$(\mathbf{Q}_2)^3 = \left\{ 3 \begin{bmatrix} \cos\left(\frac{2\pi}{3}\right) & -\sin\left(\frac{2\pi}{3}\right) \\ \sin\left(\frac{2\pi}{3}\right) & \cos\left(\frac{2\pi}{3}\right) \end{bmatrix} \right\}^3$$

$$= 27 \begin{bmatrix} \cos\left(\frac{2\pi}{3}\right) & -\sin\left(\frac{2\pi}{3}\right) \\ \sin\left(\frac{2\pi}{3}\right) & \cos\left(\frac{2\pi}{3}\right) \end{bmatrix}^3$$

$$= 27 \begin{bmatrix} \cos\left(3 \times \frac{2\pi}{3}\right) & -\sin\left(3 \times \frac{2\pi}{3}\right) \\ \sin\left(3 \times \frac{2\pi}{3}\right) & \cos\left(3 \times \frac{2\pi}{3}\right) \end{bmatrix}$$

$$= 27 \begin{bmatrix} \cos(2\pi) & -\sin(2\pi) \\ \sin(2\pi) & \cos(2\pi) \end{bmatrix}$$

$$= 27 \begin{bmatrix} 1 & 0 \\ 0 & 1 \end{bmatrix} \iff 27.$$

This explains in which sense $Q_2(x = 3\cos(2\pi/3), y = 3\sin(2\pi/3))$ is a cubic root of 27.

Thus, adding 2π to the original $\theta = 0$ resulted in a second solution. What about adding another 2π to the θ? That means what about considering the point $P_3(r = 27, \theta = 4\pi)$ to obtain a third solution? To investigate this case, take the cubic root of $r = 27$ and divide by three the angle ($\theta = 4\pi$), to obtain $Q_3(r = 3, \theta = 4\pi/3)$. In Cartesian coordinate system this is the point $Q_3(x = 3\cos(4\pi/3), y = 3\sin(4\pi/3)) = Q_4(x = -3/2, y = -3\sqrt{3}/2)$. This point is obviously also no longer on the x-axis, and markedly different than $Q_1(x = 3)$ and $Q_2(x = -3/2, y = 3\sqrt{3}/2)$. Associating with the point $Q_3(x = 3\cos(4\pi/3), y = 3\sin(4\pi/3))$ the stretching and rotation matrix $\mathbf{Q}_3 = 3\mathbf{R}_2(\cos(4\pi/3), \sin(4\pi/3))$ is immediate that $\mathbf{Q}_3^3$ is equivalent with 27. Thus, $Q_1(x = 3, y = 0)$, $Q_2(x = -3/2, y = 3\sqrt{3}/2)$, and $Q_3(x = -3/2, y = -3\sqrt{3}/2)$ are three cubic roots of 27. Similar to the preceding problem, adding further multiple of 2π reproduces either of these solutions and consequently no further solution exists.

■

As the final problem (and the culmination of the present discussion) consider the square roots of -1.

Problem: Determine the square root of -1.

Solution: Embed the x-axis into the (x, y)-plane. Endow the (x, y)-plane with the (x, y)-Cartesian coordinate system and superimpose an (r, θ)-polar coordinate system onto the (x, y)-Cartesian coordinate system. Then, the point $P_1(x = -1)$, on the x-axis is at the same time the point $P_1(x = -1, y = 0)$ in the (x, y)-Cartesian coordinate system, and the point $P_1(r = 1, \theta = \pi)$ in the (r, θ)-polar coordinate system. These equivalences are represented by the notation,

$$P_1(x = -1) \iff P_1(x = -1, y = 0) \iff P_1(r = 1, \theta = \pi). \tag{1.177}$$

The point $P_1(r = 1, \theta = \pi)$ is at the distance $r = 1$ from the origin $O(x = 0, y = 0)$ of the Cartesian (and poplar) coordinate system, and builds the angle $\theta = \pi$ with the positive x-axis.

To determine the square root of -1, associated with the point $P_1(r = 1, \theta = \pi)$ (in the (r, θ) plane rather than the point $P_1(x = -1)$ on the x-axis), take the square root of (the positive radius) $r = 1$ and halve the angle $(\theta = \pi)$, to obtain the point $Q_1(r = 1, \theta = \pi/2)$. The point $Q_1(r = 1, \theta = \pi/2)$, as a point in the (x, y)-plane has the coordinates $Q_1(x = 0, y = 1)$, and is distinctly not on the x-axis, it is rather a point of the y-axis. To show that $Q_1(x = 0, y = 1)$ is a square root of -1, and in which sense, associate $Q_1(r = 1, \theta = \pi/2)$ with the rotation matrix,

$$\mathbf{Q}_1 = \mathbf{R}_1(\cos(\pi/2), \sin(\pi/2)) = \begin{bmatrix} \cos\left(\frac{\pi}{2}\right) & -\sin\left(\frac{\pi}{2}\right) \\ \sin\left(\frac{\pi}{2}\right) & \cos\left(\frac{\pi}{2}\right) \end{bmatrix} = \begin{bmatrix} 0 & -1 \\ 1 & 0 \end{bmatrix}.$$

Thus,

$$\mathbf{Q}_1^2 = \begin{bmatrix} 0 & -1 \\ 1 & 0 \end{bmatrix} \begin{bmatrix} 0 & -1 \\ 1 & 0 \end{bmatrix} = \begin{bmatrix} -1 & 0 \\ 0 & -1 \end{bmatrix} = - \begin{bmatrix} 1 & 0 \\ 0 & 1 \end{bmatrix} \iff -1.$$

Consequently, the point $Q_1(x = 0, y = 1)$ equivalent with $Q_1(r = 1, \theta = \pi/2)$ is the square root of -1, in the above sense.

Next add 2π to $\theta = \pi$ to obtain $P_2(r = 1, \theta = 3\pi)$ which coincides with $P_1(r = 1, \theta = \pi)$. Take the square root of the positive radius $r = 1$ and halve the angle $(\theta = 3\pi)$, to obtain $Q_2(r = 1, \theta = 3\pi/2)$. In the Cartesian coordinate system, this is the point $Q_2(x = 0, y = -1)$, which is a point on the y-axis. Associate this point with the rotation matrix,

$$\mathbf{Q}_2 = \mathbf{R}_1(\cos(3\pi/2), 3\sin(\pi/2)) = \begin{bmatrix} \cos\left(\frac{3\pi}{2}\right) & -\sin\left(\frac{3\pi}{2}\right) \\ \sin\left(\frac{3\pi}{2}\right) & \cos\left(\frac{3\pi}{2}\right) \end{bmatrix} = \begin{bmatrix} 0 & 1 \\ -1 & 0 \end{bmatrix}.$$

Thus,

$$\mathbf{Q}_2^2 = \begin{bmatrix} 0 & 1 \\ -1 & 0 \end{bmatrix} \begin{bmatrix} 0 & 1 \\ -1 & 0 \end{bmatrix} = \begin{bmatrix} -1 & 0 \\ 0 & -1 \end{bmatrix} = - \begin{bmatrix} 1 & 0 \\ 0 & 1 \end{bmatrix} \iff -1.$$

Consequently, the point $Q_2(x = 0, y = -1)$ equivalent with $Q_2(r = 1, \theta = 3\pi/2)$ is the square root of -1, in the above sense. No mysteries, no obscurity, and everything remains real and simple. It was merely required to introduce the notion of rotation matrix, and the matrix multiplication.

■

The result $\{-\mathbf{I}\}^{\frac{1}{2}} = \mathbf{J}$ is reminiscent of $\sqrt{-1} = j$ ($\sqrt{-1} = j$ in engineering or $\sqrt{-1} = i$ in physics literature). Indeed, a tight relationship between $\{-\mathbf{I}\}^{\frac{1}{2}} = \mathbf{J}$ and $\sqrt{-1} = j$ will be established momentarily. While (1.174b) and (1.174c) still exhibit the minus sign under the square, or within the curly brackets in $\{\cdot\}^{\frac{1}{2}}$, the equivalent formulation (1.174a) is devoid of any obscurity.

The concept of the unitary rotation matrix $\mathbf{R}_0(\theta)$, associated with the point $P_0(\theta)$ on the unit circle, with θ continuously varying in the interval $[0, 2\pi)$, is intuitive, and exploring the consequences thereof immediate. Nonetheless, generations of engineers and technologists are led to confusion by confronting them with, among others, the minus sign under the square root sign! However, reverse engineering the occurrence of the minus sign reveals, that nothing is mysterious provided relationships and mathematical objects are introduced and employed properly.

Summary:

$$\{-\mathbf{I}\}^{\frac{1}{2}} = \left\{-\begin{bmatrix} 1 & 0 \\ 0 & 1 \end{bmatrix}\right\}^{\frac{1}{2}} = \left\{\begin{bmatrix} -1 & 0 \\ 0 & -1 \end{bmatrix}\right\}^{\frac{1}{2}} = \left\{\begin{bmatrix} 0 & -1 \\ 1 & 0 \end{bmatrix}\begin{bmatrix} 0 & -1 \\ 1 & 0 \end{bmatrix}\right\}^{\frac{1}{2}}$$

$$= \left\{\begin{bmatrix} 0 & -1 \\ 1 & 0 \end{bmatrix}^2\right\}^{\frac{1}{2}} = \begin{bmatrix} 0 & -1 \\ 1 & 0 \end{bmatrix} = \mathbf{J}. \tag{1.178}$$

$\square$

Problem: Determine the eigenpairs (eigenvalues and the corresponding eigenvectors) of $\mathbf{I}$.

Solution: Consider,

$$\begin{bmatrix} 1 & 0 \\ 0 & 1 \end{bmatrix}\begin{bmatrix} u_1 \\ u_2 \end{bmatrix} = \lambda \begin{bmatrix} u_1 \\ u_2 \end{bmatrix} \implies \begin{bmatrix} 1-\lambda & 0 \\ 0 & 1-\lambda \end{bmatrix}\begin{bmatrix} u_1 \\ u_2 \end{bmatrix} = 0. \tag{1.179}$$

The zeros of the characteristic determinant are the eigenvalues,

$$\det\left\{\begin{bmatrix} 1-\lambda & 0 \\ 0 & 1-\lambda \end{bmatrix}\right\} = (1-\lambda)^2 = 0 \implies \lambda^{(1)} = \lambda^{(2)} = 1. \tag{1.180}$$

Thus, the 2×2 unity matrix $\mathbf{I}$ has the two-fold degenerate eigenvalue 1. Substituting $\lambda = 1$ into (1.179),

$$\begin{bmatrix} 0 & 0 \\ 0 & 0 \end{bmatrix}\begin{bmatrix} u_1 \\ u_2 \end{bmatrix} = 0. \tag{1.181}$$

This equation is valid for any arbitrary finite values u_1 and u_2. In particular, the canonical (standard) unit vectors

$$\mathbf{u}^{(1)} = \begin{bmatrix} 1 \\ 0 \end{bmatrix}, \quad \mathbf{u}^{(2)} = \begin{bmatrix} 0 \\ 1 \end{bmatrix}, \tag{1.182}$$

can be chosen. $\blacksquare$

Problem: Determine the eigenpairs of $\mathbf{J}$.

Solution: Consider,

$$\begin{bmatrix} 0 & -1 \\ 1 & 0 \end{bmatrix}\begin{bmatrix} v_1 \\ v_2 \end{bmatrix} = \mu \begin{bmatrix} v_1 \\ v_2 \end{bmatrix} \implies \begin{bmatrix} -\mu & -1 \\ 1 & -\mu \end{bmatrix}\begin{bmatrix} v_1 \\ v_2 \end{bmatrix} = 0. \tag{1.183}$$

The zeros of the characteristic determinant are the requested eigenvalues,

$$\det\left\{\begin{bmatrix} -\mu & -1 \\ 1 & -\mu \end{bmatrix}\right\} = \mu^2 + 1 = 0. \tag{1.184}$$

Apparently, a problem (paradox) arises: any point on the x-axis (corresponding to $\mu \in \mathbb{R}$) renders the term $\mu^2 + 1$ positive—it can never get zero. The resolution of the contradiction, paradox, requires a change of view, a different framework, and/or new mathematical objects (structures). Being restricted to the x-axis means going back and forth on one line. Many concepts are meaningless if one is constraints to move along a line. The idea of rotation, e.g., has no interpretation in a one-dimensional world. These considerations suggest posing the next problem. ∎

Problem: Determine the eigenvalues of **J** from a different perspective—revisiting the problem.

Solution: Consider the eigenvalue problem,

$$\begin{bmatrix} 0 & -1 \\ 1 & 0 \end{bmatrix}\begin{bmatrix} v_1 \\ v_2 \end{bmatrix} = \mu \begin{bmatrix} v_1 \\ v_2 \end{bmatrix}. \tag{1.185}$$

The eigenvalue μ, and presumably the components of the associated eigenvector, v_1, and v_2, are not located on the x-axis. This suggests embedding the one dimensional x-axis into the two-dimensional xy-plane. Then, being points in the xy-plane, μ, v_1, and v_2 can be represented by their characterizing 2×2 matrices,

$$\mu \iff \begin{bmatrix} \mu_c & -\mu_s \\ \mu_s & \mu_c \end{bmatrix}, \quad \mu_c, \mu_s \in \mathbb{R}, \tag{1.186}$$

and,

$$v_1 \iff \begin{bmatrix} v_{1,c} & -v_{1,s} \\ v_{1,s} & v_{1,c} \end{bmatrix} \quad v_{1,c}, v_{1,s} \in \mathbb{R}; \quad v_2 \iff \begin{bmatrix} v_{2,c} & -v_{2,s} \\ v_{2,s} & v_{2,c} \end{bmatrix} \quad v_{2,c}, v_{2,s} \in \mathbb{R}. \tag{1.187}$$

Correspondingly, the entries of the matrix in (1.185), the scalars 0 and 1, must be promoted to their corresponding matrices. To this end,

$$0 \iff \begin{bmatrix} 0 & 0 \\ 0 & 0 \end{bmatrix}, \quad 1 \iff \begin{bmatrix} 1 & 0 \\ 0 & 1 \end{bmatrix}. \tag{1.188}$$

Remark: The representation of 0 requires a slight justification. Recall that the points on the perimeter of a circle with radius r can be represented by

$$a \iff r \begin{bmatrix} \cos(\alpha) & -\sin(\alpha) \\ \sin(\alpha) & \cos(\alpha) \end{bmatrix}, \tag{1.189}$$

with $\alpha \in [0, 2\pi)$ and the matrix entries $\cos(\alpha)$ and $\sin(\alpha)$ being subject to the constraint $\cos^2(\alpha) + \sin^2(\alpha) = 1$. The question is whether the matrix representation of 0 in (1.188), with all the entries being zero, does not contradicts the constraint $\cos^2(\alpha) + \sin^2(\alpha) = 1$. The restriction $\cos^2(\alpha) + \sin^2(\alpha) = 1$ arguably holds valid for any finite r. The representation (1.189) can be written in the form,

$$a \iff \begin{bmatrix} r\cos(\alpha) & -r\sin(\alpha) \\ r\sin(\alpha) & r\cos(\alpha) \end{bmatrix}. \tag{1.190}$$

Since $-1 \leq \cos(\alpha), \sin(\alpha) \leq 1$, in the limit $r \to 0$,

$$\begin{bmatrix} 0 & 0 \\ 0 & 0 \end{bmatrix} \iff 0. \tag{1.191}$$

The distinction between the points on the perimeter of a circle with radius zero is no longer possible, meaningful—inevitably rendering the point 0 "singular": the point 0, or, equivalently, $(0,0)$, is characterized by the radius $r = 0$ and an unspecifiable angle α. Stated differently, at $r = 0$, the angle α ceases to have any meaning. □

To facilitate the discussion, it is instructive to recast (1.185),

$$\begin{bmatrix} 0 & -1 \\ 1 & 0 \end{bmatrix}\begin{bmatrix} v_1 \\ v_2 \end{bmatrix} = \begin{bmatrix} \mu & 0 \\ 0 & \mu \end{bmatrix}\begin{bmatrix} v_1 \\ v_2 \end{bmatrix}. \tag{1.192}$$

Embedding the scalar quantities, 0, 1, v_1, v_2, μ, in the xy-space, i.e., representing them by their corresponding 2×2-matrices,

$$\begin{bmatrix} \begin{bmatrix} 0 & 0 \\ 0 & 0 \end{bmatrix} & -\begin{bmatrix} 1 & 0 \\ 0 & 1 \end{bmatrix} \\ \begin{bmatrix} 1 & 0 \\ 0 & 1 \end{bmatrix} & \begin{bmatrix} 0 & 0 \\ 0 & 0 \end{bmatrix} \end{bmatrix}\begin{bmatrix} \begin{bmatrix} v_{1,c} & -v_{1,s} \\ v_{1,s} & v_{1,c} \end{bmatrix} \\ \begin{bmatrix} v_{2,c} & -v_{2,s} \\ v_{2,s} & v_{2,c} \end{bmatrix} \end{bmatrix}$$
$$= \begin{bmatrix} \begin{bmatrix} \mu_c & -\mu_s \\ \mu_s & \mu_c \end{bmatrix} & \begin{bmatrix} 0 & 0 \\ 0 & 0 \end{bmatrix} \\ \begin{bmatrix} 0 & 0 \\ 0 & 0 \end{bmatrix} & \begin{bmatrix} \mu_c & -\mu_s \\ \mu_s & \mu_c \end{bmatrix} \end{bmatrix}\begin{bmatrix} \begin{bmatrix} v_{1,c} & -v_{1,s} \\ v_{1,s} & v_{1,c} \end{bmatrix} \\ \begin{bmatrix} v_{2,c} & -v_{2,s} \\ v_{2,s} & v_{2,c} \end{bmatrix} \end{bmatrix}. \tag{1.193}$$

Carrying out the block-matrix multiplications and simplifying,

$$-\begin{bmatrix} 1 & 0 \\ 0 & 1 \end{bmatrix}\begin{bmatrix} v_{2,c} & -v_{2,s} \\ v_{2,s} & v_{2,c} \end{bmatrix} = \begin{bmatrix} \mu_c & -\mu_s \\ \mu_s & \mu_c \end{bmatrix}\begin{bmatrix} v_{1,c} & -v_{1,s} \\ v_{1,s} & v_{1,c} \end{bmatrix} \tag{1.194a}$$

$$\begin{bmatrix} v_{1,c} & -v_{1,s} \\ v_{1,s} & v_{1,c} \end{bmatrix} = \begin{bmatrix} \mu_c & -\mu_s \\ \mu_s & \mu_c \end{bmatrix}\begin{bmatrix} v_{2,c} & -v_{2,s} \\ v_{2,s} & v_{2,c} \end{bmatrix}. \tag{1.194b}$$

The reason for keeping the identity matrix in (1.194a) will become clear momentarily. Substituting (1.194b) into (1.194a),

$$-\begin{bmatrix} 1 & 0 \\ 0 & 1 \end{bmatrix}\begin{bmatrix} v_{2,c} & -v_{2,s} \\ v_{2,s} & v_{2,c} \end{bmatrix} = \begin{bmatrix} \mu_c & -\mu_s \\ \mu_s & \mu_c \end{bmatrix}^2\begin{bmatrix} v_{2,c} & -v_{2,s} \\ v_{2,s} & v_{2,c} \end{bmatrix}. \tag{1.195}$$

Assuming non-trivial solution $v_2 \neq 0$ ($v_{2,c}^2 + v_{2,s}^2 > 0$),

$$\begin{bmatrix} \mu_c & -\mu_s \\ \mu_s & \mu_c \end{bmatrix}^2 = -\begin{bmatrix} 1 & 0 \\ 0 & 1 \end{bmatrix}. \tag{1.196}$$

Referring to the results obtained in the preceding problem set, there are two solutions,

$$\begin{bmatrix} \mu_c & -\mu_s \\ \mu_s & \mu_c \end{bmatrix} = \begin{bmatrix} 0 & -1 \\ 1 & 0 \end{bmatrix} \qquad (\Longrightarrow \quad \mu_c = 0, \quad \mu_s = 1), \tag{1.197}$$

and

$$\begin{bmatrix} \mu_c & -\mu_s \\ \mu_s & \mu_c \end{bmatrix} = -\begin{bmatrix} 0 & -1 \\ 1 & 0 \end{bmatrix} = \begin{bmatrix} 0 & 1 \\ -1 & 0 \end{bmatrix} \qquad (\Longrightarrow \quad \mu_c = 0, \quad \mu_s = -1). \tag{1.198}$$

For easy reference, and reaffirmation,

$$\begin{bmatrix} 0 & -1 \\ 1 & 0 \end{bmatrix}^2 = \begin{bmatrix} 0 & -1 \\ 1 & 0 \end{bmatrix}\begin{bmatrix} 0 & -1 \\ 1 & 0 \end{bmatrix} = \begin{bmatrix} -1 & 0 \\ 0 & -1 \end{bmatrix} = -\begin{bmatrix} 1 & 0 \\ 0 & 1 \end{bmatrix} \tag{1.199a}$$

$$\begin{bmatrix} 0 & 1 \\ -1 & 0 \end{bmatrix}^2 = \begin{bmatrix} 0 & 1 \\ -1 & 0 \end{bmatrix}\begin{bmatrix} 0 & 1 \\ -1 & 0 \end{bmatrix} = \begin{bmatrix} -1 & 0 \\ 0 & -1 \end{bmatrix} = -\begin{bmatrix} 1 & 0 \\ 0 & 1 \end{bmatrix} \tag{1.199b}$$

In order to determine the eigenvector corresponding to the eigenvalue (1.197), substitute (1.197) into (1.194a),

$$-\begin{bmatrix} v_{2,c} & -v_{2,s} \\ v_{2,s} & v_{2,c} \end{bmatrix} = \begin{bmatrix} 0 & -1 \\ 1 & 0 \end{bmatrix}\begin{bmatrix} v_{1,c} & -v_{1,s} \\ v_{1,s} & v_{1,c} \end{bmatrix}. \tag{1.200}$$

Or, equivalently,

$$\begin{bmatrix} -v_{2,c} & v_{2,s} \\ -v_{2,s} & -v_{2,c} \end{bmatrix} = \begin{bmatrix} -v_{1,s} & -v_{1,c} \\ v_{1,c} & -v_{1,s} \end{bmatrix}. \tag{1.201}$$

A term-by-term comparison reveals,

$$v_{2,c} = v_{1,s} \tag{1.202a}$$
$$v_{2,s} = -v_{1,c}. \tag{1.202b}$$

The choices $v_{1,c} = 0$ and $v_{1,s} = 1$ are legitimate and lead to $v_{2,c} = 1$ and $v_{2,s} = 0$. Consequently,

$$\begin{bmatrix} \begin{bmatrix} v_{1,c} & -v_{1,s} \\ v_{1,s} & v_{1,c} \end{bmatrix} \\ \begin{bmatrix} v_{2,c} & -v_{2,s} \\ v_{2,s} & v_{2,c} \end{bmatrix} \end{bmatrix} = \begin{bmatrix} \begin{bmatrix} 0 & -1 \\ 1 & 0 \end{bmatrix} \\ \begin{bmatrix} 1 & 0 \\ 0 & 1 \end{bmatrix} \end{bmatrix}. \tag{1.203}$$

This result can be interpreted as follows: The matrix $\begin{bmatrix} v_{1,c} & -v_{1,s} \\ v_{1,s} & v_{1,c} \end{bmatrix}$ being equal to $\begin{bmatrix} 0 & -1 \\ 1 & 0 \end{bmatrix}$ represents a point in the xy-plane. The equality $\cos(\theta) = 0$ and consistently $\sin(\theta) = 1$ imply $\theta = \pi/2$. Thus, the matrix $\begin{bmatrix} v_{1,c} & -v_{1,s} \\ v_{1,s} & v_{1,c} \end{bmatrix}$ refers to a point on the y-axis at a distance 1 from the origin. Similarly, the matrix $\begin{bmatrix} v_{2,c} & -v_{2,s} \\ v_{2,s} & v_{2,c} \end{bmatrix}$ being equal to $\begin{bmatrix} 1 & 0 \\ 0 & 1 \end{bmatrix}$ represents a point in the xy-plane. The equality $\cos(\theta) = 1$ and consistently $\sin(\theta) = 0$ imply $\theta = 0$. Thus, the matrix $\begin{bmatrix} v_{2,c} & -v_{2,s} \\ v_{2,s} & v_{2,c} \end{bmatrix}$ refers to a point on the x-axis at a distance 1 from the origin. This is the interpretation of the eigenvector in (1.203).

In order to determine the eigenvector corresponding to the eigenvalue (1.198), substitute (1.198) into (1.194a),

$$-\begin{bmatrix} v_{2,c} & -v_{2,s} \\ v_{2,s} & v_{2,c} \end{bmatrix} = \begin{bmatrix} 0 & 1 \\ -1 & 0 \end{bmatrix}\begin{bmatrix} v_{1,c} & -v_{1,s} \\ v_{1,s} & v_{1,c} \end{bmatrix}. \tag{1.204}$$

Or, equivalently,

$$\begin{bmatrix} -v_{2,c} & v_{2,s} \\ -v_{2,s} & -v_{2,c} \end{bmatrix} = \begin{bmatrix} v_{1,s} & v_{1,c} \\ -v_{1,c} & v_{1,s} \end{bmatrix}. \tag{1.205}$$

A term-by-term comparison reveals,

$$v_{2,c} = -v_{1,s} \tag{1.206a}$$

$$v_{2,s} = v_{1,c}. \tag{1.206b}$$

The choices $v_{1,c} = 1$ and $v_{1,s} = 0$ are legitimate and lead to $v_{2,c} = 0$ and $v_{2,s} = 1$. Consequently,

$$\begin{bmatrix} \begin{bmatrix} v_{1,c} & -v_{1,s} \\ v_{1,s} & v_{1,c} \end{bmatrix} \\ \begin{bmatrix} v_{2,c} & -v_{2,s} \\ v_{2,s} & v_{2,c} \end{bmatrix} \end{bmatrix} = \begin{bmatrix} \begin{bmatrix} 1 & 0 \\ 0 & 1 \end{bmatrix} \\ \begin{bmatrix} 0 & -1 \\ 1 & 0 \end{bmatrix} \end{bmatrix}. \tag{1.207}$$

An argument along the lines stated above shows that the matrix $\begin{bmatrix} v_{1,c} & -v_{1,s} \\ v_{1,s} & v_{1,c} \end{bmatrix}$ represents a point on the x-axis at a unit distance from the origin. Similarly, $\begin{bmatrix} v_{2,c} & -v_{2,s} \\ v_{2,s} & v_{2,c} \end{bmatrix}$ stands for a point on the y-axis at a unit distance from the origin.

□

Problem: Given the matrix relationship

$$e^{\mathbf{J}\theta} = \cos(\theta)\mathbf{I} + \sin(\theta)\mathbf{J}, \tag{1.208}$$

along with the set of eigenvalues $\{\lambda^{(1)} = 1, \mu^{(1)} = i\}$ and $\{\lambda^{(2)} = 1, \mu^{(2)} = -i\}$, obtained above. Replace the matrices $\mathbf{I}$ and $\mathbf{J}$ with their respective eigenvalues.

Solution: While the task in this problem is utterly easy, the result is disproportionately satisfying. Replacing $\mathbf{I}$ and $\mathbf{J}$ with $\lambda^{(1)} = 1$ and $\mu^{(1)} = i$, respectively, (1.208) yields,

$$e^{i\theta} = \cos(\theta) + i\sin(\theta), \tag{1.209}$$

which is the celebrated Euler's formula.

Similarly, replacing $\mathbf{I}$ and $\mathbf{J}$ with $\lambda^{(1)} = 1$ and $\mu^{(1)} = -i$, respectively, (1.208) results in,

$$e^{-i\theta} = \cos(\theta) - i\sin(\theta), \tag{1.210}$$

a results which can be obtained from (1.209) by substituting θ with $-\theta$.

■

1.12 Simultaneous "coupling" of $f_{\text{odd}}(x), f_{\text{even}}(x)$ with their mutual first derivatives

Consider $f_{\text{odd}}(x)$ and $f_{\text{even}}(x)$ alongside their first derivatives,

$$f_{\text{odd}}(x) = c_1 x + c_3 x^3 + c_5 x^5 + c_7 x^7 + \cdots \tag{1.211a}$$

$$f_{\text{odd}}^{(1)}(x) = c_1 + 3c_3 x^2 + 5c_5 x^4 + 7c_7 x^6 + \cdots \tag{1.211b}$$

$$f_{\text{even}}(x) = c_0 + c_2 x^2 + c_4 x^4 + c_6 x^6 + c_8 x^8 + \cdots \tag{1.211c}$$

$$f_{\text{even}}^{(1)}(x) = 2c_2 x + 4c_4 x^3 + 6c_6 x^5 + 8c_8 x^7 + \cdots . \tag{1.211d}$$

In virtue of the fact that $f_{\text{odd}}^{(1)}(x)$ and $f_{\text{even}}(x)$ are even, and $f_{\text{even}}^{(1)}(x)$ and $f_{\text{odd}}(x)$ odd functions, respectively, it is justifiable to ask which coefficients c_n ($n \in \mathbb{N}_0$) would simultaneously render the equations,

$$f_{\text{odd}}^{(1)}(x) = f_{\text{even}}(x) \tag{1.212a}$$

$$f_{\text{even}}^{(1)}(x) = f_{\text{odd}}(x) , \tag{1.212b}$$

valid. Employing (1.211) and (1.212) yields,

$$c_1 + 3c_3 x^2 + 5c_5 x^4 + 7c_7 x^6 + \cdots$$
$$= c_0 + c_2 x^2 + c_4 x^4 + c_6 x^6 + c_8 x^8 + \cdots \tag{1.213a}$$

$$2c_2 x + 4c_4 x^3 + 6c_6 x^5 + 8c_8 x^7 + \cdots$$
$$= c_1 x + c_3 x^3 + c_5 x^5 + c_7 x^7 + \cdots \tag{1.213b}$$

Problem: Show that (1.212a) alone, implying (1.213a), would be insufficient for the determination of the coefficients. ∎

Problem: Show that (1.212b) alone, implying (1.213b), would be insufficient for the determination of the coefficients. ∎

A term-by-term comparison in (1.213a) results in,

$$c_1 = c_0 \tag{1.214a}$$

$$3c_3 = c_2 \tag{1.214b}$$

$$5c_5 = c_4 \tag{1.214c}$$

$$7c_7 = c_6. \tag{1.214d}$$

A term-by-term comparison in (1.213b) results in

$$2c_2 = c_1 \tag{1.215a}$$

$$4c_4 = c_3 \tag{1.215b}$$

$$6c_6 = c_5 \tag{1.215c}$$

$$8c_8 = c_7. \tag{1.215d}$$

Observe the "interwoven" manner in which the coefficients are related to each other. Adequately interlacing the equations in (1.214) and (1.215),

$$c_1 = c_0 \tag{1.216a}$$

$$2c_2 = c_1 \tag{1.216b}$$

$$3c_3 = c_2 \tag{1.216c}$$

$$4c_4 = c_3 \tag{1.216d}$$

$$5c_5 = c_4 \tag{1.216e}$$

$$6c_6 = c_5 \tag{1.216f}$$

$$7c_7 = c_6 \tag{1.216g}$$

$$8c_8 = c_7. \tag{1.216h}$$

This equations can be solved successively,

$$c_1 = c_0 \tag{1.217a}$$

$$2c_2 = c_1 \implies c_2 = \frac{1}{2}c_1 = \frac{1}{2!}c_0 \tag{1.217b}$$

$$3c_3 = c_2 \implies c_3 = \frac{1}{3}c_2 = \frac{1}{3!}c_0 \tag{1.217c}$$

$$4c_4 = c_3 \implies c_4 = \frac{1}{4}c_3 = \frac{1}{4!}c_0 \tag{1.217d}$$

$$5c_5 = c_4 \implies c_5 = \frac{1}{5}c_4 = \frac{1}{5!}c_0 \tag{1.217e}$$

$$6c_6 = c_5 \implies c_6 = \frac{1}{6}c_5 = \frac{1}{6!}c_0 \tag{1.217f}$$

$$7c_7 = c_6 \implies c_7 = \frac{1}{7}c_6 = \frac{1}{7!}c_0 \tag{1.217g}$$

$$8c_8 = c_7 \implies c_8 = \frac{1}{8}c_7 = \frac{1}{8!}c_0 \tag{1.217h}$$

$$\vdots$$

$$c_n = \frac{1}{n!}c_0 \ . \tag{1.217i}$$

Setting $c_0 = 1$, for $f_{\text{even}}(x)$ and $f_{\text{odd}}(x)$, respectively, the functions $\cosh(x)$ and $\sinh(x)$ can be identified in terms of the coefficients in (1.217),

$$\cosh(x) = 1 + \frac{1}{2!}x^2 + \frac{1}{4!}x^4 + \frac{1}{6!}x^6 + \cdots \tag{1.218a}$$

$$\sinh(x) = x + \frac{1}{3!}x^3 + \frac{1}{5!}x^5 + \frac{1}{7!}x^7 + \cdots \ . \tag{1.218b}$$

Further above, it was established that $f_{\text{even}}(x) = \cosh(x)$ and $f_{\text{odd}}(x) = \sinh(x)$. Considering these relationships in (1.218),

$$\cosh^{(1)}(x) = \sinh(x) \tag{1.219a}$$

$$\sinh^{(1)}(x) = \cosh(x). \tag{1.219b}$$

Explicating the differential operators in (1.219),

$$\frac{d}{dx}\cosh(x) = \sinh(x) \tag{1.220a}$$

$$\frac{d}{dx}\sinh(x) = \cosh(x), \tag{1.220b}$$

or, more compactly,

$$\frac{d}{dx}\begin{bmatrix} \cosh(x) \\ \sinh(x) \end{bmatrix} = \begin{bmatrix} \sinh(x) \\ \cosh(x) \end{bmatrix}. \tag{1.221}$$

"Transforming" the vector at the R.H.S. into the vector at the L.H.S.,

$$\frac{d}{dx}\begin{bmatrix} \cosh(x) \\ \sinh(x) \end{bmatrix} = \begin{bmatrix} 0 & 1 \\ 1 & 0 \end{bmatrix}\begin{bmatrix} \cosh(x) \\ \sinh(x) \end{bmatrix}. \tag{1.222}$$

The interpretation and significance of the matrix at the R.H.S. will be made clear momentarily. Differentiating the terms on both sides with respect to x,

$$\frac{d^2}{dx^2}\begin{bmatrix} \cosh(x) \\ \sinh(x) \end{bmatrix} = \begin{bmatrix} 0 & 1 \\ 1 & 0 \end{bmatrix}\frac{d}{dx}\begin{bmatrix} \cosh(x) \\ \sinh(x) \end{bmatrix}. \tag{1.223}$$

Substituting (1.222) into (1.223),

$$\frac{d^2}{dx^2}\begin{bmatrix} \cosh(x) \\ \sinh(x) \end{bmatrix} = \begin{bmatrix} 0 & 1 \\ 1 & 0 \end{bmatrix}\begin{bmatrix} 0 & 1 \\ 1 & 0 \end{bmatrix}\begin{bmatrix} \cosh(x) \\ \sinh(x) \end{bmatrix} \tag{1.224a}$$

$$\frac{d^2}{dx^2}\begin{bmatrix} \cosh(x) \\ \sinh(x) \end{bmatrix} = \begin{bmatrix} 0 & 1 \\ 1 & 0 \end{bmatrix}^2\begin{bmatrix} \cosh(x) \\ \sinh(x) \end{bmatrix}. \tag{1.224b}$$

By induction,

$$\frac{d^n}{dx^n}\begin{bmatrix} \cosh(x) \\ \sinh(x) \end{bmatrix} = \begin{bmatrix} 0 & 1 \\ 1 & 0 \end{bmatrix}^n\begin{bmatrix} \cosh(x) \\ \sinh(x) \end{bmatrix}, \tag{1.225}$$

which is hypothesized to be valid for arbitrary $n \in \mathbb{N}_0$.

Remark: Noting that

$$\begin{bmatrix} 0 & 1 \\ 1 & 0 \end{bmatrix}\begin{bmatrix} 0 & 1 \\ 1 & 0 \end{bmatrix} = \begin{bmatrix} 1 & 0 \\ 0 & 1 \end{bmatrix}, \tag{1.226}$$

(1.224) can be written as

$$\frac{d^2}{dx^2}\begin{bmatrix} \cosh(x) \\ \sinh(x) \end{bmatrix} = \begin{bmatrix} 1 & 0 \\ 0 & 1 \end{bmatrix}\begin{bmatrix} \cosh(x) \\ \sinh(x) \end{bmatrix}, \tag{1.227}$$

which decouples into the defining equations for cosh (x) and sinh (x),

$$\frac{d^2}{dx^2}\cosh(x) = \cosh(x) \tag{1.228a}$$

$$\frac{d^2}{dx^2}\sinh(x) = \sinh(x)\,, \tag{1.228b}$$

in agreement with the results established earlier.

□

Problem: Apply the Formal Taylor Series Expansion to the function cosh $(x + \Delta x)$ and draw conclusions.

Solution: Consider, the Formal Taylor Series Expansion of cosh $(x + \Delta x)$ at x,

$$\cosh(x + \Delta x) = \sum_{n=0}^{\infty}\left[\frac{d^n}{dx^n}\cosh(x + \Delta x)\right]_{\Delta x=0}\frac{1}{n!}(\Delta x)^n. \tag{1.229}$$

Expanding the series,

$$\begin{aligned}
\cosh(x + \Delta x) = {}& \left[\frac{d^0}{dx^0}\cosh(x + \Delta x)\right]_{\Delta x=0}\frac{1}{0!}(\Delta x)^0 \\
&+ \left[\frac{d^1}{dx^1}\cosh(x + \Delta x)\right]_{\Delta x=0}\frac{1}{1!}(\Delta x)^1 \\
&+ \left[\frac{d^2}{dx^2}\cosh(x + \Delta x)\right]_{\Delta x=0}\frac{1}{2!}(\Delta x)^2 \\
&+ \left[\frac{d^3}{dx^3}\cosh(x + \Delta x)\right]_{\Delta x=0}\frac{1}{3!}(\Delta x)^3 \\
&+ \left[\frac{d^4}{dx^4}\cosh(x + \Delta x)\right]_{\Delta x=0}\frac{1}{4!}(\Delta x)^4 \\
&+ \left[\frac{d^5}{dx^5}\cosh(x + \Delta x)\right]_{\Delta x=0}\frac{1}{5!}(\Delta x)^5 \\
&+ \left[\frac{d^6}{dx^6}\cosh(x + \Delta x)\right]_{\Delta x=0}\frac{1}{6!}(\Delta x)^6 \\
&+ \left[\frac{d^7}{dx^7}\cosh(x + \Delta x)\right]_{\Delta x=0}\frac{1}{7!}(\Delta x)^7 \\
&\ \ \vdots\ .
\end{aligned} \tag{1.230}$$

Employing the coupled equations,

$$\frac{d}{dx}\cosh(x) = \sinh(x) \tag{1.231a}$$

$$\frac{d}{dx}\sinh(x) = \cosh(x), \tag{1.231b}$$

explicit expressions for the higher-order derivatives of cosh (x) can be derived successively. Results are listed below for easy reference,

$$\frac{d^0}{dx^0}\cosh(x) = \cosh(x) \tag{1.232a}$$

$$\frac{d}{dx}\cosh(x) = \sinh(x) \tag{1.232b}$$

$$\frac{d^2}{dx^2}\cosh(x) = \frac{d}{dx}\left[\frac{d}{dx}\cosh(x)\right] = \frac{d}{dx}\sinh(x) = \cosh(x) \tag{1.232c}$$

$$\frac{d^3}{dx^3}\cosh(x) = \frac{d}{dx}\left[\frac{d^2}{dx^2}\cosh(x)\right] = \frac{d}{dx}\cosh(x) = \sinh(x) \tag{1.232d}$$

$$\frac{d^4}{dx^4}\cosh(x) = \frac{d}{dx}\left[\frac{d^3}{dx^3}\cosh(x)\right] = \frac{d}{dx}\sinh(x) = \cosh(x) \tag{1.232e}$$

$$\frac{d^5}{dx^5}\cosh(x) = \frac{d}{dx}\left[\frac{d^4}{dx^4}\cosh(x)\right] = \frac{d}{dx}\cosh(x) = \sinh(x) \tag{1.232f}$$

$$\frac{d^6}{dx^6}\cosh(x) = \frac{d}{dx}\left[\frac{d^5}{dx^5}\cosh(x)\right] = \frac{d}{dx}\sinh(x) = \cosh(x) \tag{1.232g}$$

$$\frac{d^7}{dx^7}\cosh(x) = \frac{d}{dx}\left[\frac{d^6}{dx^6}\cosh(x)\right] = \frac{d}{dx}\cosh(x) = \sinh(x) \tag{1.232h}$$

$$\vdots$$

Thus, $(n \in \mathbb{N}_0)$,

$$\frac{d^{2n}}{dx^{2n}}\cosh(x) = \cosh(x) \tag{1.233a}$$

$$\frac{d^{2n+1}}{dx^{2n+1}}\cosh(x) = \sinh(x). \tag{1.233b}$$

Consequently,

$$\begin{aligned}
\cosh(x + \Delta x) = &\left[\cosh(x + \Delta x)|_{\Delta x=0}\right] \\
&+ \left[\sinh(x + \Delta x)|_{\Delta x=0}\right](\Delta x) \\
&+ \left[\cosh(x + \Delta x)|_{\Delta x=0}\right]\frac{1}{2!}(\Delta x)^2 \\
&+ \left[\sinh(x + \Delta x)|_{\Delta x=0}\right]\frac{1}{3!}(\Delta x)^3 \\
&+ \left[\cosh(x + \Delta x)|_{\Delta x=0}\right]\frac{1}{4!}(\Delta x)^4 \ldots
\end{aligned}$$

$$\ldots + \left[\sinh{(x + \triangle x)}|_{\triangle x=0}\right]\frac{1}{5!}(\triangle x)^5$$

$$+ \left[\cosh{(x + \triangle x)}|_{\triangle x=0}\right]\frac{1}{6!}(\triangle x)^6$$

$$+ \left[\sinh{(x + \triangle x)}|_{\triangle x=0}\right]\frac{1}{7!}(\triangle x)^7$$

$$\vdots \quad . \tag{1.234}$$

Simplifying the terms in the square brackets,

$$\cosh{(x + \triangle x)} = \cosh{(x)}$$
$$+ \sinh{(x)}(\triangle x)$$
$$+ \cosh{(x)}\frac{1}{2!}(\triangle x)^2$$
$$+ \sinh{(x)}\frac{1}{3!}(\triangle x)^3$$
$$+ \cosh{(x)}\frac{1}{4!}(\triangle x)^4$$
$$+ \sinh{(x)}\frac{1}{5!}(\triangle x)^5$$
$$+ \cosh{(x)}\frac{1}{6!}(\triangle x)^6$$
$$+ \sinh{(x)}\frac{1}{7!}(\triangle x)^7$$
$$\vdots \quad . \tag{1.235}$$

The terms at the R.H.S. are multiplied by either $\cosh{(x)}$ or $\sinh{(x)}$. Grouping the terms in accordance with their association with $\cosh{(x)}$ or $\sinh{(x)}$,

$$\cosh{(x + \triangle x)} = \cosh{(x)}\left[1 + \frac{1}{2!}(\triangle x)^2 + \frac{1}{4!}(\triangle x)^4 + \frac{1}{6!}(\triangle x)^6 + \ldots\right]$$
$$+ \sinh{(x)}\left[\triangle x + \frac{1}{3!}(\triangle x)^3 + \frac{1}{5!}(\triangle x)^5 + \frac{1}{7!}(\triangle x)^7 + \ldots\right]. \tag{1.236}$$

Identifying the series in the first- and the second square brackets, as the Formal Taylor Series Expansions of $\cosh{(\triangle x)}$ and $\sinh{(\triangle x)}$, respectively, the elementary hyperbolic identity

$$\cosh{(x + \triangle x)} = \cosh{(x)}\cosh{(\triangle x)} + \sinh{(x)}\sinh{(\triangle x)}, \tag{1.237}$$

manifests itself. Note that nowhere in the derivation it was necessary to require that $\triangle x$ must be small. Consequently, $\triangle x$ can be replaced by any angle of finite magnitude, if the discussion requires.

Problem: Apply the Formal Taylor Series Expansion to the function $\sinh{(x + \triangle x)}$ and draw conclusions.

Solution:

$$\sinh(x + \triangle x) = \left[\frac{d^0}{dx^0} \sinh(x + \triangle x)\right]_{\triangle x=0} \frac{1}{0!}(\triangle x)^0$$

$$+ \left[\frac{d^1}{dx^1} \sinh(x + \triangle x)|\right]_{\triangle x=0} \frac{1}{1!}(\triangle x)^1$$

$$+ \left[\frac{d^2}{dx^2} \sinh(x + \triangle x)\right]_{\triangle x=0} \frac{1}{2!}(\triangle x)^2$$

$$+ \left[\frac{d^3}{dx^3} \sinh(x + \triangle x)\right]_{\triangle x=0} \frac{1}{3!}(\triangle x)^3$$

$$+ \left[\frac{d^4}{dx^4} \sinh(x + \triangle x)\right]_{\triangle x=0} \frac{1}{4!}(\triangle x)^4$$

$$+ \left[\frac{d^5}{dx^5} \sinh(x + \triangle x)\right]_{\triangle x=0} \frac{1}{5!}(\triangle x)^5$$

$$+ \left[\frac{d^6}{dx^6} \sinh(x + \triangle x)\right]_{\triangle x=0} \frac{1}{6!}(\triangle x)^6$$

$$+ \left[\frac{d^7}{dx^7} \sinh(x + \triangle x)\right]_{\triangle x=0} \frac{1}{7!}(\triangle x)^7$$

$$\vdots \quad . \tag{1.238}$$

Employing (1.88),

$$\sinh(x + \triangle x) = [\sinh(x + \triangle x)]_{\triangle x=0}$$

$$+ [\cosh(x + \triangle x)]_{\triangle x=0}(\triangle x)$$

$$+ [\sinh(x + \triangle x)]_{\triangle x=0} \frac{1}{2!}(\triangle x)^2$$

$$+ [\cosh(x + \triangle x)]_{\triangle x=0} \frac{1}{3!}(\triangle x)^3$$

$$+ [\sinh(x + \triangle x)]_{\triangle x=0} \frac{1}{4!}(\triangle x)^4$$

$$+ [\cosh(x + \triangle x)]_{\triangle x=0} \frac{1}{5!}(\triangle x)^5$$

$$+ [\sinh(x + \triangle x)]_{\triangle x=0} \frac{1}{6!}(\triangle x)^6$$

$$+ [\cosh(x + \triangle x)]_{\triangle x=0} \frac{1}{7!}(\triangle x)^7$$

$$\vdots \quad . \tag{1.239}$$

Simplifying the terms in square brackets,

$$\sinh(x + \Delta x) = \sinh(x)$$
$$+ \cosh(x)(\Delta x)$$
$$+ \sinh(x)\frac{1}{2!}(\Delta x)^2$$
$$+ \cosh(x)\frac{1}{3!}(\Delta x)^3$$
$$+ \sinh(x)\frac{1}{4!}(\Delta x)^4$$
$$+ \cosh(x)\frac{1}{5!}(\Delta x)^5$$
$$+ \sinh(x)\frac{1}{6!}(\Delta x)^6$$
$$+ \cosh(x)\frac{1}{7!}(\Delta x)^7$$
$$\vdots \quad . \tag{1.240}$$

Grouping the terms in accordance with their association with $\sinh(x)$ or $\cosh(x)$,

$$\sinh(x + \Delta x) = \sinh(x)\left[1 + \frac{1}{2!}(\Delta x)^2 + \frac{1}{4!}(\Delta x)^4 + \frac{1}{6!}(\Delta x)^6 + \cdots\right]$$
$$+ \cosh(x)\left[\Delta x + \frac{1}{3!}(\Delta x)^3 + \frac{1}{5!}(\Delta x)^5 + \frac{1}{7!}(\Delta x)^7 + \cdots\right]. \tag{1.241}$$

Identifying the terms in the first and the second square brackets as the Formal Taylor Series Expansions of $\cosh(\Delta x)$ and $\sinh(\Delta x)$, respectively, the elementary hyperbolic identity

$$\sinh(x + \Delta x) = \sinh(x)\cosh(\Delta x) + \cosh(x)\sinh(\Delta x), \tag{1.242}$$

manifests itself. ∎

Problem: Obtain an equivalent representation for the column vector $\begin{bmatrix} \cosh(x + \Delta x) \\ \sinh(x + \Delta x) \end{bmatrix}$.

Solution: Utilizing the results obtained in the preceding two problems,

$$\begin{bmatrix} \cosh(x + \Delta x) \\ \sinh(x + \Delta x) \end{bmatrix} = \begin{bmatrix} \cosh(x)\cosh(\Delta x) + \sinh(x)\sinh(\Delta x) \\ \sinh(x)\cosh(\Delta x) + \cosh(x)\sinh(\Delta x) \end{bmatrix}. \tag{1.243}$$

Substitute the Formal Taylor Series Expansion for $\cosh(\Delta x)$ and $\sinh(\Delta x)$,

$$\cosh(\Delta x) = 1 + \frac{1}{2!}(\Delta x)^2 + \frac{1}{4!}(\Delta x)^4 + \cdots \tag{1.244a}$$
$$\sinh(\Delta x) = \Delta x + \frac{1}{3!}(\Delta x)^3 + \frac{1}{5!}(\Delta x)^5 + \cdots, \tag{1.244b}$$

at the R.H.S. of (1.243),

$$\begin{bmatrix} \cosh(x+\Delta x) \\ \sinh(x+\Delta x) \end{bmatrix} = \begin{bmatrix} \cosh(x)\left\{1+\frac{1}{2!}(\Delta x)^2 + \cdots\right\} + \sinh(x)\left\{\Delta x + \frac{1}{3!}(\Delta x)^3 + \cdots\right\} \\ \sinh(x)\left\{1+\frac{1}{2!}(\Delta x)^2 + \cdots\right\} + \cosh(x)\left\{\Delta x + \frac{1}{3!}(\Delta x)^3 + \cdots\right\} \end{bmatrix}. \tag{1.245}$$

Separating the terms associated with $(\Delta x)^0 = 1, (\Delta x)^1 = \Delta x, (\Delta x)^2, ...,$ at the R.H.S.,

$$\begin{bmatrix} \cosh(x+\Delta x) \\ \sinh(x+\Delta x) \end{bmatrix} = \begin{bmatrix} \cosh(x) \\ \sinh(x) \end{bmatrix} + (\Delta x)\begin{bmatrix} \sinh(x) \\ \cosh(x) \end{bmatrix} + \frac{1}{2!}(\Delta x)^2 \begin{bmatrix} \cosh(x) \\ \sinh(x) \end{bmatrix}$$
$$+ \frac{1}{3!}(\Delta x)^3 \begin{bmatrix} \sinh(x) \\ \cosh(x) \end{bmatrix} + \cdots . \tag{1.246}$$

Transforming the vectors at the R.H.S. into the "canonical" standard form,

$$\begin{bmatrix} \cosh(x+\Delta x) \\ \sinh(x+\Delta x) \end{bmatrix} = \begin{bmatrix} \cosh(x) \\ \sinh(x) \end{bmatrix} + (\Delta x)\begin{bmatrix} 0 & 1 \\ 1 & 0 \end{bmatrix}\begin{bmatrix} \cosh(x) \\ \sinh(x) \end{bmatrix}$$
$$+ \frac{1}{2!}(\Delta x)^2 \begin{bmatrix} 1 & 0 \\ 0 & 1 \end{bmatrix}\begin{bmatrix} \cosh(x) \\ \sinh(x) \end{bmatrix}$$
$$+ \frac{1}{3!}(\Delta x)^3 \begin{bmatrix} 0 & 1 \\ 1 & 0 \end{bmatrix}\begin{bmatrix} \cosh(x) \\ \sinh(x) \end{bmatrix} + \cdots . \tag{1.247}$$

It is immediate that the second and the third matrices at the R.H.S. are the second and the third powers of the "fundamental" matrix $\begin{bmatrix} 0 & 1 \\ 1 & 0 \end{bmatrix}$. Observe that,

$$\begin{bmatrix} 1 & 0 \\ 0 & 1 \end{bmatrix} = \begin{bmatrix} 0 & 1 \\ 1 & 0 \end{bmatrix}\begin{bmatrix} 0 & 1 \\ 1 & 0 \end{bmatrix}$$
$$= \begin{bmatrix} 0 & 1 \\ 1 & 0 \end{bmatrix}^2 . \tag{1.248}$$

Furthermore, employing (1.248),

$$\begin{bmatrix} 0 & 1 \\ 1 & 0 \end{bmatrix} = \begin{bmatrix} 1 & 0 \\ 0 & 1 \end{bmatrix}\begin{bmatrix} 0 & 1 \\ 1 & 0 \end{bmatrix}$$

$$= \begin{bmatrix} 0 & 1 \\ 1 & 0 \end{bmatrix}^2 \begin{bmatrix} 0 & 1 \\ 1 & 0 \end{bmatrix}$$

$$= \begin{bmatrix} 0 & 1 \\ 1 & 0 \end{bmatrix}^3. \tag{1.249}$$

Considering the above results along with the relationship,

$$\begin{bmatrix} \cosh(x) \\ \sinh(x) \end{bmatrix} = \begin{bmatrix} 1 & 0 \\ 0 & 1 \end{bmatrix}\begin{bmatrix} \cosh(x) \\ \sinh(x) \end{bmatrix}, \tag{1.250}$$

(1.247) reads

$$\begin{bmatrix} \cosh(x+\Delta x) \\ \sinh(x+\Delta x) \end{bmatrix} = \begin{bmatrix} 1 & 0 \\ 0 & 1 \end{bmatrix}\begin{bmatrix} \cosh(x) \\ \sinh(x) \end{bmatrix}$$

$$+ (\Delta x)\begin{bmatrix} 0 & 1 \\ 1 & 0 \end{bmatrix}\begin{bmatrix} \cosh(x) \\ \sinh(x) \end{bmatrix}$$

$$+ \frac{1}{2!}(\Delta x)^2 \begin{bmatrix} 0 & 1 \\ 1 & 0 \end{bmatrix}^2 \begin{bmatrix} \cosh(x) \\ \sinh(x) \end{bmatrix}$$

$$+ \frac{1}{3!}(\Delta x)^3 \begin{bmatrix} 0 & 1 \\ 1 & 0 \end{bmatrix}^3 \begin{bmatrix} \cosh(x) \\ \sinh(x) \end{bmatrix} + \cdots, \tag{1.251}$$

Factoring out $\begin{bmatrix} \cosh(x) \\ \sinh(x) \end{bmatrix}$ at the R.H.S.,

$$\begin{bmatrix} \cosh(x+\Delta x) \\ \sinh(x+\Delta x) \end{bmatrix} = \left\{ \begin{bmatrix} 1 & 0 \\ 0 & 1 \end{bmatrix} \right.$$

$$+(\Delta x) \begin{bmatrix} 0 & 1 \\ 1 & 0 \end{bmatrix}$$

$$+\frac{1}{2!}(\Delta x)^2 \begin{bmatrix} 0 & 1 \\ 1 & 0 \end{bmatrix}^2$$

$$\left. +\frac{1}{3!}(\Delta x)^3 \begin{bmatrix} 0 & 1 \\ 1 & 0 \end{bmatrix}^3 + \cdots \right\} \begin{bmatrix} \cosh(x) \\ \sinh(x) \end{bmatrix}. \tag{1.252}$$

The expression in the curly brackets is the Formal Taylor Series Expansion of the exponential function $e^{\begin{bmatrix} 0 & 1 \\ 1 & 0 \end{bmatrix}\Delta x}$:

$$e^{\begin{bmatrix} 0 & 1 \\ 1 & 0 \end{bmatrix}\Delta x} = \begin{bmatrix} 1 & 0 \\ 0 & 1 \end{bmatrix} + (\Delta x)\begin{bmatrix} 0 & 1 \\ 1 & 0 \end{bmatrix} + \frac{1}{2!}(\Delta x)^2 \begin{bmatrix} 0 & 1 \\ 1 & 0 \end{bmatrix}^2$$

$$+ \frac{1}{3!}(\Delta x)^3 \begin{bmatrix} 0 & 1 \\ 1 & 0 \end{bmatrix}^3 + \cdots . \tag{1.253}$$

Thus, (1.252) can be written in the form,

$$\begin{bmatrix} \cosh(x+\Delta x) \\ \sinh(x+\Delta x) \end{bmatrix} = e^{\begin{bmatrix} 0 & 1 \\ 1 & 0 \end{bmatrix}\Delta x} \begin{bmatrix} \cosh(x) \\ \sinh(x) \end{bmatrix} \tag{1.254}$$

Denoting the matrix at the R.H.S. by $\mathbf{K}$,

$$\mathbf{K} = \begin{bmatrix} 0 & 1 \\ 1 & 0 \end{bmatrix}, \tag{1.255}$$

establishes the transformation equation,

$$\begin{bmatrix} \cosh(x+\Delta x) \\ \sinh(x+\Delta x) \end{bmatrix} = e^{\mathbf{K}\Delta x} \begin{bmatrix} \cosh(x) \\ \sinh(x) \end{bmatrix}. \tag{1.256}$$

∎

Problem: Establish an alternative (equivalent) representation for the column vector $\begin{bmatrix} \cosh(x + \triangle x) \\ \sinh(x + \triangle x) \end{bmatrix}$.

Solution: The starting point is

$$\begin{bmatrix} \cosh(x + \triangle x) \\ \sinh(x + \triangle x) \end{bmatrix} = \begin{bmatrix} \cosh(x)\cosh(\triangle x) + \sinh(x)\sinh(\triangle x) \\ \sinh(x)\cosh(\triangle x) + \cosh(x)\sinh(\triangle x) \end{bmatrix}. \tag{1.257}$$

It is an easy exercise to write the vector at the R.H.S. of (1.256) as the product of a matrix (the entries of which are merely functions of $\triangle x$) and a vector (the components of which exclusively involve x),

$$\begin{bmatrix} \cosh(x)\cosh(\triangle x) + \sinh(x)\sinh(\triangle x) \\ \sinh(x)\cosh(\triangle x) + \cosh(x)\sinh(\triangle x) \end{bmatrix} = \begin{bmatrix} \cosh(\triangle x) & \sinh(\triangle x) \\ \sinh(\triangle x) & \cosh(\triangle x) \end{bmatrix} \begin{bmatrix} \cosh(x) \\ \sinh(x) \end{bmatrix}. \tag{1.258}$$

Thus, (1.257) can be written as,

$$\begin{bmatrix} \cosh(x + \triangle x) \\ \sinh(x + \triangle x) \end{bmatrix} = \begin{bmatrix} \cosh(\triangle x) & \sinh(\triangle x) \\ \sinh(\triangle x) & \cosh(\triangle x) \end{bmatrix} \begin{bmatrix} \cosh(x) \\ \sinh(x) \end{bmatrix}, \tag{1.259}$$

which is the desired representation alternative to (1.256).

∎

Euler-type identity in matrix form: Comparing the results from the previous two problems, (1.256) and (1.259),

$$e^{\mathbf{K}\triangle x} = \begin{bmatrix} \cosh(\triangle x) & \sinh(\triangle x) \\ \sinh(\triangle x) & \cosh(\triangle x) \end{bmatrix}. \tag{1.260}$$

Writing the R.H.S. in the form

$$\begin{bmatrix} \cosh(\triangle x) & \sinh(\triangle x) \\ \sinh(\triangle x) & \cosh(\triangle x) \end{bmatrix} = \begin{bmatrix} \cosh(\triangle x) & 0 \\ 0 & \cosh(\triangle x) \end{bmatrix} + \begin{bmatrix} 0 & \sinh(\triangle x) \\ \sinh(\triangle x) & 0 \end{bmatrix}. \tag{1.261}$$

Or, equivalently,

$$\begin{bmatrix} \cosh(\triangle x) & \sinh(\triangle x) \\ \sinh(\triangle x) & \cosh(\triangle x) \end{bmatrix} = \cosh(\triangle x)\underbrace{\begin{bmatrix} 1 & 0 \\ 0 & 1 \end{bmatrix}}_{\mathbf{I}} + \sinh(\triangle x)\underbrace{\begin{bmatrix} 0 & 1 \\ 1 & 0 \end{bmatrix}}_{\mathbf{K}} \tag{1.262a}$$

$$= \cosh(\triangle x)\mathbf{I} + \sinh(\triangle x)\mathbf{K}, \tag{1.262b}$$

where the 2×2 "placeholder," "scaffolding" matrices $\mathbf{K}$, and $\mathbf{I}$ have "naturally" offered themselves. Equations (1.260) and (1.262) imply,

$$e^{\mathbf{K}\triangle x} = \cosh(\triangle x)\mathbf{I} + \sinh(\triangle x)\mathbf{K}, \tag{1.263}$$

Since $\triangle x$ is arbitrary (not necessarily small), it can be replaced with any angle, say, x,

$$e^{\mathbf{K}x} = \cosh(x)\mathbf{I} + \sinh(x)\mathbf{K}. \tag{1.264}$$

Problem: Consider, the Formal Taylor Series Expansion of $\cosh(x)$ and $\sinh(x)$,

$$\cosh(x) = 1 + \frac{1}{2!}x^2 + \frac{1}{4!}x^4 + \frac{1}{6!}x^6 + \frac{1}{8!}x^8 + \mathcal{O}(x^{10}) \tag{1.265a}$$

$$\sinh(x) = x + \frac{1}{3!}x^3 + \frac{1}{5!}x^5 + \frac{1}{7!}x^7 + \frac{1}{9!}x^9 + \mathcal{O}(x^{11}). \tag{1.265b}$$

The symbol $\mathcal{O}(x^n)$ denotes the collection of the terms of the order x^n or higher. Verify the validity of the relationship,

$$\cosh^2(x) - \sinh^2(x) = 1. \tag{1.266}$$

Solution: Simple calculations reveal that,

$$\cosh^2(x) = 1 + \left(\frac{1}{2!} + \frac{1}{2!}\right)x^2 + \left(\frac{1}{4!} + \frac{1}{2!}\frac{1}{2!} + \frac{1}{4!}\right)x^4 + \left(\frac{1}{6!} + \frac{1}{2!}\frac{1}{4!} + \frac{1}{4!}\frac{1}{2!} + \frac{1}{6!}\right)x^6$$
$$+ \left(\frac{1}{8!} + \frac{1}{2!}\frac{1}{6!} + \frac{1}{4!}\frac{1}{4!} + \frac{1}{6!}\frac{1}{2!} + \frac{1}{8!}\right)x^8 + \mathcal{O}(x^{10}) \tag{1.267a}$$

$$\sinh^2(x) = x^2 + \left(\frac{1}{3!} + \frac{1}{3!}\right)x^4 + \left(\frac{1}{5!} + \frac{1}{3!}\frac{1}{3!} + \frac{1}{5!}\right)x^6$$
$$+ \left(\frac{1}{7!} + \frac{1}{3!}\frac{1}{5!} + \frac{1}{5!}\frac{1}{3!} + \frac{1}{7!}\right)x^8 + \mathcal{O}(x^{10}). \tag{1.267b}$$

Further above, in connection with $\cos^2(x)$ and $\sin^2(x)$, it was shown that the coefficients associated with the coefficients x^2, x^4, x^6, and x^8 in the two series are equal. Reasoning by mathematical induction shows that this statement is true for general terms x^{2n} ($n \in \mathbb{N}$). Consequently, subtracting (1.267b) from (1.267a), all terms cancel out, except the constant term 1, which lacks a "balancing" counterpart. This sketches the derivation of (1.266) from the Formal Taylor Series Expansions in (1.265). ∎

1.13 Equation of plane

- Consider the (x_1, x_2, x_3)-Cartesian coordinate system with its origin $O(0,0,0)$.
- Consider an arbitrary plane $\mathcal{P}$ not containing the origin O.
- Let $\mathcal{S}$ denote the sphere having the origin O as its center and touching the plane $\mathcal{P}$ at the tangent point T.
- Since the plane $\mathcal{P}$ is tangential to the sphere $\mathcal{S}$, the vector stretching from the origin O to the tangent point T is normal to the plane $\mathcal{P}$. Denote the length of this vector by d.
- Let $\mathbf{n} = (n_1, n_2, n_3)$ denote the unit vector emanating from O in the direction of the tangent point T.
- The vector stretching from the origin O to the tangent point T, having the length d, and being specified by the unit vector $\mathbf{n}$ has the representation $d\mathbf{n}$.
- Since the vector $d\mathbf{n}$ is by definition normal to the tangent plane $\mathcal{P}$, the unit vector $\mathbf{n}$ is also normal to the tangent plane $\mathcal{P}$: $\mathbf{n}$ is perpendicular to any vector in the tangent plane $\mathcal{P}$.
- Let P be an arbitrary point on the plane $\mathcal{P}$ not coinciding with the tangent point T.
- Let the point P on the plane $\mathcal{P}$ be specified by the position vector $\mathbf{r} = (x_1, x_2, x_3)$.
- Denote the vector stretching from the tangent point T to the arbitrary point P in the plane $\mathcal{P}$ by $\mathbf{t}$.
- Since $\mathbf{n}$ is perpendicular to any vector in the plane $\mathcal{P}$, and $\mathbf{t} \in \mathcal{P}$, $\mathbf{n}$ is perpendicular to $\mathbf{t}$, i.e., the scalar (dot) product of $\mathbf{n}$ with $\mathbf{t}$ is zero,

$$\mathbf{n} \cdot \mathbf{t} = 0. \tag{1.268}$$

- On the other hand,

$$d\mathbf{n} + \mathbf{t} = \mathbf{r} \qquad \Longrightarrow \qquad \mathbf{t} = \mathbf{r} - d\mathbf{n}. \tag{1.269}$$

- Substituting $\mathbf{t}$ from (1.269) into (1.268),

$$\mathbf{n} \cdot (\mathbf{r} - d\mathbf{n}) = 0 \qquad \Longrightarrow \qquad \mathbf{n} \cdot \mathbf{r} - d\mathbf{n} \cdot \mathbf{n} = 0. \tag{1.270}$$

- Since $\mathbf{n}$ is a unit vector $\mathbf{n} \cdot \mathbf{n} = 1$:

$$\mathbf{n} \cdot \mathbf{r} = d \tag{1.271}$$

This intuitive outcome is the equation of the plane $\mathcal{P}$ in the Cartesian coordinate system (x_1, x_2, x_3). The plane $\mathcal{P}$ is specified in the Cartesian coordinate system (x_1, x_2, x_3) by its distance $d \in \mathbb{R}^+$ from the origin O, and the unit normal vector $\mathbf{n}$. It states that the projection onto the unit normal vector $\mathbf{n}$ of the position vector $\mathbf{r}(x_1, x_2, x_3)$, specifying the arbitrary point P being located on the plane $\mathcal{P}$, is equal to the distance d of the origin O from the plane $\mathcal{P}$. Described graphically, for any $P \in \mathcal{P}$, OTP is the right triangle with OP being its hypotenuse.

- Writing (1.271) in coordinate form,

$$n_1 x_1 + n_2 x_2 + n_3 x_3 = d. \tag{1.272}$$

Problem: Consider the Cartesian coordinate system (x_1, x_2, x_3) and the plane $\mathcal{P}$ specified by the unit normal vector $\mathbf{n}$ and the distance d of the coordinate origin O. Assume that Ray emanating from O and running in the direction of $\mathbf{n}$ intersects the plane $\mathcal{P}$ at the point N.

Assume the points $P(x_1^p, x_2^p, x_3^p) \in \mathcal{P}$ and $Q(x_1^q, x_2^q, x_3^q) \in \mathcal{P}$ on the plane $\mathcal{P}$ and non-coinciding with the point N. Show that the line (the vector) connecting P to Q is orthogonal to $\mathbf{n}$.

Solution: Denote the position vector connecting the origin O to the point $P \in \mathcal{P}$ by $\mathbf{p}$. Denote the position vector connecting the origin O to the point $Q \in \mathcal{P}$ by $\mathbf{q}$. Denote the difference vector connecting the point P to the point Q by $\boldsymbol{\tau}$, i.e.,

$$\boldsymbol{\tau} = \mathbf{q} - \mathbf{p}. \tag{1.273}$$

Project the vectors on both sides of (1.273) onto the normal vector $\mathbf{n}$,

$$\mathbf{n} \cdot \boldsymbol{\tau} = \mathbf{n} \cdot \mathbf{q} - \mathbf{n} \cdot \mathbf{p}. \tag{1.274}$$

Above it was demonstrated that the projection of any arbitrary position vector $\mathbf{r}$ onto $\mathbf{n}$ is equal to d, the distance of the origin O from the plane $\mathcal{P}$. Since $\mathbf{p}$ and $\mathbf{q}$ are position vectors,

$$\mathbf{n} \cdot \mathbf{p} = 0, \qquad \mathbf{n} \cdot \mathbf{q} = 0. \tag{1.275}$$

Considering these relationships in (1.274),

$$\mathbf{n} \cdot \boldsymbol{\tau} = 0. \tag{1.276}$$

∎

Remark: Planes and waviness.

A firm understanding of the equation of the plane is essential in visualizing what undulates when one speaks of propagating planewaves. One must have a clear understanding of the plane and the waviness when one utters a planewave. In other words, one must have a clear mental picture of where the plane is when one paints by hand a wave in air.

Having established the equation of a plane in the preceding discussion, one more ingredient is missing before delving into the planewaves: what is the genesis of the constant π and why does it feature so prominently in describing planewaves. The constant π did not appear in the description of the equation of the plane. It can be suspected that its occurrence is related to the undulation, to what which waves. The next section clarifies the underlying ideas.

□

1.14 The universal geometric constant π

- Consider the (ξ, η)-Cartesian coordinate and the (r, φ)-polar coordinate system being superimposed. Consider the circle with the center at $(0,0)$ and the radius $r \in \mathbb{R}^+$. Denote the circumference of the circle by $C(r)$ and its area by $A(r)$.
- It turns out that the ratio $A(r)/r^2$ is independent of the radius. The ratio $A(r)/r^2$ is a universal geometric constant. The value of this universal geometric constant is slightly larger than 3. Considering the first few decimal figures it reads $3.1415926\ldots$. The universal geometric constant is an irrational, in fact, a transcendental number. There are infinitely many figures after the decimal point.
- It also turns out that the ratio $C(r)/r$ is independent of the radius.
- Quite astonishingly, the r-independent value of $C(r)/r$ is twice the same universal geometric constant, which specifies the ratio $A(r)/r^2$.
- Denoting the diameter of the circle by $d = 2r$, it is obvious that $C(r)/d$ is equal to the universal geometric constant.
- Historically, the identified universal geometric constant has been denoted by π. Since $C(r)/r$ is twice the universal geometric constant,

$$C(r)/r = 2\pi.$$

- Thus, measuring $C(r)$ in units of r results in 2π. Or, in words, measuring the circumference in units of the radius gives 2π. Using the suffix-ian allows to form from radius the adjective radian (typical of or resembling radius).
- An arc of a circle with the same length as the radius of that circle subtends an angle of 1 radian.
- One radian is defined as the angle subtended from the center of a circle which intercepts an arc equal in length to the radius of the circle.
- The radian is customarily abbreviated by rad.
- The radian is the dimensionless unit of angular measure.
- Thus, measuring the circumference $C(r)$ in radians results in 2π.
- The circumference of a circle subtends an angle of 2π radians.
- What is the genesis of the number 360 when it is said that the circumference of a circle subtends an angle of 360° (360 degrees)? In contrast to the universal geometric constant π, the numerical value 360 is rather arbitrary. It is a nice number in the sense that it is divisible by several divisors: 2, 3, 4, 5, 6, 8, 9, 10, 12, 15, 18, 20, 30, 40, 60, 90, 120, and 180. Dividing the year into 12 months and each month into 30 days might be a plausible explanation for the number 360. The correctness of this historical account is, however, immaterial to the conclusion that 360 is arbitrary and has been chosen cogently. Once a choice has been made that the circumference of a circle subtends an angle of 360°, it is consequential. Accepting this choice, it can be said that,

$$2\pi \text{ rad} \equiv 360°,$$

And thus,

$$1\pi \text{ rad} \equiv 180°.$$

Conversely, dividing both sides by $\pi = 3.1415926\ldots$,

$$1 \text{ rad} \equiv \frac{180°}{\pi} = 57.296°.$$

- Consider the point P at $(1,0)$ (the intersection of the ξ-axis and the unit circle). Let P move counterclockwise. Any position of P on the unit circle specifies a positive angle φ with $0 \le \varphi < 2\pi$.
- Consider the x-axis with the position variable x running from $-\infty$ to ∞.
- Consider the t-axis with the time variable t running from $-\infty$ to ∞.

This is all what is needed to start the discussion of the planewaves. The plane is a two-dimensional surface. However, in a Cartesian coordinate system, the unit normal vector of the plane may have one component, $(n_1, 0, 0)$, $(0, n_2, 0)$ or $(0, 0, n_3)$, or two components, $(n_1, n_2, 0)$, $(n_1, 0, n_3)$ or $(0, n_2, n_3)$, or three components, (n_1, n_2, n_3). This variety gives rise to planewaves in one-, two-, and three dimensions, which are treated next consecutively.

1.15   Planewaves in one dimension

- In the preceding sections the function $e^{i\varphi}$ was introduced to designate a point on the unit circle. The function $e^{i\varphi}$ was also meant to identify a vector with its tail at the center $(0,0)$ and its head on the perimeter of the circle while the angle between the vector and the positive x-axis being φ.
- Any point $e^{i\varphi}$ on the unit circle can be specified by assigning a φ with $0 \leq \varphi < 2\pi$. Note that $e^{i0} = e^{i2\pi} = 1$.
- Adding multiple of 2π to φ does not change the value of $e^{i\varphi}$:

$$e^{i(\varphi+n2\pi)} = e^{i\varphi}e^{in2\pi} = e^{i\varphi}(e^{i2\pi})^n = e^{i\varphi}$$

- The condition $0 \leq \varphi < 2\pi$ can also be expressed by saying that the dimensionless φ is a fraction of 2π, i.e., $\varphi = 2\pi \frac{x}{\lambda}$ with $0 \leq x < \lambda$. Or, equivalently, $0 \leq \frac{x}{\lambda} < 1$. Note that adding n multiples of λ to x is equivalent to adding n multiples of 2π to φ :

$$2\pi \frac{x+n\lambda}{\lambda} = 2\pi \frac{x}{\lambda} + 2\pi n = \varphi + n2\pi$$

Thus, considering a characteristic length λ, any point x on the x-axis, with $0 \leq x < \lambda$, corresponds to $\exp\left(i2\pi \frac{x}{\lambda}\right)$ on the unit circle. Note that for any $x \in \mathbb{R}$, an $n \in \mathbb{Z}$ can be found such that $x = n2\pi + \xi$ with $0 \leq \xi < \lambda$. Consequently, $\exp\left(i2\pi \frac{x}{\lambda}\right) = \exp\left(i2\pi \frac{\xi}{\lambda}\right)$.

- Similarly, the condition $0 \leq \varphi < 2\pi$ can also be expressed by considering the dimensionless φ to be a fraction of 2π in the form $\varphi = 2\pi \frac{t}{T}$ with $0 \leq t < T$. Or, equivalently, $0 \leq \frac{t}{T} < 1$. Obviously, adding m multiples of T to t is equivalent to adding m multiples of 2π to φ :

$$2\pi \frac{t+mT}{T} = 2\pi \frac{t}{T} + 2\pi m = \varphi + m2\pi$$

Thus, considering a characteristic time T, any point t on the t-axis, with $0 \leq t < T$, corresponds to $\exp\left(i2\pi \frac{t}{T}\right)$ on the unit circle. Furthermore, for any $t \in \mathbb{R}$, an $m \in \mathbb{Z}$ can be found such that $t = m2\pi + \tau$ with $0 \leq \tau < T$. Consequently, $\exp\left(i2\pi \frac{t}{T}\right) = \exp\left(i2\pi \frac{\tau}{T}\right)$.

- The preceding considerations lead to

$$\exp\left(i2\pi \frac{x}{\lambda}\right) \quad \text{and} \quad \exp\left(i2\pi \frac{t}{T}\right). \tag{1.277}$$

- The ratio $\frac{1}{T}$ gives the multiplicity with which T can be accommodated in one unit of time, say, a second. Analogously, the ratio $\frac{1}{\lambda}$ gives the multiplicity with which λ can be accommodated in one unit of length, say, a meter. The choice of multiplicity, as a figure of speech, is meant to eliminate the particular reference to time. The periodicity exhibited in (1.277) refers equally to space as well as time. In this sense, $\frac{1}{T}$ can be said to be the frequency in time and $\frac{1}{\lambda}$ the frequency in space. And since generally a wave induces undulation, oscillation in time and in space, $\frac{1}{T}$ can be said to be the wavenumber in time and $\frac{1}{\lambda}$ the wavenumber in space.
- It is customary to refer to $\frac{1}{T}$ as the frequency. No particular name has been suggested for $\frac{1}{\lambda}$. From the preceding discussion, $\frac{1}{\lambda}$ can be referred to as the frequency in space.

- It is customary to refer to $\frac{2\pi}{T}$ as the radian frequency. It is also an established convention to refer to $\frac{2\pi}{\lambda}$ as the wavenumber. Thus, since $\frac{2\pi}{T}$ already has an accepted name (radian frequency) there is no need to call it wavenumber in time. In summary, the following notation will be used:

$$\frac{1}{T} = f \quad \cdots\cdots\cdots \quad \text{frequency}$$

$$\frac{1}{\lambda} \quad \cdots\cdots\cdots \quad \text{frequency in space}$$

$$\frac{2\pi}{T} = \omega \quad \cdots\cdots\cdots \quad \text{radian frequency}$$

$$\frac{2\pi}{\lambda} = k \quad \cdots\cdots\cdots \quad \text{wavenumber}$$

- Utilizing the wavenumber k and the radian frequency ω just introduced, (1.277) reads,

$$\exp(ikx) \quad \text{and} \quad \exp(i\omega t). \tag{1.278}$$

Question: Given the pair (k, ω) with $k, \omega \in \mathbb{R}^+$, consider,

$$\exp(ikx)\exp(-i\omega t) = \exp(i(kx - \omega t)). \tag{1.279}$$

The function $\exp(i(kx - \omega t))$ is λ-periodic in x and simultaneously T-periodic in t. The question is what relationship between x and t ensures that the phase $\varphi_c = kx - \omega t$ is a constant.

Answer: To investigate this question consider,

$$kx - \omega t = \varphi_c = \text{const.}, \tag{1.280}$$

which can be cast in the form,

$$x = \frac{1}{k}\varphi_c + \frac{\omega}{k}t. \tag{1.281}$$

Taking the derivative of the terms with respect to t, and considering that $\frac{1}{k}\varphi_c$ is a constant,

$$v_{\text{ph}} \overset{\text{def.}}{=} \frac{dx}{dt} = \frac{\omega}{k}. \tag{1.282}$$

The phase velocity v_{ph}, as defined in (1.282), turns out to be equal to ω/k, which is in virtue of $k, \omega \in \mathbb{R}^+$, positive. ∎

Interpretation: Note that the radian phase φ_c can be assumed to be restricted to the interval, $0 \leq \varphi_c < 2\pi$, since as discussed further above, $\exp[i(\varphi_c + n2\pi)] = \exp(i\varphi_c)$. Furthermore, note that a given pair (k, ω) imposes λ-periodicity in space, and T-periodicity in time, with $k = 2\pi/\lambda$ and $\omega = 2\pi/T$. Finally, considering (1.281) note that at $t = 0$, $x|_{t=0} = \frac{1}{k}\varphi_c$. This implies that all points in space having the coordinates $(\frac{1}{k}\varphi_c, y, z)$ (their x-coordinate being $\frac{1}{k}\varphi_c$ and irrespective of their y- and z-coordinates) moving at the constant phase velocity $v_{\text{ph}} = \omega/k$, render the exponential function $\exp(i(kx - \omega t))$, equal to $\exp(i\varphi_c)$. The initial ($t = 0$) collection of points $(\frac{1}{k}\varphi_c, y, z)$ constitutes a plane perpendicular to the x-axis. The collection of the points on this plane relocates, moves, propagates, to the plane $(\frac{1}{k}\varphi_c + \frac{\omega}{k}t, y, z)$ at time t. This is what is meant if one speaks of the notion of the planewave and the planewave propagation. Note that there is nothing special about the x-axis.

$\square$

Interpretation: In which sense should be λ-periodicity understood? To answer this question, replace x with $x + \lambda$ in (1.281) and investigate the consequences thereof.

$$x = \frac{1}{k}\varphi_c + \frac{\omega}{k}t \quad \overset{x \to x+\lambda}{\Longrightarrow} \quad x + \lambda = \frac{1}{k}\varphi_c + \frac{\omega}{k}t. \tag{1.283}$$

Multiply through by k,

$$kx + k\lambda = \varphi_c + \omega t. \tag{1.284}$$

Remembering the definition $k = 2\pi/\lambda$ and thus $k\lambda = 2\pi$, and subtracting ωt from both sides,

$$kx - \omega t + 2\pi = \varphi_c. \tag{1.285}$$

Earlier it was shown that adding or subtracting multiples of 2π to or from the radian angle does not have any implications on the value of $\exp(i\varphi_c)$.

$\square$

Interpretation: A similar question arises with respect to the time. In which sense should be T-periodicity understood? To answer this question, replace t with $t + T$ in (1.281) and investigate the consequences thereof.

$$x = \frac{1}{k}\varphi_c + \frac{\omega}{k}t \quad \overset{t \to t+T}{\Longrightarrow} \quad x = \frac{1}{k}\varphi_c + \frac{\omega}{k}(t + T). \tag{1.286}$$

Multiply through by k,

$$kx = \varphi_c + \omega t + \omega T. \tag{1.287}$$

Remembering the definition $\omega = 2\pi/T$ and thus $\omega T = 2\pi$, and subtracting $\omega t + 2\pi$ from both sides,

$$kx - \omega t - 2\pi = \varphi_c. \tag{1.288}$$

However, adding or subtracting multiples of 2π to or from the radian angle does not have any impact on the value of $\exp(i\varphi_c)$.

$\square$

Question: Given the pair (k, ω) with $k, \omega \in \mathbb{R}^+$, consider,

$$\exp(-ikx)\exp(i\omega t) = \exp(i(-kx + \omega t)). \tag{1.289}$$

What relationship between x and t ensures that the phase $\varphi_c = -kx + \omega t$ is a constant.

Answer:

$$-kx + \omega t = \varphi_c = \text{const.}, \tag{1.290}$$

or, equivalently

$$x = -\frac{1}{k}\varphi_c + \frac{\omega}{k}t. \tag{1.291}$$

Taking the derivative of the terms with respect to t and considering that $-\frac{1}{k}\varphi_c$ is a constant,

$$v_{ph} \overset{\text{def.}}{=} \frac{dx}{dt} = \frac{\omega}{k}. \tag{1.292}$$

The phase velocity $v_{ph} = \omega/k$ is in virtue of $k, \omega \in \mathbb{R}^+$, positive.

■

Question: Given the pair (k, ω) with $k, \omega \in \mathbb{R}^+$, consider,

$$\exp(ikx)\exp(i\omega t) = \exp\left(i(kx + \omega t)\right). \tag{1.293}$$

What relationship between x and t ensures that the phase $\varphi_c = kx + \omega t$ is a constant.

Answer:

$$kx + \omega t = \varphi_c = \text{const.}, \tag{1.294}$$

or, equivalently,

$$x = \frac{1}{k}\varphi_c - \frac{\omega}{k}t. \tag{1.295}$$

Taking the derivative of the terms with respect to t, and considering that $\frac{1}{k}\varphi_c$ is a constant,

$$v_{ph} \overset{\text{def.}}{=} \frac{dx}{dt} = -\frac{\omega}{k}. \tag{1.296}$$

The phase velocity $v_{ph} = -\omega/k$ is in virtue of $k, \omega \in \mathbb{R}^+$, negative.

■

Question: Given the pair (k, ω) with $k, \omega \in \mathbb{R}^+$, consider,

$$\exp(-ikx)\exp(-i\omega t) = \exp\left(i(-kx - \omega t)\right). \tag{1.297}$$

What relationship between x and t ensures that the phase $\varphi_c = -kx - \omega t$ is a constant.

Answer:

$$-kx - \omega t = \varphi_c = \text{const.}, \tag{1.298}$$

or, equivalently,

$$x = -\frac{1}{k}\varphi_c - \frac{\omega}{k}t. \tag{1.299}$$

Taking the derivative of the terms with respect to t, and considering that $-\frac{1}{k}\varphi_c$ is a constant,

$$v_{ph} \overset{\text{def.}}{=} \frac{dx}{dt} = -\frac{\omega}{k}. \tag{1.300}$$

The phase velocity $v_{ph} = -\omega/k$ is in virtue of $k, \omega \in \mathbb{R}^+$, negative.

■

Remark: Forward propagating waves versus backward propagating waves.

Consider the pair (k, ω) with $k, \omega \in \mathbb{R}^+$. While the exponential functions $\exp(i(kx - \omega t))$ and $\exp(i(-kx + \omega t))$ have the phase velocity $v_{\text{ph}} = \omega/k > 0$, the exponential functions $\exp(i(-kx - \omega t))$ and $\exp(i(kx + \omega t))$ possess the phase velocity $v_{\text{ph}} = -\omega/k < 0$. As shall be made more clear shortly the former are referred to as the forward propagating planewaves, while the latter are called backward propagating planewaves.

□

Remark: On the notion of slowness.

Perhaps paradoxically, the notion of slowness can be explained more intuitively in two and three dimensions. Nonetheless, completeness commands to introduce it at this stage.
Assuming $k, \omega \in \mathbb{R}^+$, consider

$$kx - \omega t = \varphi_c = \text{const.} \tag{1.301}$$

Take the derivative with respect to t,

$$k\frac{dx}{dt} - \omega = 0. \tag{1.302}$$

Divide through by ω, and rearrange,

$$\frac{k}{\omega}\frac{dx}{dt} = 1. \tag{1.303}$$

The notion of the phase velocity $(v_{\text{ph}} = dx/dt)$ was introduced above. Since dx/dt is a velocity, and the R.H.S. is unity, k/ω must be an inverse velocity, a fact which can also be verified directly:

$$\frac{k}{\omega} = \frac{\frac{2\pi}{\lambda}}{\frac{2\pi}{T}} = \frac{T}{\lambda}. \tag{1.304}$$

Thus, k/ω having the dimension sec/m is an inverse velocity, which might more suitably called slowness and denoted by s. Since $dx/dt = v_{\text{ph}}$, (1.303) is be written as,

$$s v_{\text{ph}} = 1. \tag{1.305}$$

□

1.16   Planewaves in two dimensions

Given the triple (k_1, k_2, ω) with $k_1, k_2, \omega \in \mathbb{R}^+$, consider,

$$e^{jk_1 x_1} e^{jk_2 x_2} e^{-j\omega t} = e^{j(k_1 x_1 + k_2 x_2 - \omega t)}. \tag{1.306}$$

In view of the definitions $k_1 = 2\pi/\lambda_1$, $k_2 = 2\pi/\lambda_2$, and $\omega = 2\pi/T$, the exponential function at the R.H.S. of (1.306) reads,

$$\exp\left[j2\pi\left(\frac{x_1}{\lambda_1} + \frac{x_2}{\lambda_2} - \frac{t}{T}\right)\right]. \tag{1.307}$$

This exponential function describes a function which is λ_1-periodic in the spatial x_1-direction, λ_2-periodic in the spatial x_2-direction, and T-periodic in time,

$$\exp\left[j2\pi\left(\frac{x_1 + m_1\lambda_1}{\lambda_1} + \frac{x_2 + m_2\lambda_2}{\lambda_2} - \frac{t + nT}{T}\right)\right] = \exp\left[j2\pi\left(\frac{x_1}{\lambda_1} + \frac{x_2}{\lambda_2} - \frac{t}{T}\right)\right]. \tag{1.308}$$

Before posing a few questions and answering them, one remark is in order.

Remark: The wavenumber vector in two dimensions.

Define the position vector **x** and the wavenumber vector **k**,

$$\mathbf{x} = \begin{pmatrix} x_1 \\ x_2 \end{pmatrix}, \qquad \mathbf{k} = \begin{pmatrix} k_1 \\ k_2 \end{pmatrix}. \tag{1.309}$$

The dot-product $\mathbf{k} \cdot \mathbf{x}$, gives,

$$\mathbf{k} \cdot \mathbf{x} = \mathbf{k}^T \mathbf{x} = \begin{pmatrix} k_1 \\ k_2 \end{pmatrix}^T \begin{pmatrix} x_1 \\ x_2 \end{pmatrix} = \begin{pmatrix} k_1 & k_2 \end{pmatrix} \begin{pmatrix} x_1 \\ x_2 \end{pmatrix} = k_1 x_1 + k_2 x_2. \tag{1.310}$$

Utilizing the wavenumber vector **k** and the position vector **x**, (1.306) takes on the compact form,

$$e^{j(\mathbf{k} \cdot \mathbf{x} - \omega t)}, \tag{1.311}$$

which proves to be advantageous in the present discussion.

□

Question: Given the triple (k_1, k_2, ω) with $k_1, k_2, \omega \in \mathbb{R}^+$, what relationships between the spatial variables x_1 and x_2 and the time variable t ensure that the phase (radian angle) $k_1 x_1 + k_2 x_2 - \omega t$ in (1.306) is a constant, say, φ_c.

Answer: Consider,

$$k_1 x_1 + k_2 x_2 - \omega t = \varphi_c = \text{const.} \tag{1.312}$$

Differentiating both sides with respect to time,

$$k_1 \frac{dx_1}{dt} + k_2 \frac{dx_2}{dt} - \omega = 0. \tag{1.313}$$

Introducing the phase velocity vector $\mathbf{v}_{\text{ph}}$ with components $v_{\text{ph},1}$ and $v_{\text{ph},2}$ according to,

$$v_{\text{ph},1} = \frac{dx_1}{dt}, \qquad v_{\text{ph},2} = \frac{dx_2}{dt}, \tag{1.314}$$

and substituting into (1.313),

$$k_1 v_{\text{ph},1} + k_2 v_{\text{ph},2} - \omega = 0. \tag{1.315}$$

Dividing through by ω, and rearranging,

$$\frac{k_1}{\omega} v_{\text{ph},1} + \frac{k_2}{\omega} v_{\text{ph},2} = 1. \tag{1.316}$$

Introducing the slowness vector **s** with the components,

$$s_1 = \frac{k_1}{\omega}, \qquad s_2 = \frac{k_2}{\omega}, \tag{1.317}$$

$$s_1 v_{\text{ph},1} + s_2 v_{\text{ph},2} = 1. \tag{1.318}$$

Employing the dot-product of **s** and $\mathbf{v}_{\text{ph}}$,

$$\mathbf{s} \cdot \mathbf{v}_{\text{ph}} = 1. \tag{1.319}$$

Note that having specified the triple (k_1, k_2, ω) with $k_1, k_2, \omega \in \mathbb{R}^+$, the slowness vector $\mathbf{s} = (k_1/\omega, k_2/\omega)$ is uniquely determined.

■

1.17 Planewaves in three dimensions

Given the quadruple (k_1, k_2, k_3, ω) with $k_1, k_2, k_3, \omega \in \mathbb{R}^+$, consider,

$$e^{jk_1x_1} e^{jk_2x_2} e^{jk_3x_3} e^{-j\omega t} = e^{j(k_1x_1+k_2x_2+k_3x_3-\omega t)}. \tag{1.320}$$

With $k_1 = 2\pi/\lambda_1$, $k_2 = 2\pi/\lambda_2$, $k_3 = 2\pi/\lambda_3$, and $\omega = 2\pi/T$, the exponential function at the R.H.S. of (1.320) reads,

$$\exp\left[j2\pi\left(\frac{x_1}{\lambda_1} + \frac{x_2}{\lambda_2} + \frac{x_3}{\lambda_3} - \frac{t}{T}\right)\right]. \tag{1.321}$$

This exponential function is λ_1-periodic in the spatial x_1-direction, λ_2-periodic in the spatial x_2-direction, λ_3-periodic in the spatial x_3-direction, and T-periodic in time,

$$\exp\left[j2\pi\left(\frac{x_1 + m_1\lambda_1}{\lambda_1} + \frac{x_2 + m_2\lambda_2}{\lambda_2} + \frac{x_2 + m_3\lambda_3}{\lambda_3} - \frac{t + nT}{T}\right)\right]$$
$$= \exp\left[j2\pi\left(\frac{x_1}{\lambda_1} + \frac{x_2}{\lambda_2} + \frac{x_3}{\lambda_3} - \frac{t}{T}\right)\right]. \tag{1.322}$$

Remark: The wavenumber vector in three dimensions

Define the position vector $\mathbf{x}$ and the wavenumber vector $\mathbf{k}$,

$$\mathbf{x} = \begin{pmatrix} x_1 \\ x_2 \\ x_3 \end{pmatrix}, \qquad \mathbf{k} = \begin{pmatrix} k_1 \\ k_2 \\ k_3 \end{pmatrix}. \tag{1.323}$$

The dot-product $\mathbf{k} \cdot \mathbf{x}$, gives,

$$\mathbf{k} \cdot \mathbf{x} = \mathbf{k}^T\mathbf{x} = \begin{pmatrix} k_1 \\ k_2 \\ k_3 \end{pmatrix}^T \begin{pmatrix} x_1 \\ x_2 \\ x_3 \end{pmatrix} = \begin{pmatrix} k_1 & k_2 & k_3 \end{pmatrix} \begin{pmatrix} x_1 \\ x_2 \\ x_3 \end{pmatrix} = k_1x_1 + k_2x_2 + k_3x_3. \tag{1.324}$$

Utilizing the wavenumber vector $\mathbf{k}$ and the position vector $\mathbf{x}$, (1.320) takes on the compact form,

$$e^{j(\mathbf{k}\cdot\mathbf{x}-\omega t)}, \tag{1.325}$$

which has the same formal structure as in two dimensions.

□

Question: Given the quadruple (k_1, k_2, k_3, ω) with $k_1, k_2, k_3, \omega \in \mathbb{R}^+$, what relationships between the spatial variables x_1, x_2, x_3, and the time variable t ensure that the phase (radian angle) $k_1x_1 + k_2x_2 + k_3x_3 - \omega t$ in (1.320) is a constant, say, φ_c?

Answer: Consider,

$$k_1x_1 + k_2x_2 + k_3x_3 - \omega t = \varphi_c = \text{const.} \tag{1.326}$$

Differentiating both sides with respect to time,

$$k_1 \frac{dx_1}{dt} + k_2 \frac{dx_2}{dt} + k_3 \frac{dx_3}{dt} - \omega = 0. \tag{1.327}$$

Introducing the phase velocity vector $\mathbf{v}_{ph}$ with components $v_{ph,1}$, $v_{ph,2}$, and $v_{ph,3}$ according to,

$$v_{ph,1} = \frac{dx_1}{dt}, \qquad v_{ph,2} = \frac{dx_2}{dt} \qquad v_{ph,3} = \frac{dx_3}{dt}, \tag{1.328}$$

and substituting into (1.327),

$$k_1 v_{ph,1} + k_2 v_{ph,2} + k_3 v_{ph,3} - \omega = 0. \tag{1.329}$$

Dividing through by ω, and rearranging,

$$\frac{k_1}{\omega} v_{ph,1} + \frac{k_2}{\omega} v_{ph,2} + \frac{k_3}{\omega} v_{ph,3} = 1. \tag{1.330}$$

Introducing the slowness vector $\mathbf{s}$ with the components,

$$s_1 = \frac{k_1}{\omega}, \qquad s_2 = \frac{k_2}{\omega}, \qquad s_3 = \frac{k_3}{\omega}, \tag{1.331}$$

$$s_1 v_{ph,1} + s_2 v_{ph,2} + s_3 v_{ph,3} = 1. \tag{1.332}$$

Employing the dot-product of $\mathbf{s}$ and $\mathbf{v}_{ph}$,

$$\mathbf{s} \cdot \mathbf{v}_{ph} = 1. \tag{1.333}$$

Having specified the quadruple (k_1, k_2, k_3, ω) with $k_1, k_2, k_3, \omega \in \mathbb{R}^+$, the slowness vector $\mathbf{s} = (k_1/\omega, k_2/\omega, k_3/\omega)$ is uniquely determined.

■

Remark: Concerning the methodology.

The discussions in the last three sections, particularly in the last one, serve alluding to an important methodological aspect of content presentation in general. Contents must be made obvious and presented clearly. The presented content must be self-explanatory. The gradual concision of the intervening provisional formula leading to the final compact space-saving and abstract result is very often self-driven or even self-propelling. Writing down (1.333) and merely mentioning that $\mathbf{s}$ and $\mathbf{v}_{ph}$, respectively, refer to the slowness vector and the phase velocity vector, while leaving out the intermediate steps might be a reasonable way of communication. It may encourage the reader to get involved in identifying what lies behind this equation. The axiomatic style of communication based on gradually enhancing the intuition of the reader is one of the techniques which has been followed in this text. Presenting the final formula in (1.333) must be accompanied by all intermediate steps which had led to (1.333). The compact symbolic representation in (1.333) should be perceived as a name referring to a content, the name tag of a certain Pandora's box, or the title of a book. The reader should be able to reverse the steps from (1.333) to the set of assumptions at the beginning of the section. Only then it makes a good sense to operate with (1.333). The ability to open the Pandora's box, take a critical look at the content, evaluate the interrelationships, form an opinion, and close the box, reinforces understanding, eliminates possible misconceptions, motivates uncluttering the redundancies, and above all, facilitates creative thinking and innovative problem solving. After going through the text the reader must be able to generate their own compact version of the text to any degree of compactness they desire. The reader has the capacity to fill the gaps and read between the lines of their own version of the text.

Every though which has been thought must be communicable, irrespective of its perceived complexity.

□

Remark: The Formal Taylor Series Expansion looms large in this chapter and will continue playing an important role throughout the text. The adjective formal refers to the property that in employing the method there are no concerns related to the convergence of the series. The utility of the method stems from the fact that it allows to breaking down functions into monomials, which are readily amenable to manipulations. Thus, the great facility of the Formal Taylor Series Expansion resides in the canonical standard monomial building block. The planewave is another high performing figure in the discussion and presents itself as a most versatile tool. In particular it is proportional to its derivative. This property results from the fact that the planewave involves both the cos- and the sin-functions in an intricate manner. planewaves allow analyzing and synthesizing functions as the originating monomials do, planewaves constitute a basis.

□

1.18   Concluding remarks

Alfred Whitehead's and Bertrand Russell's joint grand vision to build the entire mathematics from a few axioms and rules of inference continues to inspire minds caring for rigor and foundational concepts. So was the zeitgeist, the intellectual ambience, in the transition from the 19th to 20th centuries, and during the first few decades then after. This ambitious project which finds its genesis perhaps in Euclid's elements, turned out to be a categorical impossibility as Kurt Gödel powerfully demonstrated in his dissertation. One of Albert Einstein's postulates in his theory of special relativity, sets the speed of light as the limit for transporting energy or moving any material body. Werner Heisenberg's uncertainty principle puts a limit onto the precision of our knowledge about certain conjugate physical quantities, e.g., position–momentum, or, time–energy. Max Planck's postulate of the smallest possible quanta of action defined yet another limit on how the microscopic world is governed. Philosophers of science spearheaded by Karl Popper and Thomas Kuhn aimed at explaining how the growth of knowledge takes place and whether or not the growth of knowledge is subject to evolutionary or revolutionary mechanisms. The philosopher Ludwig Wittgenstein and others probed the very structure of logic and the language themselves. In early decades of the 20th century quantum physics was born. It is an astonishingly successful theory and its applications are ubiquitous. The many interpretations of the theory create a sense of mystery implying that the theory is not complete yet. Leaving aside the controversies about the interpretation questions, and limiting the scope of inquiry, the axiomatic principle offers itself as a reasonable starting point to learn about the mathematics of quantum physics, despite its limitations. It turns out that most of this mathematics is in fact applied mathematics in traditional sense, and up to a few tweaks and twists, it is the same mathematics that engineers and technologists are familiar with, of course to varying degree of mastery depending on their specialties.

This chapter epitomizes the structure of the entire book by singling out a few facts and principles and developing what is needed for understanding the mathematics of quantum physics. Starting from 1 and x, the abstract notion of the planewave was developed. The intervening steps constituted the body of Chapter 1. The eminent role of Formal Taylor Series Expansion utilizing monomials was emphatically pointed out. The significance of (x, k) and (t, ω) manifested itself automatically. That Heisenberg uncertainty principle concerns the simultaneous uncertainties of x and k, or, of t and ω signals that this principle can be understood as a constraint dictated by the Fourier transform and inverse Fourier transform. These insights in turn suggest the significance of transforms and inverse transforms, and the idea underlying the resolution of identity, which is the subject matter of the next chapter.

Consequently, resorting merely to 1, x, lim, and d/dx one must be able to construct all the tools needed in this text. Along the way a few unexpected novelties motivate the refinement and extension of established ideas. Examples are, as will be witnessed in the next few chapters, novel Dirac Delta functions, generalized creation and annihilations operators, and the Taylor Transform and Inverse Transform.

The list of references [1–10] below is not representative let alone complete. They are meant to explain that Popper, Kuhn, Tarski, Wittgenstein, Gödel, and a dozen of other philosophers of science, logicians, mathematicians, and physicists play significant roles in the author's quest for pursuing fundamentals, as far as an electrical engineer and self-taught amateur mathematician can go. Interested reader should search for relevant literature on Gottlob Frege, Henri Poincare, and Alfred Whitehead. Among all these intellectual heroes and many more, Gottfried Wilhelm Leibniz, Leonhard Euler, and P.A.M. Dirac remain inspiring figures throughout one's lifetime.

References

[1]   Karl R.P., *Conjecture and Refutation: The Growth of Scientific Knowledge*, Routledge & Kegan Paul, 1963.

[2]   Karl R.P., *Objective Knowledge: An Evolutionary Approach*, Oxford University Press, 1972.

[3]   Herbert K., *The Philosophy of Karl Popper*, Cambridge University Press, 2005.

[4]   Thomas S.K., *The Structure of Scientific Revolutions*, The University of Chicago Press, 1962.

[5]   John P., *Kuhn's The Structure of Scientific Revolutions*, Continuum Reader's Guide, 2008.

[6]   Alfred T., *Introduction to Logic: And to the Methodology of Deductive Sciences*, Dover Publications Inc., 1995.

[7]   Ludwig W., *Tractatus Logico-Philosophicus*, Routledge & Kegan Paul, 1974.

[8]   Imre L., *Proofs and Refutations: The Logic of Mathematical Discovery*, Cambridge University Press, 1976.

[9]   Raymond S., *A Beginner's Further Guide to Mathematical Logic*, World Scientific, 2017.

[10] Ernest N. and James R., *Gödel's Proof*, New York University Press, 2001.

Chapter 2
The resolution of identity

2.1 A brief guide through the chapter

While the Kronecker delta symbol is a simple and useful bookkeeping vehicle, the Dirac delta function is not even a function despite what its name suggests. It is rather a functional: its application onto an arbitrarily smooth test function followed by an integration establishes the sifting property of the Dirac delta function. In virtue of its sifting property the Dirac delta function is defined. In engineering community the Dirac delta function is associated with the sampling of a function at an intended value of the independent variable. The Dirac delta function and its allied topics play prominent roles in this chapter. The discussion builds primarily upon planewaves and generalizes the underlying ideas in many ways. It is shown that customized Dirac delta functions can be constructed. Original problem-specific Dirac delta functions in one- and two spatial dimensions, designed by the author, are briefly touched upon. While Dirac delta functions in real domain are generally not factorizable, their symbolic integral representations in spectral domain permit factorization. These ideas motivate connecting to exciting developments in signal processing (more generally, functional analysis) encompassing topics such as multiresolution analysis, the theory of wavelets and dual wavelets, and the theory of frames and dual frames. The discussion in this chapter does not delve into these areas. It, however, prepares the reader to do so if they wanted to. The fact that considerations in engineering and quantum physics contributed significantly to the development of the latter theories, and the fact that these theories in turn promise to contribute to quantum physics can be viewed as a triumph of applied mathematics. Epistemologically, the formation and further development of the applied mathematics is largely inspired by applications. Its ontology is however independent of physics, be quantum physics, or for this matter, theory of general relativity, quantum electrodynamics, or string theory. This insight is revealing. The applied mathematics can be acquired as a collection of tools. The tools, however, must be distinctively separated from the interpretations of the theory they are aiming to provide. The tools and techniques in applied mathematics can be developed based on toy models. This idea has been taken seriously in this chapter. Based on simplest possible toy models, the notions of orthonormal bases and their dual bases, non-normal and dual non-normal bases, non-orthogonal and dual non-orthogonal bases, frames and dual frames, and the notion of generalized transform and inverse transform have been introduced and made plausible. The Dirac bracket notation, the inner product, the exterior product, the resolution of identity, and abstract Hilbert space have been employed to represent Fourier- and inverse Fourier transform, complete or over-complete bases, the Nyquist–Shannon sampling theorem, and the interpolation of functions. A brief discussion of the author's recently developed Discrete Taylor Transform and Inverse Transform (D-TTIT) has also been included. The proper design of elementary toy models promises to convey a host of intriguing ideas in simplest possible ways.

2.2 Elementary notions and definitions

The Kronecker delta symbol: The Kronecker delta symbol δ_{mn} is defined as,

$$\delta_{mn} \stackrel{\text{def.}}{=} \begin{cases} 1 & m = n \\ 0 & m \neq n \end{cases}. \tag{2.1}$$

One application in which δ_{mn} is routinely used, involves a sum, and has the following generic form:

$$a_m = \sum_{n=1}^{\infty} \delta_{mn} a_n, \qquad m \in \mathbb{N} \qquad \left(\begin{array}{l}\text{the quintessential} \\ \text{sifting property of } \delta_{mn}\end{array}\right). \tag{2.2}$$

Here, $\mathbb{N}$ stands for the set of positive integers, $\{1, 2, 3, \ldots\}$. The index $m \in \mathbb{N}$ is fixed in this equation. The dummy sum index n, in contrast, runs through all values in $\mathbb{N}$. At some instance, and necessarily, the running index n assumes the value m. Only in that occasion, the Kronecker delta symbol becomes unity, $\delta_{mn}|_{n=m} = \delta_{mm} = 1$; in all other cases $\delta_{mn}|_{n\neq m} = 0$. This is referred to as the sifting property of the Kronecker delta symbol. The Kronecker delta symbol is widely used in the engineering community, in particular among signal processing professionals. Defined for discrete indices, the Kronecker delta symbol has a counterpart in the continuum.

The Dirac delta function: The Dirac delta function denoted by $\delta(x)$ is also widely used in the engineering community and understood, rather casually, as,

$$\delta(x) = \begin{cases} \infty & x = 0 \\ 0 & x \neq 0 \end{cases} \qquad \left(\begin{array}{l}\text{a symbolic representation, an} \\ \text{engineering perception of } \delta(x)\end{array}\right). \tag{2.3}$$

While this symbolic designation works well for many applications, it is not quite correct. The symbol $\delta(x)$ does not in fact represent any ordinary function at all, contrary to what its name suggests. It is a distribution, a generalized function. Rather than resorting to (2.3), the Dirac delta function $\delta(x)$ must be endowed with the sifting property, which is its defining quintessential feature,

$$\int_{-\infty}^{\infty} dx\, \delta(x - x') f(x) = f(x') \qquad \left(\begin{array}{l}\text{the quintessential} \\ \text{sifting property of } \delta(x)\end{array}\right). \tag{2.4}$$

The function $f(x)$, which is referred to as the test function is assumed to be smooth to any desired degree. In particular, if $f(x) \equiv \text{const} = 1$, then,

$$\int_{-\infty}^{\infty} dx\, \delta(x - x') = 1, \tag{2.5}$$

which is the normalization condition. In order to render the Dirac delta function accessible to analysis, and recognize its immense utility, it is defined as the limit $\eta \to 0^+$ of an η-parametrized sequence of smooth functions $\delta_\eta(x)$, to be specified momentarily. The η-parametrized representation is not unique. However, all representations are required to converge to $\delta(x)$,

$$\lim_{\eta \to 0} \delta_\eta(x) = \delta(x) \qquad \left(\begin{array}{l}\delta(x) \text{ as the limit of a} \\ \text{sequence of } \eta\text{-dependent} \\ \text{proper functions } \delta_\eta(x)\end{array}\right), \tag{2.6}$$

also in a sense to be explained shortly.

Remark: It is the property in (2.6) that makes the Dirac delta functions $\delta(x)$ an indispensable tool in signal processing. In the mathematical quantum physics, $\delta(x)$ is, however, incomparably more significant. Thus, it is justified to examine distinct features of $\delta(x)$, as relevant to the current text. In the following series of solved problems and remarks several physics-based custom-tailored representations of Dirac delta functions in one-dimension $\delta(x)$, and two-dimensions $\delta(x, y)$ will be introduced. Explaining intricate relationships in abstract Hilbert space from this vantage point might be regarded as an original contribution. It will momentarily be established that the geneses of the Dirac delta function $\delta(x)$ are identity operators. It is presumably no exaggeration to compare the significance of the identity operator in quantum physics to 1 in number theory and the unity matrix in linear algebra. This insight has been one of the guiding principle in the Pseudo Axiomatic development of the current text. It will be shown that the so-called resolution of identity operator is key in understanding transforms and their associated inverse transforms, and generalizations thereof. A further novelty is the recognition that Dirac delta functions can be constructed from the governing partial differential equations in mathematical physics, justifying the qualifier physics-based. In fact, the ability to construct problem-specific Dirac delta functions strengthens its position as a fundamental tool, considerably beyond its utility as a convenient and powerful symbol in signal processing. The following solved problems and remarks are aimed at substantiating the aforementioned statements and claims.

□

Remark: In the engineering community, the Dirac delta function $\delta(\cdot)$ is predominantly employed in signal processing applications, where time-varying functions are dealt with. In solving boundary values problems, on the other hand, both spatial- and temporal-spatial varying functions arise. Since the focus in this text is field analysis, the independent variable is mostly the position variable x rather than the time variable t. Furthermore, the fact that there are three spatial dimensions offers greater flexibility in designing problems and investigating their solutions.

□

Problem: Let $\delta_\eta(x - x')$ be defined as,

$$\delta_\eta(x - x') = \frac{1}{\pi} \frac{\eta}{(x - x')^2 + \eta^2}. \tag{2.7}$$

Show that the sequence of the well-behaved arbitrarily differentiable functions $\delta_\eta(x - x')$ converges to the discontinuous generalized function $\delta(x - x')$ in the limit $\eta \to 0^+$. Thus, show that,

$$\lim_{\eta \to 0^+} \delta_\eta(x - x') = \delta(x - x'). \tag{2.8}$$

Two remarks might be in order before engaging in the solution process.

Remark: Assume (2.8) has been established. Reading (2.8) from the right to the left is revealing. The discontinuous generalized function $\delta(x - x')$ reveals itself as an accessible mathematical entity in virtue of being defined in terms of the $\eta \to 0^+$ limit of the sequence of well-behaved arbitrarily differentiable ordinary functions $\delta_\eta(x - x')$. The sequence of the functions $\delta_\eta(x - x')$ permits both theoretical scrutiny and numerical evaluation, as shall be witnessed momentarily. Stated differently, the intractable symbolic entity $\delta(x - x')$ in (2.8) has been made perfectly accessible in terms of the ordinary functions $\delta_\eta(x - x')$. The main calculations, which involve $\delta_\eta(x - x')$, will be followed by the limit process $\eta \to 0^+$. This enabling strategy is the nucleus of a variety of manipulations in quantum physics as well as whenever one is confronted with infinities and divergences (regularization, renormalization). Regularization and renormalization are not treated in this text. They constitute the subject matter of another text by this author that is exclusively dedicated to infinities arising in near-fields.

□

Remark: Without going into any further details, it should be mentioned that the genesis of (2.7) can be traced back to the Laplace operator. In view of the facts that the Laplace operator describes the electrostatic potential, and that the electrostatic Coulomb potential energy plays an important role when solving Schrödinger equation in materials and nano-electronic device applications, these observations are critically important.

□

Solution: Let $f(x)$ be an arbitrary smooth (text) function. Using (2.7), consider,

$$\lim_{\eta \to 0^+} \int_{-\infty}^{\infty} dx \{\delta_\eta(x - x')\} f(x) = \lim_{\eta \to 0^+} \int_{-\infty}^{\infty} dx \left\{ \frac{1}{\pi} \frac{\eta}{(x - x')^2 + \eta^2} \right\} f(x). \tag{2.9}$$

Focus on the integral at the R.H.S. In this integral, x is the (dummy) integration variable, x' is a constant, and η is a parameter. Consider the variable substitution $x - x' = \eta u$ ($\eta > 0$, in compliance with $\eta \to 0^+$). Thus, $x = x' + \eta u$, and with x' being a constant, $dx = \eta du$. Consequently,

$$\lim_{\eta \to 0^+} \int_{-\infty}^{\infty} dx \{\delta_\eta(x - x')\} f(x) \stackrel{\text{var. subst. } x=x'+\eta u}{=} \lim_{\eta \to 0^+} \frac{1}{\pi} \int_{-\infty}^{\infty} du \underbrace{\left(\frac{\eta^2}{\eta^2 u^2 + \eta^2} \right)} f(x' + \eta u) \qquad (2.10a)$$

$$\stackrel{\eta^2 \text{ cancels out}}{=} \lim_{\eta \to 0^+} \frac{1}{\pi} \int_{-\infty}^{\infty} du \underbrace{\left(\frac{1}{u^2 + 1} \right)} f(x' + \eta u) \qquad (2.10b)$$

$$\underset{\eta-\text{independent}}{}$$

$$\stackrel{\lim_{\eta \to 0^+}}{=} \frac{1}{\pi} \int_{-\infty}^{\infty} du \underbrace{\left(\frac{1}{u^2 + 1} \right)} f(x'). \qquad (2.10c)$$

$$\underset{\text{an even func.}}{}$$

Transferring $f(x')$ to the front of the integral, and taking into account that the integrand is an even function of u, as indicated,

$$\lim_{\eta \to 0^+} \int_{-\infty}^{\infty} dx \{\delta_\eta(x - x')\} f(x) = f(x') \frac{2}{\pi} \int_0^{\infty} du \frac{1}{1 + u^2} \qquad (2.11a)$$

$$= f(x') \frac{2}{\pi} \{\arctan(u)\} \Big|_0^{\infty} \qquad (2.11b)$$

$$= f(x') \frac{2}{\pi} \left(\frac{\pi}{2} - 0 \right) \qquad (2.11c)$$

$$= f(x'). \qquad (2.11d)$$

Using the sifting property (2.4), the R.H.S. can be rewritten,

$$\lim_{\eta \to 0^+} \int_{-\infty}^{\infty} dx \{\delta_\eta(x - x')\} f(x) = \int_{-\infty}^{\infty} dx \delta(x - x') f(x). \qquad (2.12)$$

Since $f(x)$ is an arbitrary (test) function upon assumption,

$$\lim_{\eta \to 0^+} \delta_\eta(x - x') = \delta(x - x'). \qquad (2.13)$$

Using (2.7) to explicate the structure of $\delta_\eta(x - x')$,

$$\lim_{\eta \to 0^+} \left\{ \frac{1}{\pi} \frac{\eta}{(x - x')^2 + \eta^2} \right\} = \delta(x - x'). \qquad (2.14)$$

This completes the solution.

∎

Remark: It is important to be fully aware in what sense the equality in (2.14) must be interpreted. Apply both sides of (2.14) onto any test function $f(x)$ and integrate the resulting expressions from $-\infty$ to ∞ to obtain an expression for $f(x')$. In practice the detailed dynamics of this process reveals what goes on when one speaks of sampling the function $f(x)$ at $x = x'$.

□

Remark: Many representations for the Dirac delta function are available in mathematical physics. In particular, this author has shown that the scalar or dyadic Green's functions, associated with partial differential equations, as they arise in boundary value problems, can be utilized to generate optimally problem-specific Dirac delta functions. Details are however irrelevant for the present discussion and will thus not be pursued in this text. An upcoming book by the author (with a focus on zooming into the near-field) is exclusively dedicated to the design of customized Dirac delta functions. Due to the significance of the Dirac delta functions, and their role in the resolution of identity, which is an important feature in this book, a few novel expressions for the Dirac delta function in one- and two-dimensions will be discussed nonetheless.

□

Problem: Show that the relationships

$$I_\eta(x - x') \stackrel{\text{def.}}{=} \int_{-\infty}^{\infty} \frac{dk}{2\pi} e^{jk(x-x')} e^{-|k|\eta} \tag{2.15a}$$

$$= \frac{1}{\pi} \frac{\eta}{(x - x')^2 + \eta^2} \tag{2.15b}$$

$$\stackrel{\text{def.}}{=} \delta_\eta(x - x'), \tag{2.15c}$$

hold valid, with $\lim_{\eta \to 0^+} \delta_\eta(x - x') = \delta(x - x')$.

Solution: Since $e^{-|k|\eta}$ is even symmetric in k, the sin-part of $e^{jk(x-x')}$ does not contribute to the integral. Thus,

$$I_\eta(x - x') = \int_{-\infty}^{\infty} \frac{dk}{2\pi} e^{jk(x-x')} e^{-|k|\eta} = \frac{1}{\pi} \int_0^{\infty} dk \cos [k(x - x')] e^{-k\eta}. \tag{2.16}$$

Multiplying both sides by π and integrating by parts with respect to k,

$$\pi I_\eta(x - x') = \int_0^{\infty} dk \underbrace{\cos [k(x - x')]} \underbrace{e^{-k\eta}} \tag{2.17a}$$

$$= \frac{\sin [k(x - x')]}{x - x'} e^{-k\eta} \Big|_0^{\infty} - \int_0^{\infty} dk \left\{ \frac{\sin [k(x - x')]}{x - x'} \right\} \{(-\eta)e^{-k\eta}\}. \tag{2.17b}$$

Taking into account that $\sin [k(x - x')]$ vanishes at the lower bound $k = 0$, and $e^{-k\eta}$ is zero at the upper bound $k = \infty$, the first term at the R.H.S. does not have any contribution. Simplifying the second term, by taking into account that η and $x - x'$ are k-independent,

$$\pi I_\eta(x - x') = \frac{\eta}{x - x'} \int_0^{\infty} dk \{ \sin [k(x - x')] \} \{ e^{-k\eta} \}. \tag{2.18}$$

Integrating by parts a second time,

$$\pi I_\eta(x - x') = \frac{\eta}{x - x'} \left\{ -\frac{\cos [k(x - x')]}{x - x'} e^{-k\eta} \Big|_0^{\infty} - \int_0^{\infty} dk \left[\frac{-\cos [k(x - x')]}{x - x'} \right] [(-\eta)e^{-k\eta}] \right\}. \tag{2.19}$$

Since $\cos [k(x - x')]$ and $e^{-k\eta}$ are both unity at the lower bound $k = 0$, and $e^{-k\eta}$ is zero at the upper bound $k = \infty$, the first term at the R.H.S. results in $1/(x - x')$. Simplifying the second term,

$$\pi I_\eta(x - x') = \frac{\eta}{x - x'} \left\{ \frac{1}{x - x'} - \frac{\eta}{x - x'} \underbrace{\int_0^{\infty} dk \cos [k(x - x')] e^{-k\eta}} \right\}. \tag{2.20}$$

With reference to (2.16), the under-braced term is equal to $\pi I_\eta(x - x')$. Therefore,

$$\pi I_\eta(x - x') = \frac{\eta}{x - x'}\left\{\frac{1}{x - x'} - \frac{\eta}{x - x'}\pi I_\eta(x - x')\right\} \tag{2.21}$$

$$= \frac{\eta}{(x - x')^2} - \frac{\eta^2}{(x - x')^2}\pi I_\eta(x - x'). \tag{2.22}$$

Solving for $\pi I_\eta(x - x')$, the following expression for $I_\eta(x - x')$ results,

$$I_\eta(x - x') = \frac{1}{\pi}\frac{\eta}{(x - x')^2 + \eta^2}. \tag{2.23}$$

It is worthwhile to contrast the two representations of $\delta_\eta(t - t')$,

$$\delta_\eta(x - x') = \frac{1}{\pi}\frac{\eta}{(x - x')^2 + \eta^2} \tag{2.24a}$$

$$\delta_\eta(x - x') = \int_{-\infty}^{\infty}\frac{dk}{2\pi}e^{jk(x-x')}e^{-|k|\eta}. \tag{2.24b}$$

And, consequently,

$$\delta(x - x') = \lim_{\eta \to 0^+}\frac{1}{\pi}\frac{\eta}{(x - x')^2 + \eta^2} \tag{2.25a}$$

$$\delta(x - x') = \lim_{\eta \to 0^+}\int_{-\infty}^{\infty}\frac{dk}{2\pi}e^{jk(x-x')}e^{-|k|\eta}. \tag{2.25b}$$

This completes the solution.

∎

Remark: The (convergent) integral in (2.24b) is the spectral representation of $\delta_\eta(x - x')$. The fact that the spectral representation is convergent for $\eta \neq 0$, and the fact that the expression at the R.H.S. in (2.24a) is a well-behaved arbitrarily differentiable function for $\eta \neq 0$, are equivalent statements. Note that the brute-force action of setting $\eta = 0$ in (2.25a) or (2.25b) renders these expressions divergent and thus useless. Consequently, the representations in (2.25) should be employed subject to the condition that $\eta \neq 0$, calculations carried out, and then the limit $\eta \to 0^+$ considered. Thus, by writing,

$$\int_{-\infty}^{\infty}\frac{dk}{2\pi}e^{jk(x-x')} = \delta(x - x'), \tag{2.26}$$

the integral expression is merely a symbol as $\delta(x - x')$ is a symbol, and concerning numerical calculations, it is useless. For theoretical considerations, however, the integral in (2.26) can be employed on par with the symbolic representation $\delta(x - x')$, and it offers great utility. The exposition in this chapter makes ample use of this symbolic integral representation and arrives at valuable theoretical conclusions. For numerical calculations, however, convergent integrals of the form (2.25b) must be used.

☐

Remark: Yet, another consistent approach would proceed as follows: set $\eta = 2\epsilon$ to obtain $e^{-|k|\eta} = e^{-|k|2\epsilon} = e^{-|k|\epsilon}e^{-|k|\epsilon}$. Assign one decaying exponential function $e^{-|k|\epsilon}$ to each of the oscillating non-decaying harmonic functions e^{jkx} and $e^{-jkx'}$, to obtain $e^{jkx}e^{-|k|\epsilon}$ and $e^{-jkx'}e^{-|k|\epsilon}$. A theory development along this line of thinking would be clear and clean at every step of the theoretical and computational work. For the purposes propounded in this text, however, it would not add much benefit to the overall understanding. It would rather render the presentation considerably more messy.

□

Remark: Engineers and practitioners are familiar with the symbolic representation in (2.26), which is standard and routinely used in the textbooks. The prevailing wisdom is that since this type of integrals do not converge, a judicious-chosen (yet *ad hoc*) multiplicative exponential damping term is needed to ensure that the oscillatory (non-decaying) term $e^{jk(x-x')}$ becomes integrable. Incidentally, the appearance of the exponential $e^{-|k|\eta}$ in the preceding analysis is on physical grounds rather than being *ad hoc*. Rigorously constructed physics-inspired correction terms turn out to be feature rich and reveal significant information about the hidden structure of the space where singularities and divergences happen to exist. When proceeded properly, conservation laws of some sort (physical or mathematical) take care of the divergences, provided the governing equations describe what is observed adequately, and the mathematics is followed respectfully. To summarize, it is helpful to associate with the symbolic singular form in (2.26), the corresponding regularized η-parametrized representation in (2.25b). The η-dependent forms render the expressions theoretically interpretable and numerically accessible.

□

Remark: Preparations for the design of physics-inspired Dirac delta functions.

This author has shown that modeling and simulation of materials and micro- and nano-scopic devices offer a fertile ground for generating and designing utterly new and feature-rich Dirac delta functions beyond what is known in standard textbooks and mathematical physics literature. Here a glimpse of the results will be shared with interested readers. Consider a dielectric material characterized by a 3×3 positive definite symmetric dielectric matrix $\boldsymbol{\varepsilon}$,

$$\boldsymbol{\varepsilon} = \begin{pmatrix} \varepsilon_{11} & \varepsilon_{12} & \varepsilon_{13} \\ \varepsilon_{12} & \varepsilon_{22} & \varepsilon_{23} \\ \varepsilon_{13} & \varepsilon_{23} & \varepsilon_{33} \end{pmatrix}. \tag{2.27}$$

In particular, the positive definiteness of $\boldsymbol{\varepsilon}$ implies that the sub-matrix $\boldsymbol{\varepsilon}_s$,

$$\boldsymbol{\varepsilon}_s = \begin{pmatrix} \varepsilon_{11} & \varepsilon_{13} \\ \varepsilon_{13} & \varepsilon_{33} \end{pmatrix}. \tag{2.28}$$

is also positive definite. The requirements,

$$\varepsilon_{11} > 0, \quad \varepsilon_{33} > 0, \quad \det\{\boldsymbol{\varepsilon}_s\} = \varepsilon_{11}\varepsilon_{33} - \varepsilon_{13}^2$$
$$\stackrel{\text{def.}}{=} \varepsilon_{\text{eff}}^2 > 0, \tag{2.29}$$

are necessary and sufficient for the positive-definiteness of $\boldsymbol{\varepsilon}_s$. To simplify the notation, the symbol ε_{eff} has been introduced with $\varepsilon_{\text{eff}}^2 = \varepsilon_{11}\varepsilon_{33} - \varepsilon_{13}^2$. Stated differently, these conditions imply that for an arbitrary 2×1 column vector $\mathbf{a}$ with components u and w, the quadratic form $\mathbf{a}^T \boldsymbol{\varepsilon}_s \mathbf{a}$ is strictly positive,

$$\mathbf{a}^T \boldsymbol{\varepsilon}_s \mathbf{a} = \begin{pmatrix} u & w \end{pmatrix} \begin{pmatrix} \varepsilon_{11} & \varepsilon_{13} \\ \varepsilon_{13} & \varepsilon_{33} \end{pmatrix} \begin{pmatrix} u \\ w \end{pmatrix} \tag{2.30}$$

$$= \varepsilon_{11} u^2 + 2\varepsilon_{13} uw + \varepsilon_{33} w^2 > 0. \tag{2.31}$$

Here, $\mathbf{a}^T$ is the transpose of $\mathbf{a}$. Observe that the inverse matrix $\boldsymbol{\varepsilon}_s^{-1}$,

$$\boldsymbol{\varepsilon}_s^{-1} = \frac{1}{\det\{\boldsymbol{\varepsilon}_s\}} \begin{pmatrix} \varepsilon_{33} & -\varepsilon_{13} \\ -\varepsilon_{13} & \varepsilon_{11} \end{pmatrix}, \tag{2.32}$$

inherits the properties of $\boldsymbol{\varepsilon}_s$. In particular, the positive-definiteness of $\boldsymbol{\varepsilon}_s^{-1}$ implies that for any judiciously chosen 2×1 vector $\mathbf{a}$, for example, $\mathbf{a}^T = (x - x', \eta)$, with $\eta > 0$, the quadratic form,

$$\mathbf{a}^T \boldsymbol{\varepsilon}_s^{-1} \mathbf{a} = \left(x - x', \ \eta \right) \left[\frac{1}{\det\{\boldsymbol{\varepsilon}_s\}} \begin{pmatrix} \varepsilon_{33} & -\varepsilon_{13} \\ -\varepsilon_{13} & \varepsilon_{11} \end{pmatrix} \right] \begin{pmatrix} x - x' \\ \eta \end{pmatrix} \tag{2.33a}$$

$$\mathbf{a}^T \boldsymbol{\varepsilon}_s^{-1} \mathbf{a} = \frac{1}{\det\{\boldsymbol{\varepsilon}_s\}} \left[\varepsilon_{33}(x - x')^2 - 2\varepsilon_{13}(x - x')\eta + \varepsilon_{11}\eta^2 \right] > 0, \tag{2.33b}$$

is strictly positive. Consequently, employing $\det\{\boldsymbol{\varepsilon}_s\} = \varepsilon_{\text{eff}}^2$,

$$\frac{1}{\varepsilon_{\text{eff}} \left(\mathbf{a}^T \boldsymbol{\varepsilon}_s^{-1} \mathbf{a} \right)} = \frac{\varepsilon_{\text{eff}}}{\varepsilon_{33}(x - x')^2 - 2\varepsilon_{13}(x - x')\eta + \varepsilon_{11}\eta^2} > 0. \tag{2.34}$$

Inspired by the expression at the R.H.S., define $\delta_\eta(x - x')$,

$$\delta_\eta(x - x') \stackrel{\text{def.}}{=} \frac{1}{\pi} \frac{\varepsilon_{\text{eff}}\,\eta}{\varepsilon_{33}(x - x')^2 - 2\varepsilon_{13}(x - x')\eta + \varepsilon_{11}\eta^2} > 0. \tag{2.35}$$

The following Lemma shows that in the limit $\eta \to 0^+$, the η-parametrized sequence of functions $\delta_\eta(x - x')$ converges to $\delta(x - x')$. The genesis of $\delta_\eta(x - x')$ can be traced back to near-fields in electrostatics. The plausibilization leading to (2.35) should suffice to grant the reader an idea about the origin of this expression. In particular, the occurrence of the minus sign in the denominator, and also the fact that $\delta_\eta(x - x') > 0$ for any $x - x' \in \mathbb{R}$ and $\eta \in \mathbb{R}^+$ should be pointed out.

□

Lemma: *Generalization of classical Dirac delta functions.*

Let $\boldsymbol{\varepsilon}_s$,

$$\boldsymbol{\varepsilon}_s = \begin{pmatrix} \varepsilon_{11} & \varepsilon_{13} \\ \varepsilon_{13} & \varepsilon_{33} \end{pmatrix}. \tag{2.36}$$

be a positive definite matrix. Then, the η-parametrized sequence of functions $\delta_\eta(x - x')$,

$$\delta_\eta(x - x') = \frac{1}{\pi} \frac{\varepsilon_{\text{eff}}\,\eta}{\varepsilon_{33}(x - x')^2 - 2\varepsilon_{13}(x - x')\eta + \varepsilon_{11}\eta^2} > 0, \tag{2.37}$$

converge to the Dirac delta function $\delta(x - x')$, in the $\eta \to 0^+$ limit.

Proof: To proof that the sequence of smooth arbitrarily differentiable sequence of η-parametrized functions $\delta_\eta(x - x')$ represent the Dirac delta function $\delta(x - x')$, it suffices to show that the $\eta \to 0^+$ limit of $\delta_\eta(x - x')$ possesses the sifting property. Details of the arguments are as follows.

$$\lim_{\eta \to 0^+} \int_{-\infty}^{\infty} dx \left\{ \delta_\eta(x - x') \right\} f(x)$$

$$= \lim_{\eta \to 0^+} \int_{-\infty}^{\infty} dx \left\{ \frac{1}{\pi} \frac{\varepsilon_{\text{eff}}\,\eta}{\varepsilon_{33}(x - x')^2 - 2\varepsilon_{13}(x - x')\eta + \varepsilon_{11}\eta^2} \right\} f(x). \tag{2.38}$$

Employ the variable substitution,

$$x - x' = \eta u, \quad \rightarrow \quad x = x' + \eta u, \quad \rightarrow \quad dx = \eta\,du, \tag{2.39}$$

in the integral at the R.H.S.,

$$\lim_{\eta \to 0^+} \int_{-\infty}^{\infty} dx \left\{ \delta_\eta(x - x') \right\} f(x)$$

$$= \lim_{\eta \to 0^+} \frac{1}{\pi} \int_{-\infty}^{\infty} du \left\{ \frac{\varepsilon_{\text{eff}} \eta^2}{\varepsilon_{33} u^2 \eta^2 - 2\varepsilon_{13} u \eta^2 + \varepsilon_{11} \eta^2} \right\} f(x' + \eta u). \tag{2.40}$$

It is worth noting that the argument of the test function $f(\cdot)$ records the above and the following variable substitutions with the highest fidelity, which if done systematically, allows one to reverse the order of substitutions. More importantly, it is observed that the term η^2 cancels out from the numerator and the denominator of the fraction within the curly brackets. Factoring out ε_{33} (> 0) in the denominator,

$$\int_{-\infty}^{\infty} dx \lim_{\eta \to 0^+} \delta_\eta(x - x') f(x)$$

$$= \lim_{\eta \to 0^+} \frac{1}{\pi} \int_{-\infty}^{\infty} du \frac{\varepsilon_{\text{eff}}}{\varepsilon_{33}} \left\{ \frac{1}{u^2 - 2\left(\frac{\varepsilon_{13}}{\varepsilon_{33}}\right) u + \left(\frac{\varepsilon_{11}}{\varepsilon_{33}}\right)} \right\} f(x' + \eta u). \tag{2.41}$$

The only η-dependent term in the integrand is the test function $f(\cdot)$, which is by definition smooth to any degree desired. These facts would allow, already at this stage of calculations, setting $\eta = 0$ in $f(x' + \eta u)$, i.e., replacing $f(x' + \eta u)$ with $f(x')$, getting rid of the η-dependence completely, and consequently, dismissing $\lim_{\eta \to 0^+}$. This is however not what is done here, for the purpose of continuing to record the sequence of variable substitutions, alluded to previously.

Completing the square in the denominator,

$$\text{denom.} = u^2 - 2\left(\frac{\varepsilon_{13}}{\varepsilon_{33}}\right) u + \left(\frac{\varepsilon_{11}}{\varepsilon_{33}}\right) \tag{2.42a}$$

$$= u^2 - 2\left(\frac{\varepsilon_{13}}{\varepsilon_{33}}\right) u + \left(\frac{\varepsilon_{13}}{\varepsilon_{33}}\right)^2 - \left(\frac{\varepsilon_{13}}{\varepsilon_{33}}\right)^2 + \left(\frac{\varepsilon_{11}}{\varepsilon_{33}}\right) \tag{2.42b}$$

$$= \left(u - \frac{\varepsilon_{13}}{\varepsilon_{33}}\right)^2 + \frac{\varepsilon_{11}\varepsilon_{33} - \varepsilon_{13}^2}{\varepsilon_{33}^2} \tag{2.42c}$$

$$= \left(u - \frac{\varepsilon_{13}}{\varepsilon_{33}}\right)^2 + \frac{\varepsilon_{\text{eff}}^2}{\varepsilon_{33}^2}. \tag{2.42d}$$

$$= \left(u - \frac{\varepsilon_{13}}{\varepsilon_{33}}\right)^2 + \left(\frac{\varepsilon_{\text{eff}}}{\varepsilon_{33}}\right)^2 > 0. \tag{2.42e}$$

Thus,

$$\lim_{\eta \to 0^+} \int_{-\infty}^{\infty} dx \left\{ \delta_\eta(x - x') \right\} f(x)$$

$$= \lim_{\eta \to 0^+} \frac{1}{\pi} \int_{-\infty}^{\infty} du \frac{\varepsilon_{\text{eff}}}{\varepsilon_{33}} \left\{ \frac{1}{\left(u - \frac{\varepsilon_{13}}{\varepsilon_{33}}\right)^2 + \left(\frac{\varepsilon_{\text{eff}}}{\varepsilon_{33}}\right)^2} \right\} f(x' + \eta u). \tag{2.43}$$

Using the variable substitution,

$$u - \frac{\varepsilon_{13}}{\varepsilon_{33}} = v \quad \rightarrow \quad u = \frac{\varepsilon_{13}}{\varepsilon_{33}} + v \quad \rightarrow \quad du = dv, \tag{2.44}$$

$$\lim_{\eta \to 0^+} \int_{-\infty}^{\infty} dx \{\delta_\eta(x - x')\} f(x)$$

$$= \lim_{\eta \to 0^+} \frac{1}{\pi} \int_{-\infty}^{\infty} dv \frac{\varepsilon_{\text{eff}}}{\varepsilon_{33}} \left\{ \frac{1}{v^2 + \left(\frac{\varepsilon_{\text{eff}}}{\varepsilon_{33}}\right)^2} \right\} f \left[x' + \eta \left(\frac{\varepsilon_{13}}{\varepsilon_{33}} + v \right) \right]. \tag{2.45}$$

Using the variable substitution,

$$v = \frac{\varepsilon_{\text{eff}}}{\varepsilon_{33}} w \quad \rightarrow \quad dv = \frac{\varepsilon_{\text{eff}}}{\varepsilon_{33}} dw, \tag{2.46}$$

into (2.45)

$$\lim_{\eta \to 0^+} \int_{-\infty}^{\infty} dx \{\delta_\eta(x - x')\} f(x)$$

$$= \lim_{\eta \to 0^+} \frac{1}{\pi} \int_{-\infty}^{\infty} dw \left(\frac{\varepsilon_{\text{eff}}}{\varepsilon_{33}}\right)^2 \left\{ \frac{1}{w^2 \left(\frac{\varepsilon_{\text{eff}}}{\varepsilon_{33}}\right)^2 + \left(\frac{\varepsilon_{\text{eff}}}{\varepsilon_{33}}\right)^2} \right\} f \left[x' + \eta \left(\frac{\varepsilon_{13}}{\varepsilon_{33}} + \frac{\varepsilon_{\text{eff}}}{\varepsilon_{33}} w \right) \right]. \tag{2.47}$$

The term $(\varepsilon_{\text{eff}}/\varepsilon_{33})^2$ cancels out,

$$\lim_{\eta \to 0^+} \int_{-\infty}^{\infty} dx \{\delta_\eta(x - x')\} f(x)$$

$$= \lim_{\eta \to 0^+} \frac{1}{\pi} \int_{-\infty}^{\infty} dw \frac{1}{w^2 + 1} f \left[x' + \eta \left(\frac{\varepsilon_{13}}{\varepsilon_{33}} + \frac{\varepsilon_{\text{eff}}}{\varepsilon_{33}} w \right) \right]. \tag{2.48}$$

Before completing the proof it is worthwhile to examine the argument of $f(\cdot)$: the three variable substitutions have left their footprints in the final argument of $f(\cdot)$, which can be reversed, as it is schematically shown next,

$$\text{The final argument of } f(\cdot) = x' + \eta \underbrace{\left(\frac{\varepsilon_{13}}{\varepsilon_{33}} + \underbrace{\frac{\varepsilon_{\text{eff}}}{\varepsilon_{33}} w}_{v} \right)}_{u} . \tag{2.49}$$

In the limit $\eta \to 0^+$, the argument of f simplifies to x', leading to,

$$\lim_{\eta \to 0^+} \int_{-\infty}^{\infty} dx \{\delta_\eta(x - x')\} f(x) = \frac{1}{\pi} \int_{-\infty}^{\infty} dw \frac{1}{1 + w^2} f(x'). \tag{2.50}$$

Since $f(x')$ is independent of the dummy integration variable w, it can be transferred to the front of the integral,

$$\lim_{\eta \to 0^+} \int_{-\infty}^{\infty} dx \{\delta_\eta(x - x')\} f(x) = f(x') \frac{1}{\pi} \int_{-\infty}^{\infty} dw \frac{1}{1 + w^2}. \tag{2.51}$$

As shown in an earlier problem,

$$\int\limits_{-\infty}^{\infty} dw \frac{1}{1+w^2} = \arctan(w)\Big|_{-\infty}^{\infty} = \frac{\pi}{2} - \left(-\frac{\pi}{2}\right) = \pi. \tag{2.52}$$

Considering this result in (2.51),

$$\lim_{\eta \to 0^+} \int\limits_{-\infty}^{\infty} dx \{\delta_\eta(x - x')\} f(x) = f(x') = \int\limits_{-\infty}^{\infty} dx \delta(x - x') f(x). \tag{2.53}$$

In the last transition the sifting property of the Dirac delta function has been used. Since $f(x)$ is a Test function, and with reference to (2.37),

$$\lim_{\eta \to 0^+} \left\{ \frac{1}{\pi} \frac{\varepsilon_{\text{eff}} \eta}{\varepsilon_{33}(x - x')^2 - 2\varepsilon_{13}(x - x')\eta + \varepsilon_{11}\eta^2} \right\} = \delta(x - x'). \tag{2.54}$$

This completes the proof.

∎

The above considerations are not limited to the physics-inspired problem-tailored construction of Dirac delta function in one-dimension only. The following lemma provides an example of a sequence of η-parametrized functions, $\delta_\eta(x - x', y - y')$, which approach to the Dirac delta function in two-dimensions in the limit $\eta \to 0^+$. The simplest possible example has been considered. As a matter of fact, it can be shown that for any physically realizable boundary value problem in mathematical physics corresponding Dirac delta functions can be constructed. This insight enables zooming into the near-field. Details of this principle will be explained elsewhere.

Lemma: *Dirac delta function in two dimensions.*

Consider the η-parametrized sequence of functions $\delta_\eta(x - x', y - y')$,

$$\delta_\eta(x - x', y - y') = \frac{1}{2\pi} \frac{\eta}{\left[(x - x')^2 + (y - y')^2 + \eta^2\right]^{3/2}}. \tag{2.55}$$

Show that,

$$\lim_{\eta \to 0^+} \delta_\eta(x - x', y - y') = \delta(x - x', y - y'). \tag{2.56}$$

Proof: The proof aims at showing that the limit $\eta \to 0^+$ of the proposed $\delta_\eta(x - x', y - y')$ possesses the expected sifting property. Consider the text function $f(x, y)$ and proceed as follows:

$$\lim_{\eta \to 0^+} \int\limits_{-\infty}^{\infty} \int\limits_{-\infty}^{\infty} dx dy \delta_\eta(x - x', y - y') f(x, y)$$

$$= \lim_{\eta \to 0^+} \frac{1}{2\pi} \int\limits_{-\infty}^{\infty} \int\limits_{-\infty}^{\infty} dx dy \frac{\eta}{\left[(x - x')^2 + (y - y')^2 + \eta^2\right]^{3/2}} f(x, y). \tag{2.57}$$

Use the variable substitutions,

$$x - x' = u \quad \rightarrow \quad x = x' + u \quad \rightarrow \quad dx = du \tag{2.58a}$$
$$y - y' = v \quad \rightarrow \quad y = y' + v \quad \rightarrow \quad dy = dv \tag{2.58b}$$

in the integral at the R.H.S. in (2.57),

$$\lim_{\eta \to 0^+} \int_{-\infty}^{\infty} \int_{-\infty}^{\infty} dxdy \delta_\eta(x - x', y - y')f(x,y)$$

$$= \lim_{\eta \to 0^+} \frac{1}{2\pi} \int_{-\infty}^{\infty} \int_{-\infty}^{\infty} dudv \frac{\eta}{\left[u^2 + v^2 + \eta^2\right]^{3/2}} f(x' + u, y' + v). \tag{2.59}$$

Consider the variable substitutions,

$$u = p\eta \quad \to \quad du = dp\eta \tag{2.60a}$$
$$v = q\eta \quad \to \quad dv = dq\eta \tag{2.60b}$$

in the integral at the R.H.S. in (2.59),

$$\lim_{\eta \to 0^+} \int_{-\infty}^{\infty} \int_{-\infty}^{\infty} dxdy \delta_\eta(x - x', y - y')f(x,y)$$

$$= \lim_{\eta \to 0^+} \frac{1}{2\pi} \int_{-\infty}^{\infty} \int_{-\infty}^{\infty} dpdq \underbrace{\frac{\eta^3}{\left(p^2\eta^2 + q^2\eta^2 + \eta^2\right)^{3/2}}} f(x' + p\eta, y' + q\eta). \tag{2.61}$$

The under-braced term is independent of η,

$$\frac{\eta^3}{\left[\eta^2(p^2 + q^2 + 1)\right]^{3/2}} = \frac{\eta^3}{\eta^3(p^2 + q^2 + 1)^{3/2}} = \frac{1}{(p^2 + q^2 + 1)^{3/2}}. \tag{2.62}$$

Thus,

$$\lim_{\eta \to 0^+} \int_{-\infty}^{\infty} \int_{-\infty}^{\infty} dxdy \delta_\eta(x - x', y - y')f(x,y)$$

$$= \lim_{\eta \to 0^+} \frac{1}{2\pi} \int_{-\infty}^{\infty} \int_{-\infty}^{\infty} dpdq \frac{1}{(p^2 + q^2 + 1)^{3/2}} f(x' + p\eta, y' + q\eta). \tag{2.63}$$

Since the test function $f(\cdot,\cdot)$, is arbitrarily smooth, the limit $\eta \to 0^+$ can be carried out. At the limit $\eta \to 0^+$, $f(x' + p\eta, y' + q\eta)$ becomes $f(x', y')$. Transferring $f(x', y')$ to the front of the integral,

$$\lim_{\eta \to 0^+} \int_{-\infty}^{\infty} \int_{-\infty}^{\infty} dxdy \delta_\eta(x - x', y - y')f(x,y)$$

$$= f(x', y') \frac{1}{2\pi} \underbrace{\int_{-\infty}^{\infty} \int_{-\infty}^{\infty} dpdq \frac{1}{(p^2 + q^2 + 1)^{3/2}}}_{I_{aux}}. \tag{2.64}$$

To calculate the auxiliary integral I_{aux}, as indicated, adopt the (r, ϕ)-polar coordinate system,

$$\begin{cases} p = r\cos\phi \\ q = r\sin\phi \end{cases} \implies dpdq = rdrd\phi . \tag{2.65}$$

Considering $p^2 + q^2 = r^2$,

$$I_{\text{aux}} = \frac{1}{2\pi} \int\limits_0^\infty drr \int\limits_0^{2\pi} d\phi \frac{1}{(r^2 + 1)^{3/2}}. \tag{2.66}$$

Since the integrand is independent of ϕ, the ϕ-integral gives 2π, which cancels the $1/(2\pi)$ in front of the integral. Consequently,

$$I_{\text{aux}} = \int\limits_0^\infty drr \frac{1}{(r^2 + 1)^{3/2}}. \tag{2.67}$$

Employing the variable substitution,

$$r^2 + 1 = w \qquad \rightarrow \qquad 2rdr = dw, \tag{2.68}$$

$$I_{\text{aux}} = \frac{1}{2} \int\limits_1^\infty dw \frac{1}{w^{3/2}} = \frac{1}{2} \int\limits_1^\infty dw w^{-3/2} = \frac{1}{2} \left(-2w^{-1/2} \right) \Big|_1^\infty = - \left(\frac{1}{w^{1/2}} \right) \Big|_1^\infty = -(0 - 1) = 1. \tag{2.69}$$

Considering this result in (2.64),

$$\lim_{\eta \to 0^+} \int\limits_{-\infty}^\infty \int\limits_{-\infty}^\infty dx dy \delta_\eta(x - x', y - y') f(x, y) = f(x', y'). \tag{2.70}$$

This completes the proof of the sifting property.

∎

Remark: Without providing any proof, it is merely stated that the following spectral representation for $\delta_\eta(x - x', y - y')$ holds valid,

$$\delta_\eta(x - x', y - y') = \int\limits_{-\infty}^\infty \int\limits_{-\infty}^\infty \frac{dk_1}{2\pi} \frac{dk_2}{2\pi} e^{jk_1(x-x')} e^{jk_2(y-y')} e^{-|k|\eta}, \tag{2.71}$$

with $|k| = \sqrt{k_1^2 + k_2^2}$ and $\eta \in \mathbb{R}^+$.

□

The comparatively detailed discussion of the Dirac delta function in this section was meant to assist the reader to associate with the elusive non-computable symbols $\delta(x - x')$ and $\delta(x - x', y - y')$, the corresponding tamed and well-behaved functions $\delta_\eta(x - x')$ and $\delta_\eta(x - x', y - y')$, respectively, which facilitate interpretations and enable computations.

In preparing for the next step in the development of ideas, consider the simple and at the same time far-reaching multiplicative factorizations $e^{jk(x-x')} = e^{-jkx'} e^{jkx}$ and $2\pi = \sqrt{2\pi} \sqrt{2\pi}$ in (2.26),

$$\delta(x - x') = \int\limits_{-\infty}^\infty dk \left\{ \frac{e^{jkx}}{\sqrt{2\pi}} \right\} \left\{ \frac{e^{-jkx'}}{\sqrt{2\pi}} \right\}. \tag{2.72}$$

This seemingly mundane rearrangement will serve the discussions disproportionately well. For further manipulation of (2.72), experience shows that a few surprisingly simple if not trivial examples will support the reader to gain a strong intuition for deeply understanding the subject matter.

Orthonormal bases and their corresponding dual bases: Consider the orthonornal column vectors $\mathbf{e}_1 = \begin{bmatrix} 1 \\ 0 \end{bmatrix}$ and $\mathbf{e}_2 = \begin{bmatrix} 0 \\ 1 \end{bmatrix}$ in Euclidean $\mathbb{R}^2$ space, satisfying,

$$< e_i | e_j > = \mathbf{e}_i^T \mathbf{e}_j = \mathbf{e}_i \cdot \mathbf{e}_j = \delta_{ij}, \quad i,j = 1,2. \tag{2.73}$$

The dot in $\mathbf{e}_i \cdot \mathbf{e}_j$ encodes a recipe for manipulating the components of the two column vectors in a certain way: given the column vectors $\mathbf{e}_i$ and $\mathbf{e}_j$, multiply the corresponding components of $\mathbf{e}_i$ and $\mathbf{e}_j$, and add them together. Thus,

$$\mathbf{e}_1 \cdot \mathbf{e}_1 = \begin{bmatrix} 1 \\ 0 \end{bmatrix} \cdot \begin{bmatrix} 1 \\ 0 \end{bmatrix} = 1 \times 1 + 0 \times 0 = 1 \tag{2.74a}$$

$$\mathbf{e}_1 \cdot \mathbf{e}_2 = \begin{bmatrix} 1 \\ 0 \end{bmatrix} \cdot \begin{bmatrix} 0 \\ 1 \end{bmatrix} = 1 \times 0 + 0 \times 1 = 0 \tag{2.74b}$$

$$\mathbf{e}_2 \cdot \mathbf{e}_1 = \begin{bmatrix} 0 \\ 1 \end{bmatrix} \cdot \begin{bmatrix} 1 \\ 0 \end{bmatrix} = 0 \times 1 + 1 \times 0 = 0 \tag{2.74c}$$

$$\mathbf{e}_2 \cdot \mathbf{e}_2 = \begin{bmatrix} 0 \\ 1 \end{bmatrix} \cdot \begin{bmatrix} 0 \\ 1 \end{bmatrix} = 0 \times 0 + 1 \times 1 = 1. \tag{2.74d}$$

The notation $\mathbf{e}_i^T \mathbf{e}_j$ alludes to another scheme of manipulating the components of the column vectors $\mathbf{e}_i$ and $\mathbf{e}_j$. The procedure is second nature to engineers and technologists: given the column vectors $\mathbf{e}_i$ and $\mathbf{e}_j$, build the transpose of $\mathbf{e}_i$ denoted by $\mathbf{e}_i^T$, creating a row vector. (If the components of the originating column vector are complex-valued, the components must be additionally replaced by their complex conjugates.) Leave the vector $\mathbf{e}_j$ as it is, i.e., as a column vector. The purpose is to generate a 1×2 matrix and a 2×1 matrix which are compatible and thus allow a matrix multiplication, an idea engineers are perfectly familiar with. Thus,

$$\mathbf{e}_1^T \mathbf{e}_1 = [\,1 \ 0\,] \begin{bmatrix} 1 \\ 0 \end{bmatrix} = 1 \times 1 + 0 \times 0 = 1 \tag{2.75a}$$

$$\mathbf{e}_1^T \mathbf{e}_2 = [\,1 \ 0\,] \begin{bmatrix} 0 \\ 1 \end{bmatrix} = 1 \times 0 + 0 \times 1 = 0 \tag{2.75b}$$

$$\mathbf{e}_2^T \mathbf{e}_1 = [\,0 \ 1\,] \begin{bmatrix} 1 \\ 0 \end{bmatrix} = 0 \times 1 + 1 \times 0 = 0 \tag{2.75c}$$

$$\mathbf{e}_2^T \mathbf{e}_2 = [\,0 \ 1\,] \begin{bmatrix} 0 \\ 1 \end{bmatrix} = 0 \times 0 + 1 \times 1 = 1. \tag{2.75d}$$

The inner product symbol $< e_i | e_j >$ is an abbreviation of $< e_i || e_j >$. This notation mimics Dirac bracket notation in abstract Hilbert space, to be explored in this chapter in greater detail. The symbols $|e_1 >$ and $|e_2 >$ stand for the column vectors $\mathbf{e}_1$ and $\mathbf{e}_2$, respectively. On the other hand, the symbols $< e_1 |$ and $< e_2 |$ represent the row vectors $\mathbf{e}_1^T$ and $\mathbf{e}_2^T$, respectively. Thereby, e.g., e_1 in $|e_1 >$ or even 1 in $|1 >$ suffices to indicate $\mathbf{e}_1$. Similarly, e.g., e_1 in $< e_1 |$ or even 1 in $< 1|$ suffices to indicate $\mathbf{e}_1^T$. Dirac referred to $|a >$ as the ket a, and to the corresponding $< a|$ as the bra a. Note that due to $\mathbf{a} \cdot \mathbf{b} = \mathbf{a}^T \mathbf{b} = < a|b >$, the angled bra-ket $< a|b >$ denotes the inner product of $|a >$ and $|b >$. The bra $< a|$ is the dual counterpart of the ket $|a >$. The bra $< a|$ is also referred to as the dual conjugate of the ket $|a >$. The latter convention alludes to the fact that if the components of $|a >$ are complex, then the components of $< a|$ are complex conjugate counterparts of the components of $|a >$. To emphasis this decisive content, consider, the 3×1 ket $|a >$ with complex-valued components. Let the nth component $a_n \in \mathbb{C}$ ($n = 1, 2, 3$) have the real part a_n^r and the imaginary part a_n^i. Then,

$$|a > = \begin{bmatrix} a_1^r + ia_1^i \\ a_2^r + ia_2^i \\ a_3^r + ia_3^i \end{bmatrix} \quad \Longleftrightarrow \quad < a| = \begin{bmatrix} a_1^r - ia_1^i & a_2^r - ia_2^i & a_3^r - ia_3^i \end{bmatrix}. \tag{2.76}$$

For completeness, let $|0>$ refer to the ket with all its components being zero. The utility of the bra–ket notation manifests itself already in toy models, as shown next.

Problem: Consider the ket $|a>$, defined in (2.76). Assume that not all of the terms a_1^r, a_1^i, a_2^r, a_2^i, a_3^r, and a_3^i are zero. Stated differently, assume that at least one of the preceding elements is nonzero. Or, equivalently, assume $|a> \neq |0>$. Build the inner product $< a|a >$.

Solution:

$$< a|a > = \left[\, \left(a_1^r + ia_1^i\right)^* \quad \left(a_2^r + ia_2^i\right)^* \quad \left(a_3^r + ia_3^i\right)^* \, \right] \begin{bmatrix} a_1^r + ia_1^i \\ a_2^r + ia_2^i \\ a_3^r + ia_3^i \end{bmatrix} \tag{2.77a}$$

$$= \left[\, a_1^r - ia_1^i \quad a_2^r - ia_2^i \quad a_3^r - ia_3^i \, \right] \begin{bmatrix} a_1^r + ia_1^i \\ a_2^r + ia_2^i \\ a_3^r + ia_3^i \end{bmatrix} \tag{2.77b}$$

$$= \underbrace{(a_1^r - ia_1^i)(a_1^r + ia_1^i)} + \underbrace{(a_2^r - ia_2^i)(a_2^r + ia_2^i)} + \underbrace{(a_3^r - ia_3^i)(a_3^r + ia_3^i)} \tag{2.77c}$$

$$= \Big[\underbrace{(a_1^r)^2}_{\geq 0} + \underbrace{(a_1^i)^2}_{\geq 0} \Big] + \Big[\underbrace{(a_2^r)^2}_{\geq 0} + \underbrace{(a_2^i)^2}_{\geq 0} \Big] + \Big[\underbrace{(a_3^r)^2}_{\geq 0} + \underbrace{(a_3^i)^2}_{\geq 0} \Big]. \tag{2.77d}$$

$$\underbrace{}_{\geq 0} \quad \underbrace{}_{\geq 0} \quad \underbrace{}_{\geq 0}$$

$$\underbrace{}_{>0 \text{ since } |a> \neq |0>}$$

Observe that the individual terms in (2.77c) are ≥ 0. However, not all six terms are allowed to be simultaneously zero, since $|a> \neq |0>$. At least one of the real- or imaginary parts must be non-zero to satisfy the condition $|a> \neq |0>$. ∎

The above solution shows that the ideas developed here are valid irrespective of the components of the kets being real- or complex-valued. Furthermore, it should be pointed out that exclusively all of the results are obtained from processing the components of the vectors, by following a recipe. Consequently, to ease the discussion, real-valued vectors will be considered in the following. Resort to complex-valued vectors will only be made if further clarification can be achieved.

Problem: Consider the ket $|a> \neq |0>$, $(a_1, a_2, a_3 \in \mathbb{R}$, with $|a_1| + |a_2| + |a_3| \neq 0)$, along with its dual bra $< a|$,

$$|a> = \begin{bmatrix} a_1 \\ a_2 \\ a_3 \end{bmatrix} \qquad \Longleftrightarrow \qquad < a| = \left[\, a_1 \quad a_2 \quad a_3 \, \right]. \tag{2.78}$$

Build the inner product $< a|a >$. Determine the length (the norm) of $|a>$.

Solution:

$$< a|a > = \begin{bmatrix} a_1 & a_2 & a_3 \end{bmatrix} \begin{bmatrix} a_1 \\ a_2 \\ a_3 \end{bmatrix} \tag{2.79a}$$

$$= a_1^2 + a_2^2 + a_3^2 > 0. \tag{2.79b}$$

The conclusion $a_1^2 + a_2^2 + a_3^2 > 0$ follows from $|a_1| + |a_2| + |a_3| \neq 0$, which is a formal way of saying $|a> \neq |0>$. In the Euclidean coordinate system, the term $a_1^2 + a_2^2 + a_3^2$ is equal to the magnitude squared (the length squared) of the vector $\mathbf{a} = (a_1, a_2, a_3)$. Consequently, it seems reasonable to let $a_1^2 + a_2^2 + a_3^2$ which is equal to $< a|a >$ stand for $||a>|^2$, the length squared of $|a>$. Or, perhaps in a slightly fancier fashion, for $|||a>||^2$, the norm squared of $|a>$. Thus,

$$a_1^2 + a_2^2 + a_3^2 = < a|a > = |||a>||^2 \quad \Longrightarrow \quad |||a>|| = \sqrt{< a|a >}. \tag{2.80}$$

Thus, to determine the norm of $|a>$, first build $< a|a >$, and then take the square root of $< a|a >$.

The inner product $< a|b >$ of the kets $|a>$ and $|b>$ serves to a further significant idea. It is a measure of the similarity between $|a>$ and $|b>$. If $< a|b > = 0$, $|a>$ and $|b>$ are said to orthogonal, to have no similarity at all. Provided the kets $|a>$ and $|b>$ have unit norm (length), i.e., $|||a>|| = 1$ and $|||b>|| = 1$, the condition $< a|b > = 1$ conveys the fact that they are identical.

The following problem summarizes these ideas.

◼

Problem: Given the unit kets $|e_1>$ and $|e_2>$,

$$|e_1> = \begin{bmatrix} 1 \\ 0 \end{bmatrix}, \qquad |e_2> = \begin{bmatrix} 0 \\ 1 \end{bmatrix}, \tag{2.81}$$

build $< e_i|e_j > (i, j = 1, 2)$.

Solution:

$$< e_1|e_1 > = \begin{bmatrix} 1 & 0 \end{bmatrix} \begin{bmatrix} 1 \\ 0 \end{bmatrix} = 1 \qquad < e_1|e_2 > = \begin{bmatrix} 1 & 0 \end{bmatrix} \begin{bmatrix} 0 \\ 1 \end{bmatrix} = 0$$

$$\tag{2.82}$$

$$< e_2|e_1 > = \begin{bmatrix} 0 & 1 \end{bmatrix} \begin{bmatrix} 1 \\ 0 \end{bmatrix} = 0 \qquad < e_2|e_2 > = \begin{bmatrix} 0 & 1 \end{bmatrix} \begin{bmatrix} 0 \\ 1 \end{bmatrix} = 1.$$

Utilizing the Kronecker delta symbol, these relationships can be written compactly,

$$< e_i|e_j > = \delta_{ij} \qquad i, j = 1, 2. \tag{2.83}$$

◼

Remark: Generalizations.

- Bases in higher dimensions: Obviously, there is no reason to limit i, j to 1 and 2. Considering any set of $N \in \mathbb{N}$ unit vectors $\mathbf{e}_n \in \mathbb{R}^N$, the corresponding canonical kets are orthonormal,

$$< e_i|e_j > = \delta_{ij} \qquad i, j = 1, \cdots, N, \tag{2.84}$$

  and build a basis.

- Violating the normality requirement: There is even no reason to limit the set of vectors to be unit vectors, to constitute a basis. Any orthogonal ("perpendicular" and not necessarily normal) set of $N \in \mathbb{N}$ vectors $\mathbf{b}_n \in \mathbb{R}^N$,

$$< b_i | b_j > \quad = \quad \delta_{ij} \big| \| b_i > \|^2 \qquad i,j = 1, \cdots, N, \tag{2.85}$$

with $\big| \| b_i > \|^2$ being not necessarily equal to unity, constitute a basis.
- Violating the orthogonality requirement while keeping the linear independence:
  As will be demonstrated in greater length, even the condition of orthogonality of $N \in \mathbb{N}$ vectors $\mathbf{f}_n \in \mathbb{R}^N$, can be relaxed, as long as the vectors are linearly independent. Such an instance leads to the notion of bases and the construction of the associated dual bases are required.
- Finally, by considering $N_1 \in \mathbb{N}$ vectors $\mathbf{f}_n \in \mathbb{R}^N$, with $N_1 > N$, the condition of over-completeness will be permitted (relaxing the condition of linear independence). Such an instance leads to the notion of frames and requires the construction of the associated dual frames.

Many examples will shed light onto these powerful concepts which fully deserve a prominent place in computational engineering.

□

From the above, any set of orthonormal kets satisfying (2.84) is said to constitute a basis for $\mathbb{R}^N$, an ortho-normal basis (ONB). The discussion continues now by considering kets in two dimensions. In $\mathbb{R}^2$, e.g., $|e_1 >$ and $|e_2 >$ constitute a basis. This implies that any ket $|a > \in \mathbb{R}^2$ can be synthesized from $|e_1 >$ and $|e_2 >$, i.e., it can be written as linear combination of $|e_1 >$ and $|e_2 >$,

$$|a > \quad = \quad a_1 |e_1 > + a_2 |e_2 >. \tag{2.86}$$

For reasons which becomes clear momentarily, (2.86) is preferably cast in the form,

$$|a > \quad = \quad |e_1 > a_1 + |e_2 > a_2 \qquad \text{(synthesis formula)}. \tag{2.87}$$

The question is the determination of the coefficients a_1 and a_2. To this end, build the inner product of $|a >$ with $|e_1 >$ and subsequently with $|e_2 >$, using (2.87),

$$< e_1 | a > \quad = \quad < e_1 | \big\{ |e_1 > a_1 + |e_2 > a_2 \big\} \tag{2.88a}$$

$$= \quad \underbrace{< e_1 | e_1 >}_{=1} a_1 + \underbrace{< e_1 | e_2 >}_{=0} a_2 \tag{2.88b}$$

$$= \quad a_1 \qquad \text{(analysis formula)}, \tag{2.88c}$$

and,

$$< e_2 | a > \quad = \quad < e_2 | \big\{ |e_1 > a_1 + |e_2 > a_2 \big\} \tag{2.89a}$$

$$= \quad \underbrace{< e_2 | e_1 >}_{=0} a_1 + \underbrace{< e_2 | e_2 >}_{=1} a_2 \tag{2.89b}$$

$$= \quad a_2 \qquad \text{(analysis formula)}. \tag{2.89c}$$

Substituting the resulting,

$$a_1 \quad = \quad < e_1 | a > \tag{2.90a}$$

$$a_2 \quad = \quad < e_2 | a >, \tag{2.90b}$$

into (2.87),

$$|a> \ = \ |e_1><e_1|a> +|e_2><e_2|a>. \tag{2.91}$$

Recall the brackets $<e_1|a>$ and $<e_2|a>$ were abbreviations of $<e_1||a>$ and $<e_1||a>$, respectively. Thus, (2.91) can be written as,

$$|a> \ = \ |e_1><e_1||a> +|e_2><e_2||a>. \tag{2.92}$$

The ket $|a>$ at the R.H.S. can be factored out,

$$|a> \ = \ \big(|e_1><e_1| + |e_2><e_2|\big)|a>. \tag{2.93}$$

Since the choice of the test ket $|a>$ has been arbitrary, and the addition of the two terms in the round bracket is not dependent on the choice of $|a>$, the sum of the two terms must be a universal entity, and merely dependent on the choice of $|e_1>$ and $|e_2>$. These considerations suggest examining the structure of $|e_1><e_1| + |e_2><e_2|$.

Problem: Given the unit kets $|e_1>$ and $|e_2>$ in $\mathbb{R}^2$, examine the structure of the sum of the exterior products $|e_1><e_1|$ and $|e_2><e_2|$.

Solution: Simple calculation reveals that,

$$|e_1><e_1| + |e_2><e_2| = \begin{bmatrix} 1 \\ 0 \end{bmatrix}\begin{bmatrix} 1 & 0 \end{bmatrix} + \begin{bmatrix} 0 \\ 1 \end{bmatrix}\begin{bmatrix} 0 & 1 \end{bmatrix} \tag{2.94a}$$

$$= \begin{bmatrix} 1 & 0 \\ 0 & 0 \end{bmatrix} + \begin{bmatrix} 0 & 0 \\ 0 & 1 \end{bmatrix} \tag{2.94b}$$

$$= \begin{bmatrix} 1 & 0 \\ 0 & 1 \end{bmatrix} \tag{2.94c}$$

$$= \mathbb{I}_{2\times 2}. \tag{2.94d}$$

Whenever it is clear from the context, subscripts of the form $N \times N$ will be omitted. Consequently,

$$<e_i|e_j> = \delta_{ij} \ \ (i,j=1,2) \implies \sum_{n=1}^{2} |e_n><e_n| = \mathbb{I}_{2\times 2}. \tag{2.95}$$

■

As it can readily be demonstrated, there is no reason to limit the discussion to $\mathbb{R}^2$. Assuming the orthonormal canonical kets $|e_n>$, representing $\mathbf{e}_n \in \mathbb{R}^3$ ($n=1,2,3$), it is immediate that,

$$<e_i|e_j> = \delta_{ij} \ \ (i,j=1,2,3) \implies \sum_{n=1}^{3} |e_n><e_n| = \mathbb{I}_{3\times 3}. \tag{2.96}$$

Or, even for any $N \in \mathbb{N}$, with the orthonormal canonical kets $|e_n>$, representing $\mathbf{e}_n \in \mathbb{R}^N$ ($n=1,\cdots,N$),

$$<e_i|e_j> = \delta_{ij} \ \ (i,j=1,\cdots,N) \implies \sum_{n=1}^{N} |e_n><e_n| = \mathbb{I}_{N\times N}. \tag{2.97}$$

Remark: Preliminary summary.

Let the kets $|e_n>$ ($n=1,\cdots,N$) and $N \in \mathbb{N}$, constitute an ONB. Let $<e_n|$ ($n=1,\cdots,N$) denote dual bras associated with $|e_n>$. Then, the following results hold valid.

- Orthonormality (upon assumption),

$$< e_i|e_j >= \delta_{ij} \quad (i,j = 1, \cdots , N) \qquad \text{(orthonormality condition)} \tag{2.98}$$

- Implies the existence and validity of the resolution of identity (completeness).

$$\sum_{n=1}^{N} |e_n >< e_n| = \mathbb{I}_{N \times N} \qquad \text{(resolution of identity)}. \tag{2.99}$$

- The resolution of identity (2.99) embodies both the analysis step and the synthesis step. To demonstrate this, apply $\mathbb{I}_{N \times N}$ onto an arbitrary test ket $|a >\in \mathbb{R}^N$:

The analysis step:

$$\mathbb{I}|a >= |a >= \sum_{n=1}^{N} |e_n > \underbrace{< e_n|a >}_{a_n}. \tag{2.100}$$

The analysis step consists of determining the coefficients a_n.

The synthesis step:

$$|a >= \sum_{n=1}^{N} a_n|e_n>. \tag{2.101}$$

The synthesis step comprises the employment of a_n to reconstruct $|a >$.

$\square$

At this stage, the discussion of the Dirac delta function initiated at the beginning of the chapter could be continued and consequences thereof investigated. However, the insights gained from elementary toy models are so far-reaching that it might be a good idea to explore the implications of relaxing the orthonormality condition. This is discussed in the next section.

2.3 Non-orthogonal bases and dual bases

In the preceding case, the basis vectors $\mathbf{e}_1$ and $\mathbf{e}_2$ were taken to be orthonormal. The condition of normality $< e_i|e_i >= 1$ $(i = 1, 2)$ is not substantial: (nonzero) non-normal vectors can always be rendered normal, by dividing them by their respective magnitudes. The orthogonality condition $< e_i|e_j >= 0$ $(i,j = 1, 2, i \neq j)$ is, however, crucial. In this section, the consequences of relaxing the orthogonality condition are investigated. As an example, consider the (non-collinear) non-orthogonal basis vectors $\mathbf{b}_1$ and $\mathbf{b}_2$,

$$|b_1 >= \mathbf{b}_1 = \begin{bmatrix} 3 \\ 0 \end{bmatrix}, \quad |b_2 >= \mathbf{b}_2 = \begin{bmatrix} 1 \\ 2 \end{bmatrix}, \tag{2.102}$$

along with a test vector $\mathbf{a}$ in the Euclidean space $\mathbb{R}^2$.

The resolution of identity: Adding the exterior products $|b_1 >< b_1|$ and $|b_2 >< b_2|$,

$$|b_1 >< b_1| + |b_2 >< b_2| = \begin{bmatrix} 3 \\ 0 \end{bmatrix} \begin{bmatrix} 3 & 0 \end{bmatrix} + \begin{bmatrix} 1 \\ 2 \end{bmatrix} \begin{bmatrix} 1 & 2 \end{bmatrix} \tag{2.103a}$$

$$= \begin{bmatrix} 9 & 0 \\ 0 & 0 \end{bmatrix} + \begin{bmatrix} 1 & 2 \\ 2 & 4 \end{bmatrix} \tag{2.103b}$$

$$= \begin{bmatrix} 10 & 2 \\ 2 & 4 \end{bmatrix}. \tag{2.103c}$$

Denoting the resulting matrix by $\mathbb{S}_{2\times2}$,

$$\mathbb{S}_{2\times2} = |b_1><b_1| + |b_2><b_2|. \tag{2.104}$$

Thus, relaxing the orthonormality condition of the basis vectors $\mathbf{b}_1$ and $\mathbf{b}_2$ implies that the sum of the exterior products $|b_1><b_1| + |b_2><b_2| = \mathbb{S}_{2\times2} \neq \mathbb{I}_{2\times2}$. As long as the chosen basis vectors $\mathbf{b}_1$ and $\mathbf{b}_2$ are not collinear, the inverse $\mathbb{S}_{2\times2}^{-1}$ exists. In the current case,

$$\mathbb{S}_{2\times2}^{-1} = \frac{1}{36}\begin{bmatrix} 4 & -2 \\ -2 & 10 \end{bmatrix} = \frac{1}{18}\begin{bmatrix} 2 & -1 \\ -1 & 5 \end{bmatrix}. \tag{2.105}$$

Multiplying both sides of (2.104) from the left by $\mathbb{S}_{2\times2}^{-1}$

$$\mathbb{I}_{2\times2} = \mathbb{S}_{2\times2}^{-1}\mathbb{S}_{2\times2} = \underbrace{\mathbb{S}_{2\times2}^{-1}|b_1>}_{|\tilde{b}_1>}<b_1| + \underbrace{\mathbb{S}_{2\times2}^{-1}|b_2>}_{|\tilde{b}_2>}<b_2|, \tag{2.106}$$

where the dual basis Kets $|\tilde{b}_1>$ and $|\tilde{b}_2>$ manifest themselves, as indicated,

$$|\tilde{b}_1> = \mathbb{S}_{2\times2}^{-1}|b_1> \tag{2.107a}$$

$$|\tilde{b}_2> = \mathbb{S}_{2\times2}^{-1}|b_2>. \tag{2.107b}$$

Or, more explicitly,

$$|\tilde{b}_1> = \frac{1}{18}\begin{bmatrix} 2 & -1 \\ -1 & 5 \end{bmatrix}\begin{bmatrix} 3 \\ 0 \end{bmatrix} = \frac{1}{6}\begin{bmatrix} 2 \\ -1 \end{bmatrix} \tag{2.108a}$$

$$|\tilde{b}_2> = \frac{1}{18}\begin{bmatrix} 2 & -1 \\ -1 & 5 \end{bmatrix}\begin{bmatrix} 1 \\ 2 \end{bmatrix} = \frac{1}{2}\begin{bmatrix} 0 \\ 1 \end{bmatrix}. \tag{2.108b}$$

Thus, given the non-orthonormal kets $|b_1>$ and $|b_2>$, for resolving the identity matrix $\mathbb{I}_{2\times2}$, the dual kets $|\tilde{b}_1>$ and $|\tilde{b}_2>$ must be constructed first, leading to (2.106), i.e.,

$$\mathbb{I}_{2\times2} = |\tilde{b}_1><b_1| + |\tilde{b}_2><b_2|. \tag{2.109}$$

Carrying out the exterior products and adding the resulting matrices is revealing,

$$|\tilde{b}_1><b_1| + |\tilde{b}_2><b_2| = \frac{1}{6}\begin{bmatrix} 2 \\ -1 \end{bmatrix}[3\ 0] + \frac{1}{2}\begin{bmatrix} 0 \\ 1 \end{bmatrix}[1\ 2] \tag{2.110a}$$

$$= \frac{1}{6}\begin{bmatrix} 6 & 0 \\ -3 & 0 \end{bmatrix} + \frac{1}{2}\begin{bmatrix} 0 & 0 \\ 1 & 2 \end{bmatrix} \tag{2.110b}$$

$$= \frac{1}{2}\begin{bmatrix} 2 & 0 \\ -1 & 0 \end{bmatrix} + \frac{1}{2}\begin{bmatrix} 0 & 0 \\ 1 & 2 \end{bmatrix} \tag{2.110c}$$

$$= \frac{1}{2}\begin{bmatrix} 2 & 0 \\ 0 & 2 \end{bmatrix} \tag{2.110d}$$

$$= \begin{bmatrix} 1 & 0 \\ 0 & 1 \end{bmatrix}. \tag{2.110e}$$

It is illuminating to observe the above intricate interplay of terms, despite the fact that dual kets $|\tilde{b}_1>$ and $|\tilde{b}_2>$ were constructed (designed) to yield this result in the first place. The underlying ideas should be reinforced: the ability to establish the relationship (2.109) is tantamount to having established a transform and the corresponding inverse transform, as readily demonstrated next.

The analysis step: Applying both sides of (2.109) onto the test ket $|a>$,

$$\mathbb{I}_{2\times 2}|a> = |\tilde{b}_1> \underbrace{<b_1|a>}_{=3a_1} + |\tilde{b}_2> \underbrace{<b_2|a>}_{=a_1+2a_2} . \tag{2.111}$$

The determination of the coefficients $3a_1$ and $a_1 + 2a_2$, via the inner products,

$$<b_1|a> = \mathbf{b}_1 \cdot \mathbf{a} = \mathbf{b}_1^T \mathbf{a} = \begin{bmatrix} 3 & 0 \end{bmatrix} \begin{bmatrix} a_1 \\ a_2 \end{bmatrix} = 3a_1 \tag{2.112a}$$

$$<b_2|a> = \mathbf{b}_2 \cdot \mathbf{a} = \mathbf{b}_2^T \mathbf{a} = \begin{bmatrix} 1 & 2 \end{bmatrix} \begin{bmatrix} a_1 \\ a_2 \end{bmatrix} = a_1 + 2a_2 , \tag{2.112b}$$

is referred to as the analysis step.

The synthesis step: Substituting (2.112) into (2.111),

$$|a> = (3a_1)|\tilde{b}_1> + (a_1 + 2a_2)|\tilde{b}_2>, \tag{2.113}$$

comprises the synthesis step. The fact that (2.113) is an identity can be readily examined:

$$(3a_1)|\tilde{b}_1> + (a_1 + 2a_2)|\tilde{b}_2> = (3a_1)\frac{1}{6}\begin{bmatrix} 2 \\ -1 \end{bmatrix} + (a_1 + 2a_2)\frac{1}{2}\begin{bmatrix} 0 \\ 1 \end{bmatrix} \tag{2.114a}$$

$$= \frac{a_1}{2}\begin{bmatrix} 2 \\ -1 \end{bmatrix} + \left(\frac{a_1}{2} + a_2\right)\begin{bmatrix} 0 \\ 1 \end{bmatrix} \tag{2.114b}$$

$$= \begin{bmatrix} a_1 \\ -\frac{a_1}{2} \end{bmatrix} + \begin{bmatrix} 0 \\ \frac{a_1}{2} + a_2 \end{bmatrix} \tag{2.114c}$$

$$= \begin{bmatrix} a_1 \\ a_2 \end{bmatrix} \tag{2.114d}$$

$$= |a> . \tag{2.114e}$$

Summary: Equations (2.112) constitute the analysis step, while (2.113) is referred to the synthesis step. The kets $|b_1>$ and $|b_2>$ are called the basis vectors and their associated dual basis vectors $|\tilde{b}_1>$ and $|\tilde{b}_2>$ needed to be constructed. The dual vectors were not simply the bras $<b_1|$ and $<b_2|$, the transposed counterparts of $|b_1>$ and $|b_2>$, respectively. Writing (2.113) in the standard form,

$$\mathbf{a} = (3a_1)\tilde{\mathbf{b}}_1 + (a_1 + 2a_2)\tilde{\mathbf{b}}_2. \tag{2.115}$$

might be illuminating as well.

$\square$

Problem: Consider the kets $|b_1>$ and $|b_2>$ in the above discussion,

$$|b_1> = \begin{bmatrix} 3 \\ 0 \end{bmatrix}, \quad |b_2> = \begin{bmatrix} 1 \\ 2 \end{bmatrix}. \tag{2.116}$$

Consider the constructed associated dual kets,

$$|\tilde{b}_1> = \frac{1}{6}\begin{bmatrix} 2 \\ -1 \end{bmatrix}, \quad |\tilde{b}_2> = \frac{1}{2}\begin{bmatrix} 0 \\ 1 \end{bmatrix} . \tag{2.117}$$

By direct calculation show that the dual kets $|\tilde{b}_1>$ and $|\tilde{b}_2>$ are mutually orthogonal to the kets $|b_1>$ and $|b_2>$.

Solution: Build the following inner products,

$$< \tilde{b}_1 | b_1 > \; = \; \frac{1}{6} \begin{bmatrix} 2 & -1 \end{bmatrix} \begin{bmatrix} 3 \\ 0 \end{bmatrix} = \frac{1}{6} [(2)(3) + (-1)(0)] = 1 \tag{2.118a}$$

$$< \tilde{b}_1 | b_2 > \; = \; \frac{1}{6} \begin{bmatrix} 2 & -1 \end{bmatrix} \begin{bmatrix} 1 \\ 2 \end{bmatrix} = \frac{1}{6} [(2)(1) + (-1)(2)] = 0 \tag{2.118b}$$

$$< \tilde{b}_2 | b_1 > \; = \; \frac{1}{2} \begin{bmatrix} 0 & 1 \end{bmatrix} \begin{bmatrix} 3 \\ 0 \end{bmatrix} = \frac{1}{2} [(0)(3) + (1)(0)] = 0 \tag{2.118c}$$

$$< \tilde{b}_2 | b_2 > \; = \; \frac{1}{2} \begin{bmatrix} 0 & 1 \end{bmatrix} \begin{bmatrix} 1 \\ 2 \end{bmatrix} = \frac{1}{2} [(0)(1) + (1)(2)] = 1. \tag{2.118d}$$

Thus,

$$< \tilde{b}_i | b_j > = \delta_{ij} , \tag{2.119}$$

with δ_{ij} being the Kronecker delta symbol.

∎

Remark: From the above problem and the preceding text, the following can be concluded.

- The given non-collinear kets $|b_1>$ and $|b_2>$ are neither normal ($< b_1|b_1 > = 9 \neq 1$, $< b_2|b_2 > = 5 \neq 1$) nor mutually orthogonal ($< b_1|b_2 > = < b_2|b_1 > = 3 \neq 0$).
- The constructed dual Kets $|\tilde{b}_1>$ and $|\tilde{b}_2>$ are neither normal ($< \tilde{b}_1|\tilde{b}_1 > = 5/36 \neq 1$, $< \tilde{b}_2|\tilde{b}_2 > = 1/4 \neq 1$) nor mutually orthogonal ($< \tilde{b}_1|\tilde{b}_2 > = < \tilde{b}_2|\tilde{b}_1 > = -1/12 \neq 0$).
- The given Kets $|b_1>$ and $|b_2>$ and the designed dual kets $|\tilde{b}_1>$ and $|\tilde{b}_2>$ are not only mutually orthonormal

$$< \tilde{b}_i | b_j > = \delta_{ij} , \qquad i,j = 1,2, \tag{2.120}$$

but also satisfy the resolution of identity,

$$\sum_{n=1}^{2} |\tilde{b}_n > < b_n| = \sum_{n=1}^{2} |b_n > < \tilde{b}_n| = \mathbb{I}_{2 \times 2}. \tag{2.121}$$

That these desirable properties can be achieved under much more stringent conditions is discussed in the next section.

□

The following problem should help solidifying the concept.

Problem: Consider the non-normal, non-orthogonal, and non-collinear kets $|b_1>$ and $|b_2>$,

$$|b_1> = \begin{bmatrix} 3 \\ 2 \end{bmatrix}, \quad |b_2> = \begin{bmatrix} 4 \\ 1 \end{bmatrix} \tag{2.122}$$

Construct the corresponding dual kets $|\tilde{b}_1>$ and $|\tilde{b}_2>$.

Solution: Determine the (symmetric) $\mathbb{S}$ operator consisting of the addition of the (symmetric) exterior products $|b_1> <b_1|$ and $|b_2> <b_2|$,

$$\mathbb{S} = |b_1> <b_1| + |b_2> <b_2| \tag{2.123a}$$

$$= \begin{bmatrix} 3 \\ -2 \end{bmatrix} [3\ -2] + \begin{bmatrix} 4 \\ 1 \end{bmatrix} [4\ 1] \tag{2.123b}$$

$$= \begin{bmatrix} 9 & -6 \\ -6 & 4 \end{bmatrix} + \begin{bmatrix} 16 & 4 \\ 4 & 1 \end{bmatrix} \tag{2.123c}$$

$$= \begin{bmatrix} 25 & -2 \\ -2 & 5 \end{bmatrix}. \tag{2.123d}$$

Note that $|b_1> <b_1|$ and $|b_2> <b_2|$ are individually symmetric. Consequently, the addition of $|b_1> <b_1|$ and $|b_2> <b_2|$, i.e., the matrix $\mathbb{S}$, is also symmetric. With the determinant of the non-singular matrix $\mathbb{S}$ being $\det\{\mathbb{S}\} = 121$, the inverse of $\mathbb{S}$ can be written down immediately,

$$\mathbb{S}^{-1} = \frac{1}{121} \begin{bmatrix} 5 & 2 \\ 2 & 25 \end{bmatrix}. \tag{2.124}$$

Multiplying (2.123a) from the left by $\mathbb{S}^{-1}$ generates the desired identity operator $\mathbb{I}_{2\times 2}$ at the left-hand side,

$$\mathbb{I} = \mathbb{S}^{-1}\mathbb{S} = \underbrace{\mathbb{S}^{-1}|b_1>}_{|\tilde{b}_1>} <b_1| + \underbrace{\mathbb{S}^{-1}|b_2>}_{|\tilde{b}_2>} <b_1| <b_2|. \tag{2.125}$$

As indicated, the calculation of the dual kets $|\tilde{b}_1>$ and $|\tilde{b}_2>$ amounts to operating $\mathbb{S}^{-1}$ onto $|b_1>$ and $|b_1>$, respectively. Thus,

$$|\tilde{b}_1> = \mathbb{S}^{-1}|b_1> = \frac{1}{121} \begin{bmatrix} 5 & 2 \\ 2 & 25 \end{bmatrix} \begin{bmatrix} 3 \\ -2 \end{bmatrix} = \frac{1}{121} \begin{bmatrix} 11 \\ -44 \end{bmatrix} = \frac{1}{11} \begin{bmatrix} 1 \\ -4 \end{bmatrix} \tag{2.126a}$$

$$|\tilde{b}_2> = \mathbb{S}^{-1}|b_2> = \frac{1}{121} \begin{bmatrix} 5 & 2 \\ 2 & 25 \end{bmatrix} \begin{bmatrix} 4 \\ 1 \end{bmatrix} = \frac{1}{121} \begin{bmatrix} 22 \\ 33 \end{bmatrix} = \frac{1}{11} \begin{bmatrix} 2 \\ 3 \end{bmatrix}. \tag{2.126b}$$

∎

The next problem demonstrates that the above set of column vectors $\{|\tilde{b}_1>, |\tilde{b}_2>\}$ do indeed constitute a dual basis corresponding to the basis $\{|b_1>, |b_2>\}$.

Problem: Show that,

$$|b_1> = \begin{bmatrix} 3 \\ -2 \end{bmatrix}, \quad |b_2> = \begin{bmatrix} 4 \\ 1 \end{bmatrix}, \tag{2.127}$$

and

$$|\tilde{b}_1> = \frac{1}{11} \begin{bmatrix} 1 \\ -4 \end{bmatrix}, \quad |\tilde{b}_2> = \frac{1}{11} \begin{bmatrix} 2 \\ 3 \end{bmatrix}, \tag{2.128}$$

are dual sets.

96 *Mathematical quantum physics for engineers and technologists*

Solution: The solution of this problems requires the demonstration of completeness (the resolution of identity) and the orthonormality.

Completeness:

$$|\tilde{b}_1><b_1| + |\tilde{b}_2><b_2| = \frac{1}{11}\begin{bmatrix}1\\-4\end{bmatrix}\begin{bmatrix}3 & -2\end{bmatrix} + \frac{1}{11}\begin{bmatrix}2\\3\end{bmatrix}\begin{bmatrix}4 & 1\end{bmatrix} \tag{2.129a}$$

$$= \frac{1}{11}\left\{\begin{bmatrix}3 & -2\\-12 & 8\end{bmatrix} + \begin{bmatrix}8 & 2\\12 & 3\end{bmatrix}\right\} \tag{2.129b}$$

$$= \frac{1}{11}\begin{bmatrix}11 & 0\\0 & 11\end{bmatrix} \tag{2.129c}$$

$$= \begin{bmatrix}1 & 0\\0 & 1\end{bmatrix} = \mathbb{I}. \tag{2.129d}$$

Note that taken individually neither $|\tilde{b}_1><b_1|$ nor $|\tilde{b}_2><b_2|$ is symmetric. However, combined, they lead to a distinctively symmetric matrix, i.e., the unity matrix.

Orthonormality:

$$<\tilde{b}_1||b_1> = \frac{1}{11}\begin{bmatrix}1 & -4\end{bmatrix}\begin{bmatrix}3\\-2\end{bmatrix} = \frac{1}{11}(3+8) = 1 \tag{2.130a}$$

$$<\tilde{b}_1||b_2> = \frac{1}{11}\begin{bmatrix}1 & -4\end{bmatrix}\begin{bmatrix}4\\1\end{bmatrix} = \frac{1}{11}(4-4) = 0 \tag{2.130b}$$

$$<\tilde{b}_2||b_1> = \frac{1}{11}\begin{bmatrix}2 & 3\end{bmatrix}\begin{bmatrix}3\\-2\end{bmatrix} = \frac{1}{11}(6-6) = 0 \tag{2.130c}$$

$$<\tilde{b}_2||b_2> = \frac{1}{11}\begin{bmatrix}2 & 3\end{bmatrix}\begin{bmatrix}4\\1\end{bmatrix} = \frac{1}{11}(8+3) = 1. \tag{2.130d}$$

For bookkeeping purposes, the values of the inner products can be arranged into a unity matrix,

$$\begin{bmatrix}<\tilde{b}_1||b_1> & <\tilde{b}_1||b_2>\\<\tilde{b}_2||b_1> & <\tilde{b}_2||b_2>\end{bmatrix} = \begin{bmatrix}1 & 0\\0 & 1\end{bmatrix}, \tag{2.131}$$

or, compactly,

$$<\tilde{b}_m||b_n> = \delta_{mn}, \tag{2.132}$$

with δ_{mn} standing for the Kronecker delta symbol.

∎

2.4   Frames and dual frames

A further relaxation of the conditions concerns the linear independence of the basis vectors. Consider the column vectors $\mathbf{f}_1$, $\mathbf{f}_2$, and $\mathbf{f}_3$, and their Ket representations,

$$|f_1> = \mathbf{f}_1 = \begin{bmatrix} 2 \\ -2 \end{bmatrix}, \quad |f_2> = \mathbf{f}_2 = \begin{bmatrix} 3 \\ 0 \end{bmatrix}, \quad |f_3> = \mathbf{f}_3 = \begin{bmatrix} 1 \\ 2 \end{bmatrix}, \tag{2.133}$$

along with a general test vector $\mathbf{a}$ in the Euclidean $\mathbb{R}^2$ space. The magnitudes of these vectors are not Unity, the vectors are not orthogonal, and they are not linearly independent. The latter condition, in particular, implies that $|f_1>$, $|f_2>$, and $|f_3>$, do not constitute a basis. They are referred to as the frame vectors, rather than the basis vectors. The task is the determination of the corresponding dual frame vectors—a task which closely parallels the procedure in the preceding section. The presented scheme is not restricted to three frame vectors. The choice of three vectors suffices to communicate the main idea without overloading the discussion with unnecessary details. The procedure applies to any finite number of frame vectors $\mathbf{f}_n \in \mathbb{R}^2$.

The resolution of identity: Add the exterior products $|f_1><f_1|$, $|f_2><f_2|$, and $|f_3><f_3|$,

$$|f_1><f_1| + |f_2><f_2| + |f_3><f_3| = \begin{bmatrix} 2 \\ -2 \end{bmatrix}[\,2\ -2\,] + \begin{bmatrix} 3 \\ 0 \end{bmatrix}[\,3\ 0\,] + \begin{bmatrix} 1 \\ 2 \end{bmatrix}[\,1\ 2\,]$$

$$= \begin{bmatrix} 4 & -4 \\ -4 & 4 \end{bmatrix} + \begin{bmatrix} 9 & 0 \\ 0 & 0 \end{bmatrix} + \begin{bmatrix} 1 & 2 \\ 2 & 4 \end{bmatrix}$$

$$= \begin{bmatrix} 14 & -2 \\ -2 & 8 \end{bmatrix}. \tag{2.134}$$

Denote the resulting matrix by $\mathbb{S}$,

$$\mathbb{S} = |f_1><f_1| + |f_2><f_2| + |f_3><f_3| \tag{2.135a}$$

$$= \begin{bmatrix} 14 & -2 \\ -2 & 8 \end{bmatrix}. \tag{2.135b}$$

The matrix $\mathbb{S}$ is called the frame operator.

Remark: According to (2.135a), the addition of the exterior products $|f_1><f_1|$, $|f_2><f_2|$, and $|f_3><f_3|$ does not amount to $\mathbb{I}_{2\times2}$, as this was the case with orthonormal bases. In contract it is equal to $\mathbb{S}$, which is explicated in (2.135b). Obviously, in order to obtain the desired $\mathbb{I}_{2\times2}$, (2.135a) must be multiplied from the left- or the right-hand side by the inverse of the frame operator $\mathbb{S}$, which is calculated next.

□

The inverse $\mathbb{S}^{-1}$ of $\mathbb{S}$ can be obtained simply-by-inspection,

$$\mathbb{S}^{-1} = \frac{1}{108}\begin{bmatrix} 8 & 2 \\ 2 & 14 \end{bmatrix} = \frac{1}{54}\begin{bmatrix} 4 & 1 \\ 1 & 7 \end{bmatrix}. \tag{2.136}$$

Multiplying (2.135a) from the left by $\mathbb{S}^{-1}$ creates the desired identity operator $\mathbb{I}$,

$$\mathbb{I} = \mathbb{S}^{-1}\mathbb{S} = \underbrace{\mathbb{S}^{-1}|f_1>}_{=|\tilde{f}_1>}<f_1| + \underbrace{\mathbb{S}^{-1}|f_2>}_{=|\tilde{f}_2>}<f_2| + \underbrace{\mathbb{S}^{-1}|f_3>}_{=|\tilde{f}_3>}<f_3|, \tag{2.137}$$

where the dual frame vectors $|\tilde{f}_1>$, $|\tilde{f}_2>$, and $|\tilde{f}_3>$ manifest themselves, as indicated,

$$|\tilde{f}_1> = \mathbb{S}^{-1}|f_1>$$ (2.138a)

$$|\tilde{f}_2> = \mathbb{S}^{-1}|f_2>$$ (2.138b)

$$|\tilde{f}_3> = \mathbb{S}^{-1}|f_3>.$$ (2.138c)

Or, more explicitly,

$$|\tilde{f}_1> = \frac{1}{54}\begin{bmatrix} 4 & 1 \\ 1 & 7 \end{bmatrix}\begin{bmatrix} 2 \\ -2 \end{bmatrix} = \frac{1}{18}\begin{bmatrix} 2 \\ -4 \end{bmatrix}$$ (2.139a)

$$|\tilde{f}_2> = \frac{1}{54}\begin{bmatrix} 4 & 1 \\ 1 & 7 \end{bmatrix}\begin{bmatrix} 3 \\ 0 \end{bmatrix} = \frac{1}{18}\begin{bmatrix} 4 \\ 1 \end{bmatrix}$$ (2.139b)

$$|\tilde{f}_3> = \frac{1}{54}\begin{bmatrix} 4 & 1 \\ 1 & 7 \end{bmatrix}\begin{bmatrix} 1 \\ 2 \end{bmatrix} = \frac{1}{18}\begin{bmatrix} 2 \\ 5 \end{bmatrix}.$$ (2.139c)

The resolution of the identity matrix $\mathbb{I}_{2\times2}$, (2.137),

$$\mathbb{I}_{2\times2} = |\tilde{f}_1><f_1| + |\tilde{f}_2><f_2| + |\tilde{f}_3><f_3|,$$ (2.140)

must now be shown to hold valid. The verification of this result is, however, immediate,

$$|\tilde{f}_1><f_1| + |\tilde{f}_2><f_2| + |\tilde{f}_3><f_3|$$

$$= \frac{1}{18}\begin{bmatrix} 2 \\ -4 \end{bmatrix}[\,2\ -2\,] + \frac{1}{18}\begin{bmatrix} 4 \\ 1 \end{bmatrix}[\,3\ 0\,] + \frac{1}{18}\begin{bmatrix} 2 \\ 5 \end{bmatrix}[\,1\ 2\,]$$

$$= \frac{1}{18}\begin{bmatrix} 4 & -4 \\ -8 & 8 \end{bmatrix} + \frac{1}{18}\begin{bmatrix} 12 & 0 \\ 3 & 0 \end{bmatrix} + \frac{1}{18}\begin{bmatrix} 2 & 4 \\ 5 & 10 \end{bmatrix}$$

$$= \frac{1}{18}\begin{bmatrix} 18 & 0 \\ 0 & 18 \end{bmatrix} = \mathbb{I}_{2\times2}.$$ (2.141)

It is reiterated that the ability to establish the relationship (2.140) is equivalent with having established a transform and the corresponding inverse transform, as is readily demonstrated next.

Remark: The analysis- and synthesis operations:

To see the analysis- and synthesis steps in action, apply $\mathbb{I}$ onto a test vector $\mathbf{a} \in \mathbb{R}^2$,

$$|a> = \mathbb{I}|a>$$

$$= \underbrace{|\widetilde{f_1}> \overbrace{<f_1|a>}^{\text{analysis}} + |\widetilde{f_2}> \overbrace{<f_2|a>}^{\text{analysis}} + |\widetilde{f_3}> \overbrace{<f_3|a>}^{\text{analysis}}}_{\text{synthesis}}$$

$$= \frac{1}{18}\begin{bmatrix} 2 \\ -4 \end{bmatrix} \underbrace{\begin{bmatrix} 2 & -2 \end{bmatrix}\begin{bmatrix} a_1 \\ a_2 \end{bmatrix}}_{\text{analysis}} + \frac{1}{18}\begin{bmatrix} 4 \\ 1 \end{bmatrix}\underbrace{\begin{bmatrix} 3 & 0 \end{bmatrix}\begin{bmatrix} a_1 \\ a_2 \end{bmatrix}}_{\text{analysis}} + \frac{1}{18}\begin{bmatrix} 2 \\ 5 \end{bmatrix}\underbrace{\begin{bmatrix} 1 & 2 \end{bmatrix}\begin{bmatrix} a_1 \\ a_2 \end{bmatrix}}_{\text{analysis}}$$

$$= \underbrace{\frac{1}{18}\begin{bmatrix} 2 \\ -4 \end{bmatrix}(2a_1 - 2a_2) + \frac{1}{18}\begin{bmatrix} 4 \\ 1 \end{bmatrix}(3a_1) + \frac{1}{18}\begin{bmatrix} 2 \\ 5 \end{bmatrix}(a_1 + 2a_2)}_{\text{synthesis}}$$

$$= \frac{1}{18}\begin{bmatrix} 4a_1 - 4a_2 \\ -8a_1 + 8a_2 \end{bmatrix} + \frac{1}{18}\begin{bmatrix} 12a_1 \\ 3a_1 \end{bmatrix} + \frac{1}{18}\begin{bmatrix} 2a_1 + 4a_2 \\ 5a_1 + 10a_2 \end{bmatrix}$$

$$= \frac{1}{18}\begin{bmatrix} 4a_1 - \underbrace{4a_2} + 12a_1 + 2a_1 + \underbrace{4a_2} \\ -8a_1 + 8a_2 + \underbrace{3a_1} + \underbrace{5a_1} + 10a_2 \end{bmatrix}$$

$$= \frac{1}{18}\begin{bmatrix} 18a_1 \\ 18a_2 \end{bmatrix}$$

$$= \begin{bmatrix} a_1 \\ a_2 \end{bmatrix}. \tag{2.142}$$

$\square$

Summary: As shown in (2.142), the evaluation of the inner products $<f_1|a> = 2a_1 - 2a_2$, $<f_2|a> = 3a_1$, and $<f_3|a> = a_1 + 2a_2$ amounts to the analysis step. Subsequently, multiplying the inner products by the corresponding Kets $|\widetilde{f_1}>$, $|\widetilde{f_2}>$, and $|\widetilde{f_3}>$ accomplishes the synthesis step. The kets $|f_1>$, $|f_2>$, and $|f_2>$ are called the frame vector and their associated dual frame vectors $|\widetilde{f_1}>$, $|\widetilde{f_2}>$, and $|\widetilde{f_3}>$ needed to be constructed. Thereby, the dual frame vectors were not just simply the Bras $<f_1|$, $<f_2|$, and $<f_3|$ (the transposed counterparts of the $|f_1>$, $|f_2>$, and $|f_3>$, respectively). Alternatively, writing (2.142) in the standard form,

$$\mathbf{a} = (2a_1 - 2a_2)\,\widetilde{\mathbf{f}}_1 + (3a_1)\,\widetilde{\mathbf{f}}_2 + (a_1 + 2a_2)\,\widetilde{\mathbf{f}}_3. \tag{2.143}$$

$\square$

The above elementary exposition introduced the notions of bras and kets, inner- and exterior (outer) products, the resolution of identity, the analysis and synthesis steps, the orthonormal bases, the non-orthogonal basis, frames and their dual counterparts, the transforms and inverse transforms. It is satisfactory that quite intricate relationships can be made obvious and rendered intuitive in terms of vectors in $\mathbb{R}^2$. The simple discussion in the preceding sections builds the minimum foundation for delving deeper.

2.5   The abstract Hilbert space and its dual space

Prelude: The aim in this section is setting up a theater stage on which intriguing dramas promise to unfold. Only certain type of operators are admitted onto the stage—the Hermitian operators. The Hermitian operators possess distinct properties. Their eigenvalues are all positive real numbers, and their associated eigenvectors (eigenfunctions) are orthogonal (figuratively speaking, mutually perpendicular). Since any multiplication by a scalar of an eigenvector is also an eigenvector of the underlying operator, it is legitimate to assume that the eigenvectors have been scaled by their magnitudes. Thus, it is presumed that the eigenvectors are normal (their magnitude is unity). The components of the eigenvectors are in general $\in \mathbb{C}$. The number of the eigenvectors can be finite ($\mathbf{v}_n$, $n \in 1, \cdots, N$) with N being any positive integer, countably infinite ($\mathbf{v}_n$, $n \in \mathbb{N}$), or uncountably infinite, ($\mathbf{v}(k)$, $k \in \mathbb{R}$).

Not surprisingly, various Hermitian operators have their own distinguishing names which are inherited by their corresponding eigenpairs (eigenvalues and the associated eigenvectors). Let $\hat{\mathbf{V}}$ denote a Hermitian operator. Let

$$\mathbf{v}_n \iff v_n, \ (n \in \mathbb{N}). \tag{2.144}$$

be the eigenpairs of $\hat{\mathbf{V}}$. Observe that $\mathbf{v}_n$ and v_n are dutifully carrying their names ("$\mathbf{v}$" and "v", respectively) from the name of the originating Hermitian operator $\hat{\mathbf{V}}$ (i.e., "$\mathbf{V}$"). In order to conduct a dialogue, in addition to the naming, a grammar is needed. Assume the eigenvectors $\mathbf{v}_n$ are column vectors. To designate them as column vectors, put their names within the symbols "|" and ">". It suffices to insert the name of the associated eigenvalue, i.e., v_n between "|" and ">" to obtain the symbol $|v_n>$. The symbol $|v_n>$ serves as a representation of the eigenvector $\mathbf{v}_n$ simultaneously conveying the idea that $\mathbf{v}_n$ is a column vector. It is said that the Hermitian operator $\hat{\mathbf{V}}$ possesses the eigenpairs,

$$|v_n> \iff v_n, \ (n \in \mathbb{N}). \tag{2.145}$$

In virtue of $\hat{\mathbf{V}}$ being a Hermitian operator, (i) the eigenvalues v_n are positive real numbers, (ii) the eigenvectors $|v_n>$ are normal, and (iii) any two distinct eigenvectors are orthogonal.

Since the components of the column vector $|v_n>$ are in general complex-valued, the idea of constructing a dual theater offers itself: each player performing in the dual theater is a row vector with a one-to-one correspondence to a column vector in the originating theater. The row vectors in the dual theater carry the same name as their cousin column vectors in the originating theater. However, the name of a row vector is sandwiched between the symbols "<" and "|." This notation should emphasize that the vectors in the dual theater are row vectors. There is one further distinction between the row- and their corresponding column vectors. The components of a row vector in the dual theater are complex conjugates of the respective components of the associated column vector in the originating theater.

These conventions have several utilitarian powers. The inner product of $|v_n>$ with itself is denoted by $<v_n||v_n>$, or more compactly by $<v_n|v_n>$, which corresponds to the magnitude-squared of the complex-valued column vector $|v_n>$. Thus, the magnitude of the complex-valued column vector $|v_n>$ can be obtained by taking the positive square root of the inner product $<v_n|v_n>$. It also makes sense to speak of the inner product of the vectors $|a>$ and $|b>$ in the originating theater by building $<a|b>$. That means: given $|a>$ and $|b>$ in the originating theater determine the dual counterpart $<a|$ in the dual theater of $|a>$ and form the inner product $<a|b>$. In particular, if $<a|b>$ is zero, it is said that $|a>$ and $|b>$ in the originating theater are orthogonal.

One more important construction suggests itself. Assume the nth column eigenvector $|v_n>$ of the Hermitian operator $\hat{\mathbf{V}}$, along with a test vector $|a>$ in the originating theater. Obviously, $|a>$ can be written as the superposition of two vectors, one parallel to $|v_n>$ and one normal (perpendicular) to $|v_n>$. The question is how to determine the two constituting parallel- and normal vectors. To answer this question, it suffices to determine the sub-vector of $|a>$ parallel to $|v_n>$.

To this end proceed as follows: consider $|v_n>$ in the originating theater and its corresponding vector $<v_n|$ in the dual theater. Build the exterior product $|v_n><v_n|$, which if viewed as a multiplication of an $N \times 1$ matrix by a $1 \times N$ matrix, is an $N \times N$ matrix. Multiplying $|v_n><v_n|$ onto $|a>$ from the left results in $|v_n><v_n|a>$. In this construction, the inner product $<v_n|a>$ is the projection of $|a>$ onto $|v_n>$, which is in general a scalar quantity $\in \mathbb{C}$. The resulting complex number multiplied by $|v_n>$, an element of the construction $|v_n><v_n|a>$, gives the desired vector parallel to $|v_n>$. On the other hand, the difference between $|a>$ and the constructed vector $|v_n><v_n|a>$ is the desired vector normal to $|v_n>$.

Remark: A unifying language.

The above disposition employs standard tools, which engineers and technologists have acquired during their formal training. The narrative is told in a slightly different language. Acquiring the proposed language not only facilitates communication with physicists and mathematicians, but also allows the practitioners of the language to express their own contents more clearly and concisely. Chances are that the engineers and technologists start viewing transforms and inverse transforms in a unifying style. Further below several examples will elaborate this concept.

□

Summary: Given a Hermitian operator $\hat{\mathbf{V}}$, the corresponding eigenpairs $\mathbf{v}_n \iff v_n$, $(n \in \mathbb{N})$ or, equivalently, $|v_n > \iff v_n$, $(n \in \mathbb{N})$ can be determined. Thereby, $v_n \in \mathbb{R}^+$ and $< v_m|v_n >= \delta_{m,n}$, with $\delta_{m,n}$ referring to the Kronecker delta symbol. That means $< v_n|v_n >= 1$ (the eigenvectors are normalized) and $< v_m|v_n >= 0$, $m \neq n$, any two different eigenvectors $|v_m >$ and $|v_n >$ are orthogonal. Most likely, the pinnacle of the above discussion is the realization that the term $|v_n >< v_n|a >$ expresses a vector, which is the projection of $|a >$ onto the vector $|v_n >$ and is oriented along the direction of $|v_n >$. Thus, it should not come as a surprise that when all the vectors $|v_n >< v_n|a > (n \in \mathbb{N})$ are added together, the vector $|a >$ is reproduced:

$$\sum_n |v_n >< v_n|a >= |a>. \tag{2.146}$$

Remember that the abbreviation $< v_n|a >$ stands for $< v_n||a >$. Thus the L.H.S can be written as,

$$\text{L.H.S. of (2.146)} = \sum_n |v_n >< v_n||a>. \tag{2.147}$$

In view of the fact that $|a >$ is independent of the sum index n,

$$\text{L.H.S. of (2.146)} = \left(\sum_n |v_n >< v_n| \right) |a>. \tag{2.148}$$

On the other hand, the R.H.S $|a >$ can be written as,

$$\text{R.H.S. of (2.146)} = \hat{\mathbb{I}}|a>, \tag{2.149}$$

with $\hat{\mathbb{I}}$ referring to the identity operator. From (2.148) and (2.149), it can be concluded that

$$\mathbb{I} = \sum_n |v_n >< v_n|. \tag{2.150}$$

□

The following items establish a connection between the above ideas and the customary (fancier) language and notion used in quantum physics. Whenever a difficulty arises in fully grasping what is stated, the reader should resort to the above prelude. The first item explains why Hermitian operators are crucially important in quantum physics.

- A postulate in quantum physics states that every measurable physical quantity can be associated with a corresponding Hermitian operator. This is the reason why Hermitian operators featured so prominently in the prelude. Recall Hermitian operators possess several desirable utilitarian properties. Two properties stand out: (1) the eigenvalues of Hermitian operators are positive real number, and (2) their corresponding column eigenvectors are mutually orthogonal. Exploiting the properties of Hermitian operators, a world view can be created which makes sense not only technically but also intuitively. The positive eigenvalues are associated with measurement results of (measurable) physical quantities. The orthogonal column eigenvectors referred to as the kets build an ONB. The corresponding dual row vectors referred to as the bras allow obtaining fundamental formulas for the decomposition of unity. Stated more clearly, the exterior product of the kets (column vectors) and their associated bras (row vectors) allow expressing the identity operator.

- Earlier in this chapter the discussion of the Dirac delta function was temporarily put on hold. Thereby, (2.72) was analyzed exhaustively, which is reproduced here for greater convenience,

$$\delta(x - x') = \int_{-\infty}^{\infty} dk \left\{ \frac{e^{jkx}}{\sqrt{2\pi}} \right\} \left\{ \frac{e^{-jkx'}}{\sqrt{2\pi}} \right\} \qquad \text{1. Property.} \qquad (2.151)$$

The reason for promoting this particular integral representation of $\delta(x - x')$ as the first Property (first Fact or the first Axiom) will manifest itself momentarily.

- The next step is a crucial one and will be discussed in greater length throughout the text. Here is the first attempt to explain it. The theater stage on which the drama unfolds will be henceforth referred to as the abstract Hilbert space. Put simply, take the pretentious expression abstract Hilbert space as the mere name of the above-introduced theater which will continue to offer its stage for performances to come. The abstract Hilbert space is a linear space. Alternatively, it is said to be a vector space. Both designations linear space and vector space refer to the same space (a collection of entities) possessing certain desirable properties. The linearity label refers to the property that the linear combination of two elements (entities) in the space remains in the space. The vector designation indicates that the elements (entities) of the space can be associated with vectors, having a finite or infinite number of components, as the case maybe. The inhabitants of the abstract Hilbert space are taken to be column vectors. The column vectors in the abstract Hilbert space are represented in a particular manner. The characterizing name of a vector is written with the "|" symbol to the left, and the ">" symbol to the right:

|the characterizing name of the column vector >

Column vectors in the abstract Hilbert space are referred to as the Kets.

- The next idea concerns describing the ordinary function $e^{jkx}/\sqrt{2\pi}$ in an abstract Hilbert space. Thereby, the requirement of having an inner-product rule turns out to be critically important.
- Let the column vector in the abstract Hilbert space, that represents the ordinary function $e^{jkx}/\sqrt{2\pi}$, be denoted by $|k>$, i.e., characterized by the wavenumber k. Associate with the symbol $|\cdot>$ a column vector. The "$\cdot$" in $|\cdot>$ represents anything which will replace the dot to specify a particular column vector; it is the designating name of the column vector. Similarly, let the ordinary position x be denoted by the column vector $|x>$, and characterized (named) by the value of x. Any given value of x specifies a corresponding column vector $|x>$.
- Analogously, let the row vector, dual to $|x>$, be denoted by $<x|$. The dual row vectors are the inhabitants of the abstract dual Hilbert space. Here too, the abstract dual Hilbert space is merely a fancy name of the dual theater introduced further above for accommodating dual row vectors, the actors of the dual theater. Consequently, to each given value $x \in \mathbb{R}$ a column vector, a Ket $|x>$ in the abstract Hilbert space, and a corresponding row vector, a Bra $<x|$ in the abstract dual Hilbert space can be assigned. In particular, for any $x, x' \in \mathbb{R}$, the corresponding Kets $|x>$ and $|x'>$ are orthogonal,

$$<x|x'> = \delta(x - x') \qquad \text{2. Property.} \qquad (2.152)$$

- The question is now how to establish a relationship between the ordinary variable x and the ordinary function $e^{jkx}/\sqrt{2\pi}$, and the Kets $|x>$ and $|k>$ in the abstract Hilbert space. Additionally, the roles of their respective counterparts $<x|$ and $<k|$, in the abstract dual Hilbert space, must be clarified and made obvious. The answer is quite simple: the inner product of the vectors $|x>$, and $|k>$, in the abstract Hilbert space produces the ordinary function $e^{jkx}/\sqrt{2\pi}$, which engineers and technologists are quite familiar with. More succinctly,

$$<x|k> = \frac{e^{jkx}}{\sqrt{2\pi}} \qquad \text{3. Property.} \qquad (2.153)$$

- The less familiar vectors $|x>$ and $|k>$, inhabiting the abstract Hilbert space, are useful mathematical objects, as will be shown momentarily. They offer great utility.

- Since $e^{jkx}/\sqrt{2\pi} \in \mathbb{C}$, it can be inferred that the inner product $< x|k >\in \mathbb{C}$. Thus, by taking the complex conjugate of $e^{jkx}/\sqrt{2\pi}$ and thus the Hermitian conjugate of $< x|k >$, (2.153) transforms to,

$$< k|x > \;=\; \frac{e^{-jkx}}{\sqrt{2\pi}} \qquad \text{4. Property.} \qquad (2.154)$$

Remark: Summary of the Properties 1–4:

1. Property: This is an expression of the resolution of identity. The first example in the next section offers a proof.

2. Property: This is the property that the column vectors $|x >$ and $|x' >$ corresponding to the positions x and x' in real space are orthonormal in the abstract Hilbert space.

3. Property: This property states that to fix the undulating function $e^{jkx}/\sqrt{2\pi}$ a measure for *Undulation* is needed, which is k. Furthermore, it must be determined at which point x in ordinary space the function $e^{jkx}/\sqrt{2\pi}$ is considered. These two pieces of information k and x fix $|k >$ and $|x >$. The inner product of $|k >$ and $|x >$, the projection of $|k >$ onto $|x >$, i.e., $< x|k >$ fixes the ordinary function $e^{jkx}/\sqrt{2\pi}$ at the ordinary position x.

4. Property: This property is a variation of the preceding property: $< k|x >$ is just the Hermitian conjugate of $< x|k >$. Considering this property as a property of its own right simplifies the exposition.

□

In the following, several consequences will be derived from these properties.

2.6　Hermitian operators, eigenfunctions, and eigenvalues

Problem: Orthonormality of $\{|x>,\ x \in \mathbb{R}\}$ implies completeness of $\{|k>,\ k \in \mathbb{R}\}$.

Consider the ordinary integral representation of the Dirac delta function $\delta(x - x')$,

$$\delta(x - x') \;=\; \int_{-\infty}^{\infty} dk \left\{ \frac{e^{jkx}}{\sqrt{2\pi}} \right\} \left\{ \frac{e^{-jkx'}}{\sqrt{2\pi}} \right\}, \qquad (2.155)$$

with

$$\delta(x - x') =< x|x' > . \qquad (2.156)$$

Furthermore, consider the relationships

$$< x|k > \;=\; \frac{e^{jkx}}{\sqrt{2\pi}}, \qquad (2.157)$$

and

$$< k|x > \;=\; \frac{e^{-jkx}}{\sqrt{2\pi}} \quad \overset{\text{replace x with x'}}{\Longrightarrow} \quad < k|x' > \;=\; \frac{e^{-jkx'}}{\sqrt{2\pi}}. \qquad (2.158)$$

These are just the four properties established above. The task is the determination of the operator in the abstract Hilbert space which corresponds to the Dirac delta function $\delta(x - x')$.

Solution: Substitute $\delta(x - x') = < x|x' >$, $e^{jkx}/\sqrt{2\pi} = < x|k >$, and $e^{-jkx'}/\sqrt{2\pi} = < k|x' >$ into (2.155),

$$< x|x' > = \int_{-\infty}^{\infty} dk < x|k > < k|x' > . \tag{2.159}$$

Recall that $< x|x' >$, $< x|k >$, and $< k|x' >$ are abbreviations of $< x||x' >$, $< x||k >$, and $< k||x' >$, respectively. Thus,

$$< x||x' > = \int_{-\infty}^{\infty} dk < x||k > < k||x' > . \tag{2.160}$$

The Bra $< x|$ and the Ket $|x' >$ are independent of the (dummy) integration variable k. Consequently, they can be transferred to the outside of the integral,

$$< x||x' > = < x| \left\{ \int_{-\infty}^{\infty} dk|k > < k| \right\} |x' > . \tag{2.161}$$

Let $\hat{\mathbb{I}}$ represent the identity operator. Then, in virtue of $|x' > = \hat{\mathbb{I}}|x' >$,

$$< x||x' > = < x|\hat{\mathbb{I}}|x' > . \tag{2.162}$$

Comparing (2.161) with (2.162), the k-integral sandwiched between the two position vector $< x|$ and $|x' >$ in (2.161) must be identified as the identity operator $\hat{\mathbb{I}}$,

$$\hat{\mathbb{I}} = \int_{-\infty}^{\infty} dk|k > < k| \qquad \text{(completeness of } |k > \text{)}. \tag{2.163}$$

This integral is referred to the resolution of the identity operator. It is an expression of the completeness of the set of basis vectors $|k >$ with $k \in (-\infty, \infty)$. The integral at the R.H.S. of (2.163) is the desired expression of the Dirac delta function $\delta(x - x')$ in the abstract Hilbert space.

∎

Remark: The transition from (2.155) to (2.159) invites a legitimate question which the reader might ask: If $< x|k >$ and $< k|x' >$ are ordinary scalar-valued functions, their positional order does not matter. Then, what is the motivation behind writing these functions in the order $< x|k > < k|x' >$ and not in the form $< k|x' > < x|k >$? The answer is that in fact nothing prevents one from writing,

$$\delta(x - x') = \int_{-\infty}^{\infty} dk < k|x' > < x|k > \qquad \text{(even though legitimate not useful)}. \tag{2.164}$$

The representation in (2.164), in contrast to (2.159), does not allow transferring $< k|$ and $|k >$ outside the integral, since $< k|$ and $|k >$, in contrast to $< x|$ and $|x' >$, are intrinsically k-dependent. Consequently, even though (2.164) is permissible, it does not possess any utility.

□

Remark: By the definition of the identity operator $\hat{\mathbb{I}}$, the application of $\hat{\mathbb{I}}$ onto any test function $|f >$ in the abstract Hilbert space must return $|f >$,

$$\hat{\mathbb{I}}|f > \overset{\text{def}}{=} |f > . \tag{2.165}$$

It should be reiterated that the significance of the identity operator $\hat{\mathbb{I}}$ can justifiably be compared with the role of one in number theory, and the unity matrix in linear algebra. Many relationships in quantum physics can be translated into the ordinary language of engineers and technologists by employing judiciously chosen representations of the identity operator.

$\square$

Remark: The preceding problem assumed orthonormality of the position eigenvectors and arrived at the completeness of the wavenumber eigenvectors. That means, $\delta(x - x') = <x|x'>$ was assumed and $\hat{\mathbb{I}} = \int_{-\infty}^{\infty} dk|k><k|$ was inferred:

$$\delta(x - x') = <x|x'> \quad \Longrightarrow \quad \hat{\mathbb{I}} = \int_{-\infty}^{\infty} dk|k><k|. \tag{2.166}$$

Conversely, the completeness of the wavenumber eigenvectors can be assumed, to arrive the orthonormality of the position eigenvectors. That means, $\hat{\mathbb{I}} = \int_{-\infty}^{\infty} dk|k><k|$ can be assumed and $\delta(x - x') = <x|x'>$ inferred:

$$\hat{\mathbb{I}} = \int_{-\infty}^{\infty} dk|k><k| \quad \Longrightarrow \quad \delta(x - x') = <x|x'> . \tag{2.167}$$

The next problem illuminates this idea.

$\square$

Problem: Completeness of $\{|k>, \ k \in \mathbb{R}\}$ implies orthonormality of $\{|x>, \ x \in \mathbb{R}\}$.

Consider the ordinary integral representation of the Dirac delta function $\delta(x - x')$,

$$\delta(x - x') = \int_{-\infty}^{\infty} dk \left\{\frac{e^{jkx}}{\sqrt{2\pi}}\right\} \left\{\frac{e^{-jkx'}}{\sqrt{2\pi}}\right\}. \tag{2.168}$$

Assume the completeness of the eigenvectors $\{|k>, \ k \in \mathbb{R}\}$, i.e.,

$$\hat{\mathbb{I}} = \int_{-\infty}^{\infty} dk|k><k|. \tag{2.169}$$

Furthermore, consider the relationships

$$<x|k> = \frac{e^{jkx}}{\sqrt{2\pi}}, \tag{2.170}$$

and

$$<k|x> = \frac{e^{-jkx}}{\sqrt{2\pi}} \quad \overset{\text{replace x with x'}}{\Longrightarrow} \quad <k|x'> = \frac{e^{-jkx'}}{\sqrt{2\pi}}. \tag{2.171}$$

Show that the position eigenvectors $\{|x>, \ x \in \mathbb{R}\}$ are orthonormal.

Solution: Using (2.170) and (2.171) into (2.168),

$$\delta(x - x') = \int_{-\infty}^{\infty} dk <x|k><k|x'> . \tag{2.172}$$

Or, equivalently,

$$\delta(x - x') = \ <x| \left\{ \int_{-\infty}^{\infty} dk |k><k| \right\} |x'> .$$
(2.173)

Substituting (2.169),

$$\delta(x - x') = \ <x|\hat{\mathbb{I}}|x'> .$$
(2.174)

Using $\hat{\mathbb{I}}|x'> = |x'>$,

$$\delta(x - x') = \ <x|x'> .$$
(2.175)

∎

Remark: From the previous two problems it can be concluded that,

$$\delta(x - x') = <x|x'> \quad \Longleftrightarrow \quad \hat{\mathbb{I}} = \int_{-\infty}^{\infty} dk |k><k|.$$
(2.176)

This correspondence can be interpreted as an expression of the Heisenberg uncertainty principle for planewaves propagating along the x-axis.

☐

Problem: Expressing $|x> \ (x \in \mathbb{R})$ in terms of $|k> \ (k \in \mathbb{R})$.

Given the completeness of the vectors $|k> \ (k \in \mathbb{R})$, i.e.,

$$\hat{\mathbb{I}} = \int_{-\infty}^{\infty} dk |k><k|,$$
(2.177)

and the relationships,

$$<x|k> \ = \ \frac{e^{jkx}}{\sqrt{2\pi}}$$
(2.178a)

$$<k|x> \ = \ \frac{e^{-jkx}}{\sqrt{2\pi}},$$
(2.178b)

express the position vector $|x> \ (x \in \mathbb{R})$ in terms of the basis vectors $|k> \ (k \in \mathbb{R})$.

Solution: Multiply (2.177) from the right by $|x>$,

$$\underbrace{\hat{\mathbb{I}}|x>}_{|x>} = \int_{-\infty}^{\infty} dk |k> \underbrace{<k|x>}_{e^{-jkx}/\sqrt{2\pi}}.$$
(2.179)

Thus,

$$|x> \ = \ \int_{-\infty}^{\infty} dk |k> \frac{e^{-jkx}}{\sqrt{2\pi}}.$$
(2.180)

This completes the solution.

∎

Remark: Interpretation.

The above result is illuminating. It states that the position vector $|x>$ in the abstract Hilbert space can be expressed in terms of a linear combination of the basis vectors $|k>$ $(k \in \mathbb{R})$ in the abstract Hilbert space. Additionally, it reveals that the coefficient of the basis vector $|k>$ is given by $e^{-jkx}/\sqrt{2\pi}$, i.e., the complex conjugate of the basis functions in ordinary space. Stated differently, it is a recipe for transforming the basis vectors $|k>$ to the position vectors $|x>$, both vectors ($|k>$ and $|x>$) inhabiting the abstract Hilbert space. In the course of transformation, one enters into the ordinary space. This property allows one to gain a glimpse into the structure of abstract Hilbert space.

□

Problem: Expressing $<x|$ $(x \in \mathbb{R})$ in terms of $<k|$ $(k \in \mathbb{R})$.

Given the completeness of the vectors $|k>$ $(k \in \mathbb{R})$, i.e.,

$$\hat{\mathbb{I}} = \int_{-\infty}^{\infty} dk\, |k><k|, \tag{2.181}$$

and the relationships,

$$<x|k> = \frac{e^{jkx}}{\sqrt{2\pi}} \tag{2.182a}$$

$$<k|x> = \frac{e^{-jkx}}{\sqrt{2\pi}}, \tag{2.182b}$$

express the dual position vector $<x|$ $(x \in \mathbb{R})$ in terms of the dual basis vectors $<k|$ $(k \in \mathbb{R})$.

Solution: Multiply (2.181) from the left by $<x|$,

$$\underbrace{<x|\hat{\mathbb{I}}}_{<x|} = \int_{-\infty}^{\infty} dk\, \underbrace{<x|k>}_{e^{jkx}/\sqrt{2\pi}} <k|. \tag{2.183}$$

Thus,

$$<x| = \int_{-\infty}^{\infty} dk\, \frac{e^{jkx}}{\sqrt{2\pi}} <k|. \tag{2.184}$$

This completes the solution.

■

Remark: Interpretation.

The above result states that the dual position vector $<x|$ in the dual abstract Hilbert space can be expressed in terms of a linear combination of the dual basis vectors $<k|$ $(k \in \mathbb{R})$ in the dual abstract Hilbert space. Furthermore, it reveals that the coefficient of the dual basis vector $<k|$ is given by $e^{jkx}/\sqrt{2\pi}$, i.e., the basis functions in ordinary space. It is a recipe for transforming dual basis vectors $<k|$ to dual position vectors $<x|$ both residing in the dual abstract Hilbert space. In the course of transformation one enters into the ordinary space. This property allows one to gain a glimpse into the dual abstract Hilbert space.

□

Remark: Compare the results obtained in the previous two problems,

$$|x> \ = \ \int_{-\infty}^{\infty} dk |k> \frac{e^{-jkx}}{\sqrt{2\pi}} \tag{2.185a}$$

$$<x| \ = \ \int_{-\infty}^{\infty} dk \frac{e^{jkx}}{\sqrt{2\pi}} <k|. \tag{2.185b}$$

These result shows that going from abstract Hilbert space to abstract dual Hilbert space (Eq. (2.185a) → Eq. (2.185b)) all column vectors (kets) must be replaced with their corresponding row vectors (bras). Furthermore, all complex number must be changed into their corresponding complex conjugate counterparts.

Conversely, when going from abstract dual Hilbert space to abstract Hilbert space (Eq. (2.185b) → Eq. (2.185a)) all row vectors (bras) must be replaced with their corresponding column vectors (kets). Additionally, all complex number must be changed into their corresponding complex conjugate counterparts.

□

Lemma: *Orthogonality of position states* $|x> \ (x \in \mathbb{R})$.

Using the relationships,

$$|x> \ = \ \int_{-\infty}^{\infty} dk |k> \frac{e^{-jkx}}{\sqrt{2\pi}} \tag{2.186a}$$

$$<x'| \ = \ \int_{-\infty}^{\infty} d\kappa \frac{e^{j\kappa x'}}{\sqrt{2\pi}} <\kappa|, \tag{2.186b}$$

and the orthogonality of the vectors $|k> \ (k \in \mathbb{R})$, *i.e.,*

$$<\kappa|k> \ = \ \delta(k - \kappa), \tag{2.187}$$

show that the position vectors $|x> \ (x \in \mathbb{R})$ *constitute a set of orthonormal basis vectors. More succinctly, show that for arbitrary* x *and* x' $(x, x' \in \mathbb{R})$,

$$<x'|x> \ = \ \delta(x' - x). \tag{2.188}$$

Proof: Long proof.

Build the inner product $<x'|x>$,

$$<x'|x> \ = \ \int_{-\infty}^{\infty} d\kappa \frac{e^{j\kappa x'}}{\sqrt{2\pi}} <\kappa| \int_{-\infty}^{\infty} dk |k> \frac{e^{-jkx}}{\sqrt{2\pi}}. \tag{2.189}$$

Since $<\kappa|$ is independent of k, it can be transferred into the k-integral,

$$<x'|x> \ = \ \int_{-\infty}^{\infty} d\kappa \frac{e^{j\kappa x'}}{\sqrt{2\pi}} \int_{-\infty}^{\infty} dk \underbrace{<\kappa|k>}_{\delta(k-\kappa)} \frac{e^{-jkx}}{\sqrt{2\pi}}. \tag{2.190}$$

Using the orthogonality of the vectors $|\kappa>$ and $|k>$, i.e., $<\kappa|k>= \delta(k-\kappa)$, as indicated,

$$< x'|x > \;=\; \int\limits_{-\infty}^{\infty} d\kappa \frac{e^{j\kappa x'}}{\sqrt{2\pi}} \underbrace{\int\limits_{-\infty}^{\infty} dk \delta(k-\kappa) \frac{e^{-jkx}}{\sqrt{2\pi}}}_{e^{-j\kappa x}/\sqrt{2\pi}}. \tag{2.191}$$

Using the sifting property of the Dirac delta function $\delta(k-\kappa)$, as indicated,

$$< x'|x > \;=\; \int\limits_{-\infty}^{\infty} d\kappa \frac{e^{j\kappa x'}}{\sqrt{2\pi}} \frac{e^{-j\kappa x}}{\sqrt{2\pi}}. \tag{2.192}$$

Simplifying,

$$< x'|x > \;=\; \int\limits_{-\infty}^{\infty} \frac{d\kappa}{2\pi} e^{j\kappa(x'-x)}. \tag{2.193}$$

The integral at the R.H.S. is equal to $\delta(x'-x)$. Thus,

$$< x'|x > \;=\; \delta(x'-x). \tag{2.194}$$

This completes the proof of orthonormality.

∎

Proof: Short proof.

Consider the inner product $< x'|x >$ and insert the identity operator $\hat{\mathbb{I}}$ between $< x'|$ and $|x >$,

$$< x'|x > \;=\; < x'|\hat{\mathbb{I}}|x>. \tag{2.195}$$

Recall the resolution of identity formula,

$$\hat{\mathbb{I}} = \int\limits_{-\infty}^{\infty} dk |k >< k|. \tag{2.196}$$

Then,

$$< x'|x > \;=\; < x'| \left\{ \int\limits_{-\infty}^{\infty} dk |k >< k| \right\} |x > \tag{2.197a}$$

$$=\; \int\limits_{-\infty}^{\infty} dk \underbrace{< x'|k >< k|x >}_{e^{jkx'}/\sqrt{2\pi}\; e^{-jkx}/\sqrt{2\pi}} \tag{2.197b}$$

$$=\; \int\limits_{-\infty}^{\infty} \frac{dk}{2\pi} e^{jk(x'-x)} \tag{2.197c}$$

$$=\; \delta(x'-x). \tag{2.197d}$$

This completes the comparatively shorter proof of orthonormality.

∎

Lemma: *Completeness of position vectors $|x> (x \in \mathbb{R})$.*

Using the relationships,

$$|x> = \int_{-\infty}^{\infty} dk|k> \frac{e^{-jkx}}{\sqrt{2\pi}} \qquad (2.198a)$$

$$<x| = \int_{-\infty}^{\infty} d\kappa \frac{e^{j\kappa x}}{\sqrt{2\pi}} <\kappa|, \qquad (2.198b)$$

show that

$$\int_{-\infty}^{\infty} dx|x><x| = \hat{\mathbb{I}}, \qquad (2.199)$$

with $\hat{\mathbb{I}}$ being an identity operator in the abstract Hilbert space.

Proof: Long proof.

Consider,

$$\int_{-\infty}^{\infty} dx|x><x| \overset{(2.198)}{=} \int_{-\infty}^{\infty} dx \left\{ \int_{-\infty}^{\infty} dk|k> \frac{e^{-jkx}}{\sqrt{2\pi}} \right\} \left\{ \int_{-\infty}^{\infty} d\kappa \frac{e^{j\kappa x}}{\sqrt{2\pi}} <\kappa| \right\} \qquad (2.200a)$$

$$= \int_{-\infty}^{\infty} dk \int_{-\infty}^{\infty} d\kappa |k><\kappa| \underbrace{\int_{-\infty}^{\infty} dx \left\{ \frac{e^{-jkx}}{\sqrt{2\pi}} \right\} \left\{ \frac{e^{j\kappa x}}{\sqrt{2\pi}} \right\}} \qquad (2.200b)$$

$$= \int_{-\infty}^{\infty} dk \int_{-\infty}^{\infty} d\kappa |k><\kappa| \underbrace{\int_{-\infty}^{\infty} \frac{dx}{2\pi} e^{j(\kappa-k)x}}_{\delta(\kappa-k)} \qquad (2.200c)$$

$$= \int_{-\infty}^{\infty} dk \underbrace{\int_{-\infty}^{\infty} d\kappa |k><\kappa|\delta(\kappa - k)}_{|k><k|} \qquad (2.200d)$$

$$= \int_{-\infty}^{\infty} dk|k><k| \qquad (2.200e)$$

$$= \hat{\mathbb{I}}. \qquad (2.200f)$$

Thus, the completeness of the basis vectors $|k>$ implies the completeness of the position vectors $|x>$ in abstract Hilbert space. ∎

Further above the position vectors $|x>$ ($x \in \mathbb{R}$) were expressed in terms of the basis vectors $|k>$ ($k \in \mathbb{R}$). This was made possible due to the availability of the resolution of identity relationship $\hat{\mathbb{I}} = \int\limits_{-\infty}^{\infty} dk |k><k|$.

The next problems investigate the possibility of expressing the basis vectors $|k>$ ($k \in \mathbb{R}$) in terms of the position vectors $|x>$ ($x \in \mathbb{R}$) utilizing the resolution of identity relationship $\hat{\mathbb{I}} = \int\limits_{-\infty}^{\infty} dx |x><x|$ just established.

Problem: Expressing $|k>$ ($k \in \mathbb{R}$) in terms of $|x>$ ($x \in \mathbb{R}$).

Given the completeness of the position vectors $|x>$ ($x \in \mathbb{R}$), i.e.,

$$\hat{\mathbb{I}} = \int\limits_{-\infty}^{\infty} dx |x><x|, \tag{2.201}$$

express basis vectors $|k>$ ($k \in \mathbb{R}$) in terms of position vectors $|x>$ ($x \in \mathbb{R}$).

Solution: This is an easy task. Multiply both sides of (2.201) from the right by $|k>$,

$$\hat{\mathbb{I}}|k> = \int\limits_{-\infty}^{\infty} dx |x> \underbrace{<x|k>}. \tag{2.202}$$

It is immediate that,

$$|k> = \int\limits_{-\infty}^{\infty} dx |x> \frac{e^{jkx}}{\sqrt{2\pi}}. \tag{2.203}$$

This completes the solution.

∎

Problem: Expressing $<k|$ ($k \in \mathbb{R}$) in terms of $<x|$ ($x \in \mathbb{R}$).

Given the completeness of the position vectors $|x>$ ($x \in \mathbb{R}$), i.e.,

$$\hat{\mathbb{I}} = \int\limits_{-\infty}^{\infty} dx |x><x|, \tag{2.204}$$

express basis vectors $<k|$ ($k \in \mathbb{R}$) in terms of position vectors $<x|$ ($x \in \mathbb{R}$).

Solution: Multiply both sides of (2.204) from the left by $<k|$,

$$<k|\hat{\mathbb{I}} = \int\limits_{-\infty}^{\infty} dx \underbrace{<k|x>} <x|. \tag{2.205}$$

It is immediate that

$$<k| = \int\limits_{-\infty}^{\infty} dx \frac{e^{-jkx}}{\sqrt{2\pi}} <x|. \tag{2.206}$$

This completes the solution.

∎

Remark: Concerning Fourier- and inverse Fourier transform.

A striking insight emerging from the previous discussion is that expressions of the resolution of identity operator in terms of a complete and orthonormal set of basis function enable clear description of transforms and their corresponding inverse transforms. The representation

$$\hat{\mathbb{I}} = \int_{-\infty}^{\infty} dk |k> < k|, \tag{2.207}$$

is particularly important. It is the quintessential content of the Fourier- and inverse Fourier transform, as will be shown in this chapter. The representation (2.207) allows describing the Fourier- and inverse Fourier transform in one stroke. The fact that the Fourier- and inverse Fourier transform are second nature to engineers renders the present discussion particularly relevant. Apply $\hat{\mathbb{I}}$ onto an arbitrary function $|f>$ in the abstract Hilbert space,

$$|f> = \hat{\mathbb{I}}|f> = \int_{-\infty}^{\infty} dk |k> < k|f> . \tag{2.208}$$

Further below it will be shown why (2.208) simultaneously involves the Fourier- and the inverse Fourier transform. Additionally, it will be seen that general transforms and their corresponding inverse transforms can simultaneously be expressed in terms of exterior product expansions of adequately chosen identity operators.

□

Next a slight generalization of relevant definitions, concepts, and ideas discussed so far.

2.7   Generalized transforms and inverse transforms

The following definitions and generalizations shall ease the discussion.

The ket: Consider the column vector **a** with $N(\in \mathbb{N})$ components a_n. It is instructive to associate with **a** the ket $|a>$,

$$\mathbf{a} = \begin{bmatrix} a_1 \\ \vdots \\ a_n \\ \vdots \\ a_N \end{bmatrix} = |a> . \tag{2.209}$$

Unless explicitly mentioned, $a_n \in \mathbb{R}$.

The bra: Consider the row vector $\mathbf{a}^T$ with $N(\in \mathbb{N})$ real-valued components a_n. Associate with $\mathbf{a}^T$ the bra $<a|$,

$$\mathbf{a}^T = \begin{bmatrix} a_1 & \cdots & a_n & \cdots & a_N \end{bmatrix} = <a|. \tag{2.210}$$

The superscript T denotes transposition. For complex-valued **a** ($a_n \in \mathbb{C}$), the bra $<a|$ stands for the Hermitian conjugate $\mathbf{a}^\dagger$,

$$\mathbf{a}^\dagger = \begin{bmatrix} a_1^* & \cdots & a_n^* & \cdots & a_N^* \end{bmatrix} = <a|, \tag{2.211}$$

with a_n^* denoting the complex conjugate of a_n.

The inner product: The inner product of the kets $|a>$ and $|b>$ is denoted by $<a|b>$,

$$<a|b> = \begin{bmatrix} a_1 & \cdots & a_n & \cdots & a_N \end{bmatrix} \begin{bmatrix} b_1 \\ \vdots \\ b_n \\ \vdots \\ b_N \end{bmatrix} = a_1 b_1 + \cdots + a_n b_n + \cdots + a_N b_N. \tag{2.212}$$

The exterior product: The exterior product of the kets $|a>$ and $|b>$ is denoted by $|a><b|$,

$$|a><b| = \begin{bmatrix} a_1 \\ \vdots \\ a_n \\ \vdots \\ a_N \end{bmatrix} \begin{bmatrix} b_1 & \cdots & b_n & \cdots & b_N \end{bmatrix} = \begin{bmatrix} a_1 b_1 & \cdots & a_1 b_n & \cdots & a_1 b_N \\ \vdots & & \vdots & & \vdots \\ a_n b_1 & \cdots & a_n b_n & \cdots & a_n b_N \\ \vdots & & \vdots & & \vdots \\ a_N b_1 & \cdots & a_N b_n & \cdots & a_N b_N \end{bmatrix}. \tag{2.213}$$

The determinant of A: The determinant of the $N \times N$ matrix $\mathbf{A}$ is denoted by $\det\{\mathbf{A}\}$.

The inverse of A 2×2 non-singular matrix: Given the 2×2 matrix $\mathbf{A}$,

$$\mathbf{A} = \begin{bmatrix} a_{11} & a_{12} \\ a_{21} & a_{22} \end{bmatrix}. \tag{2.214}$$

Assume that $\det\{\mathbf{A}\} = a_{11}a_{22} - a_{12}a_{21} \neq 0$. Then, the inverse of $\mathbf{A}$, denoted by $\mathbf{A}^{-1}$, is,

$$\mathbf{A}^{-1} = \frac{1}{\det\{\mathbf{A}\}} \begin{bmatrix} a_{22} & -a_{12} \\ -a_{21} & a_{11} \end{bmatrix}. \tag{2.215}$$

The set of monomials: The set of monomials $\{1, x, x^2, \cdots\}$, also written as $\{x^n | n \in \mathbb{N}_0\}$, constitutes a complete non-orthogonal basis on any finite interval $[a, b]$ with $a, b \in \mathbb{R}$ and $a < b$. Obviously, the set of monomials $\{1, x, x^2, \ldots\}$ can be generated from $\{1, x\}$ by successive multiplication of x by the preceding element. This fundamental property is the genesis of the integer algebra in the present theory. The implications of the latter property, which carries over to two-, three-, and higher dimensions, will be discussed in a separate text.

The nth derivative of $f(x)$: The nth derivative of the function $f(x)$ with respect to its argument is denoted by $\frac{d^n}{dx^n}f(x)$ or $f^{(n)}(x)$.

The nth derivative $f(x)$ evaluated at $x = 0$: The nth derivative of the function $f(x)$ with respect to its argument and evaluated at the point $x = 0$ is denoted by $\frac{d^n}{dx^n}f(x)|_{x=0}$, $\frac{d^n}{dx^n}f(0)$ or $f^{(n)}(0)$.

The n-factorial: The n-factorial is denoted by $n! = 1 \cdot 2 \cdots (n-1) \cdot n$ with $0! = 1$.

The Formal Taylor Series Expansion: The Formal Taylor Series Expansion of the function $f(x)$ is denoted by any of the following representations,

$$f(x) = c_0 + c_1 x + c_2 x^2 + \cdots = \sum_{n=0}^{\infty} c_n x^n = \sum_{n=0}^{\infty} x^n c_n, \tag{2.216}$$

with c_n being the nth derivative of $f(x)$, evaluated at $x = 0$, and divided by $n!$. The qualifier formal in quantum physics literature alludes to the convention that the above series is a representation of $f(x)$, irrespective of whether or not it converges, whether or not it converges to $f(x)$, and in which sense it converges.

Symmetric finite intervals and lattices: An interval of the form $[-N\Delta, N\Delta]$, with $N \in \mathbb{N}$ and $\Delta \in \mathbb{R}^+$, is referred to as a symmetric finite closed interval. An $[-N\Delta, N\Delta]$ interval is associated with a set of $2N + 1$ lattice points in the corresponding lattice. Examples are,

$$[-\Delta, \Delta] \iff \{-\Delta, 0, \Delta\} \tag{2.217a}$$

$$[-2\Delta, 2\Delta] \iff \{-2\Delta, -\Delta, 0, \Delta, 2\Delta\} \tag{2.217b}$$

$$[-3\Delta, 3\Delta] \iff \{-3\Delta, -2\Delta, -\Delta, 0, \Delta, 2\Delta, 3\Delta\} \tag{2.217c}$$

$$[-4\Delta, 4\Delta] \iff \{-4\Delta, -3\Delta, -2\Delta, -\Delta, 0, \Delta, 2\Delta, 3\Delta, 4\Delta\}. \tag{2.217d}$$

Asymmetric finite intervals and lattices: Asymmetric intervals along with their associated sets of lattice points in the corresponding lattices are defined similarly. Examples are,

$$[-\Delta_1, \Delta_2] \iff \{-\Delta_1, 0, \Delta_2\} \tag{2.218a}$$

$$[0, 2\Delta] \iff \{0, \Delta, 2\Delta\} \tag{2.218b}$$

$$[-2\Delta, 0] \iff \{-2\Delta, -\Delta, 0\}, \tag{2.218c}$$

with $\Delta, \Delta_1, \Delta_2 \in \mathbb{R}^+$.

Toy models on $[-N\Delta, N\Delta]$ intervals: Apart from associating a lattice with an interval, it is also instructive to assign a toy model to an interval. Consider the interval $[-N\Delta, N\Delta]$ along with its associated set of $2N + 1$ equidistant lattice points:

$$\{-N\Delta, -(N-1)\Delta, \ldots, -\Delta, 0, \Delta, \ldots, (N-1)\Delta, N\Delta\}$$

Let the function $f(x)$, defined on $[-N\Delta, N\Delta]$, possess derivatives $f^{(n)}(x)$ to any order desired. Let

$$f_{-N}, f_{-(N-1)}, \ldots, f_{-1}, f_0, f_1, \ldots, f_{N-1}, f_N$$

denote, respectively, the function values:

$$f(-N\Delta), f(-(N-1)\Delta), \ldots, f(-\Delta), f(0), f(\Delta), \ldots, f((N-1)\Delta), f(N\Delta)$$

Let the Ket $|f^{2N+1}>$ denote the column vector with the components:

$$f_{-N}, f_{-(N-1)}, \ldots, f_{-1}, f_0, f_1, \ldots, f_{N-1}, f_N$$

That means,

$$|f^{2N+1}> = \begin{bmatrix} f_{-N} \\ f_{-(N-1)} \\ \vdots \\ f_{-1} \\ f_0 \\ f_1 \\ \vdots \\ f_{N-1} \\ f_N \end{bmatrix}. \tag{2.219}$$

This defines the $(2N+1)$-order toy model. Several transforms and the corresponding inverse transforms will be considered in the following sections in this chapter. The discussion will culminate in a brief sketch of the Discrete Taylor Transform and Inverse Transform. The underlying analysis and synthesis of functions will be described in terms of low-order toy models. The simplicity of the chosen toy models enable to obtain closed form solutions to the problems. The constructive (inductive) approach employed for the development of the theory enables the interested reader to generalize the procedure self-evidently and simply-by-inspection. Clarity, self-evidence, and the ability to generalize the formulations constitute the hallmark of the developed theory based on the resolution of identity.

2.8   Fourier transform and inverse transform: the resolution of identity

Fourier transform and inverse transform can be viewed as an archetype example for elucidating the intricacies of the underlying ideas, the elegance and power of the Dirac notation, and the eminence role of the resolution of identity.* The following derivations are formal, in the sense that the operations which are carried out are assumed to be permissible.

Fourier transform: Given the function $f(x)$, its formal Fourier transform $F(k)$ is defined as,

$$F(k) = \int_{-\infty}^{\infty} dx e^{-ikx} f(x), \quad k \in \mathbb{R} \qquad \text{(engineering notation)}. \tag{2.220}$$

In this formula k has been fixed at the L.H.S. and x is the dummy integration variable. For "bookkeeping" purposes and guaranteeing maximum clarity in the derivations, it is instructive to express the integral in (2.220) in terms of a variable other than x, e.g., ξ, as well,

$$F(k) = \int_{-\infty}^{\infty} d\xi e^{-ik\xi} f(\xi), \quad k \in \mathbb{R}. \tag{2.221}$$

---

*The discussion in the next entry on the sinc-function-based transform and inverse transform shall further reinforce some of the pertinent ideas.

Inverse Fourier transform: Conversely, given the function $F(k)$, the original function $f(x)$ can be recovered employing the inverse Fourier transform formula,

$$f(x) = \int\limits_{-\infty}^{\infty} \frac{dk}{2\pi} e^{ikx} F(k), \quad x \in \mathbb{R} \qquad \text{(engineering notation)}. \tag{2.222}$$

In this formula x has been fixed at the L.H.S. and k is the dummy integration variable. It is useful to express the integral in (2.222) in terms of a variable other than k, e.g., κ, as well,

$$f(x) = \int\limits_{-\infty}^{\infty} \frac{d\kappa}{2\pi} e^{i\kappa x} F(\kappa), \quad x \in \mathbb{R}. \tag{2.223}$$

Equation (2.222) states that the information content in $F(k)$, $(k \in \mathbb{R})$ suffices to reconstruct the original function $f(x)$ for arbitrary $x \in \mathbb{R}$. Concerning the usage of 2π, there is an asymmetry in engineering notation, in contrast to the mathematical (physical) notation which has been adopted in this text. This distinction has been clarified in the following remark.

Remark: In engineering literature, the factors 1 and $1/(2\pi)$, respectively, are used in connection with Fourier transform (2.220), and inverse Fourier transform (2.222) integral expressions. In this text, the common factor $1/\sqrt{2\pi}$ has been employed in both the Fourier transform and inverse Fourier transform. The equations immediately following this remarks explicate why the symmetric choice of $1/\sqrt{2\pi}$ makes mathematically speaking more sense. On the other hand, leaving aside matters of aesthetics, and mathematical rigor, both notations are equivalent in the following sense. When performing the Fourier transform, and subsequently, the inverse Fourier transform, the choice of factors does not matter in the final result, since $(1)(1/(2\pi)) = (1/\sqrt{2\pi})(1/\sqrt{2\pi}) = 1/(2\pi)$, as demonstrated next.

$\square$

Remark: Superiority of the $(1/\sqrt{2\pi})(1/\sqrt{2\pi})$ convention.

Substituting (2.221) into (2.222), splitting 2π into $\sqrt{2\pi}\sqrt{2\pi}$, and rearranging,

$$f(x) = \int\limits_{-\infty}^{\infty} \frac{dk}{2\pi} e^{ikx} \left(\int\limits_{-\infty}^{\infty} d\xi e^{-ik\xi} f(\xi) \right) \tag{2.224a}$$

$$= \int\limits_{-\infty}^{\infty} dk \int\limits_{-\infty}^{\infty} d\xi \frac{e^{ikx}}{\sqrt{2\pi}} \frac{e^{-ik\xi}}{\sqrt{2\pi}} f(\xi) \tag{2.224b}$$

$$= \int\limits_{-\infty}^{\infty} dk \int\limits_{-\infty}^{\infty} d\xi \left\{ \frac{e^{ikx}}{\sqrt{2\pi}} \right\} \left\{ \frac{e^{ik\xi}}{\sqrt{2\pi}} \right\}^* f(\xi). \tag{2.224c}$$

Introducing the orthonormal basis functions,

$$e_k(x) = \frac{e^{ikx}}{\sqrt{2\pi}}, \tag{2.225}$$

The double integral in (2.224c) assumes the compact form,

$$f(x) = \int\limits_{-\infty}^{\infty} dk \int\limits_{-\infty}^{\infty} d\xi e_k(x) e_k(\xi) f(\xi). \tag{2.226}$$

$\square$

It is useful to employ Dirac's notation, introduce the abstract Hilbert space $\mathbb{H}$, and represent the function $f(x)$ by the ket $|f>\in \mathbb{H}$. Then, the relationship between $f(x)$ and $|f>$ can be stated by,

$$f(x) = <x|f> . \tag{2.227}$$

In words: projecting $|f>$ onto the basis position Ket $|x>$ gives $f(x)$. Here, $|x>$ is the Eigenket of the position operator $\hat{x}$,

$$\hat{x}|x> = x|x>, \tag{2.228}$$

subject to the orthonormalization condition of the Eigenkets $|x'>$ and $|x''>$,

$$<x'|x''> = \delta\left(x'-x''\right). \tag{2.229}$$

As will be demonstrated momentarily, the completeness property (the resolution of identity property) of the Eigenkets $|x>$, i.e.,

$$\hat{\mathbb{1}} = \int_{-\infty}^{\infty} dx|x><x| \left(= \int_{-\infty}^{\infty} d\xi|\xi><\xi| \right), \tag{2.230}$$

plays a prominent role in the manipulations presented here, both in terms of clarity, and elegance. The second equation in (2.230) expresses the fact that the dummy integration variable can be chosen freely.[†]

The introduction of the abstract ket $|f>$, inhabiting the abstract Hilbert space $\mathbb{H}$, is a convenient tool which is employed here, and does not require the knowledge of any of the mathematical intricacies in quantum physics. For the purposes in the present discussion, projecting $|f>$ onto, e.g., $|x-n>$, simply amounts to sampling $f(x)$ at $x=n$, and thus yielding $f(n)$. In countably infinite-dimensional Hilbert space $\mathbb{H}$, the preceding statement enables one to have the following associations:

$$f(x) \iff |f> \iff \begin{bmatrix} \vdots \\ f_{-N} \\ \vdots \\ f_{-1} \\ f_0 \\ f_1 \\ \vdots \\ f_N \\ \vdots \end{bmatrix}. \tag{2.231}$$

In finite dimensions, sampling $f(x)$ at the $2N+1$ lattice points $\{-N,\cdots,0,\cdots,N\}$ on the symmetric interval $[-N,N]$,

$$f(x) \iff |f> \iff \begin{bmatrix} f_{-N} \\ \vdots \\ f_{-1} \\ f_0 \\ f_1 \\ \vdots \\ f_N \end{bmatrix}. \tag{2.232}$$

[†]Any freedom permitted by the prevailing formulation should, in principle, be taken advantage of. The application of this principle enables presenting the expressions in their most general forms.

Or, more generally, sampling $f(x)$ at the $|M| + |N| + 1$ points on the asymmetric interval $[M, \ldots, N]$,

$$f(x) \iff |f> \iff \begin{bmatrix} f_M \\ f_{M+1} \\ \vdots \\ f_{N-1} \\ f_N \end{bmatrix}, \tag{2.233}$$

with $M, N \in \mathbb{Z}$ and $M < N$.

In view of (2.226), Dirac's notation suggests writing,

$$f(x) = <x|f> \tag{2.234a}$$

$$e_k(x) = <x|e_k> \tag{2.234b}$$

$$e_k^*(\xi) = \{e_k(\xi)\}^* = \{ <\xi|e_k> \}^* = <e_k|\xi> \tag{2.234c}$$

$$f(\xi) = <\xi|f> . \tag{2.234d}$$

Substituting the above expressions into (2.226),

$$<x|f> = \int_{-\infty}^{\infty} dk \int_{-\infty}^{\infty} d\xi \ <x|e_k><e_k|\xi><\xi|f> . \tag{2.235}$$

Noting that the inner product $<e_k|\xi>$ is an abbreviation of $<e_k||\xi>$, and transferring $\int_{-\infty}^{\infty} d\xi$ between $<e_k|$ and $|\xi>$ in $<e_k||\xi>$, and also viewing $<\xi|f>$ as $<\xi||f>$,

$$<x|f> = \int_{-\infty}^{\infty} dk \ <x|e_k><e_k| \underbrace{\left(\int_{-\infty}^{\infty} d\xi|\xi><\xi| \right)}_{=\hat{\mathbb{I}}} |f> . \tag{2.236}$$

The integral sandwiched between the brackets can be recognized as the identity operator $\hat{\mathbb{I}}$, as indicated. Thus, with $\hat{\mathbb{I}}|f> = |f>$,

$$<x|f> = \int_{-\infty}^{\infty} dk \ <x|e_k><e_k||f> . \tag{2.237}$$

Finally, transferring $\int_{-\infty}^{\infty} dk$ into the inner product $<x|e_k>$, which is an abbreviation for $<x||e_k>$,

$$<x|f> = <x| \left(\int_{-\infty}^{\infty} dk|e_k><e_k| \right) |f> . \tag{2.238}$$

Interpretation: The above equation conveys the following information: take the kth-basis ket vector $|e_k>$, build the exterior product $|e_k><e_k|$, and add over all possible instances (integrate over k), to obtain the identity operator $\hat{\mathbb{I}}$ in the abstract Hilbert space $\mathbb{H}^2$. The identity operator in (2.238), the brackted term at the R.H.S. can be "translated" into the language of the "ordinary" Hilbert function space of quadratically integrable functions, by multiplying the L.H.S. and the R.H.S. of the identity operator in (2.238) by the coordinate ket vectors $|x'>$ and $|x''>$, respectively,

$$<x'|\hat{\mathbb{I}}|x''> = <x'| \left(\int_{-\infty}^{\infty} dk |e_k><e_k| \right) |x''> \qquad (2.239a)$$

$$= \int_{-\infty}^{\infty} dk <x'|e_k><e_k|x''> \qquad (2.239b)$$

$$= \int_{-\infty}^{\infty} dk <x'|e_k> \left(<x''|e_k> \right)^* \qquad (2.239c)$$

$$= \int_{-\infty}^{\infty} dk \frac{e^{ikx'}}{\sqrt{2\pi}} \left(\frac{e^{ikx''}}{\sqrt{2\pi}} \right)^* \qquad (2.239d)$$

Thus, with $<x'|\hat{\mathbb{I}}|x''> = <x'|x''>$,

$$<x'|x''> = \int_{-\infty}^{\infty} dk \frac{e^{ikx'}}{\sqrt{2\pi}} \frac{e^{-ikx''}}{\sqrt{2\pi}}. \qquad (2.240)$$

Finally, with $<x'|x''> = \delta(x'-x'')$, and combining the exponential functions, the more familiar integral representation for the Dirac delta function offers itself,

$$\delta(x'-x'') = \int_{-\infty}^{\infty} \frac{dk}{2\pi} e^{ik(x'-x'')}. \qquad (2.241)$$

$\square$

In the above exposition the starting point was the Fourier transform and inverse transform formula. It was established that Fourier transform and inverse Fourier transform essentially amount to the existence of an associated resolution of identity.

Problem: Show that the integral $\int_{-\infty}^{\infty} dx' |x'><x'|$ represents the identity operator $\hat{\mathbb{I}}$.

Solution: Multiply,

$$\hat{\mathbb{I}} = \int_{-\infty}^{\infty} dx' |x'><x'|, \qquad (2.242)$$

on the Test Ket $|f> \in \mathbb{H}$,

$$\underbrace{\hat{\mathbb{I}}|f>} = \int_{-\infty}^{\infty} dx' |x'> \underbrace{<x'|f>}. \qquad (2.243)$$

With $\hat{\mathbb{I}}|f> = |f>$, and $<x'|f> = f(x')$,

$$|f> = \int_{-\infty}^{\infty} dx' |x'> f(x'). \tag{2.244}$$

Project both sides onto the arbitrary position Ket $|x>$,

$$\underbrace{<x|f>} = <x| \int_{-\infty}^{\infty} dx' |x'> f(x') \tag{2.245a}$$

$$= \int_{-\infty}^{\infty} dx' \underbrace{<x|x'>} f(x'). \tag{2.245b}$$

With $<x|f> = f(x)$ and $<x|x'> = \delta(x - x')$,

$$f(x) = \int_{-\infty}^{\infty} dx' \delta(x - x') f(x'). \tag{2.246a}$$

$$= f(x). \tag{2.246b}$$

The facts that the choice of the Ket $|f> \in \mathbb{H}$ and the coordinate basis ket $|x> \in \mathbb{H}$ were arbitrary, prove the claim in the problem.

∎

Problem: Show that applying $\hat{\mathbb{I}} = \int_{-\infty}^{\infty} dk |e_k> <e_k|$ onto $|f> \in \mathbb{H}$ and projecting the result onto $|x>$ quintessentially amount to the Fourier transform and inverse transform.

Solution: Apply $\hat{\mathbb{I}} = \int_{-\infty}^{\infty} dk |e_k> <e_k|$ onto $|f> \in \mathbb{H}$,

$$\underbrace{\hat{\mathbb{I}}|f>} = \int_{-\infty}^{\infty} dk |e_k> \underbrace{<e_k|f>}. \tag{2.247}$$

With $\hat{\mathbb{I}}|f> = |f>$, and inserting $\hat{\mathbb{I}} = \int_{-\infty}^{\infty} d\xi |\xi> <\xi|$ between the vertical bars in the expression of the inner product $<e_k||f> \ (= <e_k|f>)$,

$$|f> = \int_{-\infty}^{\infty} dk |e_k> <e_k| \left(\int_{-\infty}^{\infty} d\xi |\xi> <\xi| \right) |f>. \tag{2.248}$$

Rearranging,

$$|f> = \int_{-\infty}^{\infty} dk |e_k> \int_{-\infty}^{\infty} d\xi \underbrace{<e_k|\xi>} \underbrace{<\xi|f>}. \tag{2.249}$$

With $<e_k|\xi> = \frac{1}{\sqrt{2\pi}} e^{-jk\xi}$ and $<\xi|f> = f(\xi)$,

$$|f> = \int_{-\infty}^{\infty} dk |e_k> \int_{-\infty}^{\infty} d\xi \frac{1}{\sqrt{2\pi}} e^{-jk\xi} f(\xi). \tag{2.250}$$

Bringing $\frac{1}{\sqrt{2\pi}}$ outside the integral,

$$|f> = \int_{-\infty}^{\infty} \frac{dk}{\sqrt{2\pi}} |e_k> \underbrace{\int_{-\infty}^{\infty} d\xi e^{-jk\xi} f(\xi)}_{F(k)}. \qquad (2.251)$$

The under-braced integral is equal to the Fourier transform $F(k)$ of the function of $f(x)$ (engineering notation), thus,

$$|f> = \int_{-\infty}^{\infty} \frac{dk}{\sqrt{2\pi}} |e_k> F(k). \qquad (2.252)$$

Projecting both sides onto the coordinate ket vector $|x>$,

$$\underbrace{<x|f>} = <x| \int_{-\infty}^{\infty} \frac{dk}{\sqrt{2\pi}} |e_k> F(k) \qquad (2.253a)$$

$$= \int_{-\infty}^{\infty} \frac{dk}{\sqrt{2\pi}} \underbrace{<x|e_k>} F(k). \qquad (2.253b)$$

Considering $<x|f> = f(x)$ and $<x|e_k> = \frac{1}{\sqrt{2\pi}} e^{jkx}$,

$$f(x) = \int_{-\infty}^{\infty} \frac{dk}{\sqrt{2\pi}} \frac{1}{\sqrt{2\pi}} e^{jkx} F(k) \qquad (2.254a)$$

$$= \int_{-\infty}^{\infty} \frac{dk}{2\pi} e^{jkx} F(k), \qquad (2.254b)$$

a relationship which establishes the inverse Fourier transform (engineering notation). ∎

Problem: To further appreciate the power of the resolution of identity relationship, consider the following. Given the kets $|f>$ and $|g>$ in $\mathbb{H}$, translate $<f|g>$ into the language of the coordinate (position) space.

Solution: The following steps are self-explanatory,

$$<f|g> = <f|\hat{\mathbb{I}}|g> \qquad (2.255a)$$

$$= <f| \left(\int_{-\infty}^{\infty} dx |x><x| \right) |g> \qquad (2.255b)$$

$$= \int_{-\infty}^{\infty} dx \underbrace{<f|x>} <x|g> \qquad (2.255c)$$

$$= \int_{-\infty}^{\infty} dx (<x|f>)^* <x|g> \qquad (2.255d)$$

$$= \int_{-\infty}^{\infty} dx f^*(x) g(x). \qquad (2.255e)$$

The last integral is the interpretation of $< f|g >$ in the coordinate space. In particular, with $g(x) = f(x)$,

$$< f|f > = \int_{-\infty}^{\infty} dx f^*(x) f(x) \tag{2.256a}$$

$$= \int_{-\infty}^{\infty} dx |f(x)|^2 \tag{2.256b}$$

$$= ||f||^2, \tag{2.256c}$$

where $||f||$ is referred to as the norm of $f(x)$.

■

Remark: Demarcation between mathematical tools and the quantum physics.

This result was obtained from the Fourier- and inverse Fourier transform with which engineers and technologists are well familiar. It can be concluded that the equality in (2.238) (being exclusively a consequence of Fourier- and inverse Fourier transform) has little to nothing to do with quantum physics. It merely borrows the Dirac symbolic notation, which is a clever compact device for expressing vectors, dual vectors, inner products, and exterior products in a Hilbert space. Even though Hilbert spaces are used in quantum physics they are not derived from the quantum physics. A Hilbert space accommodates quadratically integrable functions and is equipped with a metric which is induced by a suitably chosen inner product rule. More generally, given partial differential equations in the mathematical physics customized Hilbert spaces can be constructed. And this is exactly what is done in quantum physics. Quantum physics makes ample use of Hilbert spaces. Engineers in signal processing and practitioners in field analysis and wave propagation operate in Hilbert spaces without necessarily being concerned with quantum phenomena. Consequently, they are already familiar with a plethora of mathematical apparatuses which are employed in quantum physics, perhaps only expressed in a slightly different language. It is essential to be aware of the power of the Fourier- and inverse Fourier transform, and its generalizations (wavelets, theory of frames and allied topics, generalized functions, and various distributions). The Fourier- and inverse Fourier transform and the Taylor- and inverse Taylor transform (introduced later in this chapter) are "natural" tools in quantum physics since they are intimately interwoven with the position variable x and the notion of the d/dx-derivative. The property that $\exp(ikx)/\sqrt{2\pi}$ is an eigenfunction of d/dx catapults $\exp(ikx)/\sqrt{2\pi}$ to prominence. The fact that the functions $\exp(ikx)/\sqrt{2\pi}$ ($k \in \mathbb{R}$) are complete and thus they permit the resolution of the unity operator $\hat{\mathbb{I}}$, further explains the widespread utility of the Fourier- and inverse Fourier transform. On the other hand the property that the composite operator $\frac{d}{dx}x - x\frac{d}{dx}$ is equal to the unity operator $\hat{\mathbb{I}}$ might be viewed as the genesis of the significance of the Taylor- and inverse Taylor transform.

□

Remark: Consequences of (2.238).

In obtaining (2.238), use was also made of (2.234), which is an excerpt from a dictionary for translating Dirac symbolic notation to the engineering notation, and conversely from the engineering language to the Dirac symbolic language. Several important implications can be deduced from (2.238), which is reproduced here, for easy reference,

$$< x|f > = < x| \left(\int_{-\infty}^{\infty} dk |e_k > < e_k| \right) |f > . \tag{2.257}$$

- **Omitting $< x|$ and $|f >$:** Note that $|x >$ and $|f >$ are arbitrary, and the inner product $< x|f >$ at the L.H.S. can be augmented by the identity operator, i.e., $< x|\hat{\mathbb{I}}|f >$. Obviously, the bra $< x|$ and the ket $|f >$ can be omitted from both sides of the equation. Then, it is seen that (2.257) is tantamount to the completeness of the kets $\{|e_k> | k \in \mathbb{R}\}$. Stated differently,

$$\hat{\mathbb{I}} = \int_{-\infty}^{\infty} dk |e_k> < e_k|. \tag{2.258}$$

This equation lives in the abstract Hilbert space.

- **Omitting** $< x|$: Since $< x|$ is arbitrary, $< x|$ can be omitted from both sides of (2.257). Removing the round brackets at the R.H.S. as well,

$$|f> = \int_{-\infty}^{\infty} dk |e_k> < e_k|f>. \tag{2.259}$$

This equation lives in the abstract Hilbert space. It is immediate that this result can alternatively be achieved by operating (2.258) from the left onto $|f>$. Thereby, $< e_k|f>$ (the projection of $|f>$ onto $|e_k>$, which is a measure for the similarity between $|f>$ and $|e_k>$) represents the Fourier analysis, and the integral at the R.H.S. of (2.259) is the Fourier synthesis.

- **Omitting** $|f>$: Since $|f>$ is a test function, it can be omitted from both sides of (2.257). The round brackets can be removed and since $< x|$ is independent of the integration variable k, it can be entered into the k-integral,

$$< x| = \int_{-\infty}^{\infty} dk < x|e_k> < e_k|. \tag{2.260}$$

This equation lives in the abstract Hilbert state space. This is an expression for the bra $< x|$ in terms of the basis bras $\{ < e_k|, k \in \mathbb{R}\}$.

- Recall the starting point were the Fourier- and inverse Fourier transform, as known to engineers and technologists, and the final results are equalities in the abstract Hilbert space.
- Conversely, one could stipulate the resolution of identity (2.258) as the initial point. Subsequently, by right multiplying by the test function $|f>$ one would obtain (2.259). By left multiplying by $< x|$ one would arrive at (2.260). Finally, by left multiplying by $< x|$ and right multiplying by $|f>$, the expression in (2.257) would emerge.
- It can be concluded that the expression of the resolution of identity in (2.258) plays a fundamental role. Stated more completely, the universal expression of the resolution of identity of the position Eigenbra ($< x|$) and Eigenket ($|x>$), i.e.,

$$\hat{\mathbb{I}} = \int_{-\infty}^{\infty} dx |x> < x|, \tag{2.261}$$

must be added to the excerpt of the dictionary in (2.234).

□

Remark: Axiomatic construction of a dictionary for two-way translation between the engineering notation and the abstract Hilbert space.

From the above observations summarized in the previous remark, it must be clear that a simple axiomatic system based on only a few rules can be stipulated to translate engineering statements in the functional analysis to the language in the abstract Hilbert space and in the reverse direction. In the following, an attempt is made to identify what the mathematical requirements might be.

- With reference to every desirable complete set of orthonormal kets $\{|a_k>, k \in \mathbb{R}\}$, $\{|b_k>, k \in \mathbb{R}\}$, $\{|c_k>, k \in \mathbb{R}\}$, $\cdots$, i.e.,

$$< a_k|a_l > = \delta(k - l) \qquad k, l \in \mathbb{R} \tag{2.262a}$$

$$< b_k|b_l > = \delta(k - l) \qquad k, l \in \mathbb{R} \tag{2.262b}$$

$$< c_k|c_l > = \delta(k - l) \qquad k, l \in \mathbb{R} \tag{2.262c}$$

$$\vdots$$

the corresponding resolution of identity must be considered,

$$\hat{\mathbb{I}} = \int_{-\infty}^{\infty} dk |a_k > < a_k| \tag{2.263a}$$

$$\hat{\mathbb{I}} = \int_{-\infty}^{\infty} dk |b_k > < b_k| \tag{2.263b}$$

$$\hat{\mathbb{I}} = \int_{-\infty}^{\infty} dk |c_k > < c_k|. \tag{2.263c}$$

- In cases where the desirable set of kets $\{|a_k>, k \in \mathbb{R}\}$, $\{|b_k>, k \in \mathbb{R}\}$, $\{|c_k>, k \in \mathbb{R}\}$, ..., are over-complete, the corresponding dual kets $\{|\alpha_k>, k \in \mathbb{R}\}$, $\{|\beta_k>, k \in \mathbb{R}\}$, $\{|\gamma_k>, k \in \mathbb{R}\}$, ..., respectively, must be constructed:

$$< \alpha_k|a_l > = \delta(k - l) \qquad k, l \in \mathbb{R} \tag{2.264a}$$

$$< \beta_k|b_l > = \delta(k - l) \qquad k, l \in \mathbb{R} \tag{2.264b}$$

$$< \gamma_k|c_l > = \delta(k - l) \qquad k, l \in \mathbb{R} \tag{2.264c}$$

$$\vdots$$

The associated expressions of the resolution of identity take the form,

$$\hat{\mathbb{I}} = \int_{-\infty}^{\infty} dk |\alpha_k > < a_k| \tag{2.265a}$$

$$\hat{\mathbb{I}} = \int_{-\infty}^{\infty} dk |\beta_k > < b_k| \tag{2.265b}$$

$$\hat{\mathbb{I}} = \int_{-\infty}^{\infty} dk |\gamma_k > < c_k|. \tag{2.265c}$$

- The position kets $\{|x>, k \in \mathbb{R}\}$ are compete, with

$$< x|x' >= \delta(x - x') \tag{2.266}$$

obeying the completeness relationship,

$$\hat{\mathbb{I}} = \int_{-\infty}^{\infty} dx |x > < x|. \tag{2.267}$$

- The projection of kets onto $|x>$,

$$< x|a_k >= a_k(x).\tag{2.268}$$

- The complex conjugate of (2.268),

$$< a_k|x >= a_k^*(x),\tag{2.269}$$

In the general case the components of $|a_k>$ are complex valued.

This is all what is needed to maneuver between the engineering notation and the abstract Hilbert space.

A few examples building upon each other may clarify the content further.

☐

Problem: Let the over-complete sequence of kets $|a_k>$ and the corresponding dual over-complete sequence of kets $|\alpha_k>$ be given. Express the test ket $|f>$ in terms of $|\alpha_k>$.

Solution: Applying (2.265a) from the left onto $|f>$,

$$|f> = \int_{-\infty}^{\infty} dk |\alpha_k> < a_k|f> .\tag{2.270}$$

■

Problem: Transformation of over-complete coordinate systems.

Let the representation of $|f>$ in the over-complete ket system $|a_k>$ associated with the ket system $|\alpha_k>$ be available, i.e.,

$$|f> = \int_{-\infty}^{\infty} dk |\alpha_k> < a_k|f> .\tag{2.271}$$

(This means assume that the scalar numbers $< a_k|f>$ are available, accessible.) The task is the transformation of (2.271) to the over-complete ket system $|b_l>$ associated with the ket system $|\beta_l>$. (Note that to guarantee maximum generality the index l rather k has been used. This idea becomes self-evident in the solution part.)

Solution: Consider (2.265b) in terms of the integral dummy variable l,

$$\hat{\mathbb{I}} = \int_{-\infty}^{\infty} dl |\beta_l> < b_l|.\tag{2.272}$$

Applying (2.272) from the left onto $|f>$,

$$|f> = \int_{-\infty}^{\infty} dl |\beta_l> < b_l|f> .\tag{2.273}$$

This is the expression of $|f>$ in the $|b_l>$ coordinates. However, assume $< b_l|f>$ is not accessible. What is available is $< a_k|f>$, according to the hypothesis. Expressing $< b_l|f>$ in terms of $< a_k|f>$ can be accomplished by writing $< b_l|f>$ at the R.H.S., in the form $< b_l||f>$ or even more clearly in the form $< b_l|\{|f>\}$, to obtain,

$$|f> = \int_{-\infty}^{\infty} dl |\beta_l> < b_l|\{|f>\}.\tag{2.274}$$

Employ (2.271) for $|f>$ at the R.H.S. in (2.274),

$$|f>= \int_{-\infty}^{\infty} dl |\beta_l> <b_l| \left\{ \int_{-\infty}^{\infty} dk |\alpha_k> <a_k|f> \right\}. \tag{2.275}$$

Since $<b_l|$ is independent of the integral dummy variable k, it can be transferred into the k-integral,

$$|f>= \int_{-\infty}^{\infty} dl |\beta_l> \left\{ \int_{-\infty}^{\infty} dk <b_l|\alpha_k> <a_k|f> \right\}. \tag{2.276}$$

This is the desired expression and can be interpreted as follows. The inner product $<a_k|f>$ is the projection of $|f>$ onto the ket $|a_k>$ (a scalar quantity, say f_k^a). The inner product $<b_l|\alpha_k>$ (a scalar quantity, say $\alpha_{k,l}^b$) is the translation formula which is the projection of the dual ket $|\alpha_k>$ corresponding to $|a_k>$ onto the ket $|b_l>$. Calculating $<b_l|\alpha_k> <a_k|f> = \alpha_{k,l}^b f_k^a$ for all conceivable k and integrating over k, the term in the curly bracket gives the scalar number f_l^b which is the projection of $|f>$ onto the ket $|b_l>$,

$$\int_{-\infty}^{\infty} dk <b_l|\alpha_k> <a_k|f> = \int_{-\infty}^{\infty} dk \alpha_{k,l}^b f_k^a = f_l^b = (<b_l|f>). \tag{2.277}$$

Thus,

$$|f>= \int_{-\infty}^{\infty} dl |\beta_l> f_l^b, \tag{2.278}$$

which is the synthesis formula in terms of the dual kets $|\beta_l>$.

This hypothetical example should demonstrate that one can easily joggle between coordinates, indeed between coordinates and their corresponding dual counterparts. Carrying out the entire gymnastic boils down to expressing the expression of the resolution of identity in various coordinates, and calculating the arising inner products.

Generally speaking a coordinate transformation makes sense, if the data $<a_k|f>$ in the a-coordinates available, and the coordinate transformation formula $<b_l|\alpha_k>$ is pre-calculated and readily available. Furthermore, one might assume that the direct calculation of the projections of $|f>$ onto the coordinates $|b_l>$, i.e., $<b_l|f>$ is not accessible. ∎

Remark: Further deeper insights into the above calculation method can be gained by scrutinizing the genesis of the formula (2.276) and by translating (2.276) into the real space. The following problems are devoted to this end. Simple toy models will suffice to bring out interesting insights which might help the reader to better grasp the underlying relationships. It is all about the orthonormality and the resolution of identity relationships. To demonstrate the generality of proposed method over-complete ket systems along with their associated dual ket systems are considered.

□

Problem: Gaining insight.

Construct a toy model based on the transformation formula in the abstract Hilbert space,

$$|f>= \int_{-\infty}^{\infty} dl |\beta_l> \left\{ \int_{-\infty}^{\infty} dk <b_l|\alpha_k> <a_k|f> \right\}, \tag{2.279}$$

and visualize the underlying relationships.

Solution: Project both sides of (2.279) onto the position Ket $|x>$, to convert $|f>$ and $|\beta_l>$ into scalar numbers which is presumably easier to handle and interpret,

$$< x|f >= \int_{-\infty}^{\infty} dl < x|\beta_l > \left\{ \int_{-\infty}^{\infty} dk < b_l|\alpha_k >< a_k|f > \right\}. \tag{2.280}$$

On the construction of the toy model: The toy model envisaged in this problem consists of replacing $\int_{-\infty}^{\infty} dl$ and $\int_{-\infty}^{\infty} dk$ by discrete sums and limiting the indices to merely run over $1, 2, 3$, i.e., $\int_{-\infty}^{\infty} dl \Rightarrow \sum_{l=1}^{3}$ and $\int_{-\infty}^{\infty} dk \Rightarrow \sum_{k=1}^{3}$, respectively. The resulting toy model,

$$< x|f >= \sum_{l=1}^{3} < x|\beta_l > \left\{ \sum_{k=1}^{3} < b_l|\alpha_k >< a_k|f > \right\}, \tag{2.281}$$

is much simpler and yet preserves all the essential features of the original transformation formula (2.280). As a first step expand the l-sum,

$$
\begin{aligned}
< x|f > = & < x|\beta_1 > \left\{ \sum_{k=1}^{3} < b_1|\alpha_k >< a_k|f > \right\} \\
& + < x|\beta_2 > \left\{ \sum_{k=1}^{3} < b_2|\alpha_k >< a_k|f > \right\} \\
& + < x|\beta_3 > \left\{ \sum_{k=1}^{3} < b_3|\alpha_k >< a_k|f > \right\}.
\end{aligned}
\tag{2.282}
$$

Next expand the k-sums,

$$
\begin{aligned}
< x|f > = & < x|\beta_1 > \left\{ < b_1|\alpha_1 >< a_1|f > + < b_1|\alpha_2 >< a_2|f > + < b_1|\alpha_3 >< a_3|f > \right\} \\
& + < x|\beta_2 > \left\{ < b_2|\alpha_1 >< a_1|f > + < b_2|\alpha_2 >< a_2|f > + < b_2|\alpha_3 >< a_3|f > \right\} \\
& + < x|\beta_3 > \left\{ < b_3|\alpha_1 >< a_1|f > + < b_3|\alpha_2 >< a_2|f > + < b_3|\alpha_3 >< a_3|f > \right\}.
\end{aligned}
\tag{2.283}
$$

It should not be difficult to see that the scalar quantity at the R.H.S. can be recast in a matrix format,

$$< x|f >= \begin{bmatrix} < x|\beta_1 > & < x|\beta_2 > & < x|\beta_3 > \end{bmatrix} \begin{bmatrix} < b_1|\alpha_1 > & < b_1|\alpha_2 > & < b_1|\alpha_3 > \\ < b_2|\alpha_1 > & < b_2|\alpha_2 > & < b_2|\alpha_3 > \\ < b_3|\alpha_1 > & < b_3|\alpha_2 > & < b_3|\alpha_3 > \end{bmatrix} \begin{bmatrix} < a_1|f > \\ < a_2|f > \\ < a_3|f > \end{bmatrix}. \tag{2.284}$$

Extending the sums to include any finite number ($N \in \mathbb{N}$) of terms should be self-evident.

The expression at the R.H.S. of (2.284) permits better visualizing the originating formula (2.280): the result in (2.284) can be interpreted in three steps by interpreting the column vector, the matrix, and the row vector at the R.H.S. The column vector contains the available information about the projection of the function $|f>$ onto the original ket coordinates $|a_k> (k = 1, 2, 3)$. The matrix contains the projections of the dual kets $|\alpha_k> (k = 1, 2, 3)$ (associated with the original coordinates) onto the kets of the new coordinate system $|b_l> (l = 1, 2, 3)$. The row vector contains the information about the projection of the dual kets of the new coordinates $|\beta_l> (l = 1, 2, 3)$ onto the position ket $|x>$.

Next consider the special case $< b_l|\alpha_k >= \delta_{kl}$. This means that the new ket coordinates are orthonormal to the old dual ket coordinates. On the other hand, and by construction, it is the case that the new kets are orthogonal to the new dual kets, and, the old dual ket are orthogonal to the old kets. Consequently, the new dual kets must be orthogonal to the old kets. This condition or a consequence thereof must follow from the special case $< b_l | \alpha_k >= \delta_{kl}$. The latter condition implies that the matrix in (2.284) is a unity matrix. Thus,

$$< x | f >= \left[\; < x | \beta_1 > \quad < x | \beta_2 > \quad < x | \beta_3 > \; \right] \begin{bmatrix} < a_1 | f > \\ < a_2 | f > \\ < a_3 | f > \end{bmatrix}. \tag{2.285}$$

Extracting $< x |$ from the left, and $| f >$ from the right,

$$< x | f >=< x | \left\{ \left[\; | \beta_1 > \quad | \beta_2 > \quad | \beta_3 > \; \right] \begin{bmatrix} < a_1 | \\ < a_2 | \\ < a_3 | \end{bmatrix} \right\} | f > . \tag{2.286}$$

Carrying out the multiplication of the row and column vectors at the R.H.S.,

$$< x | f >=< x | \{ | \beta_1 >< a_1 | + | \beta_2 >< a_2 | + | \beta_3 >< a_3 | \} | f > . \tag{2.287}$$

Comparing both sides,

$$\hat{\mathbb{I}} = | \beta_1 >< a_1 | + | \beta_2 >< a_2 | + | \beta_3 >< a_3 | . \tag{2.288}$$

Or, written more compactly,

$$\hat{\mathbb{I}} = \sum_{n=1}^{3} | \beta_n >< a_n | , \tag{2.289}$$

as predicted as a necessary condition in the considered special case.

∎

Problem: Gaining insight.

In the preceding problem use was made of the equality,

$$\left[\; | \beta_1 > \quad | \beta_2 > \quad | \beta_3 > \; \right] \begin{bmatrix} < a_1 | \\ < a_2 | \\ < a_3 | \end{bmatrix} = | \beta_1 >< a_1 | + | \beta_2 >< a_2 | + | \beta_3 >< a_3 | . \tag{2.290}$$

Show that this equality is valid.

Solution: The equality of the two representations in (2.290) for the identity operator $\hat{\mathbb{I}}$ deserves to be viewed as an important insight how the exterior product can be interpreted as the inner product of two super vectors, the super vectors being in fact matrices in disguise. To establish the first equality in (2.290), the Kets $|\beta_n>$ $(n = 1, 2, 3)$ and the Bras $<a_n|$ $(n = 1, 2, 3)$ are written as column and row vectors, respectively,

$$
\begin{bmatrix} (\beta_1)_1 & (\beta_2)_1 & (\beta_3)_1 \\ (\beta_1)_2 & (\beta_2)_2 & (\beta_3)_2 \\ (\beta_1)_3 & (\beta_2)_3 & (\beta_3)_3 \end{bmatrix} \begin{bmatrix} (a_1)_1 & (a_1)_2 & (a_1)_3 \\ (a_2)_1 & (a_2)_2 & (a_2)_3 \\ (a_3)_1 & (a_3)_2 & (a_3)_3 \end{bmatrix} = \begin{bmatrix} (\beta_1)_1 \\ (\beta_1)_2 \\ (\beta_1)_3 \end{bmatrix} \begin{bmatrix} (a_1)_1 & (a_1)_2 & (a_1)_3 \end{bmatrix}
$$

$$
+ \begin{bmatrix} (\beta_2)_1 \\ (\beta_2)_2 \\ (\beta_2)_3 \end{bmatrix} \begin{bmatrix} (a_2)_1 & (a_2)_2 & (a_2)_3 \end{bmatrix}
$$

$$
+ \begin{bmatrix} (\beta_3)_1 \\ (\beta_3)_2 \\ (\beta_3)_3 \end{bmatrix} \begin{bmatrix} (a_3)_1 & (a_3)_2 & (a_3)_3 \end{bmatrix}. \tag{2.291}
$$

It is an easy task to convince oneself that the two sides are equal.

■

Problem: Transformation of over-complete coordinates revisited.

Consider the transformation formula in the abstract Hilbert space,

$$
|f> = \int_{-\infty}^{\infty} dl |\beta_l> \left\{ \int_{-\infty}^{\infty} dk < b_l|\alpha_k > < a_k|f > \right\}. \tag{2.292}
$$

Shed light on the internal structure of transformation of the over-complete coordinates.

Solution: Reverse engineering (2.292) by omitting the test function $|f>$,

$$
\hat{\mathbb{I}} = \int_{-\infty}^{\infty} dl |\beta_l> \left\{ \int_{-\infty}^{\infty} dk < b_l|\alpha_k > < a_k| \right\}. \tag{2.293}
$$

Since $<b_l|$ are independent of k,

$$
\hat{\mathbb{I}} = \int_{-\infty}^{\infty} dl |\beta_l> < b_l| \int_{-\infty}^{\infty} dk |\alpha_k> < a_k|. \tag{2.294}
$$

Each integral being equal to $\hat{\mathbb{I}}$, this relationship is the trivial expression of the fact that $\hat{\mathbb{I}}$ is idempotent, i.e. $\hat{\mathbb{I}} = \hat{\mathbb{I}}^2$. On the other hand, since $|\beta_l>$ are independent of k,

$$
\hat{\mathbb{I}} = \int_{-\infty}^{\infty} dl \int_{-\infty}^{\infty} dk |\beta_l> < b_l|\alpha_k > < a_k|. \tag{2.295}
$$

■

Remark: Insights gained from the idempotence property of the identity operator.

From the preceding examples, several facts can be identified, which are summarized here.

- Fact I: Above it was shown that the genesis of (2.295) resides in the idempotence property of the identity operator, i.e., the property that when multiplying the identity operator by itself the identity operator is obtained. Obviously, this process can be repeated as often as it is needed.
- Fact II: Considering any complete orthonormal coordinates $\{|a_n >\}$ a corresponding resolution of the identity operator can be set up. The resulting resolution of identity solely dependents on the structure of the chosen coordinate system and does not depend on any function in the system. In this sense, the expression of the resulting resolution of identity is universal and it optimally reflects the structure of the underlying coordinate system.
- Fact III: Considering any over-complete coordinates $\{|a_n >\}$ along with its dual over-complete coordinates $\{|\alpha_n >\}$ a corresponding resolution of the identity operator can be set up. Here too, the resolution of identity solely dependents of the structure on the chosen over-complete coordinate system and its dual over-complete coordinate system. Furthermore, it does not depend on any function in the system. The resulting resolution of identity is universal and reflects the structure of the underlying coordinate system and its dual coordinate system.
- Fact IV: From Facts II and III, it follows that every complete or over-complete coordinate system is characterized by the expression of the corresponding resolution of identity. Thus, e.g., $\hat{\mathbb{I}}_a$ can be associated with the coordinate system $\{|a_n >\}$ and $\hat{\mathbb{I}}_{\alpha a}$ can be associated with the over-complete coordinate systems $\{|a_n >\}$ and its dual $\{|\alpha_n >\}$. Similarly, $\hat{\mathbb{I}}_b$ can be associated with the coordinate system $\{|b_n >\}$ and $\hat{\mathbb{I}}_{\beta b}$ can be associated with the over-complete coordinate systems $\{|b_n >\}$ and its dual $\{|\beta_n >\}$. This idea is not limited to two systems only. Likewise, $\hat{\mathbb{I}}_c$ can be associated with the coordinate system $\{|c_n >\}$ and $\hat{\mathbb{I}}_{\gamma c}$ can be associated with the over-complete coordinate systems $\{|c_n >\}$ and $\{|\gamma_n >\}$.
- Fact V: The idempotence property ensures that the multiplication of any number of identity operators is still an identity operator, e.g., $\mathbb{I} = \hat{\mathbb{I}}_{\gamma c}\,\hat{\mathbb{I}}_{\beta b}\,\hat{\mathbb{I}}_{\alpha a}$.

The following problem sheds light on a toy model which comprises two systems, i.e., $\mathbb{I} = \hat{\mathbb{I}}_{\beta b}\,\hat{\mathbb{I}}_{\alpha a}$.

$\square$

Problem: A toy model comprising two coordinate systems and their duals.

Consider (2.295) which resulted from considering the over-complete coordinates $\{|a_n >\}$ along with its dual over-complete coordinates $\{|\alpha_n >\}$, and the over-complete coordinates $\{|b_n >\}$ along with its dual over-complete coordinates $\{|\beta_n >\}$. Replacing the integrals with sums and letting the indices k and l run over the numbers $(1, 2, 3)$ generate a simple yet powerful enough toy model which illuminates the internal structure of (2.295). Details are summarized in the solution to follow.

A corresponding resolution of the identity operator can be set up. Here too, the resolution of identity solely depends on the structure of the chosen over-complete coordinate system and its dual over-complete coordinate system. Furthermore, it does not depend on any function in the system. The resulting resolution of identity is universal and reflects the structure of the underlying coordinate system and its dual coordinate system.

Solution: Consider the toy model,

$$\hat{\mathbb{I}} = \sum_{l=1}^{3}\sum_{k=1}^{3} |\beta_l >< b_l|\alpha_k >< a_k|. \tag{2.296}$$

Resolve the *l*-sum by letting *l* to run over (1,2,3),

$$\hat{\mathbb{I}} = \sum_{k=1}^{3} |\beta_1><b_1|\alpha_k><a_k|$$

$$+ \sum_{k=1}^{3} |\beta_2><b_2|\alpha_k><a_k|$$

$$+ \sum_{k=1}^{3} |\beta_3><b_3|\alpha_k><a_k|. \qquad (2.297)$$

Resolve the *k*-sum by letting *k* to run over (1,2,3),

$$\hat{\mathbb{I}} = |\beta_1><b_1|\alpha_1><a_1| + |\beta_1><b_1|\alpha_2><a_2| + |\beta_1><b_1|\alpha_3><a_3|$$
$$+ |\beta_2><b_2|\alpha_1><a_1| + |\beta_2><b_2|\alpha_2><a_2| + |\beta_2><b_2|\alpha_3><a_3|$$
$$+ |\beta_3><b_3|\alpha_1><a_1| + |\beta_3><b_3|\alpha_2><a_2| + |\beta_3><b_3|\alpha_3><a_3|. \qquad (2.298)$$

It can readily be verified that,

$$\hat{\mathbb{I}} = \begin{bmatrix} |\beta_1> & |\beta_2> & |\beta_3> \end{bmatrix} \begin{bmatrix} <b_1|\alpha_1> & <b_1|\alpha_2> & <b_1|\alpha_3> \\ <b_2|\alpha_1> & <b_2|\alpha_2> & <b_2|\alpha_3> \\ <b_3|\alpha_1> & <b_3|\alpha_2> & <b_3|\alpha_3> \end{bmatrix} \begin{bmatrix} <a_1| \\ <a_2| \\ <a_3| \end{bmatrix}. \qquad (2.299)$$

∎

Remark: Generalization of the result in (2.299):

Let (α, a) refer to the combined over-complete system $\{|a_k>, \ k \in \mathbb{N}\}$ and its dual over-complete coordinates $\{|\alpha_k>, \ k \in \mathbb{N}\}$. Let the identity operator $\hat{\mathbb{I}}_{(\alpha,a)}$ be associated with the system (α, a),

$$\hat{\mathbb{I}}_{(\alpha,a)} = \sum_{k=1}^{\infty} |\alpha_k><a_k|. \qquad (2.300)$$

Let (β, b) refer to the alternative combined over-complete system $\{|b_l>, \ l \in \mathbb{N}\}$ and its dual over-complete coordinates $\{|\beta_l>, \ l \in \mathbb{N}\}$. Let the identity operator $\hat{\mathbb{I}}_{(\beta,b)}$ be associated with the system (β, b),

$$\hat{\mathbb{I}}_{(\beta,b)} = \sum_{l=1}^{\infty} |\beta_l><b_l|. \qquad (2.301)$$

Let $[(\beta, b)(\alpha, a)]$ refer to the transformation of the system (α, a) to the system (β, b). Let the identity operator $\hat{\mathbb{I}}_{[(\beta,b)(\alpha,a)]}$ be associated with the transformation system $[(\beta, b)(\alpha, a)]$,

$$\hat{\mathbb{I}}_{[(\beta,b)(\alpha,a)]} = \sum_{l=1}^{\infty} \sum_{k=1}^{\infty} |\beta_l><b_l|\alpha_k><a_k|. \qquad (2.302)$$

This procedure can be generalized conveniently and straightforwardly. As an example, the accommodation of a second transformation via a complete system (c, c), with the corresponding identity operator $\hat{\mathbb{I}}_{[(c,c)]}$,

$$\hat{\mathbb{I}}_{[(c,c)]} = \sum_{m=1}^{\infty} |c_m><c_m|, \qquad (2.303)$$

or, an over-complete system (γ, c), with the corresponding identity operator $\hat{\mathbb{I}}_{[(\gamma,c)]}$,

$$\hat{\mathbb{I}}_{[(\gamma,c)]} = \sum_{m=1}^{\infty} |\gamma_m \rangle \langle c_m|, \tag{2.304}$$

is quite self-explanatory. It amounts to applying $\hat{\mathbb{I}}_{[(c,c)]}$ onto $\hat{\mathbb{I}}_{[(\beta,b)(\alpha,a)]}$ leading to $\hat{\mathbb{I}}_{[(c,c)(\beta,b)(\alpha,a)]}$,

$$\hat{\mathbb{I}}_{[(c,c)(\beta,b)(\alpha,a)]} = \hat{\mathbb{I}}_{[(c,c)]}\hat{\mathbb{I}}_{[(\beta,b)(\alpha,a)]} \tag{2.305}$$

$$= \sum_{m=1}^{\infty} |c_m \rangle \langle c_m| \left\{ \sum_{l=1}^{\infty} \sum_{k=1}^{\infty} |\beta_l \rangle \langle b_l|\alpha_k \rangle \langle a_k| \right\} \tag{2.306}$$

$$= \sum_{m=1}^{\infty} \sum_{l=1}^{\infty} \sum_{k=1}^{\infty} |c_m \rangle \langle c_m|\beta_l \rangle \langle b_l|\alpha_k \rangle \langle a_k|, \tag{2.307}$$

respectively, $\hat{\mathbb{I}}_{[(\gamma,c)]}$ onto $\hat{\mathbb{I}}_{[(\beta,b)(\alpha,a)]}$ leading to $\hat{\mathbb{I}}_{[(\gamma,c)(\beta,b)(\alpha,a)]}$,

$$\hat{\mathbb{I}}_{[(\gamma,c)(\beta,b)(\alpha,a)]} = \hat{\mathbb{I}}_{[(\gamma,c)]}\hat{\mathbb{I}}_{[(\beta,b)(\alpha,a)]} \tag{2.308}$$

$$= \sum_{m=1}^{\infty} |\gamma_m \rangle \langle c_m| \left\{ \sum_{l=1}^{\infty} \sum_{k=1}^{\infty} |\beta_l \rangle \langle b_l|\alpha_k \rangle \langle a_k| \right\} \tag{2.309}$$

$$= \sum_{m=1}^{\infty} \sum_{l=1}^{\infty} \sum_{k=1}^{\infty} |\gamma_m \rangle \langle c_m|\beta_l \rangle \langle b_l|\alpha_k \rangle \langle a_k|. \tag{2.310}$$

The following toy models may serve to elucidate the underlying ideas. The key is the realization that the exterior products constituting identity operators can formally be viewed as the product of a horizontal sequence of Kets by a vertical sequence of Bras. The introduction of super Bras and super Kets promise to fully clarify how transformations can be carried. The first of the following toy models explains the idea underlying super Bras and super Kets. The second toy model explains the generalization from one coordinate transformation to two coordinate transformations.

□

Remark: Super bras and super kets and coordinate transformation.

Consider the following toy model. The introduction of super bras and super kets offers itself,

$$\hat{\mathbb{I}}_{(\alpha,a)} = \sum_{n=1}^{3} |\alpha_n \rangle \langle a_n|$$

$$= |\alpha_1 \rangle \langle a_1| + |\alpha_2 \rangle \langle a_2| + |\alpha_3 \rangle \langle a_3|$$

$$= \begin{bmatrix} |\alpha_1 \rangle & |\alpha_2 \rangle & |\alpha_3 \rangle \end{bmatrix} \begin{bmatrix} \langle a_1| \\ \langle a_2| \\ \langle a_3| \end{bmatrix}. \tag{2.311}$$

The super bra $\begin{bmatrix} |\alpha_1 \rangle & |\alpha_2 \rangle & |\alpha_3 \rangle \end{bmatrix}$ is referred to as a bra since it is a collection of mathematical entities that are arranged horizontally. It happens that the mathematical entities, which constitute the components of the super bra, are kets. Consequently, super bras are in essence matrices. Viewing them as super row vectors offers great utility in the manipulations. Similarly, the super ket,

$$
\begin{bmatrix}
< a_1 | \\
< a_2 | \\
< a_3 |
\end{bmatrix},
$$

is referred to as a ket since it is a collection of mathematical entities that are arranged vertically. It happens that the mathematical entities, which constitute the components of the super kets are bras. Consequently, super kets are in essence matrices as well. Viewing them as super column vectors provides significant utility.

Next consider a second toy model, where the super bra and the super ket have been introduced in the obvious manner,

$$
\begin{aligned}
\hat{\mathbb{I}}_{(\beta,b)} &= \sum_{n=1}^{3} |\beta_n> < b_n| \\
&= |\beta_1> < b_1| + |\beta_2> < b_2| + |\beta_3> < b_3| \\
&= \begin{bmatrix} |\beta_1> & |\beta_2> & |\beta_3> \end{bmatrix} \begin{bmatrix} < b_1| \\ < b_2| \\ < b_3| \end{bmatrix}.
\end{aligned} \tag{2.312}
$$

Thus

$$
\begin{aligned}
\hat{\mathbb{I}}_{[(\beta,b)(\alpha,a)]} &= \hat{\mathbb{I}}_{(\beta,b)}\hat{\mathbb{I}}_{(\alpha,a)} \\
&= \left\{ \sum_{n=1}^{3} |\beta_n> < b_n| \right\} \left\{ \sum_{m=1}^{3} |\alpha_m> < a_m| \right\} \\
&= \left\{ \begin{bmatrix} |\beta_1> & |\beta_2> & |\beta_3> \end{bmatrix} \begin{bmatrix} < b_1| \\ < b_2| \\ < b_3| \end{bmatrix} \right\} \left\{ \begin{bmatrix} |\alpha_1> & |\alpha_2> & |\alpha_3> \end{bmatrix} \begin{bmatrix} < a_1| \\ < a_2| \\ < a_3| \end{bmatrix} \right\} \\
&= \begin{bmatrix} |\beta_1> & |\beta_2> & |\beta_3> \end{bmatrix} \underbrace{\begin{bmatrix} < b_1| \\ < b_2| \\ < b_3| \end{bmatrix} \begin{bmatrix} |\alpha_1> & |\alpha_2> & |\alpha_3> \end{bmatrix}} \begin{bmatrix} < a_1| \\ < a_2| \\ < a_3| \end{bmatrix} \\
&= \begin{bmatrix} |\beta_1> & |\beta_2> & |\beta_3> \end{bmatrix} \begin{bmatrix} < b_1|\alpha_1> & < b_1|\alpha_2> & < b_1|\alpha_3> \\ < b_2|\alpha_1> & < b_2|\alpha_2> & < b_2|\alpha_3> \\ < b_3|\alpha_1> & < b_3|\alpha_2> & < b_3|\alpha_3> \end{bmatrix} \begin{bmatrix} < a_1| \\ < a_2| \\ < a_3| \end{bmatrix}.
\end{aligned} \tag{2.313}
$$

□

Remark: Plausibilization of arranging super bras and super kets: constructability arguments.

It is important to be aware that the components of super bras and super kets can be finite, countably infinite, or continuously infinite. To follow the arguments in this and the following remarks more easily, it is instructive to view super bras and super kets to be countably infinite. This leads to a visual aid and allows one to imagine column vectors (with infinitely many components) to be arranged in parallel, while each column vector being assigned to integers in the horizontal direction: the arrangement of infinitely many parallel columns (each having infinitely many components) can easily be imagined. In contrast, column vectors (with infinitely many components) to be arranged vertically in order to build an infinite multitude of infinitely long column vectors in the vertical direction must be rejected for reasons of unconstructability. It turns out that there is no need for (unnecessarily) venturing into the realm of infinities of higher cardinality. Stated slightly differently, infinitely many infinitely long columns on top of each other cannot be rendered meaningful, imagined, let alone realized, without introducing sophisticated logical apparatuses to deal with infinities of higher cardinality. And, luckily there is no need for their introduction in the first place. Thus,

1. Horizontal arrangement of vertical columns is permitted, and it introduces the notion of super bras. It leads to a permissible matrix structure.
2. Vertical arrangement of vertical columns is forbidden. The attempt to construct an unconstructable vertical column is an oxymoron and thus refutes itself.

Similarly, infinitely many parallel horizontal rows (each having infinitely many components) can easily be imagined. In contrast, infinitely long horizontal rows to be arranging horizontally in order to build an infinite multitude of infinitely long row vectors in the horizontal direction must be rejected for reasons of unconstructability. Expressed, more succinctly, infinitely many infinitely long horizontal rows concatenated next to each other horizontally cannot be easily imagined and rendered meaningful. And, fortunately there is no need to do so. Thus

1. Vertical arrangement of horizontal rows is permitted, and it introduces the notion of super kets. It leads to a permissible matrix structure.
2. Horizontal arrangement of horizontal rows is forbidden. The attempt to construct an unconstructable horizontal row is an oxymoron and thus rules itself out.

$\square$

Remark: Multiplication rule of super bras and super kets.

The structure of the final result in (2.313) is revealing, and it permits interesting interpretations. It will be assumed that the super bras and the super kets have countably infinite components.

- The countably infinite components of super bras are arranged horizontally, by virtue of super bras being row vectors. It only makes sense that the components of super bras are themselves kets, which are column vectors.
- The components of super kets, arranged vertically by definition, can only make sense if the components are bras.
- The genesis of transformation matrices, the last line in (2.313), are super bras and super kets, the last but one line in (2.313), the under-braced vectors.
- The matrix sandwiched between the super bra and the super ket in the last line in (2.313) is the matrix responsible for transforming the (α, a) system to the (β, b) system. It is important to note that in constructing the transformation matrix the dual over-complete coordinates (α) of the first system (α, a) and the over-complete coordinates b of the second system (β, b) are involved.
- The transformation matrix, being by definition a matrix which is necessarily built from super vectors, makes sense, if it is the product of a super ket and super bra.
- Consider the super ket in the last line. Note that the structure of the super ket only makes sense if the entities comprising the components of this vector are bras.
- The multiplication of the super ket by a ket is well defined.
- Any representation of the identity operator, e.g., in (2.313), is left-guarded by a super bra and right-guarded by a super ket.

- Any representation of the identity operator, e.g., in (2.313), can operate onto a ket or onto a super bra. For example it can apply to the test function $|f>$. A toy model should reveal the details.

$$
\begin{bmatrix} <a_1| \\ <a_2| \\ <a_3| \end{bmatrix} |f> = \begin{bmatrix} (a_1)_1 & (a_1)_2 & (a_1)_3 \\ (a_2)_1 & (a_2)_2 & (a_2)_3 \\ (a_3)_1 & (a_3)_2 & (a_3)_3 \end{bmatrix} \begin{bmatrix} f_1 \\ f_2 \\ f_3 \end{bmatrix}
$$

$$
= \begin{bmatrix} (a_1)_1 f_1 + (a_1)_2 f_2 + (a_1)_3 f_3 \\ (a_2)_1 f_1 + (a_2)_2 f_2 + (a_2)_3 f_3 \\ (a_3)_1 f_1 + (a_3)_2 f_2 + (a_3)_3 f_3 \end{bmatrix}
$$

$$
= \begin{bmatrix} <a_1|f> \\ <a_2|f> \\ <a_3|f> \end{bmatrix}. \tag{2.314}
$$

Considering the first and the last terms, i.e.,

$$
\begin{bmatrix} <a_1| \\ <a_2| \\ <a_3| \end{bmatrix} |f>, \qquad \begin{bmatrix} <a_1|f> \\ <a_2|f> \\ <a_3|f> \end{bmatrix},
$$

it can be concluded that the multiplication of the super ket a and the test ket $|f>$ is akin to the multiplication of a vector by a constant. It can also be observed that the product of a super ket and a ket leads to an ordinary column vector.

- Left multiplication of $<x|$ onto a super bra.

$$
<x| \left[|\beta_1> \ |\beta_2> \ |\beta_3> \right] = \begin{bmatrix} x_1 & x_2 & x_3 \end{bmatrix} \begin{bmatrix} (\beta_1)_1 & (\beta_2)_1 & (\beta_3)_1 \\ (\beta_1)_2 & (\beta_2)_2 & (\beta_3)_2 \\ (\beta_1)_3 & (\beta_2)_3 & (\beta_3)_3 \end{bmatrix}
$$

$$
= \begin{bmatrix} \underbrace{x_1(\beta_1)_1 + x_2(\beta_1)_2 + x_3(\beta_1)_3} & \underbrace{x_1(\beta_2)_1 + x_2(\beta_2)_2 + x_3(\beta_2)_3} & \underbrace{x_1(\beta_3)_1 + x_2(\beta_3)_2 + x_3(\beta_3)_3} \end{bmatrix}
$$

$$
= \begin{bmatrix} <x|\beta_1> & <x|\beta_2> & <x|\beta_3> \end{bmatrix}. \tag{2.315}
$$

Considering the first and the last term, i.e.,

$$
<x| \left[|\beta_1> \ |\beta_2> \ |\beta_3> \right], \qquad \begin{bmatrix} <x|\beta_1> & <x|\beta_2> & <x|\beta_3> \end{bmatrix}, \tag{2.316}
$$

it can be concluded that the left multiplication of the super bra β onto the position bra $<x|$ is akin to the multiplication of a vector by a constant. It can also be observed that the product of a bra and a super bra leads to an ordinary row vector.

□

The following two successive transformation based on a simplest possible toy model should exhaust cases of potential interest.

Remark: Super bras and super kets and two successive coordinate transformations.

Consider the following simplest possible toy models each consisting of two exterior products only:

$$\hat{\mathbb{1}}_{(\alpha,a)} = \sum_{n=1}^{2} |\alpha_n><a_n| = |\alpha_1><a_1| + |\alpha_2><a_2| = \begin{bmatrix} |\alpha_1> & |\alpha_2> \end{bmatrix} \begin{bmatrix} <a_1| \\ <a_2| \end{bmatrix} \tag{2.317a}$$

$$\hat{\mathbb{1}}_{(\beta,b)} = \sum_{n=1}^{2} |\beta_n><b_n| = |\beta_1><b_1| + |\beta_2><b_2| = \begin{bmatrix} |\beta_1> & |\beta_2> \end{bmatrix} \begin{bmatrix} <b_1| \\ <b_2| \end{bmatrix} \tag{2.317b}$$

$$\hat{\mathbb{1}}_{(\gamma,c)} = \sum_{n=1}^{2} |\gamma_n><c_n| = |\gamma_1><c_1| + |\gamma_2><c_2| = \begin{bmatrix} |\gamma_1> & |\gamma_2> \end{bmatrix} \begin{bmatrix} <c_1| \\ <c_2| \end{bmatrix}. \tag{2.317c}$$

Thus,

$$\hat{\mathbb{1}}_{[(\gamma,c)(\beta,b)(\alpha,a)]} = \hat{\mathbb{1}}_{(\gamma,c)}\hat{\mathbb{1}}_{(\beta,b)}\hat{\mathbb{1}}_{(\alpha,a)}$$

$$= \left\{ \sum_{n=1}^{2} |\gamma_n><c_n| \right\} \left\{ \sum_{n=1}^{2} |\beta_n><b_n| \right\} \left\{ \sum_{m=1}^{2} |\alpha_m><a_m| \right\}$$

$$= \left\{ \begin{bmatrix} |\gamma_1> & |\gamma_2> \end{bmatrix} \begin{bmatrix} <c_1| \\ <c_2| \end{bmatrix} \right\} \left\{ \begin{bmatrix} |\beta_1> & |\beta_2> \end{bmatrix} \begin{bmatrix} <b_1| \\ <b_2| \end{bmatrix} \right\} \left\{ \begin{bmatrix} |\alpha_1> & |\alpha_2> \end{bmatrix} \begin{bmatrix} <a_1| \\ <a_2| \end{bmatrix} \right\}$$

$$= \begin{bmatrix} |\gamma_1> & |\gamma_2> \end{bmatrix} \underbrace{\begin{bmatrix} <c_1| \\ <c_2| \end{bmatrix} \begin{bmatrix} |\beta_1> & |\beta_2> \end{bmatrix}} \underbrace{\begin{bmatrix} <b_1| \\ <b_2| \end{bmatrix} \begin{bmatrix} |\alpha_1> & |\alpha_2> \end{bmatrix}} \begin{bmatrix} <a_1| \\ <a_2| \end{bmatrix}$$

$$= \begin{bmatrix} |\gamma_1> & |\gamma_2> \end{bmatrix} \begin{bmatrix} <c_1|\beta_1> & <c_1|\beta_2> \\ <c_2|\beta_1> & <c_2|\beta_2> \end{bmatrix} \begin{bmatrix} <b_1|\alpha_1> & <b_1|\alpha_2> \\ <b_2|\alpha_1> & <b_2|\alpha_2> \end{bmatrix} \begin{bmatrix} <a_1| \\ <a_2| \end{bmatrix}. \tag{2.318}$$

Consequently, by building $< x|\hat{\mathbb{I}}_{[(\gamma,c)(\beta,b)(\alpha,a)]}|f > = < x|f > = f(x)$, the analysis step and the synthesis step can be identified as follows:

$$f(x) =$$

$$\begin{bmatrix} < x|\gamma_1 > & < x|\gamma_2 > \end{bmatrix} \begin{bmatrix} < c_1|\beta_1 > & < c_1|\beta_2 > \\ < c_2|\beta_1 > & < c_2|\beta_2 > \end{bmatrix} \begin{bmatrix} < b_1|\alpha_1 > & < b_1|\alpha_2 > \\ < b_2|\alpha_1 > & < b_2|\alpha_2 > \end{bmatrix} \underbrace{\begin{bmatrix} < a_1|f > \\ < a_2|f > \end{bmatrix}}_{\text{the analysis step}}. \tag{2.319}$$

$$\underbrace{\qquad\qquad\qquad\qquad\qquad\qquad\qquad\qquad\qquad}_{\text{the synthesis step}}$$

□

The above toy models should suffice to explain the underlying intricacies when any number of complete (over-complete) systems are involved. In the remaining part of this chapter, a few further transforms have been discussed using the machinery developed so far.

2.9 Nyquist–Shannon sampling theorem: the resolution of identity

It is not the purpose here to review the Nyquist–Shannon theory of sampling and the Whittaker–Shannon interpolation of bounded band-limited signals, nor to introduce the relevant spaces (Bernstein Spaces, Paley–Wiener Spaces) and investigate their properties. Neither problems related to non-equidistant sampling have been addressed. The modest objective here is rather to view these wide-ranging ideas, concepts and methods in the context of the resolution of identity, and demonstrate the simplicity, elegance and the power of the Dirac bracket notation.

Consider the real-valued sinc(x)-function, defined by,

$$\text{sinc}(x) = \begin{cases} \frac{\sin \pi x}{\pi x} & x \neq 0 \\ 1 & x = 0. \end{cases} \tag{2.320}$$

Furthermore, define $\text{sinc}_n(x) = \text{sinc}(x - n)$ for $n \in \mathbb{Z}$. Finally, consider the Ket $|\text{sinc}_n > (n \in \mathbb{Z})$ in the abstract Hilbert space $\mathbb{H}$, with

$$< x|\text{sinc}_n >= \text{sinc}_n(x) = \text{sinc}(x - n). \tag{2.321}$$

Orthonormality of $\{|\text{sinc}_n >, n \in \mathbb{Z}\}$**:** Utilizing the expression of the resolution of identity,

$$\hat{\mathbb{I}} = \int_{-\infty}^{\infty} dx |x > < x| \tag{2.322}$$

138   *Mathematical quantum physics for engineers and technologists*

the following steps are immediate,

$$< \text{sinc}_m|\text{sinc}_n > = < \text{sinc}_m|\hat{\mathbb{1}}|\text{sinc}_n > \tag{2.323a}$$

$$= < \text{sinc}_m| \left\{ \int_{-\infty}^{\infty} dx|x >< x| \right\} |\text{sinc}_n > \tag{2.323b}$$

$$= \int_{-\infty}^{\infty} dx \underbrace{< \text{sinc}_m|x >} < x|\text{sinc}_n > \tag{2.323c}$$

$$= \int_{-\infty}^{\infty} dx < x|\text{sinc}_m >< x|\text{sinc}_n > \tag{2.323d}$$

$$= \int_{-\infty}^{\infty} dx \, \text{sinc}_m(x)\text{sinc}_n(x) \tag{2.323e}$$

$$= \delta_{mn}. \tag{2.323f}$$

The validity of the last transition can be established by resorting to the Parseval theorem: let $f(x) \Longleftrightarrow F(k)$ and $g(x) \Longleftrightarrow G(k)$ be (real- or complex-valued) Fourier pairs. Then,

$$\int_{-\infty}^{\infty} dx f^*(x)g(x) \overset{\text{Par.Th.}}{=} \int_{-\infty}^{\infty} dk F^*(k)G(k). \tag{2.324}$$

If $f(x)$ and $g(x)$ are real-valued functions, the integral at the L.H.S. meaningfully involves $f(x)g(x)$.

The reader might recall that the Fourier transform of the box-function (constant over a finite interval and zero otherwise) in real-space is the sinc-function. The following problem designs a box-function in the spectral domain such that its inverse Fourier transform renders the sinc-function in real space.

Problem: Let the function BOX(k) be defined in spectral domain in terms of two *a priori* unknown parameters $K > 0$ and $\alpha > 0$,

$$\text{BOX}(k) = \begin{cases} \alpha, & -K < k < K \\ 0, & \text{otherwise} \end{cases} . \tag{2.325}$$

Calculate the inverse Fourier transform of BOX(k), denoted by box(x). Determine the parameters $K > 0$ and $\alpha > 0$ by casting box(x) into the canonical form, to be discussed in the course of the solution. Employ the inverse Fourier transform in symmetric form.

Solution:

$$\text{box}(x) = \int_{-\infty}^{\infty} dk \frac{e^{ikx}}{\sqrt{2\pi}} \text{BOX}(k) \tag{2.326a}$$

$$\overset{(2.325)}{=} \int_{-K}^{K} dk \frac{e^{ikx}}{\sqrt{2\pi}} \alpha \tag{2.326b}$$

$$= \frac{\alpha}{\sqrt{2\pi}} \int_{-K}^{K} dk e^{ikx} \tag{2.326c}$$

$$= \frac{\alpha}{\sqrt{2\pi}} \frac{e^{iKx} - e^{-iKx}}{ix} \tag{2.326d}$$

$$= \frac{\alpha}{\sqrt{2\pi}} \frac{2i\sin(Kx)}{ix}. \tag{2.326e}$$

Focus on (2.326e). Canceling i from the numerator and the denominator, extending the denominator by K, and simplifying,

$$\text{box}(x) = \frac{2K\alpha}{\sqrt{2\pi}} \frac{\sin(Kx)}{Kx}. \tag{2.327}$$

Setting the design parameter $K = \pi$ renders $\frac{\sin(Kx)}{Kx} = \frac{\sin(\pi x)}{\pi x} = \text{sinc}(x)$. Thus, with $K = \pi$,

$$\text{box}(x) = \frac{2\pi\alpha}{\sqrt{2\pi}} \text{sinc}(x). \tag{2.328}$$

Setting the design parameter $\alpha = 1/\sqrt{2\pi}$, the coefficient $2\pi\alpha/\sqrt{2\pi}$ becomes unity. Thus

$$\text{box}(x) = \text{sinc}(x), \tag{2.329}$$

which is the desired canonical form. Consequently, the following Fourier pair can be established,

$$\text{BOX}(k) = \begin{cases} \frac{1}{\sqrt{2\pi}}, & -\pi < k < \pi \\ 0, & \text{otherwise} \end{cases} \quad \Longleftrightarrow \quad \text{sinc}(x). \tag{2.330}$$

■

It is known that a shift in spatial domains results in a corresponding phase change in spectral domain. The following problem is motivated by this idea.

Problem: Investigate the implication of multiplying $\text{BOX}(k)$ by e^{-ikn}. Stated more precisely, determine the inverse Fourier transform of

$$\text{BOX}(k)e^{-ikn} = \begin{cases} \frac{1}{\sqrt{2\pi}} e^{-ikn}, & -\pi < k < \pi \\ 0, & \text{otherwise} \end{cases}. \tag{2.331}$$

Solution:

$$\int_{-\infty}^{\infty} dk \frac{e^{ikx}}{\sqrt{2\pi}} \left[\text{BOX}(k)e^{-ikn} \right] \overset{(2.331)}{=} \int_{-\pi}^{\pi} dk \frac{e^{ikx}}{\sqrt{2\pi}} \left[\frac{1}{\sqrt{2\pi}} e^{-ikn} \right] \tag{2.332a}$$

$$= \frac{1}{2\pi} \int_{-\pi}^{\pi} dk e^{ik(x-n)} \tag{2.332b}$$

$$= \frac{1}{2\pi} \frac{e^{i\pi(x-n)} - e^{-i\pi(x-n)}}{i(x-n)} \tag{2.332c}$$

$$= \frac{1}{2\pi} \frac{2i \sin \pi(x-n)}{i(x-n)} \tag{2.332d}$$

$$= \frac{\sin \pi(x-n)}{\pi(x-n)} \tag{2.332e}$$

$$\overset{\text{def.}}{=} \text{sinc}(x-n) \tag{2.332f}$$

$$\overset{\text{def.}}{=} \text{sinc}_n(x). \tag{2.332g}$$

Consequently, the following Fourier pair can be established,

$$\text{BOX}(k)e^{-ikn} = \begin{cases} \frac{1}{\sqrt{2\pi}} e^{-ikn} & -\pi < k < \pi \\ 0 & \text{otherwise} \end{cases} \qquad \Longleftrightarrow \qquad \text{sinc}_n(x). \tag{2.333}$$

∎

Lemma: *Orthonormality of* $\{\text{sinc}_n(x),\ n \in \mathbb{Z}\}$

Prove the orthonormality of functions $\{\text{sinc}_n(x),\ n \in \mathbb{Z}\}$

Proof: It is convenient to make use of the Parseval theorem:

$$\int_{-\infty}^{\infty} dx\ \text{sinc}_m(x)\text{sinc}_n(x) \overset{\text{Par. Th.}}{=} \int_{-\pi}^{\pi} dk \left\{ \frac{e^{ikm}}{\sqrt{2\pi}} \right\}^* \left\{ \frac{e^{ikn}}{\sqrt{2\pi}} \right\} \tag{2.334a}$$

$$= \int_{-\pi}^{\pi} dk \frac{e^{-ikm}}{\sqrt{2\pi}} \frac{e^{ikn}}{\sqrt{2\pi}} \tag{2.334b}$$

$$= \frac{1}{2\pi} \int_{-\pi}^{\pi} dk e^{ik(n-m)} \tag{2.334c}$$

$$= \frac{1}{2\pi} \frac{e^{i\pi(n-m)} - e^{-i\pi(n-m)}}{i(n-m)} \tag{2.334d}$$

$$= \frac{1}{2\pi} \frac{2i \sin \pi(n-m)}{i(n-m)} \tag{2.334e}$$

$$= \frac{\sin \pi(n-m)}{\pi(n-m)} \tag{2.334f}$$

$$\overset{\text{def.}}{=} \text{sinc}(n-m) \tag{2.334g}$$

$$= \delta_{mn}. \tag{2.334h}$$

Note that the inverse Fourier transforms of $\text{sinc}_m(x)$ and $\text{sinc}_n(x)$ are phased boxes defined on the interval $[-\pi, \pi]$, a fact which was established in the preceding problem. This explains the appearance of the windowed integral in (2.334a). Next consider (2.334f). In the case $n \neq m$ the denominator is non-zero, while the numerator vanishes. Thus for $n \neq m$ the function value $\text{sinc}(n-m)$ is zero. In the case $n = m$, $\text{sinc}(0) = 1$. The latter two facts justify the transition from (2.334g) to (2.334h). Consequently, the infinite set of functions $\{\text{sinc}_n(x), \ n \in \mathbb{Z}\}$ are orthonormal; they constitute an orthonormal basis (ONB).

∎

Problem: Show that the integral $\int\limits_{-\infty}^{\infty} dx \ \text{sinc}_m(x)\text{sinc}_n(x)$ is tantamount to using the continuous resolution of the position

identity operator $\int\limits_{-\infty}^{\infty} dx \ |x><x|$ in continuous-space inner-product $<\text{sinc}_m|\text{sinc}_n>$.

Solution:

$$\int\limits_{-\infty}^{\infty} dx \ \text{sinc}_m(x)\text{sinc}_n(x) = \int\limits_{-\infty}^{\infty} dx \ <x|\text{sinc}_m><x|\text{sinc}_n> \tag{2.335a}$$

$$= \int\limits_{-\infty}^{\infty} dx \ <\text{sinc}_m|x><x|\text{sinc}_n> \tag{2.335b}$$

$$= \ <\text{sinc}_m| \left\{ \int\limits_{-\infty}^{\infty} dx \ |x><x| \right\} |\text{sinc}_n> \tag{2.335c}$$

$$= \ <\text{sinc}_m|\hat{\mathbb{I}}|\text{sinc}_n> \tag{2.335d}$$

$$= \ <\text{sinc}_m|\text{sinc}_n>. \tag{2.335e}$$

∎

Problem: Show that the set of functions $\{\text{sinc}_n(x), \ n \in \mathbb{Z}\}$ in the discrete subspace characterized by the resolution of the identity operator $\hat{\mathbb{I}} = \sum\limits_{l=-\infty}^{\infty} |l><l|$ are also orthonormal.

Solution: Using the discrete representation $\sum\limits_{l=-\infty}^{\infty} |l><l|$ rather than the continuous representation $\int\limits_{-\infty}^{\infty} dx \ |x><x|$ of the position identity operator $\hat{\mathbb{I}}$, the following steps are self-explanatory:

$$<\text{sinc}_m|\text{sinc}_n> \ = \ <\text{sinc}_m|\hat{\mathbb{I}}|\text{sinc}_n> \tag{2.336a}$$

$$= \ <\text{sinc}_m| \left\{ \sum\limits_{l=-\infty}^{\infty} |l><l| \right\} |\text{sinc}_n> \tag{2.336b}$$

$$= \sum\limits_{l=-\infty}^{\infty} <\text{sinc}_m|l><l|\text{sinc}_n> \tag{2.336c}$$

$$= \sum\limits_{l=-\infty}^{\infty} \underbrace{<l|\text{sinc}_m>}_{\delta_{ml}} \underbrace{<l|\text{sinc}_n>}_{\delta_{ln}} \tag{2.336d}$$

$$= \sum\limits_{l=-\infty}^{\infty} \delta_{ml}\delta_{ln} \tag{2.336e}$$

$$= \delta_{mn}. \tag{2.336f}$$

As indicated in (2.336d), sampling $\text{sinc}_m(x)$ at any integer other than m ($x = l \neq m$) is zero. Whereas, sampling $\text{sinc}_m(x)$ at $x = l = m$ results in the unity. This means that for any $x = l \neq m$ the function $\text{sinc}_m(x)$ crosses the x-axis. Similarly, the function $\text{sinc}_n(x)$ is zero at all integers except $x = l = n$, where it has the value one.

Conversely, it can be stated that the sampling values of $\text{sinc}_n(x)$ at integers can be viewed as a realization (model) for the Kronecker delta symbol.

∎

Remark: The preceding two problems allude to two distinguishing properties of the set of functions $\{\text{sinc}_n(x), n \in \mathbb{Z}\}$, namely, the orthogonality on $\mathbb{R}$ and the orthogonality on (much smaller space) $\mathbb{Z}$. The property that the Fourier transform of $\text{sinc}_n(x)$ is a (modulated, phased) box-function, in spectral domain, and the property that the Fourier transform of integer translate of box-functions in real space are sinc-functions in spectral domain, catapult the sequence of the sinc-functions to extreme heights. The next problem demonstrates the utilization of the completeness of the $\text{sinc}_n(x)$-functions.

□

Problem: On the completeness of $\{\text{sinc}_n(x), n \in \mathbb{Z}\}$

$$\hat{\mathbb{I}} = \sum_{n=-\infty}^{\infty} |\text{sinc}_n><\text{sinc}_n|. \tag{2.337}$$

Demonstrate that the expression of the resolution of identify operator (2.337) incorporates both the sinc-transform and the sinc-inverse transform.

Solution: Applying both sides of (2.337) onto the test function $|f>$, readily accomplishes the sinc-transform (the analysis step) and the sinc-inverse transform (the synthesis step) in the abstract Hilbert space,

$$|f> = \sum_{n=-\infty}^{\infty} \overbrace{|\text{sinc}_n>}^{\text{the synthesis step}} \underbrace{<\text{sinc}_n|f>}_{\text{the analysis step}}. \tag{2.338}$$

where the $\hat{\mathbb{I}}|f> = |f>$ has been used at the L.H.S..

Translate (2.338) into ordinary function space: Project both sides onto the position Ket $|x>$,

$$<x|f> = \sum_{n=-\infty}^{\infty} <x|\text{sinc}_n><\text{sinc}_n|f>. \tag{2.339}$$

The L.H.S. is equal to $f(x)$. Furthermore, $<x|\text{sinc}_n> = \text{sinc}_n(x)$. Consequently,

$$f(x) = \sum_{n=-\infty}^{\infty} \text{sinc}_n(x) <\text{sinc}_n|f>. \tag{2.340}$$

In order to translate $<\text{sinc}_n|f>$ in (2.340) into ordinary function space both ingredients $|f>$ and $<\text{sinc}_n|$ must be projected onto position Kets. To achieve this objective, write $<\text{sinc}_n|f> = <\text{sinc}_n||f> = <\text{sinc}_n|\hat{\mathbb{I}}|f>$ and replace $\hat{\mathbb{I}}$ with the resolution of identity operator $\mathbb{I} = \int_{-\infty}^{\infty} d\xi |\xi><\xi|$,

$$f(x) = \sum_{n=-\infty}^{\infty} \text{sinc}_n(x) <\text{sinc}_n| \left\{ \int_{-\infty}^{\infty} d\xi |\xi><\xi| \right\} |f>. \tag{2.341}$$

Rearrange,

$$f(x) = \sum_{n=-\infty}^{\infty} \text{sinc}_n(x) \int_{-\infty}^{\infty} d\xi < \text{sinc}_n|\xi> <\xi|f>. \tag{2.342}$$

Since $|\text{sinc}_n>$ is real, $<\text{sinc}_n|\xi> = <\xi|\text{sinc}_n>$,

$$f(x) = \sum_{n=-\infty}^{\infty} \text{sinc}_n(x) \int_{-\infty}^{\infty} d\xi \underbrace{<\xi|\text{sinc}_n>}_{\text{sinc}_n(\xi)} \underbrace{<\xi|f>}_{f(\xi)}. \tag{2.343}$$

Thus,

$$f(x) = \overbrace{\sum_{n=-\infty}^{\infty} \text{sinc}_n(x)}^{\text{The Synthesis Step: } f(x)} \underbrace{\int_{-\infty}^{\infty} d\xi \, \text{sinc}_n(\xi) f(\xi)}_{\text{The Analysis Step: } F_n}. \tag{2.344}$$

Obtaining F_n by projecting $f(\xi)$ onto $\text{sinc}_n(\xi)$, as indicated, amounts to the analysis step (the sinc-transform). The resulting synthesis step,

$$f(x) = \sum_{n=-\infty}^{\infty} \text{sinc}_n(x) F_n, \tag{2.345}$$

corresponds to the sinc-inverse transform.

∎

Problem: sinc-Function interpolation formula for band-limited functions.

Develop sinc-function-based interpolation formula.

Solution: Consider

$$f(x) = \sum_{n=-\infty}^{\infty} \text{sinc}_n(x) < \text{sinc}_n|f>, \tag{2.346}$$

derived above (2.340), as the starting point. This formula holds valid for arbitrary square integrable functions. Assume next that $f(x)$ is band-limited. Let the support of the Fourier transform of $f(x)$ be the unit interval $\left[-\frac{1}{2}, \frac{1}{2}\right]$ in spectral domain. Insert the discrete resolution $\hat{\mathbb{I}} = \sum_{k=-\infty}^{\infty} |k> <k|$ (instead of the continuous resolution $\hat{\mathbb{I}} = \int_{-\infty}^{\infty} d\xi |\xi> <\xi|$) into $< \text{sinc}_n|f> \ (= < \text{sinc}_n||f>)$,

$$f(x) = \sum_{n=-\infty}^{\infty} \text{sinc}_n(x) < \text{sinc}_n| \left\{ \sum_{k=-\infty}^{\infty} |k> <k| \right\} |f>. \tag{2.347}$$

Rearranging,

$$f(x) = \sum_{n=-\infty}^{\infty} \text{sinc}_n(x) \sum_{k=-\infty}^{\infty} \underbrace{< \text{sinc}_n|k> <k|f>}. \tag{2.348}$$

With $< \text{sinc}_n|k> = < k|\text{sinc}_n> = \text{sinc}_n(k) = \delta_{nk}$, and $< k|f> = f(k)$,

$$f(x) = \sum_{n=-\infty}^{\infty} \text{sinc}_n(x) \underbrace{\sum_{k=-\infty}^{\infty} \delta_{nk} f(k)}_{f(n)}. \tag{2.349}$$

Introducing $f(n)$ at the R.H.S., as indicated,

$$f(x) = \sum_{n=-\infty}^{\infty} \text{sinc}_n(x) f(n), \tag{2.350}$$

the interpolation formula for band-limited functions emerges.

∎

Remark: The resolution of identity induced by the interpolation formula.

Recalling that $f(x) = < x|f>$ and $\text{sinc}_n(x) = < x|\text{sinc}_n>$, the interpolation formula, (2.350), can be cast in the form,

$$< x|f> = \sum_{n=-\infty}^{\infty} < x|\text{sinc}_n> < n|f>. \tag{2.351}$$

In virtue of the fact that the test function $|f>$ and the position Ket $|x>$ are arbitrary,

$$\hat{\mathbb{I}} = \sum_{n=-\infty}^{\infty} |\text{sinc}_n> < n|. \tag{2.352}$$

This is the resolution of identity induced by the interpolation formula. The utility of this result will be investigated momentarily. However, first two problems, to reinforce the insights gained so far.

□

Problem: Express the resolution of identity.

$$\hat{\mathbb{I}} = \sum_{n=-\infty}^{\infty} |\text{sinc}_n> < \text{sinc}_n|, \tag{2.353}$$

in real space.

Solution: Apply both sides onto $|x>$, and project the results onto $|x'>$,

$$\underbrace{< x'|\hat{\mathbb{I}}|x>}_{\delta(x-x')} = \sum_{n=-\infty}^{\infty} \underbrace{< x'|\text{sinc}_n>}_{\text{sinc}_n(x')} \underbrace{< \text{sinc}_n|x>}_{\text{sinc}_n(x)}. \tag{2.354}$$

Considering the changes, as indicated,

$$\delta(x-x') = \sum_{n=-\infty}^{\infty} \text{sinc}_n(x')\text{sinc}_n(x). \tag{2.355}$$

Since the sinc-functions are scalar,

$$\delta(x-x') = \sum_{n=-\infty}^{\infty} \text{sinc}_n(x)\text{sinc}_n(x'). \tag{2.356}$$

Stated differently, and perhaps a bit pedantically, the fact that the order of x and x' can be exchanged at the R.H.S., confirms the property $\delta(x-x') = \delta(x'-x)$.

∎

Problem: Based on the resolution of identity in real space, (2.356), establish the sinc-transform and inverse transform.

Solution: Multiply both sides by $f(x')$ and integrate over x',

$$\underbrace{\int_{-\infty}^{\infty} dx'\, \delta(x-x')f(x')}_{f(x)} = \sum_{n=-\infty}^{\infty} \mathrm{sinc}_n(x) \underbrace{\int_{-\infty}^{\infty} dx'\, \mathrm{sinc}_n(x')f(x')}_{f_n}. \tag{2.357}$$

Considering the definition of the sinc-transform of $f(x)$, as indicated,

$$f(x) = \sum_{n=-\infty}^{\infty} \mathrm{sinc}_n(x)f_n. \tag{2.358}$$

∎

In order to relate the above exposition to standard presentations, in the remaining part of this section time-varying signals are considered rather than position-dependent functions.

Interpolation: For every bounded discrete-time signal $f = \{f(k)|k \in \mathbb{Z}\}$ it is possible to construct a bounded band-limited continuous-time signal $f(t)$ which interpolates the discrete-time signals.

Shannon sampling theorem: Certain class of bounded band-limited signals are uniquely determined by their equidistant sampled values provided the samples are taken at least at the Nyquist rate, and that these signals can be perfectly reconstructed from the samples using the series,

$$f(t) = \sum_{n=-\infty}^{\infty} \frac{\sin\left[\pi(t-n)\right]}{\pi(t-n)} f(n). \tag{2.359}$$

For $t = m \in \mathbb{Z}$,

$$f(m) = \sum_{n=-\infty}^{\infty} \frac{\sin\left[\pi(m-n)\right]}{\pi(m-n)} f(n), \tag{2.360}$$

with

$$\frac{\sin\left[\pi(m-n)\right]}{\pi(m-n)} = \delta_{mn} = \begin{cases} 1 & n = m \\ 0 & n \neq m. \end{cases} \tag{2.361}$$

Substituting (2.361) into (2.360),

$$f(m) = \sum_{n=-\infty}^{\infty} \delta_{mn} f(n). \tag{2.362}$$

Thus at the L.H.S., considering the function values at integers only, the Shannon sampling theorem boils down to a fancy representation of the Kronecker delta symbol δ_{mn} in terms of $\frac{\sin[\pi(m-n)]}{\pi(m-n)}$. Writing (2.362) by utilizing Dirac delta function, i.e.,

$$\overbrace{\int_{-\infty}^{\infty} dt\delta(t-m)f(t)}^{f(m)} = \sum_{n=-\infty}^{\infty} \delta_{mn} \overbrace{\int_{-\infty}^{\infty} dt\delta(t-n)f(t)}^{f(n)} \tag{2.363a}$$

$$= \int_{-\infty}^{\infty} dt \sum_{n=-\infty}^{\infty} \delta_{mn}\delta(t-n)f(t). \tag{2.363b}$$

Since $f(t)$ is arbitrary, this expression allows deducing the resolution of identity,

$$\delta(t-m) = \sum_{n=-\infty}^{\infty} \delta_{mn}\delta(t-n), \tag{2.364a}$$

which is not an utterly exciting relationship. Perhaps a slightly more interesting result can be achieved by substituting $\int_{-\infty}^{\infty} d\tau\delta(\tau-t)f(\tau)$ and $\int_{-\infty}^{\infty} d\tau\delta(\tau-n)f(\tau)$, for $f(t)$ and $f(n)$, respectively, into (2.359),

$$\int_{-\infty}^{\infty} d\tau\delta(\tau-t)f(\tau) = \sum_{n=-\infty}^{\infty} \frac{\sin[\pi(t-n)]}{\pi(t-n)} \int_{-\infty}^{\infty} d\tau\delta(\tau-n)f(\tau). \tag{2.365}$$

Exchanging the order of summation and integration,

$$\int_{-\infty}^{\infty} d\tau\delta(\tau-t)f(\tau) = \int_{-\infty}^{\infty} d\tau \sum_{n=-\infty}^{\infty} \frac{\sin[\pi(t-n)]}{\pi(t-n)}\delta(\tau-n)f(\tau). \tag{2.366}$$

In virtue of the fact that $f(\tau)$ is arbitrary, a more interesting relationship for the resolution of identity manifests itself,

$$\delta(\tau-t) = \sum_{n=-\infty}^{\infty} \frac{\sin[\pi(t-n)]}{\pi(t-n)}\delta(\tau-n), \quad \tau, t \in \mathbb{R}. \tag{2.367}$$

To interpret this result, observe the orthonormality properties of the constituent terms at the R.H.S. of (2.367), individually, i.e.,

$$\int_{-\infty}^{\infty} dt \frac{\sin[\pi(t-m)]}{\pi(t-m)} \frac{\sin[\pi(t-n)]}{\pi(t-n)} = \delta_{mn}. \tag{2.368}$$

and

$$\int_{-\infty}^{\infty} d\tau\delta(\tau-m)\delta(\tau-n) = \delta_{mn}. \tag{2.369}$$

Furthermore, consider the mutual orthogonality property of the constituent terms,

$$\int\limits_{-\infty}^{\infty} dt \frac{\sin\left[\pi(t-m)\right]}{\pi(t-m)}\delta(t-n) = \delta_{mn}.$$ (2.370)

Thus, it can be concluded that the set of functions $\left\{\frac{\sin\left[\pi(t-n)\right]}{\pi(t-n)}\middle| n \in \mathbb{Z}\right\}$ and $\left\{\delta(\tau-n)\middle| n \in \mathbb{Z}\right\}$ are dual bases: Shannon sampling theorem amounts to resolving the identity, as stated in (2.367). Consequently, applying (2.367) onto any permissible function $f(\tau)$ leads to (2.365) which is reproduced here and furnished with additional comments,

$$\underbrace{\int\limits_{-\infty}^{\infty} d\tau\delta(\tau-t)f(\tau)}_{f(t)} = \overbrace{\sum_{n=-\infty}^{\infty} \frac{\sin\left[\pi(t-n)\right]}{\pi(t-n)} \underbrace{\int\limits_{-\infty}^{\infty} d\tau\delta(\tau-n)f(\tau)}_{\text{the analysis step leading to } f(n)}}^{\text{the synthesis step resulting in the function } f(t)}.$$ (2.371)

2.10 Design of *bona fide* transforms and the associated inverse transforms

The title of this section suggests that problem-tailored specific transforms along with their inverse transforms can be designed in response to prevailing theoretical and practical requirements. It also implies that the notion of universal concepts rather than universal tools might serve as the driving force for the designers to develop their own transforms and inverse transforms which perfectly match to their specific applications. General considerations concerning the design and optimization, subject to stringent constraints, are among daily routines of engineers, practitioners, computer scientists, and applied mathematicians alike. Research spearheaded by applied mathematicians, flanked by signal processing engineers and computational specialists, and inspired by applications, led in the mid-1980s to the development of wavelets and frames. The resulting fruitful collaborative efforts profoundly enriched the existing conceptions of transforms and the corresponding inverse transforms, and substantially expanded the scope of their applications. Broadly speaking, given a certain class of functions f, the envisioned operator (transform) $\mathcal{T}$ is expected to transform f into another function F with certain desirable features. Ordinarily, it is also expected that an associated well-posed inverse operator (inverse transform) $\mathcal{T}^{-1}$ exists which recovers the originating function f from F. The widespread as well as less prominent Classical transforms all follow this scheme—they are associated with a transform $\mathcal{T}$ and the corresponding $\mathcal{T}^{-1}$, such that, $\mathcal{T}\mathcal{T}^{-1} = \mathcal{T}^{-1}\mathcal{T} = \hat{\mathbb{I}}$, with $\hat{\mathbb{I}}$ standing for the identity operator. It turns out that the requirement of decomposability of the identity operator $\mathbb{I}$ into $\mathcal{T}$ and $\mathcal{T}^{-1}$ is too severe of a constraint. As briefly touched upon in the preceding sections, standard transforms $\mathcal{T}$ and their corresponding inverse transforms $\mathcal{T}^{-1}$ rely on a single set of ONB functions $\{b_n\}$. It turns out that this requirement is too restrictive for the purpose of designing transforms and inverse transforms with desirable flexibility. The powerful theory of frames and dual frames comes to the rescue, and offers immensely enriched and extremely flexible new perspectives. The theory of frames requires a set of complete or over-complete, and non-orthonormal frames $\{f_n\}$ (generalization of the notion of basis) and enables the design of a set of dual frames $\{\widetilde{f_n}\}$. The factorization of the identity operator into transform and inverse transform operators takes place by utilizing the frame and dual frames. This idea which was introduced in the previous sections will be refined and discussed further in the remaining part of this chapter. The theory of frames and dual frames can be seen as a treasure trove of many important results to come.

The anatomy of the Formal Taylor Series Expansion: Consider the Formal Taylor Series Expansion of the function $f(x)$,

$$f(x) = c_0 + c_1 x + c_2 x^2 + \cdots = \sum_{n=0}^{\infty} c_n x^n = \sum_{n=0}^{\infty} x^n c_n.$$ (2.372)

The set of the monomials $\{1, x, x^2, \ldots\}$ is complete and constitutes a non-orthogonal basis. The scalar expansion coefficients c_n ($n \in \mathbb{N}_0$) must be determined (the analysis step) such that the resulting series expansion is, in some sense, equal to the originating function $f(x)$ (the synthesis step). The three representations of $f(x)$ in (2.372) will be used interchangeably.

Notation: As a reminder, in this text the set of positive integers is denoted by $\mathbb{N}$, whereas $\mathbb{N}_0$ signifies the set of nonnegative integers ($\mathbb{N}_0 = \{0\} \cup \mathbb{N}$).

$\square$

Standard determination of the expansion coefficients c_n: The nth derivative of $f(x)$ will be denoted by $\frac{d^n}{dx^n}f(x)$ or $f^{(n)}(x)$, whichever more appropriate at any given instance. The standard recipe for the determination of c_n comprises the following steps. Take the nth derivative of $f(x)$, evaluate the resulting expression at $x = 0$, and divide it by the n-factorial ($n! = 1 \cdot 2 \cdots (n-1) \cdot n$),

$$c_n = \frac{1}{n!}\left\{\frac{d^n}{dx^n}f(x)\right\}\Big|_{x=0} = \frac{1}{n!}\left\{f^{(n)}(x)\right\}\Big|_{x=0}. \tag{2.373}$$

Adopting an alternative language: The first step in developing the Taylor transform and the corresponding inverse transform is expressing the statement "the nth-derivative of $f(x)$ evaluated at $x = 0$," slightly differently. In virtue of the sifting property of the Dirac's delta function $\delta(x)$, the last term in (2.373), which is an expression for c_n, can be written more formally as,

$$c_n = \int_{-\infty}^{\infty} dx \delta(x) \frac{1}{n!} f^{(n)}(x). \tag{2.374}$$

This step is crucial in expressing c_n in terms of a (yet-to-be-determined) operator which must be applied to $f(x)$ to yield c_n. Recall, in designing transforms and their associated inverse transforms, an operator $\mathcal{A}$ must first be identified (constructed). In the present case, while taking the nth derivative is immediate (d^n/dx^n), the statement "evaluate at $x = 0$, written as $|_{x=0}$," must be further refined, and expressed in terms of the application of an operator. Integrating by parts,

$$\int_{-\infty}^{\infty} dx \delta(x) f^{(n)}(x) = \delta(x) f^{(n-1)}(x)\Big|_{-\infty}^{\infty} - \int_{-\infty}^{\infty} dx \delta^{(1)}(x) f^{(n-1)}(x). \tag{2.375}$$

Considering any finite or infinite interval which contains the origin ($x = 0$), the Dirac delta function vanishes everywhere in the interval except at $x = 0$, where the Dirac delta function is infinite. Consequently, the first term at the R.H.S. vanishes, leading to,

$$\int_{-\infty}^{\infty} dx \delta(x) f^{(n)}(x) = - \int_{-\infty}^{\infty} dx \delta^{(1)}(x) f^{(n-1)}(x). \tag{2.376}$$

Considering the equation for the determination of c_n, (2.374), and successively applying (2.376), enable "rolling over" the derivations of $f(x)$ onto the derivations of $\delta(x)$, and thus expressing the coefficient c_n in terms of an operator applied to $f(x)$. Proceeding formally,

$$c_n = \int_{-\infty}^{\infty} dx \frac{1}{n!}\delta(x)f^{(n)}(x)$$

$$= -\int_{-\infty}^{\infty} dx \frac{1}{n!}\delta^{(1)}(x)f^{(n-1)}(x)$$

$$\vdots$$

$$= (-1)^n \int_{-\infty}^{\infty} dx \frac{1}{n!}\delta^{(n)}(x)f^{(0)}(x)$$

$$= \int_{-\infty}^{\infty} dx(-1)^n \frac{1}{n!}\delta^{(n)}(x)f(x). \tag{2.377}$$

Thus, applying the distributional functional $\int_{-\infty}^{\infty} dx(-1)^n\frac{1}{n!}\delta^{(n)}(x)$ onto $f(x)$ yields the scalar coefficient c_n. In the last transition, the standard definition $f^{(0)}(x) \equiv f(x)$ has been used. Since the choice of the (dummy) integration variable is arbitrary,

$$c_n = \int_{-\infty}^{\infty} d\xi(-1)^n \frac{1}{n!}\delta^{(n)}(\xi)f(\xi). \tag{2.378}$$

Considering this integral expression for the constant c_n in the second sum in (2.372),

$$f(x) = \sum_{n=0}^{\infty} x^n \int_{-\infty}^{\infty} d\xi(-1)^n \frac{1}{n!}\delta^{(n)}(\xi)f(\xi). \tag{2.379}$$

Formal Taylor Series Expansion of the Dirac delta function: Focus on (2.379). In virtue of the fact that x^n is independent of the integration variable, after exchanging the order of integration and summation,

$$f(x) = \int_{-\infty}^{\infty} d\xi \sum_{n=0}^{\infty} x^n(-1)^n \frac{1}{n!}\delta^{(n)}(\xi)f(\xi). \tag{2.380}$$

Merge x^n and $(-1)^n$ together,

$$f(x) = \int_{-\infty}^{\infty} d\xi \sum_{n=0}^{\infty} (-x)^n \frac{1}{n!}\delta^{(n)}(\xi)f(\xi). \tag{2.381}$$

Use the sifting property of the Dirac delta function for expressing $f(x)$ at the L.H.S.,

$$f(x) = \int_{-\infty}^{\infty} d\xi \delta(\xi - x)f(\xi). \tag{2.382}$$

Equating the two expressions of $f(x)$ in (2.381) and (2.382),

$$\int_{-\infty}^{\infty} d\xi \underbrace{\delta(\xi - x)}f(\xi) = \int_{-\infty}^{\infty} d\xi \underbrace{\sum_{n=0}^{\infty}(-x)^n \frac{1}{n!}\delta^{(n)}(\xi)}f(\xi). \tag{2.383}$$

Since the choice of the test function f is arbitrary,

$$\delta(\xi - x) = \sum_{n=0}^{\infty}(-x)^n \frac{1}{n!}\delta^{(n)}(\xi). \tag{2.384}$$

The Formal Taylor Series Expansion of the shifted Dirac delta function $\delta(\xi - x)$ manifestly emerges from the formulation. More appropriately, this representation can be recognized as the Formal Taylor Series Expansion of $\delta(\xi + (-x))$ at ξ. Note that the Dirac delta function, and thus its derivatives, in virtue of being generalized functions, distributions, make sense after being multiplied by a test function and integrated over entire space. Note also that the notion of "being evaluated at ξ" has been rendered rigorous by incorporating it into the definition of the Dirac delta function and its derivatives.

Considering (2.384), the obviously equivalent representation,

$$\delta(\xi - x) = \sum_{n=0}^{\infty} x^n (-1)^n \frac{1}{n!}\delta^{(n)}(\xi), \tag{2.385}$$

shall prove advantageous, whenever the nth derivative is expected to roll back from $\delta^{(n)}(\xi)$ onto $f(\xi)$.

Remark: As a side note, the factorization (decomposition) of the generalized function $\delta(x - \xi)$ into the sum of the products of two index-dependent functions, x^n and $\delta^{(n)}(\xi)$, is fundamentally crucial. It constitutes the quintessential foundation of addition theorems in developing fast or accelerated computational algorithms.

□

Remark: Formal Taylor Series Expansion and Factorization

Equation (2.384) motivates viewing the Formal Taylor Series Expansion as a special kind of factorization. Consider,

$$f(\xi - x) = \sum_{n=0}^{\infty}(-x)^n \frac{1}{n!}f^{(n)}(\xi), \tag{2.386}$$

or, alternatively,

$$f(\xi + x) = \sum_{n=0}^{\infty} x^n \frac{1}{n!}f^{(n)}(\xi). \tag{2.387}$$

The concept of factorization is usually tacit, because the Taylor Series Expansion is ordinarily stated in a different context. Given $f(x)$, and the small constant a an approximation of $f(x + a)$ is sought. Thus,

$$f(x + a) = \sum_{n=0}^{\infty} \frac{1}{n!}\left\{\frac{d^n}{d(x + a)^n}f(x + a)\right\}_{a=0} a^n. \tag{2.388}$$

The term in the curly brackets is the nth derivative of $f(\cdot)$ with respect to its argument evaluated at $(x+a)|_{a=0}$, i.e., x. Written in the form

$$f(x+a) = \sum_{n=0}^{\infty} \frac{1}{n!} \left\{ \frac{d^n}{d(\cdot)^n} f(\cdot) \right\}_x a^n, \tag{2.389}$$

explicates x and a and separates them from each other, while conveying exactly the same content. For further clarification, let the function $f(\cdot)$ depend on the addition of the variables ξ and η, i.e., $f(\xi + \eta)$. Introduce $u = \xi + \eta$. Determine the nth derivative of $f(u)$ with respect to u, which merely depends on the functional form of f, which is $\frac{d^n}{du^n} f(u)$. Next evaluate $\frac{d^n}{du^n} f(u)$ at $u|_{\eta=0}$, i.e., at ξ, to obtain $f^{(n)}(\xi)$, which is a function of ξ. Divide by $n!$ and multiply by nth power of the variable which was set to zero to calculate $\frac{d^n}{du^n} f(u)$, i.e., multiply by $(\eta)^n$, to obtain,

$$f(\xi + \eta) = \sum_{n=0}^{\infty} \frac{1}{n!} \left\{ \frac{d^n}{du^n} f(u) \right\}_{\xi} \eta^n. \tag{2.390}$$

Writing (2.390) in the form,

$$f(\xi + \eta) = \sum_{n=0}^{\infty} \frac{1}{n!} f^{(n)}(\xi) \eta^n, \tag{2.391}$$

makes the separation of the variables explicit. Perhaps, even more clearly, by introducing $g_n(\xi) = \frac{1}{n!} f^{(n)}(\xi)$ and $h_n(\eta) = \eta^n$,

$$f(\xi + \eta) = \sum_{n=0}^{\infty} g_n(\xi) h_n(\eta). \tag{2.392}$$

$\square$

Remark: Introduce $\delta_x(\xi)$ for the x-shifted version of $\delta(\xi)$,

$$\delta_x(\xi) = \delta(\xi - x). \tag{2.393}$$

Thus,

$$\delta_0(\xi) = \delta(\xi). \tag{2.394}$$

Similarly (and consistently), introduce $\delta_x^{(n)}(\xi)$ for the x-shifted version of $\delta^{(n)}(\xi)$,

$$\delta_x^{(n)}(\xi) = \delta^{(n)}(\xi - x). \tag{2.395}$$

Thus,

$$\delta_0^{(n)}(\xi) = \delta^{(n)}(\xi). \tag{2.396}$$

Employing the above definitions, the representation (2.385) can be written clearly and succinctly,

$$\delta_x(\xi) = \sum_{n=0}^{\infty} x^n (-1)^n \frac{1}{n!} \delta_0^{(n)}(\xi). \tag{2.397}$$

$\square$

Problem: Corroborating the derived formula by an example.

Recall the representation,

$$f(x) = \sum_{n=0}^{\infty} x^n \int_{-\infty}^{\infty} d\xi (-1)^n \frac{1}{n!} \delta_0^{(n)}(\xi) f(\xi).$$
(2.398)

The integral expression at the R.H.S. stands for the coefficient c_n in the Formal Taylor Series Expansion of the function $f(x)$. Apply this formula to the special case of $f(x) = x^m$, with $m = 0, \cdots, 2N$, and $N \in \mathbb{N}$.

Solution: Substituting x^m for $f(x)$ at the L.H.S., and correspondingly substituting ξ^m for $f(\xi)$ at the R.H.S.,

$$x^m = \sum_{n=0}^{\infty} x^n \int_{-\infty}^{\infty} d\xi (-1)^n \frac{1}{n!} \delta^{(n)}(\xi) \xi^m.$$
(2.399)

Considering first the integral $\int_{-\infty}^{\infty} d\xi \delta^{(n)}(\xi)\xi^m$, and reversing the rolling over operation,

$$\int_{-\infty}^{\infty} d\xi \delta^{(n)}(\xi)\xi^m = (-1) \int_{-\infty}^{\infty} d\xi \delta^{(n-1)}(\xi) [\xi^m]^{(1)}$$
(2.400a)

$$= (-1)^2 \int_{-\infty}^{\infty} d\xi \delta^{(n-2)}(\xi) [\xi^m]^{(2)}$$
(2.400b)

$$\vdots$$

$$= (-1)^n \int_{-\infty}^{\infty} d\xi \delta^{(n-n)}(\xi) [\xi^m]^{(n)}$$
(2.400c)

$$= (-1)^n \int_{-\infty}^{\infty} d\xi \delta(\xi) [\xi^m]^{(n)}.$$
(2.400d)

Here, $[\xi^m]^{(n)}$, e.g., stands for the nth derivative of ξ^m. Consider the following relationships:

$$n > m: \quad [\xi^m]^{(n)} = 0 \qquad\qquad \implies \int_{-\infty}^{\infty} d\xi \delta(\xi) [\xi^m]^{(n)} = 0$$

$$n = m: \quad [\xi^n]^{(n)} = n[\xi^{n-1}]^{(n-1)} = \cdots = n \cdot (n-1) \cdots 1 = n! \implies \int_{-\infty}^{\infty} d\xi \delta(\xi) [\xi^m]^{(n)} = n!$$
(2.401)

$$n < m: \quad [\xi^m]^{(n)} = m(m-1)\cdots(m-n)\xi^{m-n} \qquad \implies \int_{-\infty}^{\infty} d\xi \delta(\xi) [\xi^m]^{(n)} = 0$$

Thus

$$\int_{-\infty}^{\infty} d\xi \delta(\xi) [\xi^m]^{(n)} = n! \delta_{mn},$$
(2.402)

with δ_{mn} being the Kronecker delta symbol. Substituting (2.402) into (2.400d),

$$\int_{-\infty}^{\infty} d\xi\, \delta^{(n)}(\xi)\xi^m = (-1)^n n!\, \delta_{mn}. \tag{2.403}$$

Multiplying both sides by $(-1)^n$ and dividing by $n!$, and considering $(-1)^n(-1)^n = (-1)^{2n} = 1$ at the R.H.S.,

$$\int_{-\infty}^{\infty} d\xi(-1)^n \frac{1}{n!}\delta^{(n)}(\xi)\xi^m = \delta_{mn}. \tag{2.404}$$

Considering this result in (2.399)

$$x^m = \sum_{n=0}^{\infty} x^n \delta_{mn} \tag{2.405a}$$

$$= x^m, \tag{2.405b}$$

as expected: The representation at the R.H.S. is expected to provide a power series expansion for the function at the L.H.S., which in the present case is the monomial x^m. Since the monomials are linearly independent, there can only be one term at the R.H.S. as well. The result in (2.405) confirmed this to be the case. ∎

Problem: Further corroboration of the derived formula.

Apply $\int_{-\infty}^{\infty} dx(-1)^m \frac{1}{m!}\delta_0^{(m)}(x)$ to both sides of

$$f(x) = \sum_n x^n \int_{-\infty}^{\infty} d\xi(-1)^n \frac{1}{n!}\delta_0^{(n)}(\xi)f(\xi). \tag{2.406}$$

Solution:

$$\int_{-\infty}^{\infty} dx(-1)^m \frac{1}{m!}\delta_0^{(m)}(x)f(x) = \sum_n \underbrace{\int_{-\infty}^{\infty} dx(-1)^m \frac{1}{m!}\delta_0^{(m)}(x)x^n}_{\text{Eq. (2.404)} \Longrightarrow \delta_{mn}} \int_{-\infty}^{\infty} d\xi(-1)^n \frac{1}{n!}\delta_0^{(n)}(\xi)f(\xi) \tag{2.407a}$$

$$= \sum_n \delta_{mn} \int_{-\infty}^{\infty} d\xi(-1)^n \frac{1}{n!}\delta_0^{(n)}(\xi)f(\xi) \tag{2.407b}$$

$$= \int_{-\infty}^{\infty} d\xi(-1)^m \frac{1}{m!}\delta_0^{(m)}(\xi)f(\xi). \tag{2.407c}$$

∎

Remark: Interpretation.

The above problem presents an opportunity for the reader to better familiarize themselves with the intricacies of the developed formalism. The following observations might be helpful.

- The integral at the L.H.S. in (2.407) is an expression for c_m,

$$c_m = \int_{-\infty}^{\infty} dx(-1)^m \frac{1}{m!}\delta_0^{(m)}(x)f(x).$$
(2.408)

Since the choice of the (dummy) integral variable is immaterial, replacing x with, e.g., ξ does not make any difference,

$$c_m = \int_{-\infty}^{\infty} d\xi(-1)^m \frac{1}{m!}\delta_0^{(m)}(\xi)f(\xi).$$
(2.409)

This is, however, the last integral expressions in (2.407c).
- Simply-by-inspecting (2.408) or (2.409) it must be self-evidently clear that the integrals produce $\frac{1}{m!}f^{(m)}(0)$ which is equal to c_m. In fact the type of the integral was designed to exactly produce $\frac{1}{m!}f^{(m)}(0)$.
- The derivation also required the key relationship,

$$\delta_{mn} = \int_{-\infty}^{\infty} dx(-1)^m \frac{1}{m!}\delta_0^{(m)}(x)x^n.$$
(2.410)

$\square$

Equidistant sampling: Consider,

$$f(x) = \sum_n x^n \int_{-\infty}^{\infty} d\xi(-1)^n \frac{1}{n!}\delta_0^{(n)}(\xi)f(\xi),$$
(2.411)

where the notation $\sum_n$ indicates $\sum_{n=0}^{\infty}$ or $\sum_{n=0}^{2N}$, with $N \in \mathbb{N}$. Sampling $f(x)$ at the points $m\Delta$, with $m = -N,\ldots,N$, and $\Delta \in \mathbb{R}^+$,

$$\underbrace{\int_{-\infty}^{\infty} dx\delta_{m\Delta}(x)f(x)}_{f(m\Delta)=f_m} = \sum_n \underbrace{\int_{-\infty}^{\infty} dx\delta_{m\Delta}(x)x^n}_{(m\Delta)^n=X_{mn}} \underbrace{\int_{-\infty}^{\infty} d\xi(-1)^n \frac{1}{n!}\delta_0^{(n)}(\xi)f(\xi)}_{f^{(n)}(0)}.$$
(2.412)

Recall that $\delta_{m\Delta}(x)$ stands for $\delta(x - m\Delta)$. Introducing column vectors (f_m) and $(f^{(n)}(0))$ along with the matrix (X_{mn}),

$$(f_m) = (X_{mn})(f^{(n)}(0)).$$
(2.413)

Denoting the inverse of (X_{mn}) by $((X_{mn}))^{-1}$,

$$(f^{(n)}(0)) = (X_{mn})^{-1}(f_m).$$
(2.414)

Interpretation of (2.416):

- $< \delta_{m\Delta}|f > = f(m\Delta)$: The values of the function $f(x)$ sampled at the equidistant points $m\Delta$, with $m = -N,\ldots,N$, constitute the vector $|f >$.
- $< \delta_{m\Delta}|x^n > = (m\Delta)^n = a_{mn}$: The matrix entries with $m = -N,\ldots,N$, and $n = 0,\ldots,2N+1$, constitute the "system" matrix $\mathcal{A}$.
- $< (-1)^n \frac{1}{n!}\delta_0^{(n)}|f > = F_0^{(n)}$: The nth derivative of the function $f(x)$ evaluated at $x = 0$, with $n = 0,\ldots,2N$, constitute the vector $|D >$.

With these definitions,

$$|F > = \mathcal{A}|D>.$$
(2.415)

Reading from the right to the left,

$$\mathcal{A}|D> = |F>. \tag{2.416}$$

Given the sampled function values enables the determination of the derivatives at $x = 0$ by solving (2.416), i.e.,

$$|D> = \mathcal{A}^{-1}|F>, \tag{2.417}$$

with $\mathcal{A}^{-1}$ denoting the inverse of $\mathcal{A}$.

Implication of (2.416): Inserting the identity operator $\mathcal{I}$ in the space of functions f, between $<\delta_{m\Delta}|$ and $|f>$,

$$<\delta_{m\Delta}|\underbrace{\mathcal{I}}|f> = \sum_n <\delta_{m\Delta}||x^n> \underbrace{<(-1)^n \frac{1}{n!}\delta_0^{(n)}||f>}. \tag{2.418}$$

Since $|f>$ is arbitrary,

$$\mathcal{I} = \sum_n |x^n> <(-1)^n \frac{1}{n!}\delta_0^{(n)}|. \tag{2.419}$$

Since $\mathcal{I}$ equals to its transpose,

$$\mathcal{I} = \sum_n |(-1)^n \frac{1}{n!}\delta_0^{(n)}> <x^n|. \tag{2.420}$$

Apply (2.420) to $|F>$,

$$\mathcal{I}|F> = |F> = \sum_n |(-1)^n \frac{1}{n!}\delta_0^{(n)}> <x^n|F> . \tag{2.421}$$

Project (2.421) onto $|x^m>$,

$$<x^m|F> = \sum_n <x^m|(-1)^n \frac{1}{n!}\delta_0^{(n)}> <x^n|F> . \tag{2.422}$$

The discrete case: Consider (2.388). Setting $x = m\Delta$ with $m \in \{-N, \ldots, N\}$ for any $N \in \mathbb{N}$, along with an arbitrary $\Delta \in \mathbb{R}^+$,

$$\delta(\xi - m\Delta) = \sum_{n=0}^{2N} (m\Delta)^n \frac{1}{n!}(-1)^n \delta^{(n)}(\xi). \tag{2.423}$$

Discrete Taylor transform and inverse transform: Here an attempt is made to provide a plausible explanation for the ideas underlying the proposed Taylor transform and inverse transform. The simplest possible toy model suffices to illustrate the relationships. Consider the symmetric interval $[-\Delta, \Delta]$ and the associated lattice consisting of three lattice points $\{-\Delta, 0, \Delta\}$. Let $f(x)$ be defined on $[-\Delta, \Delta]$ and possess derivatives to any order required. Consider the expansion

$$f(x) = c_0 + c_1 x + c_2 x^2. \tag{2.424}$$

In this synthesis formula, the basis monomials 1, x, and x^2 are universal while the coefficients c_0, c_1, and c_2 are linearly dependent on $f(x)$. To explicate this dependence the expression at the R.H.S. is first written as an inner product,[‡]

$$f(x) = \begin{bmatrix} 1 & x & x^2 \end{bmatrix} \begin{bmatrix} c_0 \\ c_1 \\ c_2 \end{bmatrix}. \tag{2.425}$$

Standard determination of the expansion coefficients: Taking the 0th, 1st, and the 2nd derivatives of $f(x)$, evaluating the derivatives at $x = 0$, and denoting the nth derivative of $f(x)$ by $f^{(n)}(x)$,

$$f^{(0)}(0) = \begin{bmatrix} 1 & x & x^2 \end{bmatrix}_{x=0} \begin{bmatrix} c_0 \\ c_1 \\ c_2 \end{bmatrix} = \begin{bmatrix} 1 & 0 & 0 \end{bmatrix} \begin{bmatrix} c_0 \\ c_1 \\ c_2 \end{bmatrix} = c_0 \tag{2.426a}$$

$$f^{(1)}(0) = \begin{bmatrix} 0 & 1 & 2x \end{bmatrix}_{x=0} \begin{bmatrix} c_0 \\ c_1 \\ c_2 \end{bmatrix} = \begin{bmatrix} 0 & 1 & 0 \end{bmatrix} \begin{bmatrix} c_0 \\ c_1 \\ c_2 \end{bmatrix} = c_1 \tag{2.426b}$$

$$f^{(2)}(0) = \begin{bmatrix} 0 & 0 & 2 \end{bmatrix}_{x=0} \begin{bmatrix} c_0 \\ c_1 \\ c_2 \end{bmatrix} = \begin{bmatrix} 0 & 0 & 2 \end{bmatrix} \begin{bmatrix} c_0 \\ c_1 \\ c_2 \end{bmatrix} = 2c_2. \tag{2.426c}$$

These few initial steps allow to hypothesize the relationship between the expansion coefficient c_n and the nth derivative of the function $f(x)$, i.e., $c_n = \frac{1}{n!}\frac{d^n}{dx^n}f(0)$. This is the quintessential statement in the Formal Taylor Series Expansion: The nth expansion coefficient c_n can be determined by evaluating the nth order derivative of $f(x)$ at zero (or for this matter at a given point x_0).

However, the key observation leading to the Taylor transform and inverse transform (and thus to the required resolution of identity) is that the coefficients c_n ($n = 0, 1, 2$) are expressible as linear superpositions of the function values sampled at the lattice points, i.e., in terms of linear combinations of $f(-\Delta), f(0)$, and $f(\Delta)$, in the present toy model. This content is illustrated next.

Lemma: *Alternative determination of the expansion coefficients.*

[‡]The rationale for writing the inner product in terms of an x-dependent row vector and a c-dependent column vector, rather than a c-dependent row vector and an x-dependent column vector, will become clear momentarily: the objective is to express the R.H.S. of (2.425) as an operator that acts, from the left, on the function $f(x)$, respectively, its discrete sampled values.

The coefficients in the Formal Taylor Series Expansions can be expressed in terms of linear superpositions of function values sampled at the lattice points, e.g.,

$$c_0 = c_{0,-1}f(-\Delta) + c_{0,0}f(0) + c_{0,1}f(\Delta) \tag{2.427a}$$

$$c_1 = c_{1,-1}f(-\Delta) + c_{1,0}f(0) + c_{1,1}f(\Delta) \tag{2.427b}$$

$$c_2 = c_{2,-1}f(-\Delta) + c_{2,0}f(0) + c_{2,1}f(\Delta), \tag{2.427c}$$

or, compactly,

$$\begin{bmatrix} c_0 \\ c_1 \\ c_2 \end{bmatrix} = \begin{bmatrix} c_{0,-1} & c_{0,0} & c_{0,1} \\ c_{1,-1} & c_{1,0} & c_{1,1} \\ c_{2,-1} & c_{2,0} & c_{2,1} \end{bmatrix} \begin{bmatrix} f(-\Delta) \\ f(0) \\ f(\Delta) \end{bmatrix}. \tag{2.428}$$

Furthermore, the rows in (2.428) are the components of the difference bra vectors $< \mathcal{D}^{(0)}|$, $< \mathcal{D}^{(1)}|$, and $< \mathcal{D}^{(2)}|$, i.e.,

$$< \mathcal{D}^{(0)}| = \begin{bmatrix} c_{0,-1} & c_{0,0} & c_{0,1} \end{bmatrix} \tag{2.429a}$$
$$< \mathcal{D}^{(1)}| = \begin{bmatrix} c_{1,-1} & c_{1,0} & c_{1,1} \end{bmatrix} \tag{2.429b}$$
$$< \mathcal{D}^{(2)}| = \begin{bmatrix} c_{2,-1} & c_{2,0} & c_{2,1} \end{bmatrix}. \tag{2.429c}$$

Proof: The minimalistic toy model in this section readily permits to demonstrate that the expansion coefficients c_n can be determined as linear superpositions of the sampled values of $f(x)$ at the lattice points $\{-\Delta, 0, \Delta\}$. Evaluating $f(x)$, (2.425), successively at $-\Delta$, 0, and Δ,

$$f(-\Delta) = \begin{bmatrix} 1 & (-\Delta) & (-\Delta)^2 \end{bmatrix} \begin{bmatrix} c_0 \\ c_1 \\ c_2 \end{bmatrix} \tag{2.430a}$$

$$f(0) = \begin{bmatrix} 1 & 0 & 0 \end{bmatrix} \begin{bmatrix} c_0 \\ c_1 \\ c_2 \end{bmatrix} \tag{2.430b}$$

$$f(\Delta) = \begin{bmatrix} 1 & (\Delta) & (\Delta)^2 \end{bmatrix} \begin{bmatrix} c_0 \\ c_1 \\ c_2 \end{bmatrix}. \tag{2.430c}$$

Writing (2.430) compactly,

$$
\begin{bmatrix} f(-\Delta) \\ f(0) \\ f(\Delta) \end{bmatrix} = \begin{bmatrix} 1 & (-\Delta) & (-\Delta)^2 \\ 1 & 0 & 0 \\ 1 & (\Delta) & (\Delta)^2 \end{bmatrix} \begin{bmatrix} c_0 \\ c_1 \\ c_2 \end{bmatrix}. \tag{2.431}
$$

Remark: The column vectors of the matrix at the R.H.S. can be recognized as the components of the kets $|1>$, $|x>$, $|x^2>$ representing the constant 1, the linear function x and the quadratic function x^2, respectively,

$$
|1>= \begin{bmatrix} 1 \\ 1 \\ 1 \end{bmatrix}, \quad |x>= \begin{bmatrix} -\Delta \\ 0 \\ \Delta \end{bmatrix}, \quad |x^2>= \begin{bmatrix} \Delta^2 \\ 0 \\ \Delta^2 \end{bmatrix}. \tag{2.432}
$$

□

Focus on (2.431). Multiplying both sides of this equation by the inverse of the matrix at the R.H.S. and expressing the resulting equation in terms of the expansion coefficients,

$$
\begin{bmatrix} c_0 \\ c_1 \\ c_2 \end{bmatrix} = \begin{bmatrix} 1 & -\Delta & \Delta^2 \\ 1 & 0 & 0 \\ 1 & \Delta & \Delta^2 \end{bmatrix}^{-1} \begin{bmatrix} f(-\Delta) \\ f(0) \\ f(\Delta) \end{bmatrix}. \tag{2.433}
$$

Problem: Show that,

$$
\begin{bmatrix} 1 & -\Delta & \Delta^2 \\ 1 & 0 & 0 \\ 1 & \Delta & \Delta^2 \end{bmatrix}^{-1} = \begin{bmatrix} 0 & 1 & 0 \\ -\frac{1}{2\Delta} & 0 & \frac{1}{2\Delta} \\ \frac{1}{2\Delta^2} & -\frac{2}{2\Delta^2} & \frac{1}{2\Delta^2} \end{bmatrix}. \tag{2.434}
$$

∎

Substituting (2.434) into (2.433)

$$
\begin{bmatrix} c_0 \\ c_1 \\ c_2 \end{bmatrix} = \begin{bmatrix} 0 & 1 & 0 \\ -\frac{1}{2\Delta} & 0 & \frac{1}{2\Delta} \\ \frac{1}{2\Delta^2} & -\frac{2}{2\Delta^2} & \frac{1}{2\Delta^2} \end{bmatrix} \begin{bmatrix} f(-\Delta) \\ f(0) \\ f(\Delta) \end{bmatrix}. \tag{2.435}
$$

This completes the proof.

∎

Remark: The row vectors at the R.H.S in (2.435) are the discrete difference bra vectors $< \mathcal{D}^{(0)}|$, $< \mathcal{D}^{(1)}|$, and $< \mathcal{D}^{(2)}|$, introduced further above.

□

Substituting (2.435) into (2.431),

$$\begin{bmatrix} f(-\Delta) \\ f(0) \\ f(\Delta) \end{bmatrix} = \begin{bmatrix} 1 & -\Delta & \Delta^2 \\ 1 & 0 & 0 \\ 1 & \Delta & \Delta^2 \end{bmatrix} \begin{bmatrix} 0 & 1 & 0 \\ -\frac{1}{2\Delta} & 0 & \frac{1}{2\Delta} \\ \frac{1}{2\Delta^2} & -\frac{2}{2\Delta^2} & \frac{1}{2\Delta^2} \end{bmatrix} \begin{bmatrix} f(-\Delta) \\ f(0) \\ f(\Delta) \end{bmatrix}. \tag{2.436}$$

The product of the matrices at the R.H.S can be identified as the 3×3 identity matrix $\mathbb{I}_{3\times3}$,

$$\mathbb{I}_{3\times3} = \begin{bmatrix} 1 & -\Delta & \Delta^2 \\ 1 & 0 & 0 \\ 1 & \Delta & \Delta^2 \end{bmatrix} \begin{bmatrix} 0 & 1 & 0 \\ -\frac{1}{2\Delta} & 0 & \frac{1}{2\Delta} \\ \frac{1}{2\Delta^2} & -\frac{2}{2\Delta^2} & \frac{1}{2\Delta^2} \end{bmatrix} \tag{2.437}$$

This is the (toy model) discrete analog of the factorization of the Dirac's delta function $\delta(\xi - x)$, introduced in the previous subsection.

Problem: Convince yourself that the product at the R.H.S in (2.437), can be expressed as the sum of exterior products, given below,

$$\mathbb{I}_{3\times3} = \begin{bmatrix} 1 \\ 1 \\ 1 \end{bmatrix} \begin{bmatrix} 0 & 1 & 0 \end{bmatrix}$$

$$+ \begin{bmatrix} -\Delta \\ 0 \\ \Delta \end{bmatrix} \begin{bmatrix} -\frac{1}{2\Delta} & 0 & \frac{1}{2\Delta} \end{bmatrix}$$

$$+ \begin{bmatrix} \Delta^2 \\ 0 \\ \Delta^2 \end{bmatrix} \begin{bmatrix} \frac{1}{2\Delta^2} & -\frac{2}{2\Delta^2} & \frac{1}{2\Delta^2} \end{bmatrix}. \tag{2.438}$$

∎

Consider (2.438). Substituting for the column vectors in (2.438) the Kets $|1>$, $|x>$, and $|x^2>$, and for the row vectors in (2.438) the bras $< \mathcal{D}^{(0)}|$, $< \mathcal{D}^{(1)}|$, and $< \mathcal{D}^{(2)}|$, respectively,

$$\mathbb{I}_{3\times3} = |1><\mathcal{D}^{(0)}| + |x><\mathcal{D}^{(1)}| + |x^2><\mathcal{D}^{(2)}|. \tag{2.439}$$

Or, defining the dual frame bras $< \widetilde{1}|$, $< \widetilde{x}|$, and $< \widetilde{x^2}|$,

$$< \widetilde{1}| =< \mathcal{D}^{(0)}|, \quad < \widetilde{x}| =< \mathcal{D}^{(1)}|, \quad < \widetilde{x^2}| =< \mathcal{D}^{(2)}|, \tag{2.440}$$

$$\mathbb{I}_{3\times 3} = |1 >< \widetilde{1}| + |x >< \widetilde{x}| + |x^2 >< \widetilde{x^2}|. \tag{2.441}$$

Remark: The resolution of $\mathbb{I}_{3\times 3}$ at the R.H.S. of (2.439) is the embodiment of the Discrete Taylor Transform and Inverse Transform on the basis of the simplest possible toy model considered here. Further toy models can be found in [14]. The systematic generalization of the underlying ideas builds the corpus of an independent text by this author which is exclusively dedicated to this topic.

□

Remark: A further insight might be in order prior to concluding the discussion in this chapter. In view of (2.439) the building blocks of the identity matrix are exterior products of the universal form, $|x^n >< \mathcal{D}^{(n)}|$. This expression reveals the appeal of the Discrete Taylor Transform and Inverse Transform to quantum physics, by emphasizing the powers of x and the corresponding orders of the derivatives with respect to x. The eminent role of the position coordinate x and the derivative with respect to x, i.e., d/dx manifests itself in the fact that the next two chapters are exclusively devoted to this topic and implications thereof. The algebraic and analytic aspects are elucidated in Chapters 3 and 4, respectively.

□

Remark: Finally, before concluding this chapter a brief comment on the ontology, epistemology, and the relationship between mathematics and physics seems to be appropriate. The non-commutativity of d/dx and x as will be elaborated exhaustively in this text and the associated Heisenberg uncertainty principle can be established in terms of the expressions of the resolution of identity underlying Discrete Taylor and Inverse Transform and the Fourier- and inverse Fourier transform. Both transforms were touched upon in this chapter. Neither of these transforms has anything to do with quantum physics. They are manifestations of the properties of the position variable x the differential operator d/dx and certain properties of d/dx. Foremost, the property that forward and backward propagating planewaves are eigenfunctions of the differential operator d/dx. These are all tools that engineers and technologists are very well familiar with. These tools are ubiquitous in signal processing, antennas design, electromagnetic field analysis, and generally speaking in theoretical and computational engineering. It is important to be aware that these tools and techniques can be acquired by engineers and technologists and even refined by them, without any reference to quantum physics. It is a healthy approach to do so. Quantum physics is no doubt the most successful theoretical framework which has been developed so far. At the same time the existence of several interpretations to quantum physics indicates that the current formulation of quantum physics is not and cannot be the final conception, even beyond the fact that no theory can be viewed as the final word. The theory seems to be incomplete. The approach taken in this text has been to separate the mathematical tools which are employed in quantum physics from quantum physics and its various interpretations. This approach offers several advantages against conventional text books and treatises. (i) The discussion of the mathematical tools and their development is not obscured by the interpretations of quantum physics which may unnecessarily cast unwanted shadows on the object of study and investigation. (ii) The acquired skills can be utilized in any areas of applied mathematics and computational engineering. (iii) The prevailing mathematical methods or sections of them can be refined by engineers and technologists due to their resourceful experiences from many other technical and engineering areas.

Is mathematics independent of physics and has a reality of its own? Why is mathematics so successful in describing and predicting what is observed out there?

□

2.11 Concluding remarks

The concepts of the identity operator and the resolution of identity loom large in this chapter. The identity operator and the resolution of identity characteristically and exclusively depend on the features of spaces and the corresponding dual spaces that they connect. The identity operator and the resolution of identity unify and simplify the transformation of coordinate systems. The notions of covariant and contravariant coordinates can also be made clear and obvious by utilizing the resolution of identity, even though this property was not explicated in this chapter. (Readers familiar with the subject matter might recognize this fact.) The resolution of identity unifies the standard and not well-known transforms and inverse transforms. The engineers who wish to solve boundary value problems in connection with device modeling and simulation, and technologists concerned with materials science, need to solve the Schrödinger equation in two- or three spatial dimensions. The discussion in this chapter presents itself as an excellent introduction to the challenging topics related to the regularization of singularities and taming infinities, by rendering the divergent sums and integrals, respectively, summable and integrable.

The following list of references is by no means complete. They may help the reader to find their own ways for thinking and solving problems [1–11]. The mathematical treatment in this chapter differs from them in many respect, and yet the genesis of most thoughts and presentations in this chapter originate from them and many other references which have not been included. A glimpse on the author's work on the regularization of infinities and the resolution of identity can be found in the works further below and the references therein [12,13]. The author's ongoing book projects are concerned with further elaboration and extension of the ideas, which could only be touched upon in this chapter [14].

References

[1] Ram P.K., *Generalized Functions: Theory and Technique, Mathematics in Science and Engineering*, Vol. 1771, Academic Press, 1983.

[2] Hoskins R.F. and Sousa Pinto J., *Distributions, Ultradistributions and Other Generalised Functions, Ellis Horwood Series in Mathematics and its Applications*, 1994.

[3] Zemanian A.H., *Distribution Theory and Transform Analysis*, Dover Publications, Inc., 1965

[4] Dirac P.A.M., *The Principles of Quantum Mechanics*, Oxford at the Clarendon Press, 1958.

[5] Chester M., *Primer of Quantum Mechanics*, Dover Publications, Inc., 2003.

[6] Debnath L. and Mikusinski P., *Hilbert Spaces with Applications*, Academic Press, 1999.

[7] Christensen O., *An Introduction to Frames and Riesz Bases*, Brikhäuser, 2002.

[8] Sidney Burrus C., Gopinath R.A., and Guo H., *Introduction to Wavelets and Wavelet Transform: A Primer*, Prentice Hall, 1998.

[9] Hernandez E. and Weiss G., *A First Course on Wavelets*, CRC Press, 1996.

[10] Cohen L., *Time–Frequency Analysis, Prentice Hall Signal Processing Series*, 1995.

[11] Stenger F., *Numerical Methods Based on Sinc and Analytic Functions*, Springer Verlag, 1993.

[12] Baghai-Wadji A., 3-D electrostatic charge distribution on finitely thick busbars in micro-acoustic devices: combined regularization in the near- and far-field, *IEEE Transactions on Ultrasonics, Ferroelectrics, and Frequency Control*, Vol. 62, No. 6, 2015.

[13] Baghai-Wadji A., The boundary element method applied to micro-acoustic devices: zooming into the near field, in: K. Nakamura, Ed., *Ultrasonics Transducers, Materials and Design for Sensors, Actuators and Medical Applications*, Sawston, UK: Woodhad Publishing, pp. 220–263, 2012.

[14] Baghai-Wadji A., Discrete Taylor transform and inverse transform, in: K.V.L. Narayana and G.K. Rajini, Eds., *Lecture Notes in Electrical Engineering Book Series*, Springer Verlag, pp. 1–18, 2021, Chapter no. 322.

Chapter 3
Operator gymnastic

3.1 A brief guide through the chapter

This chapter promises to be particularly appealing to engineers and technologists, and at the same time extremely empowering. The procedures are algebraic and thus easy to follow. Not unlike the processes involved in iterative techniques, self-similarity, and fractal analysis, simple rules upon repeated application lead to the emergence of many aesthetic results. Given the humble beginning of each category of examples, it is satisfying to observe the variety of powerful outcomes. In spite of the many superlatives listed, the entire discussion is built upon three major pillars. (i) Formal Taylor Series Expansion, (ii) canonical commutator relationship, and (iii) proof by mathematical induction. The reader is expected to be, by now, fully familiar with (i) and (iii) as these ideas were practiced dozens of times in the previous two chapters. The Formal Taylor Series Expansion in (i) helps to eliminate the complexity of functions of operators $f(\mathbf{A})$ and temporarily replaces $f(\mathbf{A})$ by powers of their operators $\mathbf{A}^n$ (opening the box procedure). After the required manipulations have been carried out, and $\mathbf{A}$ has been transformed to, say, $\mathbf{Q}$, the resulting $\mathbf{Q}$ replaces $\mathbf{A}$ leading to $f(\mathbf{Q})$. The method of proof by mathematical induction in (iii) takes advantage of humans' ability to build intuition and generalize. Merely by experimenting with a few initial steps, the mind recognizes the underlying pattern, which in turn enables one to state the induction hypothesis. Ordinarily, the skills acquired during the initial steps suffice to perform the induction step, and complete the proof. While (i) and (iii) are impressively powerful tools, the impact of the property in (ii) is astonishingly aesthetic. The canonical commutator relationship of two operators $\mathbf{A}$ and $\mathbf{B}$, i.e., $[\mathbf{A}, \mathbf{B}] = \mathbf{I}$, with $\mathbf{I}$ standing for the identity operator, is the genesis of far reaching implications. Several types of composite operators and functions of operators will be considered. In a few instances, problems have been solved in two ways, contrasting short, insightful, and elegant solutions against long, atomistic and algorithmic solutions. Both skills must be honed, as an interplay between the two leads to mastery. Furthermore, first and higher derivatives of operators with respect to a parameter $\mathbf{A}(\lambda)$, and first and higher derivatives of composite operators with respect to a parameter are investigated. Also, derivatives of functions $f(\mathbf{A})$ with respect to operators are scrutinized. Several relationships involving commutators of composite operators are analyzed. The chapter concludes with studying the properties of an important category of commutators which involves the position and differential operators. Being truthful to the title, each section starts with simple exercises to warm up, proceeds to rigorous training, and concludes with the satisfaction of what has been achieved.

3.2 Notations and conventions

In this chapter, operators are represented by bold characters rather that by capitalized characters carrying a "hat." Thus, e.g., $\mathbf{A}$ rather than $\hat{A}$ is used to indicate an operator.

The symbol $f(\cdot)$ is used to refer to a function f of its argument, which is what replaces the "dot" within the round brackets. This notation is important and shall prove useful in a variety of instances when, e.g., the derivative of the function f with respect to its argument, the "dot" in $(\cdot)$ is needed. It is useful to explain this important content further. There will be ample opportunities in this chapter to practice the idea. Nonetheless, one example might be in order already at the start. Consider $f(\mathbf{A}^m\mathbf{B}^n)$. To determine the derivative of f with respect to its argument (which is here the composite operator $\mathbf{A}^m\mathbf{B}^n$) one must proceed as follows. Set $\mathbf{A}^m\mathbf{B}^n = \mathbf{Z}$ to obtain $f(\mathbf{Z})$. This simple step simplifies the arguments considerably. Given $f(\mathbf{Z})$ it is an easy task to talk about the derivative of f with respect to its argument, i.e., $df(\mathbf{Z})/d\mathbf{Z}$ which makes clear what is meant using the general symbolic form $df(\cdot)/d\cdot$ with the "dot" being a placeholder. Thus, e.g., $d\sin(\cdot)/d(\cdot) = \cos(\cdot)$.

Repetitions with small variations should assist the reader to acknowledge the powers of reinforcement and rehearsal in the learning process, and furthermore, as an effective way to create windows to hidden treasures which might otherwise remain undiscovered. As in the other chapters in this book, the text is sprinkled with remarks, initiated by Remark: and closed with a □. A studious hardworking reader may decide to read all the Remarks as they are encountered, and build a strong intuition about the story told. At the same time, sandwiching a few sentences between the symbols Remark: and □ offers the possibility to safely jump over a certain passage without much loss and to the benefit of not interrupting the flow of the reading. This view might seem attractive to other groups of readers. The readers may wish to mark a Remark: as unread to return to it at a later point of time, after having gained a bird's view of the subject matter discussed in the section or covered in the chapter. A third opportunity which the Remarks offer is that they can be read whenever the reader randomly opens the book. Very often the Remarks summarize the untold (unwritten) stories between the lines. The ability to read between the lines is an art and a rare skill which can be honed and nurtured. The enhanced ability to read between the lines and the capacity to skillfully fill the existing gaps are windows to creative problem solving. Consequently, and finally, just going through the sequence of the Remarks from the beginning to the end may present itself as a subtext within the whole text.

3.3   Warming up

Exercise: Consider the operators $\mathbf{A}$ and $\mathbf{B}$. Let $\mathbf{A}^{-1}$ denote the inverse of $\mathbf{A}$. Then,

$$\mathbf{A}\mathbf{B}^2\mathbf{A}^{-1} \stackrel{(1)}{=} \mathbf{A}\mathbf{B}\mathbf{B}\mathbf{A}^{-1} \tag{3.1a}$$

$$\stackrel{(2)}{=} \mathbf{A}\mathbf{B}\mathbf{I}\mathbf{B}\mathbf{A}^{-1} \tag{3.1b}$$

$$\stackrel{(3)}{=} \mathbf{A}\mathbf{B}\mathbf{A}^{-1}\mathbf{A}\mathbf{B}\mathbf{A}^{-1} \tag{3.1c}$$

$$\stackrel{(4)}{=} \left(\mathbf{A}\mathbf{B}\mathbf{A}^{-1}\right)\left(\mathbf{A}\mathbf{B}\mathbf{A}^{-1}\right) \tag{3.1d}$$

$$\stackrel{(5)}{=} \left(\mathbf{A}\mathbf{B}\mathbf{A}^{-1}\right)^2. \tag{3.1e}$$

∎

Remark: Each of the above transitions signifies a simple yet profound definition or rule with wide-ranging implications when manipulating operators. The transition (1) is just the definition $\mathbf{B}^2 \stackrel{\text{def.}}{=} \mathbf{B}\mathbf{B}$. The transition (2) brings in the all important notions of the identity operator $\mathbf{I}$. The identity operator can, in virtue of its definition, be inserted between any two functions of operators, or be placed to the left- or right of any function of operators, without affecting the validity of given relationships. Thus, in general, $f(\mathbf{A})g(\mathbf{B}) = f(\mathbf{A})\mathbf{I}g(\mathbf{B}), f(\mathbf{A}) = \mathbf{I}f(\mathbf{A}), f(\mathbf{A}) = f(\mathbf{A})\mathbf{I}$, with $f(\cdot)$ and $g(\cdot)$ standing for arbitrary functions. It is only the representation of $\mathbf{I}$ in terms of inverse operators which enables manipulations to be carried out in desirable directions. Its facilitating power is only matched by Dirac's delta function $\delta(x)$. In particular, the so-called resolution of identity serves as an almighty manipulatory vehicle, as demonstrated in this book on countless occasions. The archetypical example, demonstrating the significance of $\mathbf{I}$, consists of expressing $\mathbf{I}$ as the product of an operator with its inverse counterpart, i.e., $\mathbf{I} = \mathbf{A}^{-1}\mathbf{A} = \mathbf{A}\mathbf{A}^{-1}$. The transition (3) employs the representation $\mathbf{I} = \mathbf{A}^{-1}\mathbf{A}$. The transition (4) indicates that the R.H.S. is the multiplication of the composite operator $\mathbf{A}\mathbf{B}\mathbf{A}^{-1}$ by itself. The transition (5) is just the usage of the definition $\mathbf{X}\mathbf{X} \stackrel{\text{def}}{=} \mathbf{X}^2$, for arbitrary operator $\mathbf{X}$.

□

Problem: Show that $\mathbf{A}\mathbf{B}^n\mathbf{A}^{-1} = \left(\mathbf{A}\mathbf{B}\mathbf{A}^{-1}\right)^n$ for arbitrary $n \in \mathbb{N}$.

Solution: Proof by mathematical induction.

The validity of the stated equation for $n = 1$ is trivially true and not revealing. The validity of the stated equation for $n = 2$ was shown in the above example. Inspired by this result, the general form,

$$\mathbf{A}\mathbf{B}^n\mathbf{A}^{-1} = \left(\mathbf{A}\mathbf{B}\mathbf{A}^{-1}\right)^n, \tag{3.2}$$

can be hypothesized to be valid for $n \in \mathbb{N}$ and $n > 2$ (the induction hypothesis). The proof amounts to demonstrating the validity of the relationship for $n + 1$, which is shown next.

$$\begin{aligned}
\mathbf{A}\mathbf{B}^{n+1}\mathbf{A}^{-1} &= \mathbf{A}\mathbf{B}^n\mathbf{B}\mathbf{A}^{-1} & \text{(3.3a)} \\
&= \mathbf{A}\mathbf{B}^n\mathbf{I}\mathbf{B}\mathbf{A}^{-1} & \text{(3.3b)} \\
&= \mathbf{A}\mathbf{B}^n\mathbf{A}^{-1}\mathbf{A}\mathbf{B}\mathbf{A}^{-1} & \text{(3.3c)} \\
&= \left(\mathbf{A}\mathbf{B}^n\mathbf{A}^{-1}\right)\left(\mathbf{A}\mathbf{B}\mathbf{A}^{-1}\right) & \text{(3.3d)} \\
&= \left(\mathbf{A}\mathbf{B}\mathbf{A}^{-1}\right)^{n+1}. & \text{(3.3e)}
\end{aligned}$$

∎

Remark: The prime role of the initial step (or, a few initial steps) is to inspire the formulation of the induction hypothesis. The proof strategy consists of rewriting $\mathbf{A}\mathbf{B}^{n+1}\mathbf{A}^{-1}$ in such a manner that the induction hypothesis (and possibly the initial step $n = 1$), and valid implications thereof, can be employed.

□

Problem: Show that $\mathbf{A}f(\mathbf{B})\mathbf{A}^{-1} = f(\mathbf{A}\mathbf{B}\mathbf{A}^{-1})$.

Remark: This is an astonishingly powerful and beautiful relationship. The preceding problem, along with the concept of Formal Taylor Series Expansion reveal the mystery behind this relationship. It prescribe a simple yet powerful recipe for calculating $\mathbf{A}f(\mathbf{B})\mathbf{A}^{-1}$. (a) Consider $\mathbf{A}f(\mathbf{B})\mathbf{A}^{-1}$. (b) In this expression, replace the function $f(\mathbf{B})$ by its argument to obtain $\mathbf{A}\mathbf{B}\mathbf{A}^{-1}$. (c) View the resulting composite operator $\mathbf{A}\mathbf{B}\mathbf{A}^{-1}$ as the new argument of the function $f(\cdot)$, i.e., $f(\mathbf{A}\mathbf{B}\mathbf{A}^{-1})$. The Verbalization of this recipe is revealing.

□

Solution: Proof by Formal Taylor Series Expansion.

Consider the Formal Taylor Series Expansion of $f(\mathbf{B})$,

$$f(\mathbf{B}) = \sum_n f_n \mathbf{B}^n, \tag{3.4}$$

which is an expression for $f(\mathbf{B})$ in terms of powers of $\mathbf{B}$. The sum extends over any finite or infinite number of terms, whatever the case maybe, depending on the nature of $f(\cdot)$.

Remark: For the arguments here details of f_n are irrelevant. This problem serves well in illuminating the command of the Formal Taylor Series Expansion. The universal utility of the Formal Taylor Series Expansion resides in the property that the possible complex form of $f(\mathbf{B})$ is temporarily put aside, by essentially replacing $f(\mathbf{B})$ with $\mathbf{B}^n$. It is much easier to operate on $\mathbf{B}^n$. After operations on $\mathbf{B}^n$ have been completed, the results are summed up. Details concerning convergence and allied topics are not relevant, when working with the Formal Taylor Series Expansion. Scenarios in which the series does not converge, or it converges slowly, or it converges to the "wrong" result, are not of any concern. These and similar problems are of considerable importance when carrying out numerical calculation. There are a few techniques available in the tool boxes of computational scientists and engineers to effectly deal with slow convergence, or divergence challenges. They can generally be subsummed under regularization techniques. They should not be mistaken with the concept of the Formal Taylor Series Expansion, which is merely a manipulatory tool.

□

Substitute (3.4) into $\mathbf{A}f(\mathbf{B})\mathbf{A}^{-1}$,

$$\mathbf{A}f(\mathbf{B})\mathbf{A}^{-1} \overset{(1)}{=} \mathbf{A}\left(\sum_n f_n \mathbf{B}^n\right)\mathbf{A}^{-1} \tag{3.5a}$$

$$\overset{(2)}{=} \sum_n f_n\left(\mathbf{A}\mathbf{B}^n\mathbf{A}^{-1}\right) \tag{3.5b}$$

$$\overset{(3)}{=} \sum_n f_n\left(\mathbf{A}\mathbf{B}\mathbf{A}^{-1}\right)^n \tag{3.5c}$$

$$\overset{(4)}{=} \sum_n f_n \mathbf{Z}^n \tag{3.5d}$$

$$\overset{(5)}{=} f(\mathbf{Z}) \tag{3.5e}$$

$$\overset{(6)}{=} f\left(\mathbf{A}\mathbf{B}\mathbf{A}^{-1}\right). \tag{3.5f}$$

∎

Remark: The transition in the above equation have the following justifications. The transition (1) employs the Formal Taylor Series Expansion of $f(\mathbf{B})$. The transition (2) utilizes the distribution law of multiplication over addition twice. This can be seen most effectively by considering a simple example. Let n to run from 1 to 3. Then, the following sequence of steps illustrates the application of the distribution law from the left followed by the application of the distribution law from the right,

$$\mathbf{A}\left(\sum_{n=1}^{3} f_n \mathbf{B}^n\right)\mathbf{A}^{-1} = \mathbf{A}\left(f_1\mathbf{B}^1 + f_2\mathbf{B}^2 + f_3\mathbf{B}^3\right)\mathbf{A}^{-1}$$

$$\overset{\text{distr. law}}{=} \left(f_1\mathbf{A}\mathbf{B}^1 + f_2\mathbf{A}\mathbf{B}^2 + f_3\mathbf{A}\mathbf{B}^3\right)\mathbf{A}^{-1}$$

$$\overset{\text{distr. law}}{=} f_1\left(\mathbf{A}\mathbf{B}^1\mathbf{A}^{-1}\right) + f_2\left(\mathbf{A}\mathbf{B}^2\mathbf{A}^{-1}\right) + f_3\left(\mathbf{A}\mathbf{B}^3\mathbf{A}^{-1}\right)$$

$$= \sum_{n=1}^{3} f_n\left(\mathbf{A}\mathbf{B}^n\mathbf{A}^{-1}\right). \tag{3.6}$$

This completes the illustration of the transition (2) in (3.5). The transition (3) takes advantage of $\mathbf{A}\mathbf{B}^n\mathbf{A}^{-1} = \left(\mathbf{A}\mathbf{B}\mathbf{A}^{-1}\right)^n$ established further above. The transition (4) introduces the auxiliary operator $\mathbf{Z} = \mathbf{A}\mathbf{B}\mathbf{A}^{-1}$. The transition (5) states that $\sum_n f_n \mathbf{Z}^n$ is the Formal Taylor Series Expansion of $f(\mathbf{Z})$. The transition (6) back-substitute the auxiliary operator $\mathbf{Z}$.

□

Problem: Show that $\mathbf{A}e^{\mathbf{B}}\mathbf{A}^{-1} = e^{\mathbf{A}\mathbf{B}\mathbf{A}^{-1}}$.

Solution: Proof by Formal Taylor Series Expansion.

The following steps must be self-explanatory by now:

$$\mathbf{A}e^{\alpha\mathbf{B}}\mathbf{A}^{-1} = \mathbf{A}\left(\sum_n \frac{1}{n!}(\alpha\mathbf{B})^n\right)\mathbf{A}^{-1} \tag{3.7a}$$

$$= \mathbf{A}\left(\sum_n \frac{\alpha^n}{n!}\mathbf{B}^n\right)\mathbf{A}^{-1} \tag{3.7b}$$

$$= \sum_n \frac{\alpha^n}{n!}\left(\mathbf{A}\mathbf{B}^n\mathbf{A}^{-1}\right) \tag{3.7c}$$

$$= \sum_n \frac{\alpha^n}{n!}\left(\mathbf{A}\mathbf{B}\mathbf{A}^{-1}\right)^n \tag{3.7d}$$

$$= \sum_n \frac{\alpha^n}{n!}\mathbf{Z}^n \tag{3.7e}$$

$$= \sum_n \frac{(\alpha\mathbf{Z})^n}{n!} \tag{3.7f}$$

$$= e^{\alpha\mathbf{Z}} \tag{3.7g}$$

$$= e^{\alpha\mathbf{A}\mathbf{B}\mathbf{A}^{-1}}. \tag{3.7h}$$

∎

Remark: The particular case, $\alpha = 1$, turns out to be of prime significance,

$$\mathbf{A}e^{\mathbf{B}}\mathbf{A}^{-1} = e^{\mathbf{A}\mathbf{B}\mathbf{A}^{-1}}. \tag{3.8}$$

□

3.4 Equations involving b and $f(\mathbf{b}^\dagger)$

The group of problems in this section involves the annihilation operator $\mathbf{b}$, and functions of the creation operator $f(\mathbf{b}^\dagger)$. The starting point is the commutation relationship,

$$\mathbf{b}\mathbf{b}^\dagger - \mathbf{b}^\dagger\mathbf{b} = \mathbf{I}, \tag{3.9}$$

with $\mathbf{I}$ standing for the identity operator. The operators $\mathbf{b}$ and $\mathbf{b}^\dagger$ will be discussed in greater detail in the next chapter which is dedicated to their construction, studying their properties, and their generalizations in original ways. Their structures can in general be astonishing complex. The present chapter focuses on one single and seemingly simple property and investigates the consequences thereof, which is the commutativity property in (3.9). The mesmerizing world that unfolds from this modest and yet overwhelmingly powerful relationship can only be grasped by studying an increasingly more complex instances, a task which is undertaken in this chapter.

The following rearrangements of (3.9) shall serve well in many manipulations:

$$\mathbf{b}\mathbf{b}^\dagger - \mathbf{b}^\dagger\mathbf{b} = \mathbf{I} \implies \begin{cases} \mathbf{b}\mathbf{b}^\dagger = \mathbf{b}^\dagger\mathbf{b} + \mathbf{I} \\ \mathbf{b}^\dagger\mathbf{b} = \mathbf{b}\mathbf{b}^\dagger - \mathbf{I} \end{cases} \tag{3.10}$$

Problem: Show the validity of the equation,

$$\mathbf{b}\left(\mathbf{b}^\dagger\right)^2 - \left(\mathbf{b}^\dagger\right)^2\mathbf{b} = 2\mathbf{b}^\dagger. \tag{3.11}$$

Solution:

$$b\left(b^\dagger\right)^2 - \left(b^\dagger\right)^2 b \quad = \quad bb^\dagger b^\dagger - b^\dagger b^\dagger b \tag{3.12a}$$

$$= \quad \left(bb^\dagger\right) b^\dagger - b^\dagger \left(b^\dagger b\right) \tag{3.12b}$$

$$\overset{\text{Eq. (3.10)}}{=} \quad \left(1 + b^\dagger b\right) b^\dagger - b^\dagger \left(bb^\dagger - 1\right) \tag{3.12c}$$

$$= \quad b^\dagger + \underbrace{b^\dagger bb^\dagger - b^\dagger bb^\dagger}_{=0} + b^\dagger \tag{3.12d}$$

$$= \quad 2b^\dagger. \tag{3.12e}$$

■

Remark: The last term at the R.H.S. is the derivative of $\left(b^\dagger\right)^2$ with respect to $b^\dagger$. The next example demonstrates that this property is a general feature.

□

Problem: Show the validity of the equation,

$$b\left(b^\dagger\right)^n - \left(b^\dagger\right)^n b = n\left(b^\dagger\right)^{n-1}. \tag{3.13}$$

Solution I: Proof by mathematical induction.

The initial commutation relationship guarantees the validity of (3.13) for $n = 1$. Assuming the validity of (3.13) for an arbitrary $n \in \mathbb{N}$, the proof using the induction hypothesis proceeds as follows.

$$\underbrace{b\left(b^\dagger\right)^{n+1}} - \underbrace{\left(b^\dagger\right)^{n+1} b} \quad = \quad \underbrace{bb^\dagger}\left(b^\dagger\right)^n - b^\dagger \left(b^\dagger\right)^n b \tag{3.14a}$$

$$\overset{\text{Eq. (3.10)}}{=} \quad \left(1 + b^\dagger b\right)\left(b^\dagger\right)^n - b^\dagger \left(b^\dagger\right)^n b \tag{3.14b}$$

$$= \quad \left(b^\dagger\right)^n + \underbrace{b^\dagger b\left(b^\dagger\right)^n - b^\dagger \left(b^\dagger\right)^n b} \tag{3.14c}$$

$$= \quad \left(b^\dagger\right)^n + b^\dagger \underbrace{\left\{b\left(b^\dagger\right)^n - \left(b^\dagger\right)^n b\right\}}_{\text{induction hypothesis}} \tag{3.14d}$$

$$= \quad \left(b^\dagger\right)^n + b^\dagger \left\{n\left(b^\dagger\right)^{n-1}\right\} \tag{3.14e}$$

$$= \quad \left(b^\dagger\right)^n + n\left(b^\dagger\right)^n \tag{3.14f}$$

$$= \quad (n+1)\left(b^\dagger\right)^n. \tag{3.14g}$$

The R.H.S. is the derivative of $\left(b^\dagger\right)^{n+1}$ with respect to $b^\dagger$.

■

Solution II: Proof by mathematical induction—an alternative way.

In view of the induction hypothesis (3.13), which is a statement for n, consider the corresponding statement for $n + 1$,

$$b\left(b^\dagger\right)^{n+1} - \left(b^\dagger\right)^{n+1} b \overset{?}{=} (n+1)\left(b^\dagger\right)^n. \tag{3.15}$$

The question mark over the equality sign signifies that at this stage it is not certain whether or not (3.15) is valid. In case the truth of (3.15) can be established, the proof by mathematical induction is completed. The relationship in (3.15), even though it has not been proven to be true, serves as a guideline how to proceed. The goal is to transition from (3.13) to (3.15), by performing a series of simple directed manipulations. To this end, multiply (3.13) from the left by $\mathbf{b}^\dagger$,

$$\underbrace{\mathbf{b}^\dagger \mathbf{b}} \left(\mathbf{b}^\dagger\right)^n - \left(\mathbf{b}^\dagger\right)^{n+1} \mathbf{b} = n \left(\mathbf{b}^\dagger\right)^n . \tag{3.16}$$

Make the replacement $\mathbf{b}^\dagger \mathbf{b} = \mathbf{b}\mathbf{b}^\dagger - 1$, as indicated in (3.16),

$$\left(\mathbf{b}\mathbf{b}^\dagger - 1\right) \left(\mathbf{b}^\dagger\right)^n - \left(\mathbf{b}^\dagger\right)^{n+1} \mathbf{b} = n \left(\mathbf{b}^\dagger\right)^n . \tag{3.17}$$

Multiply out the first term,

$$\underbrace{\mathbf{b}\mathbf{b}^\dagger \left(\mathbf{b}^\dagger\right)^n} - \left(\mathbf{b}^\dagger\right)^n - \left(\mathbf{b}^\dagger\right)^{n+1} \mathbf{b} = n \left(\mathbf{b}^\dagger\right)^n . \tag{3.18}$$

Transfer the middle term to the R.H.S. and simplify the first term as indicated,

$$\mathbf{b} \left(\mathbf{b}^\dagger\right)^{n+1} - \left(\mathbf{b}^\dagger\right)^{n+1} \mathbf{b} = (n+1) \left(\mathbf{b}^\dagger\right)^n . \tag{3.19}$$

This is the desired statement of the induction hypothesis validated for $n + 1$, as stipulated above.

∎

Remark: The observation that $(n+1)\left(\mathbf{b}^\dagger\right)^n$ is the derivative of $\left(\mathbf{b}^\dagger\right)^{n+1}$ with respect to $\mathbf{b}^\dagger$ is reaffirming. It strengthens the guess which was formed in the previous problem and thus motivates the following general problem.

□

Problem: Show that,

$$\mathbf{b}f\left(\mathbf{b}^\dagger\right) - f\left(\mathbf{b}^\dagger\right)\mathbf{b} = \frac{\partial f\left(\mathbf{b}^\dagger\right)}{\partial \mathbf{b}^\dagger}. \tag{3.20}$$

Solution I: Proof by Formal Taylor Series Expansion.

Consider the Formal Taylor Series Expansion. of $f\left(\mathbf{b}^\dagger\right)$,

$$f\left(\mathbf{b}^\dagger\right) = \sum_n f_n \left(\mathbf{b}^\dagger\right)^n . \tag{3.21}$$

Focus on the expression at the L.H.S. of (3.20). Use the expansion in (3.21) of the function $f\left(\mathbf{b}^\dagger\right)$ and proceed as indicated,

$$\underbrace{\mathbf{b}f\left(\mathbf{b}^\dagger\right)} - \underbrace{f\left(\mathbf{b}^\dagger\right)\mathbf{b}} = \mathbf{b}\left\{\underbrace{\sum_n f_n \left(\mathbf{b}^\dagger\right)^n}\right\} - \left\{\underbrace{\sum_n f_n \left(\mathbf{b}^\dagger\right)^n}\right\}\mathbf{b} \tag{3.22a}$$

$$= \sum_n f_n \underbrace{\left\{\mathbf{b}\left(\mathbf{b}^\dagger\right)^n - \left(\mathbf{b}^\dagger\right)^n \mathbf{b}\right\}}_{\text{Eq. (3.13): } \partial(\mathbf{b}^\dagger)^n / \partial \mathbf{b}^\dagger} \tag{3.22b}$$

$$= \sum_n f_n \frac{\partial \left(\mathbf{b}^\dagger\right)^n}{\partial \mathbf{b}^\dagger} \tag{3.22c}$$

$$= \frac{\partial}{\partial \mathbf{b}^\dagger}\left\{\sum_n f_n \left(\mathbf{b}^\dagger\right)^n\right\} \tag{3.22d}$$

$$\overset{\text{Eq. (3.21)}}{=} \frac{\partial}{\partial \mathbf{b}^\dagger} f(\mathbf{b}^\dagger). \tag{3.22e}$$

∎

Remark: This problem allows the solidification of a few ideas of notational and conceptual nature. At any stage in the above sequence of equations, an underbracing heralds the manipulation of the corresponding term on the other side of the equality sign. Underbracing $f(\mathbf{b}^\dagger)$ at the L.H.S. of (3.22a), for example, signals that this function is going to be replaced by its Formal Taylor Series Expansion at the R.H.S. of (3.22a). Following this interpretation, underbracing $\sum_n f_n$ at the L.H.S. of (3.22b) indicates that this term is going to be factored out at the R.H.S. of (3.22b). The vertical curly braces have their conventional bracketing function, i.e., the standard demarcation.

□

Remark: This remark is of a procedural importance. Proves and solutions involving general functions call upon the Formal Taylor Series Expansion, which is essentially tantamount to replacing functions with their arguments to the power n. In turn, relationships involving arguments to the power n, remind one of the results obtained by the method of the mathematical induction. Finally, the initial steps in each proof by mathematical induction point to a significant underlying property, which shape the body of theory developed at each stage. A good example is the commutation relation $\mathbf{bb}^\dagger - \mathbf{b}^\dagger\mathbf{b} = 1$ which has been directing the choreography of what has developed in this text. This recurring scheme is going to be a trusted accompany throughout this exposition. A little thought also reveals the importance of Formal Taylor Series Expansion. Replacing $f(\mathbf{b}^\dagger)$ with $\sum_n f_n(\mathbf{b}^\dagger)^n$ is akin to opening a box, and playing around with the elemental building blocks in the box. Going back from $\sum_n f_n(\mathbf{b}^\dagger)^n$ to $f(\mathbf{b}^\dagger)$ is similar to rapping up and closing the box. In essence, this is the role of employing transforms and inverse transforms, or, equivalently, analysis and synthesis.

□

Solution II: An alternative proof.

Further above, the relationship,

$$\mathbf{b}\left(\mathbf{b}^\dagger\right)^n - \left(\mathbf{b}^\dagger\right)^n \mathbf{b} = \underbrace{n\left(\mathbf{b}^\dagger\right)^{n-1}}, \tag{3.23}$$

was established. Express the term at the R.H.S. in terms of the derivative of $\left(\mathbf{b}^\dagger\right)^n$,

$$\mathbf{b}\left(\mathbf{b}^\dagger\right)^n - \left(\mathbf{b}^\dagger\right)^n \mathbf{b} = \frac{\partial\left(\mathbf{b}^\dagger\right)^n}{\partial\mathbf{b}^\dagger}. \tag{3.24}$$

Multiply both sides by arbitrary (summable) numbers f_n and add over n,

$$\sum_n f_n\left\{\mathbf{b}\left(\mathbf{b}^\dagger\right)^n - \left(\mathbf{b}^\dagger\right)^n \mathbf{b}\right\} = \sum_n f_n \frac{\partial\left(\mathbf{b}^\dagger\right)^n}{\partial\mathbf{b}^\dagger}. \tag{3.25}$$

Or equivalently,

$$\mathbf{b}\left\{\sum_n f_n\left(\mathbf{b}^\dagger\right)^n\right\} - \left\{\sum_n f_n\left(\mathbf{b}^\dagger\right)^n\right\}\mathbf{b} = \frac{\partial}{\partial\mathbf{b}^\dagger}\left\{\sum_n f_n\left(\mathbf{b}^\dagger\right)^n\right\}. \tag{3.26}$$

Letting the sum in the curly brackets represent the function $f(\mathbf{b}^\dagger)$,

$$\mathbf{b}f(\mathbf{b}^\dagger) - f(\mathbf{b}^\dagger)\mathbf{b} = \frac{\partial f(\mathbf{b}^\dagger)}{\partial\mathbf{b}^\dagger}. \tag{3.27}$$

■

Remark: A simple recipe.

The following steps can be viewed as a simple recipe for obtaining an expression for $\partial f(\mathbf{b}^\dagger)/\partial\mathbf{b}^\dagger$ given $f(\mathbf{b}^\dagger)$. The virtue of formulating simple recipes will manifest itself in this book repeatedly.

- Consider the order (arrangement) of the creation operator $\mathbf{b}^\dagger$ and the annihilation operator $\mathbf{b}$ in the commutation relationship, $\mathbf{bb}^\dagger - \mathbf{b}^\dagger\mathbf{b} = \mathbf{I}$. Observe that the minus sign is "guarded" from the left and the right by the daggered operator $\mathbf{b}^\dagger$. This order (arrangement) of the creation operator $\mathbf{b}^\dagger$ and the annihilation operator $\mathbf{b}$ is referred to as the "normal," "natural," or the "canonical" order.

- Observe that the derivative of the "function" $\mathbf{b}^\dagger$ with respect to the "variable" $\mathbf{b}^\dagger$ is equal to the identity operator $\mathbf{I}$.
- To obtain an expression for the derivative of the function $f\left(\mathbf{b}^\dagger\right)$ with respect to the variable $\mathbf{b}^\dagger$, i.e., $\partial f(\mathbf{b}^\dagger)/\partial \mathbf{b}^\dagger$, it is suggestive to retain the canonical order. To this effect, replace $\mathbf{b}^\dagger$ with $f(\mathbf{b}^\dagger)$ in the commutation relationship $\mathbf{bb}^\dagger - \mathbf{b}^\dagger\mathbf{b} = \mathbf{I}$. The L.H.S. yields,

$$\mathbf{b}f\left(\mathbf{b}^\dagger\right) - f\left(\mathbf{b}^\dagger\right)\mathbf{b}. \tag{3.28}$$

It was shown above that this expression is equal to $\partial f(\mathbf{b}^\dagger)/\partial \mathbf{b}^\dagger$, thus establishing (3.27).

$\square$

The following series of solved problems show that the relationship between the commutator $[\mathbf{b}, f\left(\mathbf{b}^\dagger\right)]$ and the derivative of $f(\mathbf{b}^\dagger)$, i.e., $\partial f(\mathbf{b}^\dagger)/\partial \mathbf{b}^\dagger$ is generalizable in other ways.

3.5 Equations involving $\mathbf{b}^\dagger$ and $f(\mathbf{b})$

The next group of problems involves the creation operator $\mathbf{b}^\dagger$ and functions of the annihilation operator $f(\mathbf{b})$.

Problem: Show that,

$$(\mathbf{b})^2\,\mathbf{b}^\dagger - \mathbf{b}^\dagger\,(\mathbf{b})^2 = 2\mathbf{b}. \tag{3.29}$$

Proof: Using the definition $(\mathbf{b})^2 = \mathbf{bb}$, and utilizing the canonical commutation property $\mathbf{bb}^\dagger - \mathbf{b}^\dagger\mathbf{b} = \mathbf{I}$, proceed as follows:

$$\underbrace{(\mathbf{b})^2}\mathbf{b}^\dagger - \mathbf{b}^\dagger\underbrace{(\mathbf{b})^2} = \mathbf{b}\underbrace{\mathbf{bb}^\dagger} - \underbrace{\mathbf{b}^\dagger\mathbf{b}}\,\mathbf{b} \tag{3.30a}$$

$$= \mathbf{b}\left(\mathbf{I} + \mathbf{b}^\dagger\mathbf{b}\right) - \left(\mathbf{bb}^\dagger - \mathbf{I}\right)\mathbf{b} \tag{3.30b}$$

$$= \mathbf{b} + \underbrace{\mathbf{bb}^\dagger\mathbf{b} - \mathbf{bb}^\dagger\mathbf{b}}_{=0} + \mathbf{b} \tag{3.30c}$$

$$= 2\mathbf{b}. \tag{3.30d}$$

$\blacksquare$

Remark: Note that $(\mathbf{b})^2$ and $\mathbf{b}^\dagger$ appear in (3.29) in the normal order, in the sense that the minus sign is "guarded" by the daggered operators.

$\square$

Problem: Show that,

$$(\mathbf{b})^n\,\mathbf{b}^\dagger - \mathbf{b}^\dagger\,(\mathbf{b})^n = n\,(\mathbf{b})^{n-1}. \tag{3.31}$$

Solution: Proof by mathematical induction.

Assuming the validity of the induction hypothesis (3.31) and employing the canonical commutation relationship, proceed as follows:

$$\underbrace{(\mathbf{b})^{n+1}\mathbf{b}^{\dagger} - \mathbf{b}^{\dagger}(\mathbf{b})^{n+1}} \overset{\text{def.}}{=} \mathbf{b}\underbrace{(\mathbf{b})^{n}\,\mathbf{b}^{\dagger}} - \mathbf{b}^{\dagger}(\mathbf{b})^{n+1} \tag{3.32a}$$

$$\overset{\text{Eq. (3.31)}}{=} \mathbf{b}\left\{\mathbf{b}^{\dagger}(\mathbf{b})^{n} + n\,(\mathbf{b})^{n-1}\right\} - \mathbf{b}^{\dagger}(\mathbf{b})^{n+1} \tag{3.32b}$$

$$\overset{\text{dis. law}}{=} \underbrace{\mathbf{b}\mathbf{b}^{\dagger}}\,(\mathbf{b})^{n} + n\,(\mathbf{b})^{n} - \mathbf{b}^{\dagger}(\mathbf{b})^{n+1} \tag{3.32c}$$

$$\overset{\text{can. com. rel.}}{=} \left(\mathbf{I} + \mathbf{b}^{\dagger}\mathbf{b}\right)(\mathbf{b})^{n} + n\,(\mathbf{b})^{n} - \mathbf{b}^{\dagger}(\mathbf{b})^{n+1} \tag{3.32d}$$

$$\overset{\text{dis. law}}{=} (\mathbf{b})^{n} + \underbrace{\mathbf{b}^{\dagger}(\mathbf{b})^{n+1}} + n\,(\mathbf{b})^{n} - \underbrace{\mathbf{b}^{\dagger}(\mathbf{b})^{n+1}} \tag{3.32e}$$

$$\overset{\text{cancellation}}{=} (n+1)(\mathbf{b})^{n}. \tag{3.32f}$$

■

Problem: Show that,

$$f(\mathbf{b})\mathbf{b}^{\dagger} - \mathbf{b}^{\dagger}f(\mathbf{b}) = \frac{\partial f(\mathbf{b})}{\partial \mathbf{b}}. \tag{3.33}$$

Solution: Proof by Formal Taylor Series Expansion.

Let the Formal Taylor Series Expansion $\sum_{n} f_{n}(\mathbf{b})^{n}$ represent $f(\mathbf{b})$. Substitute $\sum_{n} f_{n}(\mathbf{b})^{n}$ for $f(\mathbf{b})$ in (3.33), and proceed as follows:

$$\underbrace{f(\mathbf{b})\mathbf{b}^{\dagger} - \mathbf{b}^{\dagger}f(\mathbf{b})} = \left\{\sum_{n} f_{n}(\mathbf{b})^{n}\right\}\mathbf{b}^{\dagger} - \mathbf{b}^{\dagger}\left\{\sum_{n} f_{n}(\mathbf{b})^{n}\right\} \tag{3.34a}$$

$$= \sum_{n} f_{n}\underbrace{\left\{(\mathbf{b})^{n}\,\mathbf{b}^{\dagger} - \mathbf{b}^{\dagger}(\mathbf{b})^{n}\right\}}_{\text{Eq. (3.31): } \partial(\mathbf{b})^{n}/\partial\mathbf{b}} \tag{3.34b}$$

$$= \sum_{n} f_{n}\frac{\partial(\mathbf{b})^{n}}{\partial\mathbf{b}} \tag{3.34c}$$

$$= \frac{\partial}{\partial\mathbf{b}}\underbrace{\sum_{n} f_{n}(\mathbf{b})^{n}}_{=f(\mathbf{b})} \tag{3.34d}$$

$$= \frac{\partial f(\mathbf{b})}{\partial\mathbf{b}}. \tag{3.34e}$$

■

Remark: On the relevance of formulating simple recipes.

Having established a useful (insightful) relationship and developed a sufficiently deep understanding of the intricacies involved, one should proceed to developing an easy and plausible recipe to commit the derived relationship to the memory. This is a crucial step in problem solving, since it permits retrieving the content figuratively at once. It can be hypothesized that all contents possess shapes, forms, or stated more pretentiously geometries. Thereby, the degree of the vividness of the corresponding shapes or forms in the mind determines how well the corresponding contents have been understood and thus they can be employed in routine or creative problem solving. A crucial realization is that the proof step must precede the memorization step. A further benefit of developing a clear plausible recipe is that it forces one to verbalize the details which might otherwise remain hidden in obscurity. Developing a recipe, writing a pseudo-code, conceptualizing a fully-fletched functional algorithm, all serve to scrutinizing the anatomy of contents, and inevitably enable sharpening the mind and honing the ability to think critically.

□

Remark: A simple recipe.

- Consider the order of the creation and annihilation operators in the canonical commutation relationship,

$$\mathbf{b}\mathbf{b}^\dagger - \mathbf{b}^\dagger \mathbf{b} = \mathbf{I}. \tag{3.35}$$

- To obtain an expression for the derivative of $f(\mathbf{b})$ with respect to $\mathbf{b}$, i.e., $\partial f(\mathbf{b})/\partial \mathbf{b}$, from (3.35), the canonical order must be retained. Thus, $\mathbf{b}$ in (3.35) must be replaced by $f(\mathbf{b})$,

$$f(\mathbf{b})\,\mathbf{b}^\dagger - \mathbf{b}^\dagger f(\mathbf{b})\,. \tag{3.36}$$

As demonstrated above, this is the desired expression for the derivative $\partial f(\mathbf{b})/\partial \mathbf{b}$.
- The recipes propounded in this and the preceding remark are powerful manipulatory tools for obtaining strikingly simple and yet wide-ranging results.

□

The following comment lists four special cases which will prove to be of particular importance.

Remark: Based on the recipes provided in the previous two remarks, the following relationships are readily established. They are significant in manipulations:

$$\mathbf{b}e^{\alpha \mathbf{b}^\dagger} - e^{\alpha \mathbf{b}^\dagger}\mathbf{b} = \alpha e^{\alpha \mathbf{b}^\dagger} \tag{3.37a}$$

$$\mathbf{b}e^{-\alpha \mathbf{b}^\dagger} - e^{-\alpha \mathbf{b}^\dagger}\mathbf{b} = -\alpha e^{-\alpha \mathbf{b}^\dagger} \tag{3.37b}$$

$$e^{\alpha \mathbf{b}}\mathbf{b}^\dagger - \mathbf{b}^\dagger e^{\alpha \mathbf{b}} = \alpha e^{\alpha \mathbf{b}} \tag{3.37c}$$

$$e^{-\alpha \mathbf{b}}\mathbf{b}^\dagger - \mathbf{b}^\dagger e^{-\alpha \mathbf{b}} = -\alpha e^{-\alpha \mathbf{b}}. \tag{3.37d}$$

□

Remark: Consider the general formulae established for the derivatives of $f(\mathbf{b}^\dagger)$ and $f(\mathbf{b})$,

$$\mathbf{b}f(\mathbf{b}^\dagger) - f(\mathbf{b}^\dagger)\,\mathbf{b} = \frac{\partial f(\mathbf{b}^\dagger)}{\partial \mathbf{b}^\dagger} \tag{3.38a}$$

$$f(\mathbf{b})\,\mathbf{b}^\dagger - \mathbf{b}^\dagger f(\mathbf{b}) = \frac{\partial f(\mathbf{b})}{\partial \mathbf{b}}. \tag{3.38b}$$

Recall that the genesis for arriving at these relationships has been the canonical commutation relationship,

$$\mathbf{b}\mathbf{b}^\dagger - \mathbf{b}^\dagger \mathbf{b} = \mathbf{I}. \tag{3.39}$$

In a certain determinate sense, the structures of the general formulae in (3.38) are consistent with the structures of the originating canonical commutation relationship in (3.39). It is worth pointing out that (3.39) can be interpreted in two ways: (i) the R.H.S. is the result of taking the derivative of $\mathbf{b}^\dagger$ with respect to $\mathbf{b}^\dagger$ which leads to $\mathbf{I}$. (ii) The R.H.S. is the result of taking the derivative of $\mathbf{b}$ with respect to $\mathbf{b}$ which leads to $\mathbf{I}$.

□

Exercise: Show the commutativity property,

$$\mathbf{b}^{\dagger m}\mathbf{b}^{\dagger n} = \mathbf{b}^{\dagger n}\mathbf{b}^{\dagger m}. \tag{3.40}$$

Solution: Employing the definition of $\mathbf{b}^{\dagger m}$ and $\mathbf{b}^{\dagger n}$, the following steps are self-evident:

$$\left(\mathbf{b}^{\dagger}\right)^{m}\left(\mathbf{b}^{\dagger}\right)^{n} = \underbrace{\mathbf{b}^{\dagger}\cdots\mathbf{b}^{\dagger}}_{m-\text{times}}\cdot\underbrace{\mathbf{b}^{\dagger}\cdots\mathbf{b}^{\dagger}}_{n-\text{times}} \tag{3.41a}$$

$$= \underbrace{\mathbf{b}^{\dagger}\cdots\cdots\cdots\cdots\cdots\mathbf{b}^{\dagger}}_{(m+n)-\text{times}} \tag{3.41b}$$

$$= \underbrace{\mathbf{b}^{\dagger}\cdots\mathbf{b}^{\dagger}}_{n-\text{times}}\cdot\underbrace{\mathbf{b}^{\dagger}\cdots\mathbf{b}^{\dagger}}_{m-\text{times}} \tag{3.41c}$$

$$= \left(\mathbf{b}^{\dagger}\right)^{n}\left(\mathbf{b}^{\dagger}\right)^{m}. \tag{3.41d}$$

■

Problem: Let $f\left(\mathbf{b}^{\dagger}\right)$ and $g\left(\mathbf{b}^{\dagger}\right)$ be arbitrary functions. Show that the commutativity property,

$$f\left(\mathbf{b}^{\dagger}\right)g\left(\mathbf{b}^{\dagger}\right) = g\left(\mathbf{b}^{\dagger}\right)f\left(\mathbf{b}^{\dagger}\right), \tag{3.42}$$

holds true.

Solution: Formal Taylor Series Expansion.

Consider the Formal Taylor Series Expansions of $f\left(\mathbf{b}^{\dagger}\right)$ and $g\left(\mathbf{b}^{\dagger}\right)$,

$$f\left(\mathbf{b}^{\dagger}\right) = \sum_{m}f_{m}\left(\mathbf{b}^{\dagger}\right)^{m} \tag{3.43a}$$

$$g\left(\mathbf{b}^{\dagger}\right) = \sum_{n}g_{n}\left(\mathbf{b}^{\dagger}\right)^{n}. \tag{3.43b}$$

Build the expression of $f\left(\mathbf{b}^{\dagger}\right)g\left(\mathbf{b}^{\dagger}\right)$ as follows:

$$f\left(\mathbf{b}^{\dagger}\right)\underbrace{g\left(\mathbf{b}^{\dagger}\right)}\;\overset{\text{FTSE of g}}{=}\;f\left(\mathbf{b}^{\dagger}\right)\sum_{n}g_{n}\left(\mathbf{b}^{\dagger}\right)^{n} \tag{3.44a}$$

$$\overset{\text{dist. law}}{=}\;\sum_{n}g_{n}\underbrace{f\left(\mathbf{b}^{\dagger}\right)}\left(\mathbf{b}^{\dagger}\right)^{n} \tag{3.44b}$$

$$\overset{\text{FTSE of f}}{=}\;\sum_{n}g_{n}\left\{\sum_{m}f_{m}\left(\mathbf{b}^{\dagger}\right)^{m}\right\}\left(\mathbf{b}^{\dagger}\right)^{n} \tag{3.44c}$$

$$\overset{\text{dist. law}}{=}\;\sum_{n}g_{n}\left\{\sum_{m}f_{m}\underbrace{\left(\mathbf{b}^{\dagger}\right)^{m}\left(\mathbf{b}^{\dagger}\right)^{n}}\right\} \tag{3.44d}$$

$$\overset{(3.40)}{=}\;\sum_{n}g_{n}\left\{\sum_{m}f_{m}\left(\mathbf{b}^{\dagger}\right)^{n}\left(\mathbf{b}^{\dagger}\right)^{m}\right\} \tag{3.44e}$$

$$\overset{\text{dist. law}}{=}\;\sum_{n}g_{n}\left(\mathbf{b}^{\dagger}\right)^{n}\underbrace{\left\{\sum_{m}f_{m}\left(\mathbf{b}^{\dagger}\right)^{m}\right\}} \tag{3.44f}$$

$$\overset{\text{FTSE of f}}{=}\;\underbrace{\sum_{n}g_{n}\left(\mathbf{b}^{\dagger}\right)^{n}f\left(\mathbf{b}^{\dagger}\right)} \tag{3.44g}$$

$$\overset{\text{FTSE of g}}{=}\;g\left(\mathbf{b}^{\dagger}\right)f\left(\mathbf{b}^{\dagger}\right). \tag{3.44h}$$

In the above, the abbreviation "dist. law" stands for the distribute law. For **a**, **b**, and **c** being operators, the distributive law permits multiplying out $\mathbf{a(b + c) = ab + ac}$. And, conversely, it allows factoring out a common operator, i.e., $\mathbf{ab + ac = a(b + c)}$.

■

Problem: Show in a direct way, that the inverse of $e^{\mathbf{b}^\dagger}$ is $e^{-\mathbf{b}^\dagger}$.

Solution: Formal Taylor Series Expansion.

Consider the Formal Taylor Series Expansions of $e^{\mathbf{b}^\dagger}$ and $e^{-\mathbf{b}^\dagger}$,

$$e^{\mathbf{b}^\dagger} = \mathbf{I} + \mathbf{b}^\dagger + \frac{1}{2}\left(\mathbf{b}^\dagger\right)^2 + \frac{1}{6}\left(\mathbf{b}^\dagger\right)^3 + \frac{1}{24}\left(\mathbf{b}^\dagger\right)^4 + \cdots \tag{3.45a}$$

$$e^{-\mathbf{b}^\dagger} = \mathbf{I} - \mathbf{b}^\dagger + \frac{1}{2}\left(\mathbf{b}^\dagger\right)^2 - \frac{1}{6}\left(\mathbf{b}^\dagger\right)^3 + \frac{1}{24}\left(\mathbf{b}^\dagger\right)^4 - \cdots . \tag{3.45b}$$

Multiply the expressions of $e^{\mathbf{b}^\dagger}$ and $e^{-\mathbf{b}^\dagger}$ term-by-term and arrange the resulting terms column-wise according to the powers of $\mathbf{b}^\dagger$, as shown below,

$$
\begin{aligned}
e^{\mathbf{b}^\dagger} e^{-\mathbf{b}^\dagger} = \mathbf{I} &+ \mathbf{b}^\dagger + \tfrac{1}{2}\left(\mathbf{b}^\dagger\right)^2 + \tfrac{1}{6}\left(\mathbf{b}^\dagger\right)^3 + \tfrac{1}{24}\left(\mathbf{b}^\dagger\right)^4 + \cdots \\
&- \mathbf{b}^\dagger - \left(\mathbf{b}^\dagger\right)^2 - \tfrac{1}{6}\left(\mathbf{b}^\dagger\right)^3 - \tfrac{1}{6}\left(\mathbf{b}^\dagger\right)^4 - \cdots \\
&\quad\quad + \tfrac{1}{2}\left(\mathbf{b}^\dagger\right)^2 + \tfrac{1}{2}\left(\mathbf{b}^\dagger\right)^3 + \tfrac{1}{4}\left(\mathbf{b}^\dagger\right)^4 + \cdots \\
&\quad\quad\quad\quad\quad - \tfrac{1}{6}\left(\mathbf{b}^\dagger\right)^3 - \tfrac{1}{6}\left(\mathbf{b}^\dagger\right)^4 - \cdots \\
&\quad\quad\quad\quad\quad\quad\quad\quad + \tfrac{1}{24}\left(\mathbf{b}^\dagger\right)^4 + \cdots .
\end{aligned}
\tag{3.46}
$$

Group terms in accordance with the powers of $\mathbf{b}^\dagger$,

$$
\begin{aligned}
e^{\mathbf{b}^\dagger} e^{-\mathbf{b}^\dagger} = \ &\mathbf{I} \\
&+ (1 - 1)\,\mathbf{b}^\dagger \\
&+ \left(\frac{1}{2} - 1 + \frac{1}{2}\right)\left(\mathbf{b}^\dagger\right)^2 \\
&+ \left(\frac{1}{6} - \frac{1}{2} + \frac{1}{2} - \frac{1}{6}\right)\left(\mathbf{b}^\dagger\right)^3 \\
&+ \left(\frac{1}{24} - \frac{1}{6} + \frac{1}{4} - \frac{1}{6} + \frac{1}{24}\right)\left(\mathbf{b}^\dagger\right)^4 \\
&+ \cdots .
\end{aligned}
\tag{3.47}
$$

It is observed that the coefficients of odd-powers of $\mathbf{b}^\dagger$ cancel each other pair-wisely. On the other hand, the coefficients of even-powers of $\mathbf{b}^\dagger$, even though not canceling pair-wisely, add up to zero, taken collectively. The only surviving term is the identity operator $\mathbf{I}$.

Similarly, it can be shown that $e^{-\mathbf{b}^\dagger} e^{\mathbf{b}^\dagger} = \mathbf{I}$. Consequently,

$$e^{\mathbf{b}^\dagger} e^{-\mathbf{b}^\dagger} = e^{-\mathbf{b}^\dagger} e^{\mathbf{b}^\dagger} = \mathbf{I}. \tag{3.48}$$

■

Remark: A rigorous proof would require the involvement of general formulae for the coefficients in the round brackets in (3.47). No efforts have been made here to develop the details.

□

Remark: Further above the relationship,

$$\mathbf{A}^{-1} f\,(\mathbf{B})\,\mathbf{A} = f\left(\mathbf{A}^{-1}\mathbf{B}\mathbf{A}\right), \tag{3.49}$$

for general operators $\mathbf{A}$ and $\mathbf{B}$ was established. Obviously, the transition from $\mathbf{A}^{-1}f(\mathbf{B})\mathbf{A}$ at the L.H.S. to $f(\mathbf{A}^{-1}\mathbf{BA})$ at the R.H.S. can be accomplished by going through the following elementary steps,

- Consider $\mathbf{A}^{-1}f(\mathbf{B})\mathbf{A}$.
- Replace $f(\mathbf{B})$ by its argument to obtain $\mathbf{A}^{-1}\mathbf{BA}$.
- Consider $\mathbf{A}^{-1}\mathbf{BA}$ as the new argument of $f(\cdot)$ to obtain $f(\mathbf{A}^{-1}\mathbf{BA})$.

$\square$

In the following series of problems, this simple recipe will be utilized to establish further important relationships.

Problem: Find an equivalent representation for $e^{-\alpha\mathbf{b}}\mathbf{b}^\dagger e^{\alpha\mathbf{b}}$.

Solution: The term $e^{-\alpha\mathbf{b}}\mathbf{b}^\dagger e^{\alpha\mathbf{b}}$, in virtue of its construction, suggests defining $\mathbf{A}=e^{\alpha\mathbf{b}}$, $\mathbf{A}^{-1}=e^{-\alpha\mathbf{b}}$, and $f(\mathbf{B})=\mathbf{b}^\dagger$, rendering $e^{-\alpha\mathbf{b}}\mathbf{b}^\dagger e^{\alpha\mathbf{b}}$ in the form $\mathbf{A}^{-1}f(\mathbf{B})\mathbf{A}$, and, thus, motivating the application of the formula $\mathbf{A}^{-1}f(\mathbf{B})\mathbf{A}=f(\mathbf{A}^{-1}\mathbf{BA})$. Applying the recipe in the preceding remark reads as follows,

- Consider $e^{-\alpha\mathbf{b}}\mathbf{b}^\dagger e^{\alpha\mathbf{b}}$.
- Replace $f(\cdot)(=\mathbf{b}^\dagger)$ by its argument, i.e., $\mathbf{b}^\dagger$ to obtain $e^{-\alpha\mathbf{b}}\mathbf{b}^\dagger e^{\alpha\mathbf{b}}$.
- Consider $e^{-\alpha\mathbf{b}}\mathbf{b}^\dagger e^{\alpha\mathbf{b}}$ as the new argument of $f(\cdot)(=\mathbf{b}^\dagger)$ to obtain $e^{-\alpha\mathbf{b}}\mathbf{b}^\dagger e^{\alpha\mathbf{b}}$.

This "circular" procedure is hardly useful, since the function $f(\mathbf{b}^\dagger)=\mathbf{b}^\dagger$ is linear, and even more restrictively, is equal to its argument. Consequently, the formula $\mathbf{A}^{-1}f(\mathbf{B})\mathbf{A}=f(\mathbf{A}^{-1}\mathbf{BA})$ being, in the present case, the trivial identity $\mathbf{A}^{-1}\mathbf{BA}=\mathbf{A}^{-1}\mathbf{BA}$, does not lead to any equivalent expression (simplification) of $e^{-\alpha\mathbf{b}}\mathbf{b}^\dagger e^{\alpha\mathbf{b}}$. Luckily, there are varied other tools available, which allow dealing with this "singular" case, which is of great merit for a variety of further cases.

▲

Remark: Concluding the above solution with a black triangle rather than a black square should signify that the problem was not concluded successfully. This task will be accomplished by resorting to another technique.

$\square$

Remark: The analysis in above discussion is meant to emphasize the significance of the establishment of the initial step in pursuing proofs by mathematical induction.

$\square$

Problem: Find an equivalent representation for $e^{-\alpha\mathbf{b}}\mathbf{b}^\dagger e^{\alpha\mathbf{b}}$.

Solution: This problem can be approached in two ways, depending on the chosen strategy for the "factorization" of $e^{-\alpha\mathbf{b}}\mathbf{b}^\dagger e^{\alpha\mathbf{b}}$. The associativity property enables one to write any of the following two valid factorizations:

$$e^{-\alpha\mathbf{b}}\mathbf{b}^\dagger e^{\alpha\mathbf{b}}=\{e^{-\alpha\mathbf{b}}\mathbf{b}^\dagger\}e^{\alpha\mathbf{b}}\implies\text{Solution I}\tag{3.50a}$$

$$e^{-\alpha\mathbf{b}}\mathbf{b}^\dagger e^{\alpha\mathbf{b}}=e^{-\alpha\mathbf{b}}\{\mathbf{b}^\dagger e^{\alpha\mathbf{b}}\}\implies\text{Solution II}\tag{3.50b}$$

Solution I:
- As indicated in (3.50a), focus on the term $e^{-\alpha\mathbf{b}}\mathbf{b}^\dagger$ in the curly brackets.
- Complete $e^{-\alpha\mathbf{b}}\mathbf{b}^\dagger$ to a "derivative" formula, established earlier,

$$e^{-\alpha\mathbf{b}}\mathbf{b}^\dagger-\mathbf{b}^\dagger e^{-\alpha\mathbf{b}}=-\alpha e^{-\alpha\mathbf{b}}.\tag{3.51}$$

This construction is unique. The minus sign must be "guarded" by the "daggered" operators.
- In view of the form $\{e^{-\alpha\mathbf{b}}\mathbf{b}^\dagger\}e^{\alpha\mathbf{b}}$ in (3.50a), multiply (3.51) by $e^{\alpha\mathbf{b}}$ from the R.H.S.,

$$e^{-\alpha\mathbf{b}}\mathbf{b}^\dagger e^{\alpha\mathbf{b}}-\mathbf{b}^\dagger=-\alpha.\tag{3.52}$$

– Add $\mathbf{b}^\dagger$ to both sides,

$$e^{-\alpha \mathbf{b}}\mathbf{b}^\dagger e^{\alpha \mathbf{b}} = \mathbf{b}^\dagger - \alpha. \tag{3.53}$$

This completes the Solution I.

Solution II:
– As indicated in (3.50b) focus on the term $\mathbf{b}^\dagger e^{\alpha \mathbf{b}}$ in the curly brackets.
– Complete $\mathbf{b}^\dagger e^{\alpha \mathbf{b}}$ to a "derivative" formula, established earlier,

$$e^{\alpha \mathbf{b}}\mathbf{b}^\dagger - \mathbf{b}^\dagger e^{\alpha \mathbf{b}} = \alpha e^{\alpha \mathbf{b}}. \tag{3.54}$$

This construction is unique. The minus sign must be "guarded" by the "daggered" operators.
– In view of the form $e^{-\alpha \mathbf{b}}\left\{\mathbf{b}^\dagger e^{\alpha \mathbf{b}}\right\}$ in (3.50b), multiply (3.54) by $e^{-\alpha \mathbf{b}}$ from the L.H.S.,

$$\mathbf{b}^\dagger - e^{-\alpha \mathbf{b}}\mathbf{b}^\dagger e^{\alpha \mathbf{b}} = \alpha. \tag{3.55}$$

– Rearrange,

$$e^{-\alpha \mathbf{b}}\mathbf{b}^\dagger e^{\alpha \mathbf{b}} = \mathbf{b}^\dagger - \alpha. \tag{3.56}$$

This completes the Solution II.

∎

Problem: Find an equivalent representation for $e^{\alpha \mathbf{b}}\mathbf{b}^\dagger e^{-\alpha \mathbf{b}}$.

Solution: The solution of this problem can be obtained by substituting α with $-\alpha$ in the previous problem and its solution. However, in order to solidify the ideas, it is solved independently, merely by following the proposed recipe. The associativity property suggests writing:

$$e^{\alpha \mathbf{b}}\mathbf{b}^\dagger e^{-\alpha \mathbf{b}} = \left\{e^{\alpha \mathbf{b}}\mathbf{b}^\dagger\right\} e^{-\alpha \mathbf{b}} \quad \Longrightarrow \quad \text{Solution I} \tag{3.57a}$$

$$e^{\alpha \mathbf{b}}\mathbf{b}^\dagger e^{-\alpha \mathbf{b}} = e^{\alpha \mathbf{b}}\left\{\mathbf{b}^\dagger e^{-\alpha \mathbf{b}}\right\} \quad \Longrightarrow \quad \text{Solution II} \tag{3.57b}$$

Solution I:
– Focus on the term $e^{\alpha \mathbf{b}}\mathbf{b}^\dagger$ in the curly brackets in (3.57a).
– Complete $e^{\alpha \mathbf{b}}\mathbf{b}^\dagger$ to a "derivative" formula, established earlier,

$$e^{\alpha \mathbf{b}}\mathbf{b}^\dagger - \mathbf{b}^\dagger e^{\alpha \mathbf{b}} = \alpha e^{\alpha \mathbf{b}}. \tag{3.58}$$

This construction is unique. The minus sign is "guarded" by "daggered" operators.
– In view of the form $\left\{e^{\alpha \mathbf{b}}\mathbf{b}^\dagger\right\} e^{-\alpha \mathbf{b}}$ in (3.57a), multiply (3.58) by $e^{-\alpha \mathbf{b}}$ from the R.H.S.,

$$e^{\alpha \mathbf{b}}\mathbf{b}^\dagger e^{-\alpha \mathbf{b}} - \mathbf{b}^\dagger = \alpha. \tag{3.59}$$

– Add $\mathbf{b}^\dagger$ to both sides,

$$e^{\alpha \mathbf{b}}\mathbf{b}^\dagger e^{-\alpha \mathbf{b}} = \mathbf{b}^\dagger + \alpha. \tag{3.60}$$

This completes the Solution I.

Solution II:
– Focus on the term $\mathbf{b}^\dagger e^{-\alpha \mathbf{b}}$ in the curly brackets in (3.57b).

－ Complete $\mathbf{b}^\dagger e^{-\alpha\mathbf{b}}$ to a "derivative" formula, established earlier,

$$e^{-\alpha\mathbf{b}}\mathbf{b}^\dagger - \mathbf{b}^\dagger e^{-\alpha\mathbf{b}} = -\alpha e^{-\alpha\mathbf{b}}. \tag{3.61}$$

This construction is unique. The minus sign is "guarded" by the "daggered" operators.
－ In view of the form $e^{\alpha\mathbf{b}}\left\{\mathbf{b}^\dagger e^{-\alpha\mathbf{b}}\right\}$ in (3.57b), multiply (3.61) by $e^{\alpha\mathbf{b}}$ from the L.H.S.,

$$\mathbf{b}^\dagger - e^{\alpha\mathbf{b}}\mathbf{b}^\dagger e^{-\alpha\mathbf{b}} = -\alpha. \tag{3.62}$$

－ Rearrange,

$$e^{\alpha\mathbf{b}}\mathbf{b}^\dagger e^{-\alpha\mathbf{b}} = \mathbf{b}^\dagger + \alpha. \tag{3.63}$$

This completes the Solution II.

∎

Remark: In order to demonstrate the power of the differentiation rules in derivations, and also to emphasize the significance of the underlying canonical commutativity property $\mathbf{b}\mathbf{b}^\dagger - \mathbf{b}^\dagger\mathbf{b} = \mathbf{I}$, consider the following alternative way of establishing (3.56). Consider explicit Formal Taylor Series Expansions of $e^{-\alpha\mathbf{b}}$ and $e^{\alpha\mathbf{b}}$ up to and including the second terms in α and proceed as follows:

$$e^{-\alpha\mathbf{b}}\mathbf{b}^\dagger e^{\alpha\mathbf{b}} = \left(\mathbf{I} - \alpha\mathbf{b} + \frac{1}{2}\alpha^2\mathbf{b}^2 + \mathcal{O}(\alpha^3)\right)\mathbf{b}^\dagger\left(\mathbf{I} + \alpha\mathbf{b} + \frac{1}{2}\alpha^2\mathbf{b}^2 + \mathcal{O}(\alpha^3)\right) \tag{3.64a}$$

$$= \left(\mathbf{I} - \alpha\mathbf{b} + \frac{1}{2}\alpha^2\mathbf{b}^2 + \mathcal{O}(\alpha^3)\right)\left(\mathbf{b}^\dagger + \alpha\mathbf{b}^\dagger\mathbf{b} + \frac{1}{2}\alpha^2\mathbf{b}^\dagger\mathbf{b}^2 + \mathcal{O}(\alpha^3)\right) \tag{3.64b}$$

$$= \mathbf{b}^\dagger + \alpha\mathbf{b}^\dagger\mathbf{b} + \frac{1}{2}\alpha^2\mathbf{b}^\dagger\mathbf{b}^2 + \mathcal{O}(\alpha^3)$$
$$- \alpha\mathbf{b}\mathbf{b}^\dagger - \alpha^2\mathbf{b}\mathbf{b}^\dagger\mathbf{b} + \mathcal{O}(\alpha^3)$$
$$+ \frac{1}{2}\alpha^2\mathbf{b}^2\mathbf{b}^\dagger + \mathcal{O}(\alpha^3). \tag{3.64c}$$

Order the terms at the R.H.S. according to the powers of α,

$$e^{-\alpha\mathbf{b}}\mathbf{b}^\dagger e^{\alpha\mathbf{b}} = \mathbf{b}^\dagger - \alpha\left(\mathbf{b}\mathbf{b}^\dagger - \mathbf{b}^\dagger\mathbf{b}\right) + \alpha^2\left(\frac{1}{2}\mathbf{b}^\dagger\mathbf{b}^2 - \mathbf{b}\mathbf{b}^\dagger\mathbf{b} + \frac{1}{2}\mathbf{b}^2\mathbf{b}^\dagger\right) + \mathcal{O}(\alpha^3). \tag{3.65}$$

Consider the fact that the term associated with α is the identity operator $\mathbf{I}$, and rewrite the term associated with α^2, by utilizing the definition $\mathbf{b}^2 = \mathbf{b}\mathbf{b}$,

$$e^{-\alpha\mathbf{b}}\mathbf{b}^\dagger e^{\alpha\mathbf{b}} = \mathbf{b}^\dagger - \alpha + \alpha^2\left(\frac{1}{2}\mathbf{b}^\dagger\mathbf{b}\mathbf{b} - \mathbf{b}\mathbf{b}^\dagger\mathbf{b} + \frac{1}{2}\mathbf{b}\mathbf{b}\mathbf{b}^\dagger\right) + \mathcal{O}(\alpha^3). \tag{3.66}$$

Focus on the terms associated with $1/2$,

$$\underbrace{\mathbf{b}^\dagger\mathbf{b}\mathbf{b}} + \underbrace{\mathbf{b}\mathbf{b}\mathbf{b}^\dagger} = \left(\mathbf{b}\mathbf{b}^\dagger - \mathbf{I}\right)\mathbf{b} + \mathbf{b}\left(\mathbf{b}^\dagger\mathbf{b} + \mathbf{I}\right) \tag{3.67a}$$

$$= 2\mathbf{b}\mathbf{b}^\dagger\mathbf{b}. \tag{3.67b}$$

Thus, the bracketed term at the R.H.S. in (3.66) results in,

$$\frac{1}{2}\mathbf{b}^\dagger\mathbf{b}\mathbf{b} - \mathbf{b}\mathbf{b}^\dagger\mathbf{b} + \frac{1}{2}\mathbf{b}\mathbf{b}\mathbf{b}^\dagger = \frac{1}{2}\left(\mathbf{b}^\dagger\mathbf{b}\mathbf{b} + \mathbf{b}\mathbf{b}\mathbf{b}^\dagger\right) - \mathbf{b}\mathbf{b}^\dagger\mathbf{b} \tag{3.68}$$

$$\stackrel{(3.67b)}{=} \frac{1}{2}\left(2\mathbf{b}\mathbf{b}^\dagger\mathbf{b}\right) - \mathbf{b}\mathbf{b}^\dagger\mathbf{b} \tag{3.69}$$

$$= \mathbf{0}. \tag{3.70}$$

Similarly, it can be shown that the terms associated with with α^n ($n \geq 3$) are all zero. Consequently,

$$e^{-\alpha\mathbf{b}}\mathbf{b}^\dagger e^{\alpha\mathbf{b}} = \mathbf{b}^\dagger - \alpha. \tag{3.71}$$

$\square$

Problem: Find an equivalent representation for $e^{-\alpha\mathbf{b}}\left(\mathbf{b}^\dagger\right)^2 e^{\alpha\mathbf{b}}$.

Solution: Apply the recipe for establishing the relationship $\mathbf{A}^{-1}f\left(\mathbf{B}\right)\mathbf{A} = f\left(\mathbf{A}^{-1}\mathbf{B}\mathbf{A}\right)$:

- Consider $e^{-\alpha\mathbf{b}}\left(\mathbf{b}^\dagger\right)^2 e^{\alpha\mathbf{b}}$.
- Replace $\left(\mathbf{b}^\dagger\right)^2$ by its argument, i.e., $\mathbf{b}^\dagger$ to obtain $e^{-\alpha\mathbf{b}}\mathbf{b}^\dagger e^{\alpha\mathbf{b}}$.
- Consider $e^{-\alpha\mathbf{b}}\mathbf{b}^\dagger e^{\alpha\mathbf{b}}$ as the new argument of $\left(\mathbf{b}^\dagger\right)^2$ to obtain $\left(e^{-\alpha\mathbf{b}}\mathbf{b}^\dagger e^{\alpha\mathbf{b}}\right)^2$.

Employing (3.71),

$$e^{-\alpha\mathbf{b}}\left(\mathbf{b}^\dagger\right)^2 e^{\alpha\mathbf{b}} = \left(\underbrace{e^{-\alpha\mathbf{b}}\mathbf{b}^\dagger e^{\alpha\mathbf{b}}}\right)^2 \tag{3.72a}$$

$$\stackrel{\text{Eq. (3.71)}}{=} \left(\mathbf{b}^\dagger - \alpha\right)^2. \tag{3.72b}$$

∎

Problem: Show that

$$e^{-\alpha\mathbf{b}}\left(\mathbf{b}^\dagger\right)^n e^{\alpha\mathbf{b}} = \left(\mathbf{b}^\dagger - \alpha\right)^n. \tag{3.73}$$

Solution: Apply the recipe for establishing the relationship $\mathbf{A}^{-1}f\left(\mathbf{B}\right)\mathbf{A} = f\left(\mathbf{A}^{-1}\mathbf{B}\mathbf{A}\right)$:

$$e^{-\alpha\mathbf{b}}\left(\mathbf{b}^\dagger\right)^n e^{\alpha\mathbf{b}} \rightarrow e^{-\alpha\mathbf{b}}\mathbf{b}^\dagger e^{\alpha\mathbf{b}} \tag{3.74a}$$

$$\rightarrow \left(\underbrace{e^{-\alpha\mathbf{b}}\mathbf{b}^\dagger e^{\alpha\mathbf{b}}}\right)^n \tag{3.74b}$$

$$\stackrel{\text{Eq. (3.71)}}{=} \left(\mathbf{b}^\dagger - \alpha\right)^n. \tag{3.74c}$$

∎

Problem: Show that

$$e^{-\alpha\mathbf{b}}f\left(\mathbf{b}^\dagger\right) e^{\alpha\mathbf{b}} = f\left(\mathbf{b}^\dagger - \alpha\right). \tag{3.75}$$

Solution: Apply the recipe for establishing the relationship $\mathbf{A}^{-1}f\left(\mathbf{B}\right)\mathbf{A} = f\left(\mathbf{A}^{-1}\mathbf{B}\mathbf{A}\right)$:

$$e^{-\alpha\mathbf{b}}f\left(\mathbf{b}^\dagger\right) e^{\alpha\mathbf{b}} \rightarrow e^{-\alpha\mathbf{b}}\mathbf{b}^\dagger e^{\alpha\mathbf{b}} \tag{3.76a}$$

$$\rightarrow f\left(\underbrace{e^{-\alpha\mathbf{b}}\mathbf{b}^\dagger e^{\alpha\mathbf{b}}}\right) \tag{3.76b}$$

$$\stackrel{\text{Eq. (3.71)}}{=} f\left(\mathbf{b}^\dagger - \alpha\right). \tag{3.76c}$$

In what follows relationships dual to those established in this section will be derived, by interchanging the roles of $\mathbf{b}^\dagger$ and $\mathbf{b}$.

Problem: Find an equivalent representation for $e^{-\alpha\mathbf{b}^\dagger}\mathbf{b}e^{\alpha\mathbf{b}^\dagger}$.

Solution: In this and the following problem, the annihilation operator $\mathbf{b}$ is sandwiched between the exponential functions of the creation operator $\mathbf{b}^\dagger$. However, the solution scheme proceeds along the same lines as expressed in the proposed recipe. The associativity property suggests writing:

$$e^{-\alpha\mathbf{b}^\dagger}\mathbf{b}e^{\alpha\mathbf{b}^\dagger} = \left\{e^{-\alpha\mathbf{b}^\dagger}\mathbf{b}\right\}e^{\alpha\mathbf{b}^\dagger} \implies \text{Solution I} \tag{3.77a}$$

$$e^{-\alpha\mathbf{b}^\dagger}\mathbf{b}e^{\alpha\mathbf{b}^\dagger} = e^{-\alpha\mathbf{b}^\dagger}\left\{\mathbf{b}e^{\alpha\mathbf{b}^\dagger}\right\} \implies \text{Solution II} \tag{3.77b}$$

Solution I:
- Focus on the term $e^{-\alpha\mathbf{b}^\dagger}\mathbf{b}$ in the curly brackets in (3.77a).
- Complete $e^{-\alpha\mathbf{b}^\dagger}\mathbf{b}$ to a "derivative" formula, established earlier,

$$\mathbf{b}e^{-\alpha\mathbf{b}^\dagger} - e^{-\alpha\mathbf{b}^\dagger}\mathbf{b} = -\alpha e^{-\alpha\mathbf{b}^\dagger}. \tag{3.78}$$

  This construction is unique. The minus sign is "guarded" by the "daggered" operators.
- In view of the form $\left\{e^{-\alpha\mathbf{b}^\dagger}\mathbf{b}\right\}e^{\alpha\mathbf{b}^\dagger}$ in (3.77a), multiply (3.78) by $e^{\alpha\mathbf{b}^\dagger}$ from the R.H.S.,

$$\mathbf{b} - e^{-\alpha\mathbf{b}^\dagger}\mathbf{b}e^{\alpha\mathbf{b}^\dagger} = -\alpha. \tag{3.79}$$

- Rearrange,

$$e^{\alpha\mathbf{b}^\dagger}\mathbf{b}e^{\alpha\mathbf{b}^\dagger} = \mathbf{b} + \alpha. \tag{3.80}$$

This completes the solution procedure.

Solution II:
- Focus on the term $\mathbf{b}e^{\alpha\mathbf{b}^\dagger}$ in the curly brackets in (3.77b).
- Complete $\mathbf{b}e^{\alpha\mathbf{b}^\dagger}$ to a "derivative" formula, established earlier,

$$\mathbf{b}e^{\alpha\mathbf{b}^\dagger} - e^{\alpha\mathbf{b}^\dagger}\mathbf{b} = \alpha e^{\alpha\mathbf{b}^\dagger}. \tag{3.81}$$

  This construction is unique. The minus sign is "guarded" by the "daggered" operators.
- In view of the form $e^{-\alpha\mathbf{b}^\dagger}\left\{\mathbf{b}e^{\alpha\mathbf{b}^\dagger}\right\}$ in (3.77b), multiply (3.81) by $e^{-\alpha\mathbf{b}^\dagger}$ from the L.H.S.,

$$e^{-\alpha\mathbf{b}^\dagger}\mathbf{b}e^{\alpha\mathbf{b}^\dagger} - \mathbf{b} = \alpha. \tag{3.82}$$

- Add $\mathbf{b}$ to both sides,

$$e^{-\alpha\mathbf{b}^\dagger}\mathbf{b}e^{\alpha\mathbf{b}^\dagger} = \mathbf{b} + \alpha. \tag{3.83}$$

This completes the solution procedure.

Problem: Find an equivalent representation for $e^{\alpha\mathbf{b}^\dagger}\mathbf{b}e^{-\alpha\mathbf{b}^\dagger}$.

Solution: The associativity property suggests writing:

$$e^{\alpha \mathbf{b}^\dagger} \mathbf{b} e^{-\alpha \mathbf{b}^\dagger} = \left\{ e^{\alpha \mathbf{b}^\dagger} \mathbf{b} \right\} e^{-\alpha \mathbf{b}^\dagger} \implies \text{Solution I} \tag{3.84a}$$

$$e^{\alpha \mathbf{b}^\dagger} \mathbf{b} e^{-\alpha \mathbf{b}^\dagger} = e^{\alpha \mathbf{b}^\dagger} \left\{ \mathbf{b} e^{-\alpha \mathbf{b}^\dagger} \right\} \implies \text{Solution II} \tag{3.84b}$$

Solution I:
- Focus on the term $e^{\alpha \mathbf{b}^\dagger} \mathbf{b}$ in the curly brackets in (3.84a).
- Complete $e^{\alpha \mathbf{b}^\dagger} \mathbf{b}$ to the "derivative" formula, established earlier,

$$\mathbf{b} e^{\alpha \mathbf{b}^\dagger} - e^{\alpha \mathbf{b}^\dagger} \mathbf{b} = \alpha e^{\alpha \mathbf{b}^\dagger}. \tag{3.85}$$

This construction is unique. The minus sign is "guarded" by "daggered" operators.
- In view of the form $\left\{ e^{\alpha \mathbf{b}^\dagger} \mathbf{b} \right\} e^{-\alpha \mathbf{b}^\dagger}$ in (3.84a), multiply (3.85) by $e^{-\alpha \mathbf{b}^\dagger}$ from the R.H.S.,

$$\mathbf{b} - e^{\alpha \mathbf{b}^\dagger} \mathbf{b} e^{-\alpha \mathbf{b}^\dagger} = \alpha. \tag{3.86}$$

- Rearrange,

$$e^{\alpha \mathbf{b}^\dagger} \mathbf{b} e^{-\alpha \mathbf{b}^\dagger} = \mathbf{b} - \alpha. \tag{3.87}$$

This completes the solution procedure.

Solution II:
- Focus on the term $\mathbf{b} e^{-\alpha \mathbf{b}^\dagger}$ in the curly brackets in (3.84b).
- Complete $\mathbf{b} e^{-\alpha \mathbf{b}^\dagger}$ to the "derivative" formula, established earlier,

$$\mathbf{b} e^{-\alpha \mathbf{b}^\dagger} - e^{-\alpha \mathbf{b}^\dagger} \mathbf{b} = -\alpha e^{-\alpha \mathbf{b}^\dagger}. \tag{3.88}$$

This construction is unique. The minus sign is "guarded" by "daggered" operators.
- In view of the form $e^{\alpha \mathbf{b}^\dagger} \left\{ \mathbf{b} e^{-\alpha \mathbf{b}^\dagger} \right\}$ in (3.84b), multiply (3.88) by $e^{\alpha \mathbf{b}^\dagger}$ from the L.H.S.,

$$e^{\alpha \mathbf{b}^\dagger} \mathbf{b} e^{-\alpha \mathbf{b}^\dagger} - \mathbf{b} = -\alpha. \tag{3.89}$$

- Add $\mathbf{b}$ to both sides,

$$e^{\alpha \mathbf{b}^\dagger} \mathbf{b} e^{-\alpha \mathbf{b}^\dagger} = \mathbf{b} - \alpha. \tag{3.90}$$

This completes the solution procedure.

∎

Problem: Show that

$$e^{-\alpha \mathbf{b}^\dagger} (\mathbf{b})^2 e^{\alpha \mathbf{b}^\dagger} = (\mathbf{b} + \alpha)^2. \tag{3.91}$$

Solution:

$$e^{-\alpha \mathbf{b}^\dagger} (\mathbf{b})^2 e^{\alpha \mathbf{b}^\dagger} \quad \rightarrow \quad e^{-\alpha \mathbf{b}^\dagger} \mathbf{b} e^{\alpha \mathbf{b}^\dagger} \tag{3.92a}$$

$$\rightarrow \quad \left(\underbrace{e^{-\alpha \mathbf{b}^\dagger} \mathbf{b} e^{\alpha \mathbf{b}^\dagger}} \right)^2 \tag{3.92b}$$

$$\overset{\text{Eq. (3.83)}}{=} (\mathbf{b} + \alpha)^2. \tag{3.92c}$$

∎

Problem: Show that

$$e^{-\alpha \mathbf{b}^\dagger} (\mathbf{b})^n e^{\alpha \mathbf{b}^\dagger} = (\mathbf{b} + \alpha)^n .$$ (3.93)

Solution: Following the recipe

$$e^{-\alpha \mathbf{b}^\dagger} (\mathbf{b})^n e^{\alpha \mathbf{b}^\dagger} \quad \rightarrow \quad e^{-\alpha \mathbf{b}^\dagger} \mathbf{b} e^{\alpha \mathbf{b}^\dagger}$$ (3.94a)

$$\rightarrow \quad \left(e^{-\alpha \mathbf{b}^\dagger} \mathbf{b} e^{\alpha \mathbf{b}^\dagger} \right)^n$$ (3.94b)

$$\overset{\text{Eq. (3.83)}}{=} (\mathbf{b} + \alpha)^n .$$ (3.94c)

∎

Solution continued: Mathematical induction.

A complete proof by mathematical induction would require the following arguments:

$$e^{-\alpha \mathbf{b}^\dagger} \underbrace{(\mathbf{b})^{n+1}} e^{\alpha \mathbf{b}^\dagger} \overset{(1)}{=} e^{-\alpha \mathbf{b}^\dagger} (\mathbf{b})^n \, \mathbf{b} e^{\alpha \mathbf{b}^\dagger}$$ (3.95a)

$$\overset{(2)}{=} e^{-\alpha \mathbf{b}^\dagger} (\mathbf{b})^n \underbrace{\mathbf{I}} \mathbf{b} e^{\alpha \mathbf{b}^\dagger}$$ (3.95b)

$$\overset{(3)}{=} e^{-\alpha \mathbf{b}^\dagger} (\mathbf{b})^n \overbrace{e^{\alpha \mathbf{b}^\dagger} e^{-\alpha \mathbf{b}^\dagger}} \mathbf{b} e^{\alpha \mathbf{b}^\dagger}$$ (3.95c)

$$\overset{(4)}{=} \underbrace{e^{-\alpha \mathbf{b}^\dagger} (\mathbf{b})^n e^{\alpha \mathbf{b}^\dagger}} \underbrace{e^{-\alpha \mathbf{b}^\dagger} \mathbf{b} e^{\alpha \mathbf{b}^\dagger}}$$ (3.95d)

$$\overset{(5)}{=} \underbrace{\left(e^{-\alpha \mathbf{b}^\dagger} \mathbf{b} e^{\alpha \mathbf{b}^\dagger} \right)^n} \underbrace{\left(e^{-\alpha \mathbf{b}^\dagger} \mathbf{b} e^{\alpha \mathbf{b}^\dagger} \right)}$$ (3.95e)

$$\overset{(6)}{=} (\mathbf{b} + \alpha)^n (\mathbf{b} + \alpha)$$ (3.95f)

$$\overset{(7)}{=} (\mathbf{b} + \alpha)^{n+1} .$$ (3.95g)

∎

Remark: Mapping out the logic and psychology of the solution procedure.

Transition (1) employs the definition $\mathbf{b}^{n+1} = \mathbf{b}^n \mathbf{b}^1$. The judicious splitting of $\mathbf{b}^{n+1}$ not only creates $\mathbf{b}^n$ and $\mathbf{b}^1$ but also gives rise to the idea of the space in-between, a space between $\mathbf{b}^n$ and $\mathbf{b}$, which is a theater stage, a Platform to operate on. Judicious because $\mathbf{b}^n$ relates to the induction hypothesis, while $\mathbf{b}^1$ relates to the induction initial step. Transition (2) inserts the identity operator $\mathbf{I}$ into the space between $\mathbf{b}^n$ and $\mathbf{b}^1$. Transition (3) uses the judicious factorization of $\mathbf{I}$ in the form of $\mathbf{I} = e^{\alpha \mathbf{b}^\dagger} e^{-\alpha \mathbf{b}^\dagger}$. Judicious because the alternative equally-valid factorization $\mathbf{I} = e^{-\alpha \mathbf{b}^\dagger} e^{\alpha \mathbf{b}^\dagger}$ would not be goal oriented. The motivation for the chosen insightful factorization is explained in the next step or the sequence of following steps. This demonstrates that the individual steps are guided by the aim (expectation, anticipation) of achieving a determinate goal. In practice, this happens after a series of trial and removal of error experiments, which are to the greater part educated attempts or pseudo random experiments merely based on accumulated hunches, which are usually precursors of well-formed educated guesses. Transition (4) uses the associativity property, the ability to group operators together without changing their order, i.e., proper bracketing. As always is the case, the rationale for the chosen grouping is understood either subsequently or in the following steps, which collectively reveal the purpose of the current enacted choice. Transition (5) resorts to the formula $\mathbf{A}^{-1} f(\mathbf{B}) \mathbf{A} = f(\mathbf{A}^{-1} \mathbf{B} \mathbf{A})$. Transition (6) employs the induction hypothesis and the induction initial step. Transition (7) takes advantage of the merging property $\mathbf{b}^n \mathbf{b}^1 = \mathbf{b}^{n+1}$. As can be seen the challenge in the proof by mathematical induction is the formation of the induction hypothesis, which is itself based on the inductive reasoning, forming an idea based on a few initial steps. Some hidden structures may reveal themselves after going through several steps. After a pattern emerges the formulation of the induction hypothesis offers itself, and the steps in the proof are more or less self-guided. One of the major appealing features of the Formal

Taylor Series Expansion in quantum physics is that it introduces countable enumerable sets by letting summation indices to run from an initial finite value (usually 0) to infinity.

□

Remark: The extraordinary power of the identity operator.

It seems appropriate to reemphasize the almighty power of the identity property in manipulations. The above derivation illustrates a technique which is frequently used in derivations in quantum physics. Whenever, permissible and appropriate, the identity operator $\mathbf{I}$ is inserted, and then expressed by any of its myriad representations as respective situations dictate. In the course of developing further relationships in this text, other interesting representations of $\mathbf{I}$ suggest themselves nearly cogently. Thus, the versatility and generality of this simple, yet, profound and powerful technique manifests itself in the fact, that the respective applications suggest which representation of $\mathbf{I}$ should be chosen or even designed. This leads to the notion of the problem-specific resolution of the identity. If "0" is a placeholder for nothing, the notions of "1" (the unity), the identity matrix, the Dirac delta function, and the identity operator, represent the whole. The concept of the resolution of the identity can justifiably be viewed as a treasure trove for creative problem solving.

□

Problem: Show that

$$e^{-\alpha \mathbf{b}^\dagger} f(\mathbf{b}) \, e^{\alpha \mathbf{b}^\dagger} = f(\mathbf{b} + \alpha). \tag{3.96}$$

Solution:

$$e^{-\alpha \mathbf{b}^\dagger} f(\mathbf{b}) \, e^{\alpha \mathbf{b}^\dagger} \;\rightarrow\; e^{-\alpha \mathbf{b}^\dagger} \mathbf{b} \, e^{\alpha \mathbf{b}^\dagger} \tag{3.97a}$$

$$\rightarrow\; f\!\left(\underbrace{e^{-\alpha \mathbf{b}^\dagger} \mathbf{b} \, e^{\alpha \mathbf{b}^\dagger}}\right) \tag{3.97b}$$

$$= f(\mathbf{b} + \alpha). \tag{3.97c}$$

∎

Remark: In the preceding series of problems, an important class of relationships was introduced. The aim here is to provide a simple "pseudo rule" for expressing the results simply-by-inspection. Consider the relationships,

$$e^{-\alpha \mathbf{b}} \mathbf{b}^\dagger e^{\alpha \mathbf{b}} = \mathbf{b}^\dagger - \alpha \tag{3.98a}$$

$$e^{\alpha \mathbf{b}} \mathbf{b}^\dagger e^{-\alpha \mathbf{b}} = \mathbf{b}^\dagger + \alpha \tag{3.98b}$$

$$e^{-\alpha \mathbf{b}^\dagger} \mathbf{b} \, e^{\alpha \mathbf{b}^\dagger} = \mathbf{b} + \alpha \tag{3.98c}$$

$$e^{\alpha \mathbf{b}^\dagger} \mathbf{b} \, e^{-\alpha \mathbf{b}^\dagger} = \mathbf{b} - \alpha. \tag{3.98d}$$

The results at the R.H.S. can be readily obtained by a few "mental gymnastic moves," or by what it may be referred to as the "rationalized memorization," or, "simply-by-inspection." For the present purposes, it suffices to know that $\mathbf{b}^\dagger$ and $\mathbf{b}$ can be synthesized as a linear combination of the x and d/dx,

$$\mathbf{b}^\dagger = \frac{1}{\sqrt{2}} \left(x - \frac{d}{dx} \right) \tag{3.99a}$$

$$\mathbf{b} = \frac{1}{\sqrt{2}} \left(x + \frac{d}{dx} \right). \tag{3.99b}$$

Thus, the operator $\mathbf{b}^\dagger$ involves the independent variable x and the differential operator $-\frac{d}{dx}$ carrying a negative sign. Similarly, the operator $\mathbf{b}$ involves the independent variable x and the differential operator $\frac{d}{dx}$ carrying a positive sign. Having made these simple observations, proceed as follows, by treating the relationships in (3.98) one by one.

The term $e^{-\alpha \mathbf{b}}\mathbf{b}^{\dagger}e^{\alpha \mathbf{b}}$:

$$e^{-\alpha \mathbf{b}}\mathbf{b}^{\dagger}e^{\alpha \mathbf{b}} = e^{-\alpha \mathbf{b}}\underbrace{\mathbf{b}^{\dagger}e^{\alpha \mathbf{b}}}$$

$$= e^{-\alpha \mathbf{b}}\begin{cases} 1) \text{ view } \mathbf{b} \text{ and } \mathbf{b}^{\dagger} \text{ as variables}: \ \mathbf{b}^{\dagger}e^{\alpha \mathbf{b}} \Longrightarrow e^{\alpha \mathbf{b}}\mathbf{b}^{\dagger} \\[2mm] 2) \text{ view } \mathbf{b}^{\dagger} \text{ as a } \textbf{nag}. \text{ diff. oper.}: \ \mathbf{b}^{\dagger}e^{\alpha \mathbf{b}} \Longrightarrow -\alpha e^{\alpha \mathbf{b}} \end{cases}.$$

$$= \begin{cases} e^{-\alpha \mathbf{b}}e^{\alpha \mathbf{b}}\mathbf{b}^{\dagger} = \mathbf{b}^{\dagger} \\[2mm] -\alpha e^{-\alpha \mathbf{b}}e^{\alpha \mathbf{b}} = -\alpha \end{cases} \quad \text{(add the two contributions)}$$

$$= \mathbf{b}^{\dagger} - \alpha. \tag{3.100}$$

The term $e^{\alpha \mathbf{b}}\mathbf{b}^{\dagger}e^{-\alpha \mathbf{b}}$:

$$e^{\alpha \mathbf{b}}\mathbf{b}^{\dagger}e^{-\alpha \mathbf{b}} = e^{\alpha \mathbf{b}}\underbrace{\mathbf{b}^{\dagger}e^{-\alpha \mathbf{b}}}$$

$$= e^{\alpha \mathbf{b}}\begin{cases} 1) \text{ view } \mathbf{b} \text{ and } \mathbf{b}^{\dagger} \text{ as variables}: \ \mathbf{b}^{\dagger}e^{-\alpha \mathbf{b}} \Longrightarrow e^{-\alpha \mathbf{b}}\mathbf{b}^{\dagger} \\[2mm] 2) \text{ view } \mathbf{b}^{\dagger} \text{ as a } \textbf{nag}. \text{ diff. oper.}: \ \mathbf{b}^{\dagger}e^{-\alpha \mathbf{b}} \Longrightarrow \alpha e^{-\alpha \mathbf{b}} \end{cases}.$$

$$= \begin{cases} e^{\alpha \mathbf{b}}e^{-\alpha \mathbf{b}}\mathbf{b}^{\dagger} = \mathbf{b}^{\dagger} \\[2mm] \alpha e^{\alpha \mathbf{b}}e^{-\alpha \mathbf{b}} = \alpha \end{cases} \quad \text{(add the two contributions)}$$

$$= \mathbf{b}^{\dagger} + \alpha. \tag{3.101}$$

The term $e^{-\alpha \mathbf{b}^{\dagger}}\mathbf{b}e^{\alpha \mathbf{b}^{\dagger}}$:

$$e^{-\alpha \mathbf{b}^{\dagger}}\mathbf{b}e^{\alpha \mathbf{b}^{\dagger}} = e^{-\alpha \mathbf{b}^{\dagger}}\underbrace{\mathbf{b}e^{\alpha \mathbf{b}^{\dagger}}}$$

$$= e^{-\alpha \mathbf{b}^{\dagger}}\begin{cases} 1) \text{ view } \mathbf{b} \text{ and } \mathbf{b}^{\dagger} \text{ as variables}: \ \mathbf{b}e^{\alpha \mathbf{b}^{\dagger}} \Longrightarrow e^{\alpha \mathbf{b}^{\dagger}}\mathbf{b} \\[2mm] 2) \text{ view } \mathbf{b} \text{ as a } \textbf{pos}. \text{ diff. oper.}: \ \mathbf{b}e^{\alpha \mathbf{b}^{\dagger}} \Longrightarrow \alpha e^{\alpha \mathbf{b}^{\dagger}} \end{cases}.$$

$$= \begin{cases} e^{-\alpha \mathbf{b}^{\dagger}}e^{\alpha \mathbf{b}^{\dagger}}\mathbf{b} = \mathbf{b} \\[2mm] \alpha e^{-\alpha \mathbf{b}^{\dagger}}e^{\alpha \mathbf{b}^{\dagger}} = \alpha \end{cases} \quad \text{(add the two contributions)}$$

$$= \mathbf{b} + \alpha. \tag{3.102}$$

The term $e^{\alpha \mathbf{b}^\dagger} \mathbf{b} e^{-\alpha \mathbf{b}^\dagger}$:

$$e^{\alpha \mathbf{b}^\dagger} \mathbf{b} e^{-\alpha \mathbf{b}^\dagger} = e^{\alpha \mathbf{b}^\dagger} \underbrace{\mathbf{b} e^{-\alpha \mathbf{b}^\dagger}}$$

$$= e^{\alpha \mathbf{b}^\dagger} \begin{cases} 1) \text{ view } \mathbf{b} \text{ and } \mathbf{b}^\dagger \text{ as variables}: \ \mathbf{b} e^{-\alpha \mathbf{b}^\dagger} \implies e^{-\alpha \mathbf{b}^\dagger} \mathbf{b} \\[2mm] 2) \text{ view } \mathbf{b} \text{ as a } \textbf{pos.} \text{ diff. oper.}: \ \mathbf{b} e^{-\alpha \mathbf{b}^\dagger} \implies -\alpha e^{-\alpha \mathbf{b}^\dagger} \end{cases}.$$

$$= \begin{cases} e^{\alpha \mathbf{b}^\dagger} e^{-\alpha \mathbf{b}^\dagger} \mathbf{b} = \mathbf{b} \\[2mm] -\alpha e^{\alpha \mathbf{b}^\dagger} e^{-\alpha \mathbf{b}^\dagger} = -\alpha \end{cases} \quad \text{(add the two contributions)}$$

$$= \mathbf{b} - \alpha. \tag{3.103}$$

$\square$

Remark: By applying the above rules of games, the following results are immediate:

$$e^{-\alpha \mathbf{b}} f\left(\mathbf{b}^\dagger\right) e^{\alpha \mathbf{b}} \implies e^{-\alpha \mathbf{b}} \left(\mathbf{b}^\dagger\right) e^{\alpha \mathbf{b}} \implies e^{-\alpha \mathbf{b}} \mathbf{b}^\dagger e^{\alpha \mathbf{b}} \implies \mathbf{b}^\dagger - \alpha \implies f\left(\mathbf{b}^\dagger - \alpha\right) \tag{3.104a}$$

$$e^{\alpha \mathbf{b}} f\left(\mathbf{b}^\dagger\right) e^{-\alpha \mathbf{b}} \implies e^{\alpha \mathbf{b}} \left(\mathbf{b}^\dagger\right) e^{-\alpha \mathbf{b}} \implies e^{\alpha \mathbf{b}} \mathbf{b}^\dagger e^{-\alpha \mathbf{b}} \implies \mathbf{b}^\dagger + \alpha \implies f\left(\mathbf{b}^\dagger + \alpha\right) \tag{3.104b}$$

$$e^{-\alpha \mathbf{b}^\dagger} f\left(\mathbf{b}\right) e^{\alpha \mathbf{b}^\dagger} \implies e^{-\alpha \mathbf{b}^\dagger} \left(\mathbf{b}\right) e^{\alpha \mathbf{b}^\dagger} \implies e^{-\alpha \mathbf{b}^\dagger} \mathbf{b} e^{\alpha \mathbf{b}^\dagger} \implies \mathbf{b} + \alpha \implies f\left(\mathbf{b} + \alpha\right) \tag{3.104c}$$

$$e^{\alpha \mathbf{b}^\dagger} f\left(\mathbf{b}\right) e^{-\alpha \mathbf{b}^\dagger} \implies e^{\alpha \mathbf{b}^\dagger} \left(\mathbf{b}\right) e^{-\alpha \mathbf{b}^\dagger} \implies e^{\alpha \mathbf{b}^\dagger} \mathbf{b} e^{-\alpha \mathbf{b}^\dagger} \implies \mathbf{b} - \alpha \implies f\left(\mathbf{b} - \alpha\right) \tag{3.104d}$$

It should be reemphasized that the above pseudo rules are neither logically nor mathematically valid rules or algorithms. They are consistent recipes, or possibly, more suitably stated, consistent tricks for obtaining results at once, simply-by-inspection. They might serve as pseudo codes and employed for automatic generation of software code. For routine manipulation of formulae, they may also prove to be extraordinarily useful. Applying pseudo rules, after rigorous proofs have been carried out, enhances creativity and the ability to solve problems. Employing pseudo rules, mechanically and without understanding the underlying relationships, is akin to simply memorizing relationships and as such severs understanding, obscures logical thinking and counters creative problem solving efforts.

$\square$

Remark: Certain category of commutators involving annihilation and creation operators will not be considered in this text. Among those commutators which are not included is the commutator, $\mathbf{b}^m \mathbf{b}^{\dagger n} - \mathbf{b}^{\dagger n} \mathbf{b}^m$. There are several other constructions which are excluded considering the available space. The interested reader may wish to engage in studying the properties of this type of commutators and discover for themselves a myriad of intriguing relationships.

$\square$

Problem: Find an equivalent expression for

$$e^{-\alpha \mathbf{b}^\dagger} e^{\beta \mathbf{b}} e^{\alpha \mathbf{b}^\dagger}. \tag{3.105}$$

Solution: Focus on the function $e^{\beta \mathbf{b}}$ being sandwiched between the operator $e^{\alpha \mathbf{b}^\dagger}$ and its inverse $e^{-\alpha \mathbf{b}^\dagger}$. In the preceding discussion, it was shown exhaustively that multiplying a function (here $e^{\beta \mathbf{b}}$) by $e^{-\alpha \mathbf{b}^\dagger}$ from the L.H.S. and $e^{\alpha \mathbf{b}^\dagger}$ from the R.H.S., respectively, effects the shift $\mathbf{b} \to \mathbf{b} + \alpha$ in the argument of $e^{\beta \mathbf{b}}$, thus, leading to $e^{\beta(\mathbf{b}+\alpha)}$. Since α and β are constants, $e^{\beta(\mathbf{b}+\alpha)}$ simplifies, resulting in $e^{\beta \mathbf{b} + \beta \alpha} = e^{\beta \mathbf{b}} e^{\beta \alpha}$. Thus,

$$e^{-\alpha \mathbf{b}^\dagger} e^{\beta \mathbf{b}} e^{\alpha \mathbf{b}^\dagger} = e^{\beta(\mathbf{b}+\alpha)} = e^{\beta \mathbf{b}} e^{\beta \alpha}. \tag{3.106}$$

■

Problem: Verify the above result directly.

Solution: Formal Taylor Series Expansion.

$$e^{-\alpha \mathbf{b}^\dagger} \underbrace{e^{\beta \mathbf{b}}} e^{\alpha \mathbf{b}^\dagger} = e^{-\alpha \mathbf{b}^\dagger} \left\{ \sum_n \frac{\beta^n}{n!} \mathbf{b}^n \right\} e^{\alpha \mathbf{b}^\dagger} \tag{3.107a}$$

$$= \sum_n \frac{\beta^n}{n!} \underbrace{e^{-\alpha \mathbf{b}^\dagger} \mathbf{b}^n e^{\alpha \mathbf{b}^\dagger}} \tag{3.107b}$$

$$= \sum_n \frac{\beta^n}{n!} (\mathbf{b} + \alpha)^n \tag{3.107c}$$

$$= \sum_n \frac{[\beta (\mathbf{b} + \alpha)]^n}{n!} \tag{3.107d}$$

$$= e^{\beta (\mathbf{b} + \alpha)} \tag{3.107e}$$

$$= e^{\beta \mathbf{b}} e^{\beta \alpha}. \tag{3.107f}$$

■

3.6   Building derivatives of operators

Problem: With $\mathbf{A} = \mathbf{A}(\lambda)$ and $\mathbf{B} = \mathbf{B}(\lambda)$ calculate $\frac{d}{d\lambda}(\mathbf{AB})$.

Solution: By definition,

$$\frac{d}{d\lambda}(\mathbf{AB}) \overset{\text{def.}}{=} \lim_{\epsilon \to 0} \frac{\mathbf{A}(\lambda + \epsilon)\mathbf{B}(\lambda + \epsilon) - \mathbf{A}(\lambda)\mathbf{B}(\lambda)}{\epsilon} \tag{3.108}$$

Subtract and add $\mathbf{A}(\lambda)\mathbf{B}(\lambda + \epsilon)$ in the numerator, and group terms as indicated,

$$\frac{d}{d\lambda}(\mathbf{AB}) = \lim_{\epsilon \to 0} \frac{\overbrace{\mathbf{A}(\lambda + \epsilon)\mathbf{B}(\lambda + \epsilon) - \mathbf{A}(\lambda)\mathbf{B}(\lambda + \epsilon)} + \overbrace{\mathbf{A}(\lambda)\mathbf{B}(\lambda + \epsilon) - \mathbf{A}(\lambda)\mathbf{B}(\lambda)}}{\epsilon}. \tag{3.109}$$

Remark: As is shown further below, an alternative way is subtracting and adding $\mathbf{A}(\lambda + \epsilon)\mathbf{B}(\lambda)$, which has no implications in the final result due to the commutativity property of the operator addition. The rationale for choosing $\mathbf{A}(\lambda)\mathbf{B}(\lambda + \epsilon)$ or $\mathbf{A}(\lambda + \epsilon)\mathbf{B}(\lambda)$ and the reason for having these two equivalent options are as follows:

$$\begin{cases} \text{Consider } \mathbf{A}(\lambda + \epsilon)\mathbf{B}(\lambda + \epsilon) \overset{\text{split into}}{\Longrightarrow} \begin{cases} \mathbf{A}(\lambda + \epsilon) \\ \\ \mathbf{B}(\lambda + \epsilon) \end{cases} \overset{\text{preserve the order of } \mathbf{A} \text{ and } \mathbf{B}}{\Longrightarrow} \\ \\ \text{Consider } \mathbf{A}(\lambda)\mathbf{B}(\lambda) \overset{\text{split into}}{\Longrightarrow} \begin{cases} \mathbf{A}(\lambda) \\ \\ \mathbf{B}(\lambda) \end{cases} \end{cases} \tag{3.110}$$

$$\overset{\text{only two combinations are possible}}{\Longrightarrow} \begin{cases} \mathbf{A}(\lambda + \epsilon)\mathbf{B}(\lambda) \cdots\cdots \text{Case I} \\ \\ \mathbf{A}(\lambda)\mathbf{B}(\lambda + \epsilon) \cdots\cdots \text{Case II} \end{cases} \tag{3.111}$$

This explains why subtracting and adding of either of $\mathbf{A}(\lambda + \epsilon)\,\mathbf{B}(\lambda)$ or $\mathbf{A}(\lambda)\,\mathbf{B}(\lambda + \epsilon)$ works out and why only these two cases are possible.

$\square$

Factor out terms, while respecting the order of the operators,

$$\frac{d}{d\lambda}\,(\mathbf{AB}) = \lim_{\epsilon \to 0} \frac{[\mathbf{A}(\lambda + \epsilon) - \mathbf{A}(\lambda)]\,\mathbf{B}(\lambda + \epsilon) + \mathbf{A}(\lambda)\,[\mathbf{B}(\lambda + \epsilon) - \mathbf{B}(\lambda)]}{\epsilon}. \tag{3.112}$$

Separate terms,

$$\frac{d}{d\lambda}\,(\mathbf{AB}) = \lim_{\epsilon \to 0} \frac{[\mathbf{A}(\lambda + \epsilon) - \mathbf{A}(\lambda)]\,\mathbf{B}(\lambda + \epsilon)}{\epsilon} + \lim_{\epsilon \to 0} \frac{\mathbf{A}(\lambda)\,[\mathbf{B}(\lambda + \epsilon) - \mathbf{B}(\lambda)]}{\epsilon}. \tag{3.113}$$

Noting that $\lim_{\epsilon \to 0} \mathbf{B}(\lambda + \epsilon) = \mathbf{B}(\lambda)$ and the replacement of $\mathbf{B}(\lambda + \epsilon)$ with $\mathbf{B}(\lambda)$ causes no singularity in the first term at the R.H.S. This observation along with the facts that $\mathbf{A}(\lambda)$ and $\mathbf{B}(\lambda)$ are independent of ϵ,

$$\frac{d}{d\lambda}\,(\mathbf{AB}) = \left\{ \lim_{\epsilon \to 0} \frac{\mathbf{A}(\lambda + \epsilon) - \mathbf{A}(\lambda)}{\epsilon} \right\} \mathbf{B}(\lambda) + \mathbf{A}(\lambda) \left\{ \lim_{\epsilon \to 0} \frac{\mathbf{B}(\lambda + \epsilon) - \mathbf{B}(\lambda)}{\epsilon} \right\}, \tag{3.114}$$

and thus,

$$\frac{d}{d\lambda}\,(\mathbf{AB}) = \frac{d\mathbf{A}}{d\lambda}\mathbf{B} + \mathbf{A}\frac{d\mathbf{B}}{d\lambda}. \tag{3.115}$$

Note that respecting the order of $\mathbf{A}$ and $\mathbf{B}$ is crucially significant.

$\blacksquare$

Remark: In transition from (3.108) to (3.109), the term $\mathbf{A}(\lambda)\,\mathbf{B}(\lambda + \epsilon)$ was subtracted and added in the numerator. The analysis below chooses the alternative term $\mathbf{A}(\lambda + \epsilon)\,\mathbf{B}(\lambda)$, which is subtracted and added in the numerator,

$$\frac{d}{d\lambda}\,(\mathbf{AB}) = \lim_{\epsilon \to 0} \frac{\overbrace{\mathbf{A}(\lambda + \epsilon)\,\mathbf{B}(\lambda + \epsilon) - \mathbf{A}(\lambda + \epsilon)\,\mathbf{B}(\lambda)} + \overbrace{\mathbf{A}(\lambda + \epsilon)\,\mathbf{B}(\lambda) - \mathbf{A}(\lambda)\,\mathbf{B}(\lambda)}}{\epsilon}. \tag{3.116}$$

Factoring out within the groups, as indicated,

$$\frac{d}{d\lambda}\,(\mathbf{AB}) = \lim_{\epsilon \to 0} \frac{\mathbf{A}(\lambda + \epsilon)\,[\mathbf{B}(\lambda + \epsilon) - \mathbf{B}(\lambda)] + [\mathbf{A}(\lambda + \epsilon) - \mathbf{A}(\lambda)]\,\mathbf{B}(\lambda)}{\epsilon}. \tag{3.117}$$

Separating terms,

$$\frac{d}{d\lambda}\,(\mathbf{AB}) = \lim_{\epsilon \to 0} \frac{\mathbf{A}(\lambda + \epsilon)\,[\mathbf{B}(\lambda + \epsilon) - \mathbf{B}(\lambda)]}{\epsilon} + \lim_{\epsilon \to 0} \frac{[\mathbf{A}(\lambda + \epsilon) - \mathbf{A}(\lambda)]\,\mathbf{B}(\lambda)}{\epsilon}. \tag{3.118}$$

Considering $\lim_{\epsilon \to 0} \mathbf{A}(\lambda + \epsilon) = \mathbf{A}(\lambda)$ in the first term at the R.H.S., and slightly rearranging,

$$\frac{d}{d\lambda}\,(\mathbf{AB}) = \mathbf{A}(\lambda) \left\{ \lim_{\epsilon \to 0} \frac{[\mathbf{B}(\lambda + \epsilon) - \mathbf{B}(\lambda)]}{\epsilon} \right\} + \left\{ \lim_{\epsilon \to 0} \frac{[\mathbf{A}(\lambda + \epsilon) - \mathbf{A}(\lambda)]}{\epsilon} \right\} \mathbf{B}(\lambda). \tag{3.119}$$

Or, alternatively,

$$\frac{d}{d\lambda}(\mathbf{AB}) = \mathbf{A}\frac{d\mathbf{B}}{d\lambda} + \frac{d\mathbf{A}}{d\lambda}\mathbf{B}, \tag{3.120}$$

which is the result obtained in (3.115), except that the additive order of the terms at the R.H.S. has been reserved. However, since the additive order of terms is immaterial, both expressions are identical.

□

3.7   A key relationship

One of the key relationships, that is frequently used in manipulations, can be stated as follows: for arbitrary operators $\mathbf{A}$ and $\mathbf{B}$,

$$e^{\mathbf{A}}e^{\mathbf{B}} = e^{\mathbf{A}+\mathbf{B}+\frac{1}{2}[\mathbf{A},\mathbf{B}]}, \tag{3.121}$$

with the commutator $[\mathbf{A},\mathbf{B}] = \mathbf{AB} - \mathbf{BA}$. Further below this relationship and its implications are discussed in great detail. If $\mathbf{A}$ and $\mathbf{B}$ commute, i.e., $[\mathbf{A},\mathbf{B}] = 0$, then,

$$\begin{aligned} e^{\mathbf{A}}e^{\mathbf{B}} &= e^{\mathbf{A}+\mathbf{B}} = e^{\mathbf{B}+\mathbf{A}} = e^{\mathbf{B}}e^{\mathbf{A}} \\ &\quad ([\mathbf{A},\mathbf{B}] = 0) \end{aligned} \tag{3.122}$$

Problem: Assuming $\mathbf{A} \neq \mathbf{A}(\lambda)$ calculate $\frac{d}{d\lambda}e^{\lambda\mathbf{A}}$.

Solution: By definition,

$$\frac{d}{d\lambda}e^{\lambda\mathbf{A}} \overset{\text{def.}}{=} \lim_{\epsilon\to 0} \frac{e^{(\lambda+\epsilon)\mathbf{A}} - e^{\lambda\mathbf{A}}}{\epsilon}. \tag{3.123}$$

With $e^{(\lambda+\epsilon)\mathbf{A}} = e^{\lambda\mathbf{A}+\epsilon\mathbf{A}}$, the property $[\lambda\mathbf{A},\epsilon\mathbf{A}] = 0$, and applying (3.122), $e^{\lambda\mathbf{A}+\epsilon\mathbf{A}} = e^{\lambda\mathbf{A}}e^{\epsilon\mathbf{A}}$. Thus,

$$e^{(\lambda+\epsilon)\mathbf{A}} = e^{\lambda\mathbf{A}}e^{\epsilon\mathbf{A}} = e^{\epsilon\mathbf{A}}e^{\lambda\mathbf{A}}. \tag{3.124}$$

Substituting (3.124) into (3.123),

$$\frac{d}{d\lambda}e^{\lambda\mathbf{A}} \overset{(1)}{=} \lim_{\epsilon\to 0} \frac{e^{\epsilon\mathbf{A}}e^{\lambda\mathbf{A}} - \mathbf{I}e^{\lambda\mathbf{A}}}{\epsilon} \tag{3.125a}$$

$$\overset{(2)}{=} \lim_{\epsilon\to 0} \frac{\left(e^{\epsilon\mathbf{A}} - \mathbf{I}\right)e^{\lambda\mathbf{A}}}{\epsilon} \tag{3.125b}$$

$$\overset{(3)}{=} \left\{\lim_{\epsilon\to 0} \frac{e^{\epsilon\mathbf{A}} - \mathbf{I}}{\epsilon}\right\}e^{\lambda\mathbf{A}} \tag{3.125c}$$

$$\overset{(4)}{=} \left\{\lim_{\epsilon\to 0} \frac{\mathbf{I} + \epsilon\mathbf{A} + O(\epsilon^2) - \mathbf{I}}{\epsilon}\right\}e^{\lambda\mathbf{A}} \tag{3.125d}$$

$$\overset{(5)}{=} \left\{\lim_{0} [\mathbf{A} + O(\epsilon)]\right\}e^{\lambda\mathbf{A}} \tag{3.125e}$$

$$\overset{(6)}{=} \mathbf{A}e^{\lambda\mathbf{A}}. \tag{3.125f}$$

Elaborating the transitions in (3.125).

(1) This is a consequence of the definition of the derivative and the commutativity property of operators, as shown in the opening section of the solution. Also note the introduction of the identity operator **I**. Making use of the identity operator and its many expansions is one of the most powerful machineries in quantum physics. In the current case, its introduction serves to render the discussion more clear. In the myriad of cases, as will be demonstrated in this text, its introduction is key for solving the problem at hand.

(2) The ϵ-independent term $e^{\lambda \mathbf{A}}$ has been factored out in the numerator.

(3) This is a further implication of the fact that $e^{\lambda \mathbf{A}}$ is independent of ϵ.

(4) Application of the de l' Hospital's rule is not permitted. This would require the derivative of $e^{\epsilon \mathbf{A}}$, which is essentially the aim of the current exercise. Instead, the expansion of $e^{\epsilon \mathbf{A}}$ for small values of ϵ has been pursued, while explicitly writing down terms up to and including the linear term. Strictly speaking, there is a bit of cheating involved here. In writing the approximation of $e^{\epsilon \mathbf{A}}$ for small values of ϵ, the Formal Taylor Series Expansion has been used, which in turn requires the derivatives of $e^{\epsilon \mathbf{A}}$, evaluated at $\epsilon = 0$.

(5) The identity operator cancels out, and ϵ-dependent terms simplify, respectively, reduce.

(6) In the limit $\epsilon \rightarrow 0$, the term $O(\epsilon)$ vanishes.

∎

Remark: The preceding result can also be established by starting with the Formal Taylor Series Expansion of $e^{\lambda \mathbf{A}}$ and applying $d/d\lambda$ onto the expansion terms. A simple manipulation of the resulting terms leads to $\mathbf{A} e^{\lambda \mathbf{A}}$.

□

Problem: Assuming $\mathbf{A} \neq \mathbf{A}(\lambda)$ and $\mathbf{B} \neq \mathbf{B}(\lambda)$, calculate $\frac{d}{d\lambda} \left\{ e^{\lambda \mathbf{A}} e^{\lambda \mathbf{B}} \right\}$.

Solution:

$$\frac{d}{d\lambda} \left\{ e^{\lambda \mathbf{A}} e^{\lambda \mathbf{B}} \right\} = \left\{ \frac{d e^{\lambda \mathbf{A}}}{d\lambda} \right\} e^{\lambda \mathbf{B}} + e^{\lambda \mathbf{A}} \left\{ \frac{d e^{\lambda \mathbf{B}}}{d\lambda} \right\} \tag{3.126a}$$

$$\overset{(3.125)}{=} \left\{ \mathbf{A} e^{\lambda \mathbf{A}} \right\} e^{\lambda \mathbf{B}} + e^{\lambda \mathbf{A}} \left\{ \mathbf{B} e^{\lambda \mathbf{B}} \right\}. \tag{3.126b}$$

Consider $\mathbf{A} e^{\lambda \mathbf{A}} = e^{\lambda \mathbf{A}} \mathbf{A}$ as a special case of the commutation property $f(\mathbf{A})g(\mathbf{A}) = g(\mathbf{A})f(\mathbf{A})$, in (3.126b),

$$\frac{d}{d\lambda} \left\{ e^{\lambda \mathbf{A}} e^{\lambda \mathbf{B}} \right\} = \left\{ e^{\lambda \mathbf{A}} \mathbf{A} \right\} e^{\lambda \mathbf{B}} + e^{\lambda \mathbf{A}} \left\{ \mathbf{B} e^{\lambda \mathbf{B}} \right\}. \tag{3.127}$$

Combining the terms at the R.H.S., while respecting the order of the operators,

$$\frac{d}{d\lambda} \left\{ e^{\lambda \mathbf{A}} e^{\lambda \mathbf{B}} \right\} = e^{\lambda \mathbf{A}} (\mathbf{A} + \mathbf{B}) e^{\lambda \mathbf{B}}. \tag{3.128}$$

∎

Problem: Assuming $\mathbf{A} \neq \mathbf{A}(\lambda)$ and $\mathbf{B} \neq \mathbf{B}(\lambda)$, calculate $\frac{d}{d\lambda} \left\{ e^{\lambda \mathbf{A}} e^{\lambda \mathbf{B}} \right\}$ directly.

Solution:

$$\frac{d}{d\lambda} \left\{ e^{\lambda \mathbf{A}} e^{\lambda \mathbf{B}} \right\} \overset{\text{def.}}{=} \lim_{\epsilon \to 0} \frac{e^{(\lambda + \epsilon)\mathbf{A}} e^{(\lambda + \epsilon)\mathbf{B}} - e^{\lambda \mathbf{A}} e^{\lambda \mathbf{B}}}{\epsilon} \tag{3.129a}$$

$$= \lim_{\epsilon \to 0} \frac{e^{\lambda \mathbf{A}} e^{\epsilon \mathbf{A}} e^{\lambda \mathbf{B}} e^{\epsilon \mathbf{B}} - e^{\lambda \mathbf{A}} e^{\lambda \mathbf{B}}}{\epsilon} \tag{3.129b}$$

$$= \lim_{\epsilon \to 0} \frac{e^{\lambda \mathbf{A}} e^{\epsilon \mathbf{A}} e^{\epsilon \mathbf{B}} e^{\lambda \mathbf{B}} - e^{\lambda \mathbf{A}} \mathbf{I} e^{\lambda \mathbf{B}}}{\epsilon} \tag{3.129c}$$

$$= \lim_{\epsilon \to 0} \frac{e^{\lambda \mathbf{A}} (e^{\epsilon \mathbf{A}} e^{\epsilon \mathbf{B}} - \mathbf{I}) e^{\lambda \mathbf{B}}}{\epsilon} \tag{3.129d}$$

$$= e^{\lambda \mathbf{A}} \left\{ \lim_{\epsilon \to 0} \frac{e^{\epsilon \mathbf{A}} e^{\epsilon \mathbf{B}} - \mathbf{I}}{\epsilon} \right\} e^{\lambda \mathbf{B}} \tag{3.129e}$$

$$= e^{\lambda \mathbf{A}} \left\{ \lim_{\epsilon \to 0} \frac{[\mathbf{I} + \epsilon \mathbf{A} + O(\epsilon^2)][\mathbf{I} + \epsilon \mathbf{B} + O(\epsilon^2)] - \mathbf{I}}{\epsilon} \right\} e^{\lambda \mathbf{B}} \tag{3.129f}$$

$$= e^{\lambda \mathbf{A}} \left\{ \lim_{\epsilon \to 0} \frac{\mathbf{I} + \epsilon (\mathbf{A} + \mathbf{B}) + O(\epsilon^2) - \mathbf{I}}{\epsilon} \right\} e^{\lambda \mathbf{B}} \tag{3.129g}$$

$$= e^{\lambda \mathbf{A}} \left\{ \lim_{\epsilon \to 0} [(\mathbf{A} + \mathbf{B}) + O(\epsilon)] \right\} e^{\lambda \mathbf{B}} \tag{3.129h}$$

$$= e^{\lambda \mathbf{A}} (\mathbf{A} + \mathbf{B}) e^{\lambda \mathbf{B}}. \tag{3.129i}$$

The reader is encouraged to explain the individual steps in the above derivation. In particular, the introduction of the identity operator $\mathbf{I}$ in (3.129c).

∎

Problem: Assuming $\mathbf{A} \neq \mathbf{A}(\lambda)$ and $\mathbf{B} \neq \mathbf{B}(\lambda)$, calculate $\frac{d^2}{d\lambda^2} \{e^{\lambda \mathbf{A}} e^{\lambda \mathbf{B}}\}$. What can be implied if $\mathbf{A}$ and $\mathbf{B}$ were commutative ($[\mathbf{A}, \mathbf{B}] = 0$)?

Solution:

$$\frac{d^2}{d\lambda^2} \{e^{\lambda \mathbf{A}} e^{\lambda \mathbf{B}}\} \overset{\text{def.}}{=} \frac{d}{d\lambda} \left\{ \frac{d}{d\lambda} \{e^{\lambda \mathbf{A}} e^{\lambda \mathbf{B}}\} \right\} \tag{3.130a}$$

$$\overset{(3.129)}{=} \frac{d}{d\lambda} \{e^{\lambda \mathbf{A}} (\mathbf{A} + \mathbf{B}) e^{\lambda \mathbf{B}}\} \tag{3.130b}$$

$$= \left\{ \frac{d}{d\lambda} e^{\lambda \mathbf{A}} \right\} (\mathbf{A} + \mathbf{B}) e^{\lambda \mathbf{B}} + e^{\lambda \mathbf{A}} (\mathbf{A} + \mathbf{B}) \left\{ \frac{d}{d\lambda} e^{\lambda \mathbf{B}} \right\} \tag{3.130c}$$

$$\overset{(3.125)}{=} e^{\lambda \mathbf{A}} \mathbf{A} (\mathbf{A} + \mathbf{B}) e^{\lambda \mathbf{B}} + e^{\lambda \mathbf{A}} (\mathbf{A} + \mathbf{B}) \mathbf{B} e^{\lambda \mathbf{B}} \tag{3.130d}$$

$$= e^{\lambda \mathbf{A}} [\mathbf{A} (\mathbf{A} + \mathbf{B}) + (\mathbf{A} + \mathbf{B}) \mathbf{B}] e^{\lambda \mathbf{B}}. \tag{3.130e}$$

If $[\mathbf{A}, \mathbf{B}] = 0$, then, $(\mathbf{A} + \mathbf{B}) \mathbf{B} = \mathbf{B} (\mathbf{A} + \mathbf{B})$. Consequently,

$$\frac{d^2}{d\lambda^2} \{e^{\lambda \mathbf{A}} e^{\lambda \mathbf{B}}\} = e^{\lambda \mathbf{A}} \left[\mathbf{A} (\mathbf{A} + \mathbf{B}) + \underbrace{(\mathbf{A} + \mathbf{B}) \mathbf{B}}_{\mathbf{B}(\mathbf{A}+\mathbf{B})} \right] e^{\lambda \mathbf{B}} \tag{3.131a}$$

$$= e^{\lambda \mathbf{A}} [(\mathbf{A} + \mathbf{B})(\mathbf{A} + \mathbf{B})] e^{\lambda \mathbf{B}} \tag{3.131b}$$

$$= e^{\lambda \mathbf{A}} (\mathbf{A} + \mathbf{B})^2 e^{\lambda \mathbf{B}}. \tag{3.131c}$$

∎

Problem: Assuming $\mathbf{A} \neq \mathbf{A}(\lambda)$ and $\mathbf{B} \neq \mathbf{B}(\lambda)$, calculate $\frac{d^3}{d\lambda^3}\{e^{\lambda\mathbf{A}}e^{\lambda\mathbf{B}}\}$. What can be implied if $\mathbf{A}$ and $\mathbf{B}$ are commutative ($[\mathbf{A},\mathbf{B}]=0$)?

Solution:

$$\frac{d^3}{d\lambda^3}\{e^{\lambda\mathbf{A}}e^{\lambda\mathbf{B}}\} \stackrel{\text{def.}}{=} \frac{d}{d\lambda}\left\{\underbrace{\frac{d^2}{d\lambda^2}\{e^{\lambda\mathbf{A}}e^{\lambda\mathbf{B}}\}}\right\} \tag{3.132a}$$

$$\stackrel{(3.130)}{=} \frac{d}{d\lambda}\left\{e^{\lambda\mathbf{A}}\big[\mathbf{A}(\mathbf{A}+\mathbf{B})+(\mathbf{A}+\mathbf{B})\mathbf{B}\big]e^{\lambda\mathbf{B}}\right\} \tag{3.132b}$$

$$= \left\{\underbrace{\frac{d}{d\lambda}e^{\lambda\mathbf{A}}}\right\}\big[\mathbf{A}(\mathbf{A}+\mathbf{B})+(\mathbf{A}+\mathbf{B})\mathbf{B}\big]e^{\lambda\mathbf{B}}$$

$$+ \quad e^{\lambda\mathbf{A}}\big[\mathbf{A}(\mathbf{A}+\mathbf{B})+(\mathbf{A}+\mathbf{B})\mathbf{B}\big]\left\{\underbrace{\frac{d}{d\lambda}e^{\lambda\mathbf{B}}}\right\} \tag{3.132c}$$

$$\stackrel{(3.125)}{=} \{e^{\lambda\mathbf{A}}\mathbf{A}\}\big[\mathbf{A}(\mathbf{A}+\mathbf{B})+(\mathbf{A}+\mathbf{B})\mathbf{B}\big]e^{\lambda\mathbf{B}}$$

$$+ \quad e^{\lambda\mathbf{A}}\big[\mathbf{A}(\mathbf{A}+\mathbf{B})+(\mathbf{A}+\mathbf{B})\mathbf{B}\big]\{\mathbf{B}e^{\lambda\mathbf{B}}\} \tag{3.132d}$$

$$= \quad e^{\lambda\mathbf{A}}\big[\mathbf{A}^2(\mathbf{A}+\mathbf{B})+\mathbf{A}(\mathbf{A}+\mathbf{B})\mathbf{B}\big]e^{\lambda\mathbf{B}}$$

$$+ \quad e^{\lambda\mathbf{A}}\big[\mathbf{A}(\mathbf{A}+\mathbf{B})\mathbf{B}+(\mathbf{A}+\mathbf{B})\mathbf{B}^2\big]e^{\lambda\mathbf{B}} \tag{3.132e}$$

$$= \quad e^{\lambda\mathbf{A}}\big[\mathbf{A}^2(\mathbf{A}+\mathbf{B})+2\mathbf{A}(\mathbf{A}+\mathbf{B})\mathbf{B}+(\mathbf{A}+\mathbf{B})\mathbf{B}^2\big]e^{\lambda\mathbf{B}}. \tag{3.132f}$$

Remark: With $\mathbf{A}^0=\mathbf{B}^0=\mathbf{I}$, (3.132f) can be symmetrized,

$$\frac{d^3}{d\lambda^3}\{e^{\lambda\mathbf{A}}e^{\lambda\mathbf{B}}\} = e^{\lambda\mathbf{A}}\big[\mathbf{A}^2(\mathbf{A}+\mathbf{B})\mathbf{B}^0+2\mathbf{A}(\mathbf{A}+\mathbf{B})\mathbf{B}+\mathbf{A}^0(\mathbf{A}+\mathbf{B})\mathbf{B}^2\big]e^{\lambda\mathbf{B}}. \tag{3.133}$$

Viewing the terms in the square brackets, a pattern emerges, that can be used to formulate the induction hypothesis. In view of (3.133) the following recipe suggests itself.

- Step (1): Ignoring $(\mathbf{A}+\mathbf{B})$, the terms in the square bracket in (3.133) read: $\mathbf{A}^2\mathbf{B}^0+2\mathbf{A}\mathbf{B}+\mathbf{A}^0\mathbf{B}^2$. Note that in all terms $\mathbf{A}$ precedes $\mathbf{B}$. Respecting this order of the operators,

$$\mathbf{A}^2\mathbf{B}^0+2\mathbf{A}\mathbf{B}+\mathbf{B}^0\mathbf{B}^2 = \sum_{i=0}^{2}\binom{2}{2-i}\mathbf{A}^{2-i}\mathbf{B}^i. \tag{3.134}$$

- Step (2): Insert $(\mathbf{A}+\mathbf{B})$ between each of the terms $\mathbf{A}^{2-i}\mathbf{B}^i$ to obtain

$$\sum_{i=0}^{2}\binom{2}{2-i}\mathbf{A}^{2-i}(\mathbf{A}+\mathbf{B})\mathbf{B}^i. \tag{3.135}$$

- Step (3): Multiply this expression from the L.H.S. by $e^{\lambda\mathbf{A}}$ and from the R.H.S. by $e^{\lambda\mathbf{B}}$,

$$\frac{d^3}{d\lambda^3}\{e^{\lambda\mathbf{A}}e^{\lambda\mathbf{B}}\} = e^{\lambda\mathbf{A}}\left[\sum_{i=0}^{2}\binom{2}{2-i}\mathbf{A}^{2-i}(\mathbf{A}+\mathbf{B})\mathbf{B}^i\right]e^{\lambda\mathbf{B}}. \tag{3.136}$$

In the next example, this result will be employed to formulate the induction hypothesis.

$\square$

If $[\mathbf{A}, \mathbf{B}] = 0$, (3.132f) reads,

$$\frac{d^3}{d\lambda^3}\left\{e^{\lambda\mathbf{A}}e^{\lambda\mathbf{B}}\right\} = e^{\lambda\mathbf{A}}\left[\mathbf{A}^2\left(\mathbf{A}+\mathbf{B}\right) + \underbrace{2\mathbf{A}\left(\mathbf{A}+\mathbf{B}\right)\mathbf{B}} + \underbrace{\left(\mathbf{A}+\mathbf{B}\right)\mathbf{B}^2}\right]e^{\lambda\mathbf{B}} \tag{3.137a}$$

$$= e^{\lambda\mathbf{A}}\left[\mathbf{A}^2\left(\mathbf{A}+\mathbf{B}\right) + \underbrace{2\mathbf{A}\mathbf{B}\left(\mathbf{A}+\mathbf{B}\right)} + \underbrace{\mathbf{B}^2\left(\mathbf{A}+\mathbf{B}\right)}\right]e^{\lambda\mathbf{B}} \tag{3.137b}$$

$$= e^{\lambda\mathbf{A}}\left[\left(\mathbf{A}^2 + 2\mathbf{A}\mathbf{B} + \mathbf{B}^2\right)\left(\mathbf{A}+\mathbf{B}\right)\right]e^{\lambda\mathbf{B}} \tag{3.137c}$$

$$= e^{\lambda\mathbf{A}}\left(\mathbf{A}+\mathbf{B}\right)^3 e^{\lambda\mathbf{B}}. \tag{3.137d}$$

∎

Remark: To ease the formulation of the next problem, here is a summary of the results obtained in the above problem:

$$\frac{d^3}{d\lambda^3}\left\{e^{\lambda\mathbf{A}}e^{\lambda\mathbf{B}}\right\} = e^{\lambda\mathbf{A}}\left[\sum_{i=0}^{2}\binom{2}{2-i}\mathbf{A}^{2-i}\left(\mathbf{A}+\mathbf{B}\right)\mathbf{B}^i\right]e^{\lambda\mathbf{B}} \tag{3.138a}$$

$$\frac{d^3}{d\lambda^3}\left\{e^{\lambda\mathbf{A}}e^{\lambda\mathbf{B}}\right\} = e^{\lambda\mathbf{A}}\left(\mathbf{A}+\mathbf{B}\right)^3 e^{\lambda\mathbf{B}} \tag{3.138b}$$

$$[\mathbf{A}, \mathbf{B}] = 0.$$

□

Problem: Show that for arbitrary $n \in \mathbb{N}$,

Part I: $\qquad \dfrac{d^n}{d\lambda^n}\left\{e^{\lambda\mathbf{A}}e^{\lambda\mathbf{B}}\right\} = e^{\lambda\mathbf{A}}\left[\displaystyle\sum_{i=0}^{n-1}\binom{n-1}{(n-1)-i}\mathbf{A}^{(n-1)-i}\left(\mathbf{A}+\mathbf{B}\right)\mathbf{B}^i\right]e^{\lambda\mathbf{B}} \tag{3.139a}$

Part II: $\qquad \dfrac{d^n}{d\lambda^n}\left\{e^{\lambda\mathbf{A}}e^{\lambda\mathbf{B}}\right\} = e^{\lambda\mathbf{A}}\left(\mathbf{A}+\mathbf{B}\right)^n e^{\lambda\mathbf{B}} \tag{3.139b}$

$$[\mathbf{A}, \mathbf{B}] = 0.$$

Solution of Part I: Proof by mathematical induction.

It will be assumed that (3.139a) and (3.139b) are valid for arbitrary $n \in \mathbb{N}$ (induction hypothesis). The proof consists of showing the validity of these formulae for $n + 1$. Pat I assumes arbitrary operators $\mathbf{A}$ and $\mathbf{B}$. Given (3.139a), proceed as follows:

$$\frac{d^{n+1}}{d\lambda^{n+1}}\left(e^{\lambda\mathbf{A}}e^{\lambda\mathbf{B}}\right) \overset{\text{def.}}{=} \frac{d}{d\lambda}\left\{\frac{d^n}{d\lambda^n}\left(e^{\lambda\mathbf{A}}e^{\lambda\mathbf{B}}\right)\right\} \tag{3.140a}$$

$$\overset{(3.139a)}{=} \frac{d}{d\lambda}\left\{e^{\lambda\mathbf{A}}\left[\sum_{i=0}^{n-1}\binom{n-1}{(n-1)-i}\mathbf{A}^{(n-1)-i}\left(\mathbf{A}+\mathbf{B}\right)\mathbf{B}^i\right]e^{\lambda\mathbf{B}}\right\} \tag{3.140b}$$

$$= \left\{\frac{d}{d\lambda}e^{\lambda\mathbf{A}}\right\}\left[\sum_{i=0}^{n-1}\binom{n-1}{(n-1)-i}\mathbf{A}^{(n-1)-i}\left(\mathbf{A}+\mathbf{B}\right)\mathbf{B}^i\right]e^{\lambda\mathbf{B}}$$

$$+ \; e^{\lambda\mathbf{A}}\left[\sum_{i=0}^{n-1}\binom{n-1}{(n-1)-i}\mathbf{A}^{(n-1)-i}\left(\mathbf{A}+\mathbf{B}\right)\mathbf{B}^i\right]\left\{\frac{d}{d\lambda}e^{\lambda\mathbf{B}}\right\}. \tag{3.140c}$$

With $\frac{d}{d\lambda}e^{\lambda\mathbf{A}} = e^{\lambda\mathbf{A}}\mathbf{A}$ and $\frac{d}{d\lambda}e^{\lambda\mathbf{B}} = \mathbf{B}e^{\lambda\mathbf{B}}$,

$$\frac{d^{n+1}}{d\lambda^{n+1}}\left(e^{\lambda\mathbf{A}}e^{\lambda\mathbf{B}}\right) = \{e^{\lambda\mathbf{A}}\mathbf{A}\}\left[\sum_{i=0}^{n-1}\binom{n-1}{(n-1)-i}\mathbf{A}^{(n-1)-i}(\mathbf{A}+\mathbf{B})\mathbf{B}^{i}\right]e^{\lambda\mathbf{B}}$$

$$+ e^{\lambda\mathbf{A}}\left[\sum_{i=0}^{n-1}\binom{n-1}{(n-1)-i}\mathbf{A}^{(n-1)-i}(\mathbf{A}+\mathbf{B})\mathbf{B}^{i}\right]\{\mathbf{B}e^{\lambda\mathbf{B}}\}. \tag{3.141}$$

Consider the first term at the R.H.S. Multiply the expression within the square bracket by $\mathbf{A}$ from the L.H.S. Consider the second term at the R.H.S. Multiply the expression within the square bracket by $\mathbf{B}$ from the R.H.S. Then,

$$\frac{d^{n+1}}{d\lambda^{n+1}}\left(e^{\lambda\mathbf{A}}e^{\lambda\mathbf{B}}\right) = e^{\lambda\mathbf{A}}\left[\sum_{i=0}^{n-1}\binom{n-1}{(n-1)-i}\mathbf{A}^{n-i}(\mathbf{A}+\mathbf{B})\mathbf{B}^{i}\right]e^{\lambda\mathbf{B}}$$

$$+ e^{\lambda\mathbf{A}}\left[\sum_{i=0}^{n-1}\binom{n-1}{(n-1)-i}\mathbf{A}^{(n-1)-i}(\mathbf{A}+\mathbf{B})\mathbf{B}^{i+1}\right]e^{\lambda\mathbf{B}}. \tag{3.142}$$

Manipulate the second sum with the aim of transforming $\mathbf{A}^{(n-1)-i}(\mathbf{A}+\mathbf{B})\mathbf{B}^{i+1}$ into $\mathbf{A}^{n-i}(\mathbf{A}+\mathbf{B})\mathbf{B}^{i}$ which appears in the first sum. In more detail,

$$\sum_{i=0}^{n-1}\binom{n-1}{(n-1)-i}\mathbf{A}^{(n-1)-i}(\mathbf{A}+\mathbf{B})\mathbf{B}^{i+1}$$

$$= \sum_{i=0}^{n-1}\binom{n-1}{n-(i+1)}\mathbf{A}^{n-(i+1)}(\mathbf{A}+\mathbf{B})\mathbf{B}^{i+1} \tag{3.143a}$$

$$\overset{i+1=j}{=} \sum_{j=1}^{n}\binom{n-1}{n-j}\mathbf{A}^{n-j}(\mathbf{A}+\mathbf{B})\mathbf{B}^{j} \tag{3.143b}$$

$$\overset{j\to i}{=} \sum_{i=1}^{n}\binom{n-1}{n-i}\mathbf{A}^{n-i}(\mathbf{A}+\mathbf{B})\mathbf{B}^{i} \tag{3.143c}$$

Substituting the last expression in (3.143c) for the second term in (3.142),

$$\frac{d^{n+1}}{d\lambda^{n+1}}\left(e^{\lambda\mathbf{A}}e^{\lambda\mathbf{B}}\right) = e^{\lambda\mathbf{A}}\left[\sum_{i=0}^{n-1}\binom{n-1}{(n-1)-i}\mathbf{A}^{n-i}(\mathbf{A}+\mathbf{B})\mathbf{B}^{i}\right]e^{\lambda\mathbf{B}}$$

$$+ e^{\lambda\mathbf{A}}\left[\sum_{i=1}^{n}\binom{n-1}{n-i}\mathbf{A}^{n-i}(\mathbf{A}+\mathbf{B})\mathbf{B}^{i}\right]e^{\lambda\mathbf{B}}. \tag{3.144}$$

Factoring out the common exponential functions $e^{\lambda\mathbf{A}}$ and $e^{\lambda\mathbf{B}}$,

$$\frac{d^{n+1}}{d\lambda^{n+1}}\left(e^{\lambda\mathbf{A}}e^{\lambda\mathbf{B}}\right) = e^{\lambda\mathbf{A}}\left[\sum_{i=0}^{n-1}\binom{n-1}{(n-1)-i}\mathbf{A}^{n-i}(\mathbf{A}+\mathbf{B})\mathbf{B}^{i}\right.$$

$$+ \left.\sum_{i=1}^{n}\binom{n-1}{n-i}\mathbf{A}^{n-i}(\mathbf{A}+\mathbf{B})\mathbf{B}^{i}\right]e^{\lambda\mathbf{B}}. \tag{3.145}$$

The first sum comprises n-terms, say, s_i, with the dummy index i running from 0 to $n-1$. In detail $s_0 + s_1 + \cdots + s_{n-2} + s_{n-1}$. The second sum also comprises n-terms, say, t_i, with the dummy index i running from 1 to n. In detail

$t_1 + t_2 + \cdots + t_{n-1} + t_n$. Thus, it is seen that s_0 has no counterpart t_0. Similarly, t_n possesses no counterpart s_n. Below an attempt has been made to visualize this content:

$$
\begin{aligned}
s_0, \ s_1, \ s_2, \ s_3 \ \ldots, \ s_{n-2}, \ s_{n-1} \\
t_1, \ t_2, \ t_3, \ \ldots, \ t_{n-2}, \ t_{n-1}, \ t_n
\end{aligned}
\tag{3.146}
$$

Adding the two sums,

$$
s_0, \ s_1 + t_1, \ s_2 + t_2, \ s_3 + t_3, \ \ldots, \ s_{n-2} + t_{n-2}, \ s_{n-1} + t_{n-1}, \ t_n
\tag{3.147}
$$

On the determination of s_0:

$$
s_0 = \begin{pmatrix} n - 1 \\ (n-1) - i \end{pmatrix} \mathbf{A}^{n-i} \left(\mathbf{A} + \mathbf{B} \right) \mathbf{B}^i \Big|_{i=0}
\tag{3.148a}
$$

$$
= \begin{pmatrix} n - 1 \\ n - 1 \end{pmatrix} \mathbf{A}^n \left(\mathbf{A} + \mathbf{B} \right) \mathbf{B}^0
\tag{3.148b}
$$

$$
= \mathbf{A}^n \left(\mathbf{A} + \mathbf{B} \right) \mathbf{B}^0
\tag{3.148c}
$$

In the last transition use was made of

$$
\begin{pmatrix} n - 1 \\ n - 1 \end{pmatrix} = 1.
\tag{3.149}
$$

On the determination of t_n:

$$
t_n = \begin{pmatrix} n - 1 \\ n - i \end{pmatrix} \mathbf{A}^{n-i} \left(\mathbf{A} + \mathbf{B} \right) \mathbf{B}^i \Big|_{i=n}
\tag{3.150a}
$$

$$
= \begin{pmatrix} n - 1 \\ 0 \end{pmatrix} \mathbf{A}^0 \left(\mathbf{A} + \mathbf{B} \right) \mathbf{B}^n
\tag{3.150b}
$$

$$
= \mathbf{A}^0 \left(\mathbf{A} + \mathbf{B} \right) \mathbf{B}^n
\tag{3.150c}
$$

In the last transition use was made of

$$
\begin{pmatrix} n - 1 \\ 0 \end{pmatrix} = 1.
\tag{3.151}
$$

On the determination of $s_i + t_i$:

$$
s_i + t_i = \left\{ \begin{pmatrix} n - 1 \\ (n-1) - i \end{pmatrix} + \begin{pmatrix} n - 1 \\ n - i \end{pmatrix} \right\} \mathbf{A}^{n-i} \left(\mathbf{A} + \mathbf{B} \right) \mathbf{B}^i.
\tag{3.152}
$$

Focus on the individual terms in the curly brackets:

$$\binom{n-1}{(n-1)-i} = \binom{n-1}{n-(i+1)} \tag{3.153a}$$

$$= \frac{(n-1)!}{(n-(i+1))!((n-1)-(n-(i+1)))!} \tag{3.153b}$$

$$= \frac{(n-1)!}{(n-(i+1))!i!} \tag{3.153c}$$

$$= \frac{n-i}{n} \frac{(n-1)!n}{(n-(i+1))!(n-i)i!} \tag{3.153d}$$

$$= \frac{n-i}{n} \frac{n!}{(n-i)!i!} \tag{3.153e}$$

$$= \left(1 - \frac{i}{n}\right) \frac{n!}{(n-i)!i!}. \tag{3.153f}$$

$$\binom{n-1}{n-i} = \frac{(n-1)!}{(n-i)!((n-1)-(n-i))!} \tag{3.154a}$$

$$= \frac{(n-1)!}{(n-i)!(i-1)!} \tag{3.154b}$$

$$= \frac{i}{n} \frac{(n-1)!n}{(n-i)!(i-1)!i} \tag{3.154c}$$

$$= \frac{i}{n} \frac{n!}{(n-i)!i!} \tag{3.154d}$$

$$\tag{3.154e}$$

Adding the two terms,

$$\binom{n-1}{(n-1)-i} + \binom{n-1}{n-i} = \frac{n!}{(n-i)!i!} = \binom{n}{i}. \tag{3.155}$$

Thus,

$$s_i + t_i = \binom{n}{i} \mathbf{A}^{n-i} (\mathbf{A} + \mathbf{B}) \mathbf{B}^i. \tag{3.156}$$

In the light of the results obtained for s_0, $s_i + t_i$ (with $i = 1, \ldots, n-1$), and t_n,

$$\frac{d^{n+1}}{d\lambda^{n+1}} \left(e^{\lambda \mathbf{A}} e^{\lambda \mathbf{B}}\right) = e^{\lambda \mathbf{A}} \left[\sum_{i=0}^{n} \binom{n}{i} \mathbf{A}^{n-i} (\mathbf{A} + \mathbf{B}) \mathbf{B}^i\right] e^{\lambda \mathbf{B}}. \tag{3.157}$$

This completes the solution of the problem in cases that $\mathbf{A}$ and $\mathbf{B}$ do not commute.

Solution of Part II: The above result is general in the sense that it is valid irrespective of $\mathbf{A}$ and $\mathbf{B}$ being commutative. In the case that $[\mathbf{A}, \mathbf{B}] = 0$, the order of $\mathbf{A}^{n-i}$ and $(\mathbf{A} + \mathbf{B})$ can be exchanged,

$$\frac{d^{n+1}}{d\lambda^{n+1}} \left(e^{\lambda \mathbf{A}} e^{\lambda \mathbf{B}}\right) = e^{\lambda \mathbf{A}} \left[\sum_{i=0}^{n} \binom{n}{i} (\mathbf{A} + \mathbf{B}) \mathbf{A}^{n-i} \mathbf{B}^i\right] e^{\lambda \mathbf{B}}. \tag{3.158}$$

Since $\mathbf{A} + \mathbf{B}$ is independent of the dummy index i, it transfers to the front of the summation sign,

$$\frac{d^{n+1}}{d\lambda^{n+1}} \left(e^{\lambda\mathbf{A}} e^{\lambda\mathbf{B}} \right) = e^{\lambda\mathbf{A}} \left[(\mathbf{A}+\mathbf{B}) \underbrace{\sum_{i=0}^{n} \binom{n}{i} \mathbf{A}^{n-i}\mathbf{B}^i}_{(\mathbf{A}+\mathbf{B})^n} \right] e^{\lambda\mathbf{B}}. \tag{3.159}$$

Recognizing the under-braced term as $(\mathbf{A}+\mathbf{B})^n$,

$$\frac{d^{n+1}}{d\lambda^{n+1}} \left(e^{\lambda\mathbf{A}} e^{\lambda\mathbf{B}} \right) = e^{\lambda\mathbf{A}} (\mathbf{A}+\mathbf{B})^{n+1} e^{\lambda\mathbf{B}}. \tag{3.160}$$

∎

Problem: Assuming $\mathbf{A} \neq \mathbf{A}(\lambda)$ and $\mathbf{B} \neq \mathbf{B}(\lambda)$, and the commutativity of $\mathbf{A}$ and $\mathbf{B}$, $[\mathbf{A}, \mathbf{B}] = 0$, prove

$$\frac{d^n}{d\lambda^n} \left(e^{\lambda\mathbf{A}} e^{\lambda\mathbf{B}} \right) = e^{\lambda\mathbf{A}} (\mathbf{A}+\mathbf{B})^n e^{\lambda\mathbf{B}}, \tag{3.161}$$

by mathematical induction.

Solution: The initial step(s) were shown further above exhaustively. The demonstration of the induction step, based on the condition $[\mathbf{A}, \mathbf{B}] = 0$, is quite straightforward.

$$\frac{d^{n+1}}{d\lambda^{n+1}} \left(e^{\lambda\mathbf{A}} e^{\lambda\mathbf{B}} \right) \overset{\text{def.}}{=} \frac{d}{d\lambda} \left\{ \frac{d^n}{d\lambda^n} \left(e^{\lambda\mathbf{A}} e^{\lambda\mathbf{B}} \right) \right\} \tag{3.162a}$$

$$\overset{\text{induc. hypo.}}{=} \frac{d}{d\lambda} \left\{ e^{\lambda\mathbf{A}} (\mathbf{A}+\mathbf{B})^n e^{\lambda\mathbf{B}} \right\} \tag{3.162b}$$

$$\overset{[\mathbf{A},\mathbf{B}]=0}{=} (\mathbf{A}+\mathbf{B})^n \underbrace{\frac{d}{d\lambda} \left\{ e^{\lambda\mathbf{A}} e^{\lambda\mathbf{B}} \right\}}_{e^{\lambda\mathbf{A}} (\mathbf{A}+\mathbf{B}) e^{\lambda\mathbf{B}}} \tag{3.162c}$$

$$\overset{\text{ind. init. step}}{=} (\mathbf{A}+\mathbf{B})^n e^{\lambda\mathbf{A}} (\mathbf{A}+\mathbf{B}) e^{\lambda\mathbf{B}} \tag{3.162d}$$

$$\overset{[\mathbf{A},\mathbf{B}]=0}{=} e^{\lambda\mathbf{A}} (\mathbf{A}+\mathbf{B})^{n+1} e^{\lambda\mathbf{B}}. \tag{3.162e}$$

∎

Solution: Quick solution.

Consider $e^{\lambda\mathbf{A}} e^{\lambda\mathbf{B}}$ subject to the condition, $[\mathbf{A}, \mathbf{B}] = 0$. Commutative operators $\mathbf{A}$ and $\mathbf{B}$ can formally be viewed as "ordinary" variables. Thus, analogous to $e^{\lambda x} e^{\lambda y} = e^{\lambda(x+y)}$, the relationship $e^{\lambda\mathbf{A}} e^{\lambda\mathbf{B}} = e^{\lambda(\mathbf{A}+\mathbf{B})}$ holds valid. The following steps follow immediately:

$$\frac{d^{n+1}}{d\lambda^{n+1}} \left(e^{\lambda\mathbf{A}} e^{\lambda\mathbf{B}} \right) = \frac{d^{n+1}}{d\lambda^{n+1}} e^{\lambda(\mathbf{A}+\mathbf{B})} = (\mathbf{A}+\mathbf{B})^{n+1} e^{\lambda(\mathbf{A}+\mathbf{B})} = e^{\lambda\mathbf{A}} (\mathbf{A}+\mathbf{B})^{n+1} e^{\lambda\mathbf{B}}. \tag{3.163}$$

∎

Remark: Yet another class of relationships will be introduced next to gain further insights. Similar to learning any second or further "natural" language, it is hoped that the reader realizes the instantiation of recurring patterns that they already have been practicing, as well as acknowledges the "emergence" of new unexpected, and, at times, counterintuitive insights. In other words, it is hoped that the reader experiences the manifestation of the "learning process." The reader might also recognize the fact that no obscurity arises, provided the leaps from one step to the next are small enough, to "emulate continuity of thought." Expressed more cautiously, in cases where paradoxes and erroneous conclusions threaten to sneak into the arguments, they are easier to spot. This clearcut procedure, trying to ensure watertight reasoning, is the hallmark of inductive reasoning, enabling the move from special cases to more general instances. Intriguingly, paradigm changes in thinking, revisions of existing models, and the construction of novel models are also promoted in this style of thinking. The breaking points where the emergence and paradigm changes manifest themselves seem however to be outside the realm of the "linear" process of inductive thinking. A learning mind undergoes "nonlinear" processes which are powerful enough to accommodate the occurrences of the breaking points. It seems that the multitude and the variety of different cases play significant roles. Thus, the strategy of investigating as many categories and instances and mastering them seem to be a reasonable approach. The reader is encouraged to pause and reflect on the above anecdotal wisdoms, each time they experience an Eureka moment. Any new realization would be an indication of a genuine instantiation of "deep learning."

□

Problem: Assuming $\mathbf{A} \neq \mathbf{A}(\lambda)$ and $\mathbf{B} \neq \mathbf{B}(\lambda)$, calculate $\frac{d}{d\lambda}\left(e^{\lambda\mathbf{A}}\mathbf{B}e^{-\lambda\mathbf{A}}\right)$.

Solution: The operator $\mathbf{B}$ being independent of λ, and respecting the order of the operators,

$$\frac{d}{d\lambda}\left(e^{\lambda\mathbf{A}}\mathbf{B}e^{-\lambda\mathbf{A}}\right) = \left\{\frac{de^{\lambda\mathbf{A}}}{d\lambda}\right\}\mathbf{B}e^{-\lambda\mathbf{A}} + e^{\lambda\mathbf{A}}\mathbf{B}\left\{\frac{de^{-\lambda\mathbf{A}}}{d\lambda}\right\} \tag{3.164a}$$

$$= \left\{\mathbf{A}e^{\lambda\mathbf{A}}\right\}\mathbf{B}e^{-\lambda\mathbf{A}} + e^{\lambda\mathbf{A}}\mathbf{B}\left\{-\mathbf{A}e^{-\lambda\mathbf{A}}\right\}. \tag{3.164b}$$

$$\tag{3.164c}$$

In virtue of $f(\mathbf{A})g(\mathbf{A}) = g(\mathbf{A})f(\mathbf{A})$, and thus considering $\mathbf{A}e^{\lambda\mathbf{A}} = e^{\lambda\mathbf{A}}\mathbf{A}$ in the first term at the R.H.S.,

$$\frac{d}{d\lambda}\left(e^{\lambda\mathbf{A}}\mathbf{B}e^{-\lambda\mathbf{A}}\right) = \left\{e^{\lambda\mathbf{A}}\mathbf{A}\right\}\mathbf{B}e^{-\lambda\mathbf{A}} - e^{\lambda\mathbf{A}}\mathbf{B}\left\{\mathbf{A}e^{-\lambda\mathbf{A}}\right\}. \tag{3.165}$$

Remove the curly brackets, factor out $e^{\lambda\mathbf{A}}$ and $e^{-\lambda\mathbf{A}}$, at the L.H.S. and the R.H.S., respectively, while respecting the order of the operators, and introduce the commutator $[\mathbf{A}, \mathbf{B}]$, to obtain,

$$\frac{d}{d\lambda}\left(e^{\lambda\mathbf{A}}\mathbf{B}e^{-\lambda\mathbf{A}}\right) = e^{\lambda\mathbf{A}}\mathbf{A}\mathbf{B}e^{-\lambda\mathbf{A}} - e^{\lambda\mathbf{A}}\mathbf{B}\mathbf{A}e^{-\lambda\mathbf{A}} \tag{3.166a}$$

$$= e^{\lambda\mathbf{A}}\left(\mathbf{A}\mathbf{B} - \mathbf{B}\mathbf{A}\right)e^{-\lambda\mathbf{A}} \tag{3.166b}$$

$$= e^{\lambda\mathbf{A}}\left[\mathbf{A}, \mathbf{B}\right]e^{-\lambda\mathbf{A}}. \tag{3.166c}$$

∎

Problem: Assuming $\mathbf{A} \neq \mathbf{A}(\lambda)$ and $\mathbf{B} \neq \mathbf{B}(\lambda)$, calculate $\frac{d^2}{d\lambda^2}\left(e^{\lambda\mathbf{A}}\mathbf{B}e^{-\lambda\mathbf{A}}\right)$.

Solution:

$$\frac{d^2}{d\lambda^2}\left(e^{\lambda\mathbf{A}}\mathbf{B}e^{-\lambda\mathbf{A}}\right) = \frac{d}{d\lambda}\underbrace{\left\{\frac{d}{d\lambda}\left(e^{\lambda\mathbf{A}}\mathbf{B}e^{-\lambda\mathbf{A}}\right)\right\}}_{} \tag{3.167a}$$

$$= \frac{d}{d\lambda}\left\{e^{\lambda\mathbf{A}}\left[\mathbf{A}, \mathbf{B}\right]e^{-\lambda\mathbf{A}}\right\} \tag{3.167b}$$

$$= \left\{\frac{d}{d\lambda}e^{\lambda\mathbf{A}}\right\}\left[\mathbf{A}, \mathbf{B}\right]e^{-\lambda\mathbf{A}} + e^{\lambda\mathbf{A}}\left[\mathbf{A}, \mathbf{B}\right]\left\{\frac{d}{d\lambda}e^{-\lambda\mathbf{A}}\right\} \tag{3.167c}$$

$$= \left\{\mathbf{A}e^{\lambda\mathbf{A}}\right\}\left[\mathbf{A}, \mathbf{B}\right]e^{-\lambda\mathbf{A}} + e^{\lambda\mathbf{A}}\left[\mathbf{A}, \mathbf{B}\right]\left\{-\mathbf{A}e^{-\lambda\mathbf{A}}\right\} \tag{3.167d}$$

$$\overset{(p)}{=} e^{\lambda\mathbf{A}}\mathbf{A}\left[\mathbf{A}, \mathbf{B}\right]e^{-\lambda\mathbf{A}} - e^{\lambda\mathbf{A}}\left[\mathbf{A}, \mathbf{B}\right]\mathbf{A}e^{-\lambda\mathbf{A}} \tag{3.167e}$$

$$\overset{(q)}{=} e^{\lambda\mathbf{A}}\left\{\mathbf{A}\left[\mathbf{A}, \mathbf{B}\right] - \left[\mathbf{A}, \mathbf{B}\right]\mathbf{A}\right\}e^{-\lambda\mathbf{A}} \tag{3.167f}$$

$$\overset{(r)}{=} e^{\lambda\mathbf{A}}\left[\mathbf{A}, \left[\mathbf{A}, \mathbf{B}\right]\right]e^{-\lambda\mathbf{A}} \tag{3.167g}$$

$$\overset{(s)}{=} e^{\lambda\mathbf{A}}\left[\mathbf{A}, \mathbf{B}\right]_2 e^{-\lambda\mathbf{A}}. \tag{3.167h}$$

In the transition (p), the commutativity relationship $\mathbf{A}e^{\lambda\mathbf{A}} = e^{\lambda\mathbf{A}}\mathbf{A}$ was employed.

In the transition (q), the exponential terms were factored out, while respecting the order of operators.

In the transitions (r) and s, respectively, $[\mathbf{A}, [\mathbf{A}, \mathbf{B}]]$ was identified and $[\mathbf{A}, \mathbf{B}]_2$ introduced, according to,

$$[\mathbf{A}, [\mathbf{A}, \mathbf{B}]] = \mathbf{A}[\mathbf{A}, \mathbf{B}] - [\mathbf{A}, \mathbf{B}]\mathbf{A} \tag{3.168a}$$

$$[\mathbf{A}, \mathbf{B}]_2 = [\mathbf{A}, [\mathbf{A}, \mathbf{B}]]. \tag{3.168b}$$

∎

Problem: Assuming $\mathbf{A} \neq \mathbf{A}(\lambda)$ and $\mathbf{B} \neq \mathbf{B}(\lambda)$, show that

$$\frac{d^n}{d\lambda^n}\left(e^{\lambda\mathbf{A}}\mathbf{B}e^{-\lambda\mathbf{A}}\right) = e^{\lambda\mathbf{A}}[\mathbf{A}, \mathbf{B}]_n e^{-\lambda\mathbf{A}}, \tag{3.169}$$

with

$$[\mathbf{A}, \mathbf{B}]_n = \left[\mathbf{A}, [\mathbf{A}, \mathbf{B}]_{n-1}\right], \tag{3.170}$$

for arbitrary $n \in \mathbb{N}$. To unify the notation, the conventions

$$[\mathbf{A}, \mathbf{B}]_1 = [\mathbf{A}, \mathbf{B}], \tag{3.171}$$

and

$$[\mathbf{A}, \mathbf{B}]_0 = \mathbf{B}, \tag{3.172}$$

are introduced.

Solution: Proof by mathematical induction.

The proof by the mathematical induction offers itself. The initial steps for $n = 1$ (3.166) and $n = 2$ (3.167) were established in the preceding two problems. Employing the induction hypothesis (3.169), the induction step is immediate,

$$\frac{d^{n+1}}{d\lambda^{n+1}}\left(e^{\lambda\mathbf{A}}\mathbf{B}e^{-\lambda\mathbf{A}}\right) \stackrel{\text{def.}}{=} \frac{d}{d\lambda}\underbrace{\left\{\frac{d^n}{d\lambda^n}\left(e^{\lambda\mathbf{A}}\mathbf{B}e^{-\lambda\mathbf{A}}\right)\right\}} \tag{3.173a}$$

$$\stackrel{\text{ind. hypo.}}{=} \frac{d}{d\lambda}\left\{e^{\lambda\mathbf{A}}[\mathbf{A}, \mathbf{B}]_n e^{-\lambda\mathbf{A}}\right\} \tag{3.173b}$$

$$= \left\{\frac{d}{d\lambda}e^{\lambda\mathbf{A}}\right\}[\mathbf{A}, \mathbf{B}]_n e^{-\lambda\mathbf{A}} + e^{\lambda\mathbf{A}}[\mathbf{A}, \mathbf{B}]_n\left\{\frac{d}{d\lambda}e^{-\lambda\mathbf{A}}\right\} \tag{3.173c}$$

$$= \left\{\mathbf{A}e^{\lambda\mathbf{A}}\right\}[\mathbf{A}, \mathbf{B}]_n e^{-\lambda\mathbf{A}} + e^{\lambda\mathbf{A}}[\mathbf{A}, \mathbf{B}]_n\left\{-\mathbf{A}e^{-\lambda\mathbf{A}}\right\} \tag{3.173d}$$

$$= e^{\lambda\mathbf{A}}\mathbf{A}[\mathbf{A}, \mathbf{B}]_n e^{-\lambda\mathbf{A}} - e^{\lambda\mathbf{A}}[\mathbf{A}, \mathbf{B}]_n \mathbf{A}e^{-\lambda\mathbf{A}} \tag{3.173e}$$

$$= e^{\lambda\mathbf{A}}\underbrace{\left\{\mathbf{A}[\mathbf{A}, \mathbf{B}]_n - [\mathbf{A}, \mathbf{B}]_n\mathbf{A}\right\}}e^{-\lambda\mathbf{A}} \tag{3.173f}$$

$$\stackrel{\text{def.}}{=} e^{\lambda\mathbf{A}}[\mathbf{A}, [\mathbf{A}, \mathbf{B}]_n]e^{-\lambda\mathbf{A}} \tag{3.173g}$$

$$\stackrel{\text{def.}}{=} e^{\lambda\mathbf{A}}[\mathbf{A}, \mathbf{B}]_{n+1}e^{-\lambda\mathbf{A}}. \tag{3.173h}$$

∎

Problem: Let $\mathbf{A}$ and $\mathbf{B}$ be arbitrary linear operators with $\mathbf{A} \neq \mathbf{A}(\lambda)$ and $\mathbf{B} \neq \mathbf{B}(\lambda)$. Show the validity of the relationship,

$$e^{\lambda\mathbf{A}}\mathbf{B}e^{-\lambda\mathbf{A}} = \mathbf{B} + \sum_{n=1}^{\infty}\frac{\lambda^n}{n!}[\mathbf{A}, \mathbf{B}]_n. \tag{3.174}$$

Solution: Define the λ-dependent operator $\mathbf{F}(\lambda)$,

$$\mathbf{F}(\lambda) = e^{\lambda\mathbf{A}}\mathbf{B}e^{-\lambda\mathbf{A}}. \tag{3.175}$$

Expand $\mathbf{F}(\lambda)$ in a Formal Taylor Series Expansion at $\lambda = 0$,

$$\mathbf{F}(\lambda) = \mathbf{F}(0) + \sum_{n=1}^{\infty} \frac{\lambda^n}{n!} \left\{ \frac{d^n}{d\lambda^n}\mathbf{F}(\lambda) \right\}_{\lambda=0}. \tag{3.176}$$

From (3.175),

$$\mathbf{F}(0) = \mathbf{B}. \tag{3.177}$$

Furthermore, with reference to (3.175) and (3.169),

$$\frac{d^n}{d\lambda^n}\mathbf{F}(\lambda) = \frac{d^n}{d\lambda^n}\left(e^{\lambda\mathbf{A}}\mathbf{B}e^{-\lambda\mathbf{A}}\right) \tag{3.178}$$

$$= e^{\lambda\mathbf{A}}\left[\mathbf{A},\mathbf{B}\right]_n e^{-\lambda\mathbf{A}}. \tag{3.179}$$

In the special case $\lambda = 0$,

$$\left\{ \frac{d^n}{d\lambda^n}\mathbf{F}(\lambda) \right\}\Big|_{\lambda=0} = \left[\mathbf{A},\mathbf{B}\right]_n. \tag{3.180}$$

Substituting (3.177) and (3.180) into (3.176),

$$\mathbf{F}(\lambda) = \mathbf{B} + \sum_{n=1}^{\infty} \frac{\lambda^n}{n!}\left[\mathbf{A},\mathbf{B}\right]_n. \tag{3.181}$$

Substituting back for the auxiliary function $\mathbf{F}(\lambda)$, as introduced in (3.175),

$$e^{\lambda\mathbf{A}}\mathbf{B}e^{-\lambda\mathbf{A}} = \mathbf{B} + \sum_{n=1}^{\infty} \frac{\lambda^n}{n!}\left[\mathbf{A},\mathbf{B}\right]_n. \tag{3.182}$$

∎

Remark: Assume the special case $[\mathbf{A},\mathbf{B}]_2 = 0$. The condition $[\mathbf{A},\mathbf{B}]_2 = 0$ implies that $[\mathbf{A},\mathbf{B}]_n = 0$ for $n \geq 3$. Then, (3.182) simplifies greatly, and

$$e^{\lambda\mathbf{A}}\mathbf{B}e^{-\lambda\mathbf{A}} = \mathbf{B} + \lambda\left[\mathbf{A},\mathbf{B}\right]. \tag{3.183}$$

□

Remark: Assume the special case $[\mathbf{A},\mathbf{B}]_2 = 0$. Written explicitly this means $[\mathbf{A},[\mathbf{A},\mathbf{B}]] = 0$. In particular, $[\mathbf{A},\mathbf{B}] = \mathbf{I}$ is sufficient for $[\mathbf{A},[\mathbf{A},\mathbf{B}]] = 0$ to hold true. The canonical commutation relationship $[\mathbf{A},\mathbf{B}] = \mathbf{I}$ is satisfied by a myriad of operators, as shown in this Text exhaustively. The archetypical equation, involving the annihilation operator $\mathbf{b}$ and creation operator $\mathbf{b}^\dagger$, is $[\mathbf{b},\mathbf{b}^\dagger] = \mathbf{I}$. Thus,

$$e^{\lambda\mathbf{b}}\mathbf{b}^\dagger e^{-\lambda\mathbf{b}} = \mathbf{b}^\dagger + \lambda, \tag{3.184}$$

which was established in this chapter following a different path of reasoning.

□

3.8 A most important relationship and consequences thereof

The above title heralding a new section, and preceding the next lemma, should allude to the significance of the LEMMA, and the consequences derived from it, which will be presented as special cases.

Let ξ and η stand for "ordinary" variables (x, y) or constants (a, b). Then, the factorization $e^{\xi+\eta} = e^{\xi} e^{\eta}$ is immediate. However, if ξ and η stand for operators (or matrices representing them), the stated factorization is in general no longer valid. The following lemma analyzes this content.

Lemma: *Let $\mathbf{A}$ and $\mathbf{B}$ be arbitrary linear operators with $\mathbf{A} \neq \mathbf{A}(\lambda)$ and $\mathbf{B} \neq \mathbf{B}(\lambda)$. Let the conditions $[\mathbf{A}, [\mathbf{A}, \mathbf{B}]] = 0$, $[\mathbf{B}, [\mathbf{A}, \mathbf{B}]] = 0$ be met. Show the validity of the relationship,*

$$e^{\mathbf{A}+\mathbf{B}} = e^{\mathbf{A}} e^{\mathbf{B}} e^{-\frac{1}{2}[\mathbf{A},\mathbf{B}]} = e^{\mathbf{B}} e^{\mathbf{A}} e^{\frac{1}{2}[\mathbf{A},\mathbf{B}]}. \tag{3.185}$$

Remark: The addition of operators (matrices) if permissible, is commutative: $\mathbf{A} + \mathbf{B} = \mathbf{B} + \mathbf{A}$. Thus, $e^{\mathbf{A}+\mathbf{B}} = e^{\mathbf{B}+\mathbf{A}}$. In virtue of $[\mathbf{A}, \mathbf{B}] = -[\mathbf{B}, \mathbf{A}]$, the second equation in (3.185) is a testimony of the additive symmetry in $e^{\mathbf{A}+\mathbf{B}} = e^{\mathbf{B}+\mathbf{A}}$,

$$
\begin{array}{rcl}
e^{\mathbf{A}+\mathbf{B}} & = & e^{\mathbf{A}} e^{\mathbf{B}} e^{-\frac{1}{2}[\mathbf{A},\mathbf{B}]} \\[2mm]
= & \rightarrow & = \\[2mm]
e^{\mathbf{B}+\mathbf{A}} & = & e^{\mathbf{B}} e^{\mathbf{A}} e^{-\frac{1}{2}[\mathbf{B},\mathbf{A}]} = e^{\mathbf{B}} e^{\mathbf{A}} e^{\frac{1}{2}[\mathbf{A},\mathbf{B}]}.
\end{array}
\tag{3.186}
$$

Thus, it suffices to establish the first claim in the lemma, which implies the validity of the second claim.

$\square$

Proof: To prove this extraordinarily important classical statement, define $\mathbf{F}(\lambda)$, a family of operators depending on the (real- or complex-valued) parameter λ, according to,

$$\mathbf{F}(\lambda) = e^{\lambda\mathbf{A}} e^{\lambda\mathbf{B}}. \tag{3.187}$$

The "λ-parametrization" technique is a powerful vehicle in establishing novel relationships. The "initial condition" $\mathbf{F}(0) = \mathbf{I}$ will be utilized necessarily and advantageously. Differentiating $\mathbf{F}(\lambda)$ with respect to λ, and using (3.129i),

$$\frac{d}{d\lambda}\mathbf{F}(\lambda) = e^{\lambda\mathbf{A}} (\mathbf{A} + \mathbf{B}) e^{\lambda\mathbf{B}}. \tag{3.188}$$

Insert the identity operator $\mathbf{I}$, written in the form $\mathbf{I} = e^{-\lambda\mathbf{A}} e^{\lambda\mathbf{A}}$, between $(\mathbf{A} + \mathbf{B})$ and $e^{\lambda\mathbf{B}}$,

$$\frac{d}{d\lambda}\mathbf{F}(\lambda) = e^{\lambda\mathbf{A}} (\mathbf{A} + \mathbf{B}) \overbrace{e^{-\lambda\mathbf{A}} e^{\lambda\mathbf{A}}}^{\mathbf{I}} e^{\lambda\mathbf{B}}. \tag{3.189}$$

The rationale for choosing $\mathbf{I} = e^{-\lambda\mathbf{A}} e^{\lambda\mathbf{A}}$ rather than $\mathbf{I} = e^{\lambda\mathbf{A}} e^{-\lambda\mathbf{A}}$ is that the last two exponential functions in (3.189) give rise to $\mathbf{F}(\lambda)$,

$$\frac{d}{d\lambda}\mathbf{F}(\lambda) = \underbrace{e^{\lambda\mathbf{A}} (\mathbf{A} + \mathbf{B}) e^{-\lambda\mathbf{A}}}\mathbf{F}(\lambda). \tag{3.190}$$

Expanding the under-braced terms,

$$\frac{d}{d\lambda}\mathbf{F}(\lambda) = \left\{ e^{\lambda\mathbf{A}} \mathbf{A} e^{-\lambda\mathbf{A}} + e^{\lambda\mathbf{A}} \mathbf{B} e^{-\lambda\mathbf{A}} \right\} \mathbf{F}(\lambda). \tag{3.191}$$

Since $e^{\lambda\mathbf{A}}$ and $\mathbf{A}$ commute, $\underbrace{e^{\lambda\mathbf{A}} \mathbf{A} e^{-\lambda\mathbf{A}}} = \mathbf{A}\underbrace{e^{\lambda\mathbf{A}} e^{-\lambda\mathbf{A}}} = \mathbf{A}$. Furthermore, since upon assumption, $\mathbf{B}$ and $[\mathbf{A}, \mathbf{B}]$ commute, $e^{\lambda\mathbf{A}} \mathbf{B} e^{-\lambda\mathbf{A}} = \mathbf{B} + \lambda [\mathbf{A}, \mathbf{B}]$, following (3.183). Consequently,

$$\frac{d}{d\lambda}\mathbf{F}(\lambda) = \{ \mathbf{A} + \mathbf{B} + \lambda [\mathbf{A}, \mathbf{B}] \}\mathbf{F}(\lambda) \tag{3.192a}$$

$$= \{ (\mathbf{A} + \mathbf{B}) + \lambda [\mathbf{A}, \mathbf{B}] \}\mathbf{F}(\lambda). \tag{3.192b}$$

The two constituent terms in the curly brackets in (3.192b), i.e., $\mathbf{A} + \mathbf{B}$ and $\lambda\,[\mathbf{A}, \mathbf{B}]$ are commutative in virtue of the assumed property that $\mathbf{A}$ and $\mathbf{B}$ commute with $[\mathbf{A}, \mathbf{B}]$ individually,

$$[(\mathbf{A} + \mathbf{B}), [\mathbf{A}, \mathbf{B}]] = \underbrace{[\mathbf{A}, [\mathbf{A}, \mathbf{B}]]}_{=0} + \underbrace{[\mathbf{B}, [\mathbf{A}, \mathbf{B}]]}_{=0} \tag{3.193a}$$

$$= \mathbf{0}. \tag{3.193b}$$

The fact that $\mathbf{A} + \mathbf{B}$ commute with $[\mathbf{A}, \mathbf{B}]$ along with $\mathbf{F}(0) = \mathbf{I}$ suggests considering the following ordinary differential equation of the first order,

$$\frac{d}{d\lambda} f(\lambda) = (\alpha + \lambda\beta) f(\lambda) \tag{3.194a}$$

$$f(0) = 1, \tag{3.194b}$$

as a model for (3.192b). Finding a solution for $f(\lambda)$ is immediate,

$$f(\lambda) = e^{\alpha\lambda + \frac{1}{2}\beta\lambda^2}, \tag{3.195}$$

inspiring to consider the solution

$$\mathbf{F}(\lambda) = e^{(\mathbf{A}+\mathbf{B})\lambda + \frac{1}{2}[\mathbf{A},\mathbf{B}]\lambda^2}. \tag{3.196}$$

The above considerations allow to transform the product of two exponential functions of operators into one exponential function. For greater clarity the general and special cases are listed below, according to degree of their generality. In each case, the condition under which the equality holds valid has been explicated.

- Referring to the definition of $\mathbf{F}(\lambda)$ in (3.187),

$$e^{\lambda\mathbf{A}} e^{\lambda\mathbf{B}} = e^{(\mathbf{A}+\mathbf{B})\lambda + \frac{1}{2}[\mathbf{A},\mathbf{B}]\lambda^2} \tag{3.197a}$$

  subject to the conditions: $[\mathbf{A}, [\mathbf{A}, \mathbf{B}]] = \mathbf{0}, \quad [\mathbf{B}, [\mathbf{A}, \mathbf{B}]] = \mathbf{0}.$ \tag{3.197b}

- In the special case that $\lambda = 1$,

$$e^{\mathbf{A}} e^{\mathbf{B}} = e^{(\mathbf{A}+\mathbf{B}) + \frac{1}{2}[\mathbf{A},\mathbf{B}]} \tag{3.198a}$$

  subject to the conditions: $[\mathbf{A}, [\mathbf{A}, \mathbf{B}]] = \mathbf{0}, \quad [\mathbf{B}, [\mathbf{A}, \mathbf{B}]] = \mathbf{0}.$ \tag{3.198b}

- If $[\mathbf{A}, \mathbf{B}] = 0$, the conditions (3.198b) are met trivially, and (3.198a) assumes a particularly simple form,

$$e^{\mathbf{A}} e^{\mathbf{B}} = e^{\mathbf{A}+\mathbf{B}} \tag{3.199a}$$

  subject to the condition: $[\mathbf{A}, \mathbf{B}] = \mathbf{0}.$ \tag{3.199b}

The above results should illuminate and guide the intuition of the reader to investigate the structure of (3.197) further. The following observations can be made.

1. In view of (3.199), if any two operators $\mathbf{X}$ and $\mathbf{Y}$ commute, $[\mathbf{X}, \mathbf{Y}] = 0$, they can essentially be treated as "ordinary" variables. In particular, $e^{\mathbf{X}} e^{\mathbf{Y}} = e^{\mathbf{X}+\mathbf{Y}}$.
2. Let $\mathbf{X} = \mathbf{A} + \mathbf{B}$ and $\mathbf{Y} = \frac{1}{2}[\mathbf{A}, \mathbf{B}]$. Then

$$[\mathbf{X}, \mathbf{Y}] = \left[(\mathbf{A} + \mathbf{B}), \frac{1}{2}[\mathbf{A}, \mathbf{B}]\right] \tag{3.200a}$$

$$= \left[\mathbf{A}, \frac{1}{2}[\mathbf{A}, \mathbf{B}]\right] + \left[\mathbf{B}, \frac{1}{2}[\mathbf{A}, \mathbf{B}]\right] \tag{3.200b}$$

$$= \frac{1}{2}[\mathbf{A}, [\mathbf{A}, \mathbf{B}]] + \frac{1}{2}[\mathbf{B}, [\mathbf{A}, \mathbf{B}]]. \tag{3.200c}$$

3.   While assuming $[\mathbf{A}, \mathbf{B}] \neq 0$, let the commutativity conditions $[\mathbf{A}, [\mathbf{A}, \mathbf{B}]] = 0$ and $[\mathbf{B}, [\mathbf{A}, \mathbf{B}]] = 0$ hold. Then,

$$\begin{cases} [\mathbf{A}, \mathbf{B}] \neq 0 \\ [\mathbf{A}, [\mathbf{A}, \mathbf{B}]] = 0 \implies \left[(\mathbf{A} + \mathbf{B}), \tfrac{1}{2}[\mathbf{A}, \mathbf{B}]\right] = \tfrac{1}{2}\underbrace{[\mathbf{A}, [\mathbf{A}, \mathbf{B}]]}_{=0} + \tfrac{1}{2}\underbrace{[\mathbf{B}, [\mathbf{A}, \mathbf{B}]]}_{=0} = 0 \\ [\mathbf{B}, [\mathbf{A}, \mathbf{B}]] = 0 \end{cases}$$

$$\implies \quad [\mathbf{X}, \mathbf{Y}] = 0.$$

Consequently, the exponential function at the R.H.S. of (3.198a) can be factorized into the product of two exponential functions, involving $(\mathbf{A} + \mathbf{B})$ and $\frac{1}{2}[\mathbf{A}, \mathbf{B}]$,

$$e^{\mathbf{A}} e^{\mathbf{B}} = e^{\mathbf{A}+\mathbf{B}} e^{\frac{1}{2}[\mathbf{A},\mathbf{B}]} \qquad (3.201a)$$

subject to the conditions: $[\mathbf{A}, [\mathbf{A}, \mathbf{B}]] = 0$, $[\mathbf{B}, [\mathbf{A}, \mathbf{B}]] = 0$. $\qquad (3.201b)$

Multiplying both sides from the right by $e^{-\frac{1}{2}[\mathbf{A},\mathbf{B}]}$, and rearranging,

$$e^{\mathbf{A}+\mathbf{B}} = e^{\mathbf{A}} e^{\mathbf{B}} e^{-\frac{1}{2}[\mathbf{A},\mathbf{B}]} \qquad (3.202a)$$

subject to the conditions: $[\mathbf{A}, [\mathbf{A}, \mathbf{B}]] = 0$, $[\mathbf{B}, [\mathbf{A}, \mathbf{B}]] = 0$. $\qquad (3.202b)$

∎

Remark: Following the derivation of the relationships in (3.201) and (3.202), a mnemonics will be offered here to aid the reader to remember the structure of the formulae. In the following, the mnemonics has been sandwiched between two other remarks, which might also be helpful.

- The coefficients of $(\mathbf{A} + \mathbf{B})$ and $[\mathbf{A}, \mathbf{B}]$, respectively, are 1 and $\frac{1}{2}$. Recall a key point in the analysis was the choice of the expression $\alpha + \beta\lambda$. The integration of $\alpha + \beta\lambda$ with respect to λ led to $\alpha\lambda + \frac{1}{2}\beta\lambda^2$, and thus to the coefficients 1 and $\frac{1}{2}$ of λ and λ^2, respectively.
- Viewing the multiplication to be in general a more complex and sophisticated operation than the addition operation, a higher "score" can be assigned to the multiplication than the score associated with the addition: Thus, the score of $e^{\mathbf{A}} e^{\mathbf{B}}$ is higher than the score of $e^{\mathbf{A}+\mathbf{B}}$.
    1. In (3.201a), the term $e^{\frac{1}{2}[\mathbf{A},\mathbf{B}]}$, with the positive sign in the exponent, increases the "score" of $e^{\mathbf{A}+\mathbf{B}}$ to the level of the "score" of $e^{\mathbf{A}} e^{\mathbf{B}}$.
    2. Conversely, in (3.202a), the term $e^{-\frac{1}{2}[\mathbf{A},\mathbf{B}]}$, with the negative sign in the exponent decreases the "score" of $e^{\mathbf{A}} e^{\mathbf{B}}$ down to the "score" of $e^{\mathbf{A}+\mathbf{B}}$.
    This mnemonics should enable, simply-by-inspection, to translate the terms $e^{\mathbf{A}+\mathbf{B}}$ and $e^{\mathbf{A}} e^{\mathbf{B}}$ into each other.
- Since $[\mathbf{A}, \mathbf{B}]$ commutes with $\mathbf{A}$ and $\mathbf{B}$, and thus with $\mathbf{A} + \mathbf{B}$, the following expressions are equally valid:

$$e^{\mathbf{A}} e^{\mathbf{B}} = e^{\mathbf{A}+\mathbf{B}} e^{\frac{1}{2}[\mathbf{A},\mathbf{B}]} = e^{\frac{1}{2}[\mathbf{A},\mathbf{B}]} e^{\mathbf{A}+\mathbf{B}} \qquad (3.203a)$$

subject to the conditions: $[\mathbf{A}, [\mathbf{A}, \mathbf{B}]] = 0$, $[\mathbf{B}, [\mathbf{A}, \mathbf{B}]] = 0$. $\qquad (3.203b)$

$$e^{\mathbf{A}+\mathbf{B}} = e^{\mathbf{A}} e^{\mathbf{B}} e^{-\frac{1}{2}[\mathbf{A},\mathbf{B}]} = e^{\mathbf{A}} e^{-\frac{1}{2}[\mathbf{A},\mathbf{B}]} e^{\mathbf{B}} = e^{-\frac{1}{2}[\mathbf{A},\mathbf{B}]} e^{\mathbf{A}} e^{\mathbf{B}} \qquad (3.204a)$$

subject to the conditions: $[\mathbf{A}, [\mathbf{A}, \mathbf{B}]] = 0$, $[\mathbf{B}, [\mathbf{A}, \mathbf{B}]] = 0$. $\qquad (3.204b)$

□

3.9 Derivatives of functions of operators with respect to operators

In this brief series of problems, derivatives of functions with respect to operators rather than parameters are investigated.

Problem: Show that,

$$\frac{d}{d\mathbf{A}}\mathbf{A} = \mathbf{I}. \tag{3.205}$$

Solution:

$$\frac{d}{d\mathbf{A}}\mathbf{A} = \lim_{\epsilon \to 0} \frac{(\mathbf{A} + \epsilon\mathbf{I}) - \mathbf{A}}{\epsilon} \tag{3.206a}$$

$$= \mathbf{I}. \tag{3.206b}$$

■

Problem: Show that,

$$\frac{d}{d\mathbf{A}}\mathbf{A}^2 = 2\mathbf{A}. \tag{3.207}$$

Solution:

$$\frac{d}{d\mathbf{A}}\mathbf{A}^2 \stackrel{\text{def.}}{=} \frac{d}{d\mathbf{A}}\mathbf{A}\mathbf{A} \tag{3.208a}$$

$$= \left\{\frac{d}{d\mathbf{A}}\mathbf{A}\right\}\mathbf{A} + \mathbf{A}\left\{\frac{d}{d\mathbf{A}}\mathbf{A}\right\} \tag{3.208b}$$

$$= \mathbf{I}\mathbf{A} + \mathbf{A}\mathbf{I} \tag{3.208c}$$

$$= 2\mathbf{A}. \tag{3.208d}$$

■

Problem: Show that,

$$\frac{d}{d\mathbf{A}}\mathbf{A}^n = n\mathbf{A}^{n-1}. \tag{3.209}$$

Solution: Proof by mathematical induction.

In the preceding problems, the initial steps were shown. The case of $\mathbf{A}^2$ was in fact not necessary. However, dealing with a few initial steps is not only assuring, but also provides a guide how to carry out the induction step, which consists of assuming the validity of (3.209), the induction hypothesis, and examining the validity of the same for $n + 1$. It is seen that the proof scheme follows the same line of arguments as in the case of obtaining the derivative of $\mathbf{A}^2$.

$$\frac{d}{d\mathbf{A}}\mathbf{A}^{n+1} \stackrel{\text{def.}}{=} \frac{d}{d\mathbf{A}}\mathbf{A}^n\mathbf{A} \tag{3.210a}$$

$$= \left\{\frac{d}{d\mathbf{A}}\mathbf{A}^n\right\}\mathbf{A} + \mathbf{A}^n\left\{\frac{d}{d\mathbf{A}}\mathbf{A}\right\} \tag{3.210b}$$

$$= \left\{n\mathbf{A}^{n-1}\right\}\mathbf{A} + \mathbf{A}^n\{\mathbf{I}\} \tag{3.210c}$$

$$= n\mathbf{A}^n + \mathbf{A}^n \tag{3.210d}$$

$$= (n + 1)\mathbf{A}^n. \tag{3.210e}$$

■

Remark: Since the function considered is a monomial, there are several alternatives to the above proof. Each alternative sheds a different light on the internal structure of the problem. Consider,

$$\frac{d}{d\mathbf{A}}\mathbf{A}^n = \lim_{\epsilon \to 0} \frac{(\mathbf{A}+\epsilon)^n - \mathbf{A}^n}{\epsilon}. \tag{3.211}$$

Obviously, having replaced $\epsilon\mathbf{I}$ with ϵ in (3.211) has no implications on the arguments made.

Consider the expansion,

$$a^n - b^n = (a-b)\underbrace{\left(a^{n-1} + a^{n-2}b + a^{n-3}b^2 + \cdots + a^2b^{n-3} + ab^{n-2} + b^{n-1}\right)}_{n-\text{terms}}. \tag{3.212}$$

Set $a = \mathbf{A} + \epsilon$ and $b = \mathbf{A}$, with $a - b = \epsilon$,

$$\begin{aligned}(\mathbf{A}+\epsilon)^n - \mathbf{A}^n = \epsilon\big\{&(\mathbf{A}+\epsilon)^{n-1} + (\mathbf{A}+\epsilon)^{n-2}\mathbf{A} + (\mathbf{A}+\epsilon)^{n-3}\mathbf{A}^2 \\ &+ \cdots \\ &+ (\mathbf{A}+\epsilon)^2\mathbf{A}^{n-3} + (\mathbf{A}+\epsilon)\mathbf{A}^{n-2} + \mathbf{A}^{n-1}\big\}.\end{aligned} \tag{3.213}$$

Dividing both sides by arbitrarily small, however, finite ϵ,

$$\begin{aligned}\frac{(\mathbf{A}+\epsilon)^n - \mathbf{A}^n}{\epsilon} = &(\mathbf{A}+\epsilon)^{n-1} + (\mathbf{A}+\epsilon)^{n-2}\mathbf{A} + (\mathbf{A}+\epsilon)^{n-3}\mathbf{A}^2 \\ &+ \cdots \\ &+ (\mathbf{A}+\epsilon)^2\mathbf{A}^{n-3} + (\mathbf{A}+\epsilon)\mathbf{A}^{n-2} + \mathbf{A}^{n-1}.\end{aligned} \tag{3.214}$$

In the limit $\epsilon \to 0$ each of the n terms at the R.H.S. becomes equal to $\mathbf{A}^{n-1}$. Thus,

$$\lim_{\epsilon \to 0} \frac{(\mathbf{A}+\epsilon)^n - \mathbf{A}^n}{\epsilon} = n\mathbf{A}^{n-1}. \tag{3.215}$$

In view of (3.211),

$$\frac{d}{d\mathbf{A}}\mathbf{A}^n = n\mathbf{A}^{n-1}. \tag{3.216}$$

Thus taking the derivatives of operator monomials with respect to their operator follows exactly the same path as if they were "ordinary" variables.

$\square$

Problem: Show that,

$$\frac{d}{d\mathbf{A}}e^{\mathbf{A}} = e^{\mathbf{A}}. \tag{3.217}$$

Solution:

$$\frac{d}{d\mathbf{A}}e^{\mathbf{A}} = \lim_{\epsilon \to 0} \frac{e^{\mathbf{A}+\epsilon} - e^{\mathbf{A}}}{\epsilon} \tag{3.218a}$$

$$= \lim_{\epsilon \to 0} \frac{e^{\mathbf{A}}e^{\epsilon} - e^{\mathbf{A}}}{\epsilon} \tag{3.218b}$$

$$= \lim_{\epsilon \to 0} \frac{e^{\epsilon} - 1}{\epsilon}e^{\mathbf{A}} \tag{3.218c}$$

$$= \lim_{\epsilon \to 0} \frac{\{1 + \epsilon + O(\epsilon^2)\} - 1}{\epsilon}e^{\mathbf{A}} \tag{3.218d}$$

$$= \lim_{\epsilon \to 0} \{1 + O(\epsilon)\}e^{\mathbf{A}} \tag{3.218e}$$

$$= e^{\mathbf{A}}. \tag{3.218f}$$

In the transition from (3.218a) to (3.218b) advantage was taken of the fact that $\mathbf{A}$ and the constant ϵ, or stated more precisely, $\mathbf{A}$ and $\epsilon\mathbf{I}$ are commutative, and consequently, it was permitted to factorize the exponential function $e^{\mathbf{A}+\epsilon}$ in the form $e^{\mathbf{A}}e^{\epsilon}$. Also it was permitted to write $e^{\mathbf{A}}e^{\epsilon} = e^{\epsilon}e^{\mathbf{A}}$, which enabled to factor out $e^{\mathbf{A}}$, as shown in (3.218c). ∎

3.10   Commutator gymnastics

In this section, several general commutator identities will be introduced. The starting point are the following two identities for moving general operators inside and outside of commutator brackets:

$$[\mathbf{A}, \mathbf{BC}] = [\mathbf{A}, \mathbf{B}]\mathbf{C} + \mathbf{B}[\mathbf{A}, \mathbf{C}] \tag{3.219a}$$

$$[\mathbf{AB}, \mathbf{C}] = [\mathbf{A}, \mathbf{C}]\mathbf{B} + \mathbf{A}[\mathbf{B}, \mathbf{C}]. \tag{3.219b}$$

These relationships will be established by solving the following two problems. Thereby, each problem will be solved in two ways.

Problem: Show that,

$$[\mathbf{A}, \mathbf{BC}] = [\mathbf{A}, \mathbf{B}]\mathbf{C} + \mathbf{B}[\mathbf{A}, \mathbf{C}]. \tag{3.220}$$

Solution:

$$[\mathbf{A}, \mathbf{BC}] \stackrel{\text{def.}}{=} \mathbf{ABC} - \mathbf{BCA}. \tag{3.221}$$

The existence of the composite operator $\mathbf{BC}$ in $[\mathbf{A}, \mathbf{BC}]$ suggests the necessity for extracting either $\mathbf{B}$ or $\mathbf{C}$. Whatever choice is made is irrelevant, and the process suggests itself.

1. Extracting $\mathbf{B}$
    *   In view of (3.221), extracting $\mathbf{B}$ requires considering the term $-\mathbf{BCA}$.
    *   With $\mathbf{B}$ extracted, the remaining term $-\mathbf{CA}$ requires the companion term $\mathbf{AC}$ to constitute the commutator $[\mathbf{A}, \mathbf{C}]$. Consequently, the term $\mathbf{BAC}$ must be added.
    *   To compensate for the added term, the term $\mathbf{BAC}$ must also be subtracted. Thus,

$$[\mathbf{A}, \mathbf{BC}] = \mathbf{ABC} - \mathbf{BCA} + \underbrace{\mathbf{BAC} - \mathbf{BAC}}_{=0}. \tag{3.222}$$

    *   In view of this relationship, the reader should convince themselves, that the "necessary" addition of the (3rd) term $\mathbf{BAC}$ for producing a commutator expression together with the (2nd) term $-\mathbf{BCA}$, also "necessarily" leads to the introduction of the (4th) term $-\mathbf{BAC}$. The latter term on its part together with the (1st) term $\mathbf{ABC}$

automatically generates the missing commutator expression in the solution. Consequently, after making the decision for extracting **B**, every other step follows compellingly. Thus

$$[\mathbf{A}, \mathbf{BC}] = \overbrace{\mathbf{ABC}\underbrace{-\mathbf{BCA} + \mathbf{BAC}} - \mathbf{BAC}} \tag{3.223a}$$

$$= (\mathbf{ABC} - \mathbf{BAC}) + (\mathbf{BAC} - \mathbf{BCA}) \tag{3.223b}$$

$$= (\mathbf{AB} - \mathbf{BA})\,\mathbf{C} + \mathbf{B}\,(\mathbf{AC} - \mathbf{CA}) \tag{3.223c}$$

$$= [\mathbf{A}, \mathbf{B}]\,\mathbf{C} + \mathbf{B}\,[\mathbf{A}, \mathbf{C}]\,. \tag{3.223d}$$

By viewing these equations, one is reminded of the product rule of differentiation:

- Equation (3.219a) is reminiscent of taking the "derivative" of **BC** from the left with respect to **A**. Thereby, the "derivative operator" **A** is applied to the product (**BC**), while maintaining the order of the **B** and **C** terms.
- Equation (3.219b) is reminiscent of taking the "derivative" of **AB** from the right with respect to **A**. Thereby, the "derivative operator" **C** is applied to the product (**AB**), while maintaining the order of the **A** and **B** terms.

- To see this similarity more clearly, interpret the commutator $[\mathbf{X}, \mathbf{Y}]$ as the differentiation of **Y** with respect to **X**, and denote it by $\mathcal{D}_{\mathbf{X}}(\mathbf{Y})$:

$$\mathcal{D}_{\mathbf{X}}(\mathbf{Y}) := [\mathbf{X}, \mathbf{Y}]\,. \tag{3.224}$$

Employing this definition, (3.219a) can be cast in the form,

$$\mathcal{D}_{\mathbf{A}}(\mathbf{BC}) = \mathcal{D}_{\mathbf{A}}(\mathbf{B})\,\mathbf{C} + \mathbf{B}\mathcal{D}_{\mathbf{A}}(\mathbf{C})\,. \tag{3.225}$$

As will be shown momentarily, there is no need to separately performing the "derivation" process from the R.H.S.

2. Extracting **C**
   - In view of (3.221), extracting **C** requires considering the term **ABC**.
   - With **C** extracted, the remaining term **AB** requires the companion term $-\mathbf{BA}$ to constitute the commutator $[\mathbf{A}, \mathbf{B}]$. Consequently, the term $-\mathbf{BAC}$ is required.
   - To compensate for the subtracted term, the term **BAC** must also be added. Thus,

$$[\mathbf{A}, \mathbf{BC}] = \mathbf{ABC} - \mathbf{BCA}\underbrace{-\mathbf{BAC} + \mathbf{BAC}}_{=0}. \tag{3.226}$$

It is illuminating to compare the under-braced terms in (3.222) and (3.226).

   - As in the previous case, after making the decision for extracting **C**, every other step follows compellingly. In particular, after adding $-\mathbf{BAC} + \mathbf{BAC}$ to the R.H.S., the following steps suggest themselves,

$$[\mathbf{A}, \mathbf{BC}] = \mathbf{ABC}\underbrace{-\mathbf{BCA} - \mathbf{BAC}} + \mathbf{BAC} \tag{3.227a}$$

$$= \underbrace{\mathbf{ABC} - \mathbf{BAC}}\underbrace{-\mathbf{BCA} + \mathbf{BAC}} \tag{3.227b}$$

$$= (\mathbf{AB} - \mathbf{BA})\,\mathbf{C} + \mathbf{B}\,(\mathbf{AC} - \mathbf{CA}) \tag{3.227c}$$

$$= [\mathbf{A}, \mathbf{B}]\,\mathbf{C} + \mathbf{B}\,[\mathbf{A}, \mathbf{C}]\,. \tag{3.227d}$$

■

Problem: Show that,

$$[\mathbf{AB}, \mathbf{C}] = [\mathbf{A}, \mathbf{C}]\,\mathbf{B} + \mathbf{A}\,[\mathbf{B}, \mathbf{C}]\,. \tag{3.228}$$

Solution: The following steps are self-explanatory:

$$[\mathbf{AB}, \mathbf{C}] = -[\mathbf{C}, \mathbf{AB}] \tag{3.229a}$$

$$\stackrel{\text{def.}}{=} -\mathcal{D}_{\mathbf{C}}(\mathbf{AB}) \tag{3.229b}$$

$$= -\mathcal{D}_{\mathbf{C}}(\mathbf{A})\mathbf{B} - \mathbf{A}\mathcal{D}_{\mathbf{C}}(\mathbf{B}) \tag{3.229c}$$

$$\stackrel{\text{def.}}{=} -[\mathbf{C}, \mathbf{A}]\mathbf{B} - \mathbf{A}[\mathbf{C}, \mathbf{B}] \tag{3.229d}$$

$$= [\mathbf{A}, \mathbf{C}]\mathbf{B} + \mathbf{A}[\mathbf{B}, \mathbf{C}]. \tag{3.229e}$$

Considering the result $[\mathbf{AB}, \mathbf{C}] = [\mathbf{A}, \mathbf{C}]\mathbf{B} + \mathbf{A}[\mathbf{B}, \mathbf{C}]$ the following rule can be stated: taking the derivative of $(\mathbf{AB})$ from the right with respect to $\mathbf{C}$ is equal to taking the derivative of $\mathbf{A}$ with respect to $\mathbf{C}$ from the right, multiplied by $\mathbf{B}$, plus the derivative of $\mathbf{B}$ with respect to $\mathbf{C}$ from the right, multiplied by $\mathbf{A}$, invariably respecting the order of the operators $\mathbf{A}$ and $\mathbf{B}$. To the latter, note that in the entire discussion presented in this solution, $\mathbf{A}$ precedes $\mathbf{B}$.

∎

In order to prepare the move to the next problem, it is worthwhile to recall the definition of a function of an operator, $f(\mathbf{A})$. Given $f(\mathbf{A})$, proceed as follows:

- Replace the operator $\mathbf{A}$ in $f(\mathbf{A})$ with the "ordinary" variable x to obtain $f(x)$. This is the first "critical step" towards reducing the complexity of calculations: working with $f(x)$ is immeasurably simpler than working with $f(\mathbf{A})$. As an example, consider $\sin(\mathbf{A})$ with $\mathbf{A}$ being any square matrix of a finite or an infinite dimension, versus $\sin(x)$, which is a decisively simpler mathematical object.
- The "ordinary" function $f(x)$ permits considering the Formal Taylor Series Expansion $f(x) = \sum_i f_i x^i$. This is the second "critical step," towards reducing the complexity. Employing the Formal Taylor Series Expansion of the function $f(\cdot)$ translates the complex structure of $f(\cdot)$ into the coefficients f_i, and enables working with the monomials x^i rather than the function $f(x)$. Handling the monomials x^i ($i \in \mathbb{N}_0$), additionally enables the utilization of the powerful machinery of the mathematical induction.

Problem: Show that

$$f(\mathbf{A})[g(\mathbf{A}), \mathbf{B}] = [g(\mathbf{A}), f(\mathbf{A})\mathbf{B}]. \tag{3.230}$$

Solution: The following steps are self-explanatory:

$$f(\mathbf{A})[g(\mathbf{A}), \mathbf{B}] \stackrel{\text{def.}}{=} f(\mathbf{A})\{g(\mathbf{A})\mathbf{B} - \mathbf{B}g(\mathbf{A})\} \tag{3.231a}$$

$$= \underbrace{f(\mathbf{A})g(\mathbf{A})}_{\text{commutative}}\mathbf{B} - f(\mathbf{A})\mathbf{B}g(\mathbf{A}) \tag{3.231b}$$

$$= g(\mathbf{A})\underbrace{f(\mathbf{A})\mathbf{B}}_{} - \underbrace{f(\mathbf{A})\mathbf{B}}_{}g(\mathbf{A}) \tag{3.231c}$$

$$\stackrel{\text{def.}}{=} [g(\mathbf{A}), f(\mathbf{A})\mathbf{B}]. \tag{3.231d}$$

Going from the R.H.S. to the L.H.S. is even more straightforward,

$$[g(\mathbf{A}), f(\mathbf{A})\mathbf{B}] = f(\mathbf{A})[g(\mathbf{A}), \mathbf{B}] + \underbrace{[g(\mathbf{A}), f(\mathbf{A})]}_{\text{commutative}}\mathbf{B} \tag{3.232a}$$

$$= f(\mathbf{A})[g(\mathbf{A}), \mathbf{B}]. \tag{3.232b}$$

The indicated commutativity property can be easily verified by expanding the functions $f(\mathbf{A})$ and $g(\mathbf{A})$ into their respective Formal Taylor Series Expansions, and employing the commutativity property $[\mathbf{A}^m, \mathbf{A}^n] = 0$.

∎

Problem: Show that

$$[\mathbf{A}, g(\mathbf{B})]f(\mathbf{B}) = [\mathbf{A}f(\mathbf{B}), g(\mathbf{B})]. \tag{3.233}$$

Solution: The following steps are self-explanatory:

$$[\mathbf{A}, g(\mathbf{B})]f(\mathbf{B}) \stackrel{\text{def.}}{=} \{\mathbf{A}g(\mathbf{B}) - g(\mathbf{B})\mathbf{A}\}f(\mathbf{B}) \tag{3.234a}$$

$$= \underbrace{\mathbf{A}g(\mathbf{B})f(\mathbf{B})}_{\text{commutative}} - g(\mathbf{B})\mathbf{A}f(\mathbf{B}) \tag{3.234b}$$

$$= \underbrace{\mathbf{A}f(\mathbf{B})g(\mathbf{B})} - g(\mathbf{B})\underbrace{\mathbf{A}f(\mathbf{B})} \tag{3.234c}$$

$$= [\mathbf{A}f(\mathbf{B}), g(\mathbf{B})]. \tag{3.234d}$$

∎

3.11   Commutators involving operator monomials

Problem: Show that

$$[\mathbf{A}, \mathbf{B}^n] = \sum_{k=0}^{n-1} \mathbf{B}^k [\mathbf{A}, \mathbf{B}] \mathbf{B}^{n-1-k}, \quad n \geq 1. \tag{3.235}$$

Remark: The exponent of $\mathbf{B}$ at the L.H.S. is n. The exponent of $\mathbf{B}$ in each summand at the R.H.S. is also n: This is because one $\mathbf{B}$ is invested in $[\mathbf{A}, \mathbf{B}]$, and the exponents of the monomials $\mathbf{B}^k$ and $\mathbf{B}^{n-1-k}$, guarding $[\mathbf{A}, \mathbf{B}]$, must add up to $n - 1$, as they do.

□

Solution: Proof by mathematical induction.

The initial step ($n = 1$) can be verified straightforwardly:

$$[\mathbf{A}, \mathbf{B}^1] = \sum_{k=0}^{1-1} \mathbf{B}^k [\mathbf{A}, \mathbf{B}] \mathbf{B}^{1-1-k} \tag{3.236a}$$

$$= \mathbf{B}^0 [\mathbf{A}, \mathbf{B}] \mathbf{B}^0 \tag{3.236b}$$

$$= [\mathbf{A}, \mathbf{B}]. \tag{3.236c}$$

Assuming the validity of the induction hypothesis (3.235), the induction step consists of showing that (3.235) holds true for $n + 1$, i.e.,

$$[\mathbf{A}, \mathbf{B}^{n+1}] = \sum_{k=0}^{n} \mathbf{B}^k [\mathbf{A}, \mathbf{B}] \mathbf{B}^{n-k}, \quad n \geq 1. \tag{3.237}$$

In the following, it is fascinating to observe how the intermediate steps suggest themselves, the manipulations unfold, and the final result establishes itself. It is obvious that the steps can be automated:

$$[A, B^{n+1}] \overset{(1)}{=} [A, BB^n] \tag{3.238a}$$

$$\overset{(2)}{=} \underbrace{B[A, B^n]}_{\text{Ind.Hyp.}} + [A, B] B^n \tag{3.238b}$$

$$\overset{(3)}{=} B\left(\sum_{k=0}^{n-1} B^k [A, B] B^{n-1-k}\right) + [A, B] B^n \tag{3.238c}$$

$$\overset{(4)}{=} \left(\sum_{k=0}^{n-1} B^{k+1} [A, B] B^{n-1-k}\right) + [A, B] B^n \tag{3.238d}$$

$$\overset{i=k+1}{=} \left(\sum_{i=1}^{n} B^i [A, B] B^{n-i}\right) + \underbrace{[A, B] B^n} \tag{3.238e}$$

$$\overset{(6)}{=} \sum_{i=1}^{n} B^i [A, B] B^{n-i} + B^0 [A, B] B^n \tag{3.238f}$$

$$\overset{(7)}{=} \sum_{i=0}^{n} B^i [A, B] B^{n-i}. \tag{3.238g}$$

Since the summation index is irrelevant ($i \to k$),

$$[A, B^{n+1}] = \sum_{k=0}^{n} B^k [A, B] B^{n-k}. \tag{3.239}$$

■

Remark: The transitions between the equations in (3.238) make usage of the following properties.

(1) $B^{n+1} = BB^n$.
(2) $[A, BC] = B[A, C] + [A, B] C$.
(3) The induction hypothesis, (3.235), is utilized.
(4) Each summand is multiplied from the left by B.
(5) The summation index k is introduced via $i = k + 1$. Thereby, $k = 0$ corresponds to $i = 1$, and $k = n - 1$ to $i = n$. Consequently, $\sum_{k=0}^{n-1} \to \sum_{i=1}^{n}$.
(6) The term $[A, B] B^n$ is written in the equivalent form: $B^0 [A, B] B^n$.
(7) Since $B^0 [A, B] B^n$ is $B^i [A, B] B^{n-i}$ for $i = 0$, (3.238g) follows. In (3.238g), the summation index i runs from 0 to n with the summand corresponding to $i = 0$ included in the sum.

□

3.12 Preparatory considerations for treating $[A, B]^{(n)}$

To state the next problem, it is instructive to introduce the symbol $[A, B]^{(n)}$ for $n \in \mathbb{N}_0$. Starting with the "initial" term,

$$[A, B]^{(0)} \overset{\text{def.}}{=} A, \tag{3.240}$$

recursively introduce the nth term according to,

$$[\mathbf{A}, \mathbf{B}]^{(n)} \overset{\text{def.}}{=} \left[[\mathbf{A}, \mathbf{B}]^{(n-1)}, \mathbf{B}\right] \qquad n \in \mathbb{N}. \tag{3.241}$$

Examining a few low-order terms will allow to explicate the general formula.

- $n = 1$

$$[\mathbf{A}, \mathbf{B}]^{(1)} = \left[[\mathbf{A}, \mathbf{B}]^{(0)}, \mathbf{B}\right] \tag{3.242a}$$
$$= [\mathbf{A}, \mathbf{B}] \tag{3.242b}$$
$$= \underbrace{[}_{1-\text{time}} \underbrace{\mathbf{A}, \mathbf{B}}_{1-\text{time}}]. \tag{3.242c}$$

$$\tag{3.242d}$$

- $n = 2$

$$[\mathbf{A}, \mathbf{B}]^{(2)} = \left[[\mathbf{A}, \mathbf{B}]^{(1)}, \mathbf{B}\right] \tag{3.243a}$$
$$= [[\mathbf{A}, \mathbf{B}], \mathbf{B}] \tag{3.243b}$$
$$= \underbrace{[[}_{2-\text{times}} \mathbf{A} \underbrace{, \mathbf{B}], \mathbf{B}}_{2-\text{times}}]. \tag{3.243c}$$

$$\tag{3.243d}$$

- $n = 3$

$$[\mathbf{A}, \mathbf{B}]^{(3)} = \left[[\mathbf{A}, \mathbf{B}]^{(2)}, \mathbf{B}\right] \tag{3.244a}$$
$$= [[[\mathbf{A}, \mathbf{B}], \mathbf{B}], \mathbf{B}] \tag{3.244b}$$
$$= \underbrace{[[[}_{3-\text{times}} \mathbf{A} \underbrace{, \mathbf{B}], \mathbf{B}], \mathbf{B}}_{3-\text{times}}]. \tag{3.244c}$$

$$\tag{3.244d}$$

- n

$$[\mathbf{A}, \mathbf{B}]^{(n)} = \left[[\mathbf{A}, \mathbf{B}]^{(n-1)}, \mathbf{B}\right] \tag{3.245a}$$
$$= \underbrace{[[\cdots[}_{n-\text{times}} \mathbf{A} \underbrace{, \mathbf{B}], \mathbf{B}], \cdots, \mathbf{B}}_{n-\text{times}}]. \tag{3.245b}$$

Thus, the proposed "construction rule," the induction hypothesis, (3.245) consists of considering $\mathbf{A}$ and prefixing it with "[" (n) times and postfixing it with ", $\mathbf{B}$]" (n)-times. By employing mathematical induction, it is possible to examine whether or not the proposed "construction rule" is consistent. In fact, the course of the proof by mathematical induction itself will expose any possible (otherwise hidden) internal contradiction.

- $n + 1$

$$[\mathbf{A}, \mathbf{B}]^{(n+1)} \overset{\text{def.}}{=} \left[[\mathbf{A}, \mathbf{B}]^{(n)}, \mathbf{B}\right] \tag{3.246a}$$

$$\overset{\text{observe}}{=} \underbrace{[\quad}_{1-\text{time}} [\mathbf{A}, \mathbf{B}]^{(n)} \underbrace{, \mathbf{B}\quad]}_{1-\text{time}} \tag{3.246b}$$

$$\overset{\text{Ind. Hyp.}}{=} \underbrace{[\quad}_{1-\text{time}} \underbrace{[\,[\cdots[}_{n-\text{times}} \mathbf{A}, \mathbf{B}\,], \mathbf{B}\,], \underbrace{\cdots, \mathbf{B}\,]}_{n-\text{times}} \underbrace{, \mathbf{B}\,]}_{1-\text{time}}. \tag{3.246c}$$

$$= \underbrace{[\,[\cdots[}_{(n+1)-\text{times}} \mathbf{A}, \mathbf{B}\,], \mathbf{B}\,], \underbrace{\cdots, \mathbf{B}\,]}_{(n+1)-\text{times}}. \tag{3.246d}$$

Even the initial case $n = 0$ is included in the proposed construction scheme, provided "0-time" is interpreted as "no-time at all":

- $n = 0$

$$[\mathbf{A}, \mathbf{B}]^{(0)} = \underbrace{[\quad}_{0-\text{time}} \mathbf{A} \underbrace{, \mathbf{B}\quad]}_{0-\text{time}} \tag{3.247a}$$

$$= \mathbf{A}. \tag{3.247b}$$

This completes the definition of $[\mathbf{A}, \mathbf{B}]^{(n)}$.

3.13   On the genesis of $[\mathbf{A}, \mathbf{B}]^{(n)}$ and its generating expression

Problem: Show that

$$e^{-\mathbf{B}} \mathbf{A} e^{\mathbf{B}} = \sum_{k=0}^{\infty} \frac{1}{n!} [\mathbf{A}, \mathbf{B}]^{(n)}. \tag{3.248}$$

Solution: Define the λ-dependent function $f(\lambda)$,

$$f(\lambda) = e^{-\lambda \mathbf{B}} \mathbf{A} e^{\lambda \mathbf{B}}. \tag{3.249}$$

The reader might already guess the rationale behind defining $f(\lambda)$: The first derivative of $f(\lambda)$ with respect to λ generates $\mathbf{AB}$ and $-\mathbf{BA}$ and thus $[\mathbf{A}, \mathbf{B}]$, sandwiched between the exponentials $e^{-\lambda \mathbf{B}}$ and $e^{\lambda \mathbf{B}}$. A little thought shows, possibly less obviously, that the higher-order derivatives of $f(\lambda)$ give rise to $[\mathbf{A}, \mathbf{B}]^{(n)}$. The following analysis reveals the details.

1. Calculate the 1st derivative of $f(\lambda)$, while stringently respecting the order of operators.

$$\frac{df(\lambda)}{d\lambda} = \frac{d}{d\lambda}\left\{e^{-\lambda\mathbf{B}}\mathbf{A}e^{\lambda\mathbf{B}}\right\} \tag{3.250a}$$

$$= \left\{\frac{d}{d\lambda}e^{-\lambda\mathbf{B}}\right\}\mathbf{A}e^{\lambda\mathbf{B}} + e^{-\lambda\mathbf{B}}\mathbf{A}\left\{\frac{d}{d\lambda}e^{\lambda\mathbf{B}}\right\} \tag{3.250b}$$

$$= \left\{e^{-\lambda\mathbf{B}}(-\mathbf{B})\right\}\mathbf{A}e^{\lambda\mathbf{B}} + e^{-\lambda\mathbf{B}}\mathbf{A}\left\{\mathbf{B}e^{\lambda\mathbf{B}}\right\} \tag{3.250c}$$

$$= e^{-\lambda\mathbf{B}}[\mathbf{AB} - \mathbf{BA}]e^{\lambda\mathbf{B}} \tag{3.250d}$$

$$= e^{-\lambda\mathbf{B}}[\mathbf{A},\mathbf{B}]e^{\lambda\mathbf{B}} \tag{3.250e}$$

$$= e^{-\lambda\mathbf{B}}[\mathbf{A},\mathbf{B}]^{(1)}e^{\lambda\mathbf{B}}. \tag{3.250f}$$

In particular,

$$\frac{df(\lambda)}{d\lambda}\Big|_{\lambda=0} = e^{-\lambda\mathbf{B}}[\mathbf{A},\mathbf{B}]^{(1)}e^{\lambda\mathbf{B}}\Big|_{\lambda=0} \tag{3.251a}$$

$$= [\mathbf{A},\mathbf{B}]^{(1)}. \tag{3.251b}$$

2. Calculate the 2nd derivative of $f(\lambda)$.

$$\frac{d^2f(\lambda)}{d\lambda^2} \stackrel{\text{def.}}{=} \frac{d}{d\lambda}\left\{\frac{df(\lambda)}{d\lambda}\right\} \tag{3.252a}$$

$$\stackrel{(3.250)}{=} \frac{d}{d\lambda}\left\{e^{-\lambda\mathbf{B}}[\mathbf{A},\mathbf{B}]e^{\lambda\mathbf{B}}\right\} \tag{3.252b}$$

$$= \left\{\frac{d}{d\lambda}e^{-\lambda\mathbf{B}}\right\}[\mathbf{A},\mathbf{B}]e^{\lambda\mathbf{B}} + e^{-\lambda\mathbf{B}}[\mathbf{A},\mathbf{B}]\left\{\frac{d}{d\lambda}e^{\lambda\mathbf{B}}\right\} \tag{3.252c}$$

$$= \left\{e^{-\lambda\mathbf{B}}(-\mathbf{B})\right\}[\mathbf{A},\mathbf{B}]e^{\lambda\mathbf{B}} + e^{-\lambda\mathbf{B}}[\mathbf{A},\mathbf{B}]\left\{\mathbf{B}e^{\lambda\mathbf{B}}\right\} \tag{3.252d}$$

$$= e^{-\lambda\mathbf{B}}[[\mathbf{A},\mathbf{B}]\mathbf{B} - \mathbf{B}[\mathbf{A},\mathbf{B}]]e^{\lambda\mathbf{B}} \tag{3.252e}$$

$$= e^{-\lambda\mathbf{B}}[[\mathbf{A},\mathbf{B}],\mathbf{B}]e^{\lambda\mathbf{B}} \tag{3.252f}$$

$$= e^{-\lambda\mathbf{B}}[\mathbf{A},\mathbf{B}]^{(2)}e^{\lambda\mathbf{B}}. \tag{3.252g}$$

In particular,

$$\frac{d^2f(\lambda)}{d\lambda^2}\Big|_{\lambda=0} = e^{-\lambda\mathbf{B}}[\mathbf{A},\mathbf{B}]^{(2)}e^{\lambda\mathbf{B}}\Big|_{\lambda=0} \tag{3.253a}$$

$$= [\mathbf{A},\mathbf{B}]^{(2)}. \tag{3.253b}$$

3. Calculate the nth derivative of $f(\lambda)$. Using the method of mathematical induction, based on the induction hypothesis,

$$\frac{d^n f(\lambda)}{d\lambda^n} = e^{-\lambda \mathbf{B}} [\mathbf{A}, \mathbf{B}]^{(n)} e^{\lambda \mathbf{B}}, \tag{3.254}$$

the following steps are self-explanatory,

$$\frac{d^{n+1} f(\lambda)}{d\lambda^{n+1}} \quad \overset{\text{def.}}{=} \quad \frac{d}{d\lambda} \left\{ \frac{d^n f(\lambda)}{d\lambda^n} \right\} \tag{3.255a}$$

$$\overset{\text{ind. hypo.}}{=} \frac{d}{d\lambda} \left\{ e^{-\lambda \mathbf{B}} [\mathbf{A}, \mathbf{B}]^n e^{\lambda \mathbf{B}} \right\} \tag{3.255b}$$

$$= \quad \left\{ \frac{d}{d\lambda} e^{-\lambda \mathbf{B}} \right\} [\mathbf{A}, \mathbf{B}]^n e^{\lambda \mathbf{B}} + e^{-\lambda \mathbf{B}} [\mathbf{A}, \mathbf{B}]^n \left\{ \frac{d}{d\lambda} e^{\lambda \mathbf{B}} \right\} \tag{3.255c}$$

$$= \quad \left\{ e^{-\lambda \mathbf{B}} (-\mathbf{B}) \right\} [\mathbf{A}, \mathbf{B}]^n e^{\lambda \mathbf{B}} + e^{-\lambda \mathbf{B}} [\mathbf{A}, \mathbf{B}]^n \left\{ \mathbf{B} e^{\lambda \mathbf{B}} \right\} \tag{3.255d}$$

$$= \quad e^{-\lambda \mathbf{B}} \left[[\mathbf{A}, \mathbf{B}]^n \mathbf{B} - \mathbf{B} [\mathbf{A}, \mathbf{B}]^n \right] e^{\lambda \mathbf{B}} \tag{3.255e}$$

$$= \quad e^{-\lambda \mathbf{B}} \left[[\mathbf{A}, \mathbf{B}]^n, \mathbf{B} \right] e^{\lambda \mathbf{B}} \tag{3.255f}$$

$$= \quad e^{-\lambda \mathbf{B}} [\mathbf{A}, \mathbf{B}]^{(n+1)} e^{\lambda \mathbf{B}}. \tag{3.255g}$$

In particular,

$$\frac{d^{n+1} f(\lambda)}{d\lambda^{n+1}} \Big|_{\lambda=0} = e^{-\lambda \mathbf{B}} [\mathbf{A}, \mathbf{B}]^{(n+1)} e^{\lambda \mathbf{B}} \Big|_{\lambda=0} \tag{3.256a}$$

$$= [\mathbf{A}, \mathbf{B}]^{(n+1)}. \tag{3.256b}$$

4. Consider the Formal Taylor Series Expansion of $f(\lambda)$,

$$f(\lambda) = \sum_{n=0}^{\infty} \frac{1}{n!} \frac{d^n f(\lambda)}{d\lambda^n} \Big|_{\lambda=0} \lambda^n \tag{3.257a}$$

$$= \sum_{n=0}^{\infty} \frac{1}{n!} [\mathbf{A}, \mathbf{B}]^{(n)} \lambda^n. \tag{3.257b}$$

5. Recall the definition of $f(\lambda)$ in (3.249). Then,

$$e^{-\lambda \mathbf{B}} \mathbf{A} e^{\lambda \mathbf{B}} = \sum_{n=0}^{\infty} \frac{1}{n!} [\mathbf{A}, \mathbf{B}]^{(n)} \lambda^n. \tag{3.258}$$

In particular, setting $\lambda = 1$,

$$e^{-\mathbf{B}} \mathbf{A} e^{\mathbf{B}} = \sum_{n=0}^{\infty} \frac{1}{n!} [\mathbf{A}, \mathbf{B}]^{(n)}. \tag{3.259}$$

∎

Remark: Keeping in mind the steps which led to (3.258) and recalling the definition of $[\mathbf{A}, \mathbf{B}]^{(n)}$, the equality (3.258) can readily be made plausible. Stated slightly differently, the expression at the R.H.S. of (3.258) can be obtained simply-by-inspection. The "methodology" underlying the Simply-by-Inspection idea should decisively be contrasted against memorization. To clarify this distinction, several key intuitive relationships will be analyzed and explicated next.

1. Focus on (3.258).
2. The exponent of $\mathbf{A}$, sandwiched between $e^{-\mathbf{B}}$ and $e^{\mathbf{B}}$ is unity at the L.H.S. Recalling the definition of $[\mathbf{A}, \mathbf{B}]^{(n)}$, i.e., $[[\cdots[\mathbf{A}, \mathbf{B}], \mathbf{B}], \ldots, \mathbf{B}]$, it is seen that the exponent of $\mathbf{A}$ is also unity at the R.H.S., as expected. In contrast, all powers of $\mathbf{B}^n$ ($n \in \mathbb{N}_0$) happen to be present on both sides of the equation. This suggests the choice of $[\mathbf{A}, \mathbf{B}]^{(n)}$ in view of $e^{-\lambda\mathbf{B}}\mathbf{A}e^{\lambda\mathbf{B}}$. This observation, however, does not justify why $e^{-\lambda\mathbf{B}}\mathbf{A}e^{\lambda\mathbf{B}}$ rather than $e^{\lambda\mathbf{B}}\mathbf{A}e^{-\lambda\mathbf{B}}$ has been chosen. The criterion for this choice is the subject of the next item.
3. Differentiating $e^{-\lambda\mathbf{B}}\mathbf{A}e^{\lambda\mathbf{B}}$ with respect to λ resulted in $-\mathbf{B}\mathbf{A}$ and $\mathbf{A}\mathbf{B}$ and thus the commutator $[\mathbf{A}, \mathbf{B}]$. This suggests that $e^{-\lambda\mathbf{B}}\mathbf{A}e^{\lambda\mathbf{B}}$ is the desired order.
4. Another way to see that the derivative of $e^{-\lambda\mathbf{B}}\mathbf{A}e^{\lambda\mathbf{B}}$ with respect to λ generates $[\mathbf{A}, \mathbf{B}]$ is by considering the first two terms of the Formal Taylor Series Expansion of the exponential functions $e^{-\lambda\mathbf{B}}$ and $e^{\lambda\mathbf{B}}$. Then, $e^{-\lambda\mathbf{B}}\mathbf{A}e^{\lambda\mathbf{B}}$ becomes,

$$e^{-\lambda\mathbf{B}}\mathbf{A}e^{\lambda\mathbf{B}} = (\mathbf{I} - \lambda\mathbf{B} + \mathcal{O}(\lambda^2))\mathbf{A}(\mathbf{I} + \lambda\mathbf{B} + \mathcal{O}(\lambda^2)). \tag{3.260}$$

Ordering the terms with respect to increasing powers of λ,

$$e^{-\lambda\mathbf{B}}\mathbf{A}e^{\lambda\mathbf{B}} = \mathbf{A} + \lambda\left(-\mathbf{B}\mathbf{A} + \mathbf{A}\mathbf{B}\right) + \mathcal{O}\left(\lambda^2\right). \tag{3.261}$$

Thus the presence of $e^{-\lambda\mathbf{B}}$ to the left of $\mathbf{A}$ results in $-\mathbf{B}\mathbf{A}$ which is required in constructing the commutator $[\mathbf{A}, \mathbf{B}]$.
5. When considering terms, say, to the 3rd-order in λ,

$$e^{-\lambda\mathbf{B}}\mathbf{A}e^{\lambda\mathbf{B}}$$
$$= \left(\mathbf{I} - \lambda\mathbf{B} + \frac{1}{2!}\lambda^2\mathbf{B}^2 - \frac{1}{3!}\lambda^3\mathbf{B}^3 + \mathcal{O}(\lambda^4)\right)\mathbf{A}\left(\mathbf{I} + \lambda\mathbf{B} + \frac{1}{2!}\lambda^2\mathbf{B}^2 + \frac{1}{3!}\lambda^3\mathbf{B}^3 + \mathcal{O}(\lambda^2)\right) \tag{3.262}$$

and invariably respecting the order of operators, it is revealing, intriguing, and reassuring to find out the way how the coefficient proportional to λ^2 and λ^3 "collaboratively" result in $[[\mathbf{A}, \mathbf{B}], \mathbf{B}] = [\mathbf{A}, \mathbf{B}]^{(2)}$ and $[[[\mathbf{A}, \mathbf{B}], \mathbf{B}], \mathbf{B}] = [\mathbf{A}, \mathbf{B}]^{(3)}$, accompanied by the multiplicative factors $1/2!$ and $1/3!$, respectively.

$\square$

Remark: To emphasize the significance of the preceding analysis, the main idea has been summarized in this remark separately. It is hypothesized that the "mental post-processing" (after rigorously completing the proof), as exemplified in the above steps, can be viewed as an essential part of the psychology and perhaps neurophysiology of learning and discovery. A thorough analysis sheds light on what was done in the course of the proof, and more crucially, directs the attention to seek for generalizations. Numerous other examples in this text shall render this method self-evident.

$\square$

3.14 On the genesis of $^{(n)}[\mathbf{B}, \mathbf{A}]$ and its generating expression

Rather than using $[\mathbf{B}, \mathbf{A}]^n$, which was employed for notational simplicity further above, it is here instructive to introduce the more suggestive notation $^{(n)}[\mathbf{B}, \mathbf{A}]$,

$$^{(n)}[\mathbf{B}, \mathbf{A}] = \left[\mathbf{B}, {}^{(n-1)}[\mathbf{B}, \mathbf{A}]\right] \tag{3.263a}$$
$$= \underbrace{[\mathbf{B}, [\mathbf{B}, \ldots, [\mathbf{B},}_{n-\text{rimes}} \underbrace{\mathbf{A}\,]\ldots]]}_{n-\text{times}}. \tag{3.263b}$$

Problem: Show that,

$$^{(n)}[\mathbf{B}, \mathbf{A}] = (-1)^n\,[\mathbf{A}, \mathbf{B}]^{(n)}, \quad n \in \mathbb{N}_0. \tag{3.264}$$

Interpreting the case "0-time" as "no time at all,"

$$^{(0)}[\mathbf{B}, \mathbf{A}] = \mathbf{A}. \tag{3.265}$$

Remark: Considering (3.264), the adjacency of n to $\mathbf{B}$ in $^{(n)}[\mathbf{B}, \mathbf{A}]$ is a reminder of the property that "$[\mathbf{B},$ " and thus $\mathbf{B}$ occurs n-times. Similarly, the adjacency of n to $\mathbf{B}$ in $[\mathbf{A}, \mathbf{B}]^{(n)}$ is a reminder of the property that ", $\mathbf{B}]$" and thus $\mathbf{B}$ occurs n-times. The operator $\mathbf{A}$ occurs only once in both $^{(n)}[\mathbf{B}, \mathbf{A}]$ and $[\mathbf{A}, \mathbf{B}]^{(n)}$. Consequently, it is meaningful to seek a relationship between $^{(n)}[\mathbf{B}, \mathbf{A}]$ and $[\mathbf{A}, \mathbf{B}]^{(n)}$ (or, for this matter between $^{(n)}[\mathbf{A}, \mathbf{B}]$ and $[\mathbf{B}, \mathbf{A}]^{(n)}$), which is the subject matter of the current problem.

$\square$

Solution: The solution of the problem is obtained by applying the method of mathematical induction.

The trivial case $n = 0$ is the fundamental property of the reflexivity: $\mathbf{A} = \mathbf{A}$,

$$\mathbf{A} \Longleftarrow {}^{(0)}[\mathbf{B}, \mathbf{A}] = (-1)^0 \, [\mathbf{A}, \mathbf{B}]^0 \implies \mathbf{A}. \tag{3.266}$$

This (singular) case does not reveal much of the structures of $^{(n)}[\mathbf{B}, \mathbf{A}]$ and $[\mathbf{A}, \mathbf{B}]^n$.

The case $n = 1$ is slightly more illuminating.

$$^{(1)}[\mathbf{B}, \mathbf{A}] \overset{\text{def.}}{=} [\mathbf{B}, \mathbf{A}] \tag{3.267a}$$

$$\overset{\text{property}}{=} (-1)^1 \, [\mathbf{A}, \mathbf{B}] \tag{3.267b}$$

$$\overset{\text{def.}}{=} (-1)^1 \, [\mathbf{A}, \mathbf{B}]^{(1)}. \tag{3.267c}$$

The induction step, is however, revealing. Using the induction hypothesis (3.264),

$$^{(n+1)}[\mathbf{B}, \mathbf{A}] \overset{(3.263a)}{=} \left[\mathbf{B}, {}^{(n)}[\mathbf{B}, \mathbf{A}] \right] \tag{3.268a}$$

$$\overset{\text{def.}}{=} \mathbf{B} \left\{ {}^{(n)}[\mathbf{B}, \mathbf{A}] \right\} - \left\{ {}^{(n)}[\mathbf{B}, \mathbf{A}] \right\} \mathbf{B} \tag{3.268b}$$

$$\overset{(3.264)}{=} \mathbf{B} \left\{ (-1)^n \, [\mathbf{A}, \mathbf{B}]^{(n)} \right\} - \left\{ (-1)^n \, [\mathbf{A}, \mathbf{B}]^{(n)} \right\} \mathbf{B} \tag{3.268c}$$

$$= (-1)^n \left\{ \mathbf{B} \, [\mathbf{A}, \mathbf{B}]^{(n)} - [\mathbf{A}, \mathbf{B}]^{(n)} \, \mathbf{B} \right\} \tag{3.268d}$$

$$\overset{\text{fact. out } (-1)}{=} (-1)^{n+1} \left\{ [\mathbf{A}, \mathbf{B}]^{(n)} \, \mathbf{B} - \mathbf{B} \, [\mathbf{A}, \mathbf{B}]^{(n)} \right\} \tag{3.268e}$$

$$\overset{\text{def.}}{=} (-1)^{n+1} \left[[\mathbf{A}, \mathbf{B}]^{(n)}, \mathbf{B} \right] \tag{3.268f}$$

$$\overset{\text{def.}}{=} (-1)^{n+1} \, [\mathbf{A}, \mathbf{B}]^{(n+1)}. \tag{3.268g}$$

■

Exercise: Define the λ-dependent operator $\mathbf{G}(\lambda)$,

$$\mathbf{G}(\lambda) = e^{\lambda \mathbf{B}} \mathbf{A} e^{-\lambda \mathbf{B}}. \tag{3.269}$$

It is straightforward to establish the relationship,

$$\frac{d^n}{d\lambda^n} \mathbf{G}(\lambda) = e^{\lambda \mathbf{B}} \left({}^{(n)}[\mathbf{B}, \mathbf{A}] \right) e^{-\lambda \mathbf{B}}. \tag{3.270}$$

In particular,

$$\left\{ \frac{d^n}{d\lambda^n} \mathbf{G}(\lambda) \right\}_{\lambda=0} = {}^{(n)}[\mathbf{B}, \mathbf{A}]. \tag{3.271}$$

Consider the Formal Taylor Series Expansion of $\mathbf{G}(\lambda)$,

$$\mathbf{G}(\lambda) = \sum_{n=0}^{\infty} \frac{1}{n!} \left\{ \frac{d^n}{d\lambda^n} \mathbf{G}(\lambda) \right\}_{\lambda=0} \lambda^n. \tag{3.272}$$

Substituting (3.271), and remembering the definition of $\mathbf{G}(\lambda)$ in (3.269),

$$e^{\lambda \mathbf{B}} \mathbf{A} e^{-\lambda \mathbf{B}} = \sum_{n=0}^{\infty} \frac{1}{n!}^{(n)}[\mathbf{B}, \mathbf{A}] \lambda^n. \tag{3.273}$$

In particular, for $\lambda = 1$,

$$e^{\mathbf{B}} \mathbf{A} e^{-\mathbf{B}} = \sum_{n=0}^{\infty} \frac{1}{n!}^{(n)}[\mathbf{B}, \mathbf{A}]. \tag{3.274}$$

∎

3.15 Commutators involving $\hat{\mathbf{r}}$ and $\hat{\mathbf{p}}$

The operators $\hat{\mathbf{r}}$ and $\hat{\mathbf{p}}$, corresponding to the position vector $\mathbf{r}$ and the linear momentum $\mathbf{p}$, have the following coordinate space representations,

$$\mathbf{r} = (x_1, x_2, x_3) \iff \hat{\mathbf{r}} = (\hat{x}_1, \hat{x}_3, \hat{x}_1) \tag{3.275a}$$

$$\mathbf{p} = (p_1, p_2, p_3) \iff \hat{\mathbf{p}} = (\hat{p}_1, \hat{p}_2, \hat{p}_3) = \frac{\hbar}{i} \nabla = \frac{\hbar}{i} \left(\frac{\partial}{\partial x_1}, \frac{\partial}{\partial x_2}, \frac{\partial}{\partial x_3} \right) = \frac{\hbar}{i} \left(\hat{D}_1, \hat{D}_2, \hat{D}_3 \right). \tag{3.275b}$$

Here, $\hbar$ refers to the reduced Planck constant. Operators corresponding to physical quantities (observable) will be thoroughly discussed in Chapter 1, Volume 2.

An effective way to investigate the properties of operators is to applying them to functions in their definition domain. This is done next to elucidate the meaning and properties of the position- and linear momentum operators.

3.15.1 Position operators $\hat{x}_i$ $(i = 1, 2, 3)$

The impact of applying $\hat{x}_i$ onto the function $f(x_1, x_2, x_3)$ is just multiplying $f(x_1, x_2, x_3)$ by x_i. Thus,

$$\hat{x}_1 f(x_1, x_2, x_3) = x_1 f(x_1, x_2, x_3) \tag{3.276a}$$
$$\hat{x}_2 f(x_1, x_2, x_3) = x_2 f(x_1, x_2, x_3) \tag{3.276b}$$
$$\hat{x}_3 f(x_1, x_2, x_3) = x_3 f(x_1, x_2, x_3). \tag{3.276c}$$

3.15.2 Linear momentum operators $\hat{p}_i$ $(i = 1, 2, 3)$

The impact of applying $\hat{p}_i = (\hbar/i) \hat{D}_i$ onto the function $f(x_1, x_2, x_3)$ is the application of $\hat{D}_i$ onto the function $f(x_1, x_2, x_3)$ and multiplying the result by $\hbar/i$. Thus, it is required to differentiate the function $f(x_1, x_2, x_3)$ with respect to x_i is required:

$$\hat{p}_1 f(x_1, x_2, x_3) = \frac{\hbar}{i} \hat{D}_1 f(x_1, x_2, x_3) = \frac{\hbar}{i} \frac{\partial}{\partial x_1} f(x_1, x_2, x_3) \tag{3.277a}$$

$$\hat{p}_2 f(x_1, x_2, x_3) = \frac{\hbar}{i} \hat{D}_2 f(x_1, x_2, x_3) = \frac{\hbar}{i} \frac{\partial}{\partial x_1} f(x_1, x_2, x_3) \tag{3.277b}$$

$$\hat{p}_3 f(x_1, x_2, x_3) = \frac{\hbar}{i} \hat{D}_3 f(x_1, x_2, x_3) = \frac{\hbar}{i} \frac{\partial}{\partial x_1} f(x_1, x_2, x_3). \tag{3.277c}$$

Remark: Differentiating and integrating functions in addition to a good dose of linear algebra are all what is needed in computational quantum physics, leaving aside the various interpretations of quantum theory. Generalizations of the notions of differentiation and integration, which might be needed, are comparatively straightforward, and can be acquired after one acquaints themselves with the fundamentals in calculus, functional analysis and linear algebra. The only note of caution, at the current stage, is the careful treatment of the product rule when differentiating the product of two functions, and the rule of integration by parts. Nothing more is needed, as shown next.

$\square$.

Problem: Determine a representation for the commutator $\left[\hat{D}_x, \hat{x}\right]$, with $\hat{D}_x = d/dx$.

Solution: To investigate the properties of $\left[\hat{D}_x, \hat{x}\right]$, which is an operator, apply it to an arbitrary test function $f(x)$,

$$\left[\hat{D}_x, \hat{x}\right] f(x) \stackrel{(1)}{=} \left[\frac{d}{dx}, \hat{x}\right] f(x) \tag{3.278a}$$

$$\stackrel{(2)}{=} \left(\frac{d}{dx}\hat{x} - \hat{x}\frac{d}{dx}\right) f(x) \tag{3.278b}$$

$$\stackrel{(3)}{=} \left(\frac{d}{dx}\hat{x}\right) f(x) - \left(\hat{x}\frac{d}{dx}\right) f(x) \tag{3.278c}$$

$$\stackrel{(4)}{=} \frac{d}{dx}(\hat{x}f(x)) - \hat{x}\left(\frac{d}{dx}f(x)\right) \tag{3.278d}$$

$$\stackrel{(5)}{=} \underbrace{\frac{d}{dx}(xf(x))} - x\left(\frac{d}{dx}f(x)\right) \tag{3.278e}$$

$$\stackrel{(6)}{=} \overbrace{\left\{\left(\frac{d}{dx}x\right)f(x) + x\left(\frac{d}{dx}f(x)\right)\right\}} - x\left(\frac{d}{dx}f(x)\right) \tag{3.278f}$$

$$\stackrel{(7)}{=} 1 \cdot f(x) \tag{3.278g}$$

$$\stackrel{(8)}{=} f(x) \tag{3.278h}$$

$$\stackrel{(9)}{=} \hat{I}f(x). \tag{3.278i}$$

The transitions in (3.278) can be justified as follows:

(1) The definition $\hat{D}_x \stackrel{\text{def.}}{=} d/dx$.
(2) The definition of the commutator, while respecting the order of the operators involved.
(3) The linearity of the composite operators $\frac{d}{dx}\hat{x}$ and $\hat{x}\frac{d}{dx}$.
(4) The application of the composite operator $\left(\frac{d}{dx}\hat{x}\right)$ on $f(x)$ consists of applying $\hat{x}$ first, i.e., $\hat{x}f(x)$, followed by applying $\frac{d}{dx}$, i.e., $\frac{d}{dx}\left(\hat{x}f(x)\right)$. Similarly, the application of the composite operator $\left(\hat{x}\frac{d}{dx}\right)$ on $f(x)$ consists of applying $\frac{d}{dx}$ first, i.e., $\frac{d}{dx}f(x)$, followed by applying $\hat{x}$, i.e., $\hat{x}\left(\frac{d}{dx}f(x)\right)$.
(5) Applying $\hat{x}$ onto $f(x)$ results, by definition, in $xf(x)$, which is the product of two functions. This simple fact, the impact of which is seen momentarily, is the most relevant idea in the entire discussion in this problem. On the other hand, the application of $\frac{d}{dx}$ onto $f(x)$ leads to the function $\frac{d}{dx}f(x)$, the derivative of $f(x)$. This term does not require any further analysis, as it will be canceled out in the course of calculation, as seen in the next two steps.
(6) Taking the derivative of $xf(x)$, $\frac{d}{dx}(xf(x))$, requires, not surprisingly, the application of the product rule. Consequently, two terms are generated.
(7) The term $x\frac{d}{dx}(f(x))$ appearing with the plus and the minus sign cancel out. Furthermore, the derivative of the "function" x with respect to x results in the natural number 1.
(8) Scaling $f(x)$ by $1 \in \mathbb{N}$ results in $f(x)$.
(9) The function $f(x)$ can be written in the form $\hat{I}f(x)$ with $\hat{I}$ being the identity operator.

Equating the first term to the last term in (3.278), and in view of the fact that $f(x)$ is arbitrary,

$$\left[\hat{D}_x, \hat{x}\right] = \hat{I}. \tag{3.279}$$

■

Problem: Determine a representation for the commutator $\left[\hat{D}_x, \hat{y}\right]$, with $\hat{D}_x = \partial/\partial x$.

Solution: To investigate the properties of $[\hat{D}_x, \hat{y}]$, the commutator must be applied to an arbitrary function in its definition domain. In the present case, there are two independent variables involved, i.e., x, and y. Consequently, the test function must be a general bivariate function, $f(x, y)$:

$$\left[\hat{D}_x, \hat{y}\right] f(x, y) = \left[\frac{\partial}{\partial x}, \hat{y}\right] f(x, y) \tag{3.280a}$$

$$= \left(\frac{\partial}{\partial x}\hat{y} - \hat{y}\frac{\partial}{\partial x}\right) f(x, y) \tag{3.280b}$$

$$= \left(\frac{\partial}{\partial x}\hat{y}\right) f(x, y) - \left(\hat{y}\frac{\partial}{\partial x}\right) f(x, y) \tag{3.280c}$$

$$= \frac{\partial}{\partial x}\left(\hat{y}f(x, y)\right) - \hat{y}\left(\frac{\partial}{\partial x}f(x, y)\right) \tag{3.280d}$$

$$\overset{(5)}{=} \underbrace{\frac{\partial}{\partial x}\left(yf(x, y)\right)} - y\left(\frac{\partial}{\partial x}f(x, y)\right) \tag{3.280e}$$

$$\overset{(6)}{=} \overbrace{\left\{\left(\frac{\partial}{\partial x}y\right)f(x, y) + y\left(\frac{\partial}{\partial x}f(x, y)\right)\right\}} - y\left(\frac{\partial}{\partial x}f(x, y)\right) \tag{3.280f}$$

$$\overset{(7)}{=} 0 \cdot f(x, y) \tag{3.280g}$$

$$\overset{(8)}{=} 0 \tag{3.280h}$$

$$\overset{(9)}{=} \hat{0}f(x, y). \tag{3.280i}$$

(5) This is a crucial step whereby the definition $\hat{y}f(x, y) \overset{\text{def.}}{=} yf(x, y)$ has been employed.
(6) The product rule has been applied to the term $\frac{\partial}{\partial x}(yf(x, y))$.
(7) This step, stating that the y-derivative of the independent variable x is zero, is the quintessential reason why $\hat{D}_x$ and $\hat{y}$ are commutative.
(8) The symbol 0 stands for the zero being element of $\mathbb{R}$.
(9) The symbol $\hat{0}$ is an operator defined by projecting $f(x, y)$ onto 0: $\hat{0}f(x, y) \overset{\text{def.}}{=} 0$.

Equating the first term to the last term in (3.280), and in view of the fact that the test function $f(x, y)$ is arbitrary,

$$\left[\hat{D}_x, \hat{y}\right] = \hat{0}. \tag{3.281}$$

■

Remark: From the preceding two problems it can be concluded that,

$$\left[\hat{D}_{x_i}, x_j\right] = \delta_{ij}I, \tag{3.282}$$

with δ_{ij} standing for the Kronecker delta symbol.

Due to the paramount significance of this relationship, it is instructive to write it in a less pretentious compact form. The relationship in (3.282) is equivalent with the following equations:

$$\left[\hat{D}_x, \hat{x}\right] = \hat{I} \quad \left[\hat{D}_x, \hat{y}\right] = \hat{0} \quad \left[\hat{D}_x, \hat{z}\right] = \hat{0}$$

$$\left[\hat{D}_y, \hat{x}\right] = \hat{0} \quad \left[\hat{D}_y, \hat{y}\right] = \hat{I} \quad \left[\hat{D}_y, \hat{z}\right] = \hat{0} \qquad (3.283)$$

$$\left[\hat{D}_z, \hat{x}\right] = \hat{0} \quad \left[\hat{D}_z, \hat{y}\right] = \hat{0} \quad \left[\hat{D}_z, \hat{z}\right] = \hat{I}$$

These equations build the foundation for countless useful relationships, which are the subject matter of further analysis in this text. $\square$

Remark: From the above discussion the utility of the "^" notation must be clear. A few further comments should solidify the ideas. To this end, it is instructive to introduce the notation,

$$f^{(m,n,l)}(x,y,z) = \hat{D}_x^m \hat{D}_y^n \hat{D}_z^l f(x,y,z), \qquad (3.284)$$

with,

$$\hat{D}_x^m \hat{D}_y^n \hat{D}_z^l \stackrel{\text{def.}}{=} \frac{\partial^m}{\partial x^m} \frac{\partial^n}{\partial y^n} \frac{\partial^l}{\partial z^l} = \frac{\partial^{m+n+l}}{\partial x^m \partial y^n \partial z^l}. \qquad (3.285)$$

- $\hat{D}_x \stackrel{\text{def.}}{=} \partial/\partial x$ is a differential operator, resulting in the derivative with respect to x of the function upon which it operates. Analogous statements can be made about $\hat{D}_y \stackrel{\text{def.}}{=} \partial/\partial y$, and $\hat{D}_z \stackrel{\text{def.}}{=} \partial/\partial z$. For completeness, $\hat{D}_t \stackrel{\text{def.}}{=} \partial/\partial t$, with t referring to the time variable:

  1. $\hat{D}_x x^m = \frac{\partial}{\partial x} x^m = m x^{m-1}$.

  2. $\hat{D}_x x^m y^n = \left(\hat{D}_x x^m\right) y^n = \left(\frac{\partial}{\partial x} x^m\right) y^n = \left(m x^{m-1}\right) y^n = m x^{m-1} y^n$.

  3. $\hat{D}_x^2 \hat{D}_y x^3 y^5 = \left(\hat{D}_x^2 x^3\right)\left(\hat{D}_y y^5\right) = \left(\frac{\partial^2}{\partial x^2} x^3\right)\left(\frac{\partial}{\partial y} y^5\right) = (6x)\left(5y^4\right) = 30xy^4$.

  4. The order of the operators $\hat{D}_\xi$ and $\hat{D}_\eta$ (with ξ and η standing for x, y, or z), and consequently, of $\hat{D}_\xi^m$ and $\hat{D}_\eta^n$ is immaterial. Thus $\left[\hat{D}_\xi^m, \hat{D}_\eta^n\right] = \hat{0}$.

- $\hat{x}f(x) \stackrel{\text{def.}}{=} xf(x)$. Similarly, $\hat{y}f(y) \stackrel{\text{def.}}{=} yf(y)$ and $\hat{z}f(z) \stackrel{\text{def.}}{=} zf(z)$. For completeness, $\hat{t}f(t) \stackrel{\text{def.}}{=} tf(t)$.

  1. $\hat{x}^2 x^m = \hat{x}\left(\hat{x}x^m\right) = \hat{x}\left(xx^m\right) = x\left(xx^m\right) = x^2 x^m$.

  2. $\hat{x}^n f(x) = x^n f(x)$.

  3. $\hat{x}^m \hat{y}^n f(x,y) = \hat{x}^m\left(\hat{y}^n f(x,y)\right) = \hat{x}^m\left(y^n f(x,y)\right) = x^m\left(y^n f(x,y)\right) = x^m y^n f(x,y)$.

  4. Obviously, the order of $\hat{\xi}^m$ and $\hat{\eta}^n$ (with ξ and η standing for x, y, or z) is not relevant. Thus, $\left[\hat{\xi}^m, \hat{\eta}^n\right] = \hat{0}$.

- Instances where the order of operators matters are of the type $\hat{D}_\xi \hat{\xi}$, or more generally, $\hat{D}_\xi^m \hat{\xi}^n$. The fact that $\hat{D}_\xi \hat{\xi} \neq \hat{\xi} \hat{D}_\xi$, and more generally, $\hat{D}_\xi^m \hat{\xi}^n \neq \hat{\xi}^n \hat{D}_\xi^m$, merely means that changing the order of operators must follow certain strict, but, straightforward rules. Under the provision that certain rules (e.g., commutativity, anti-commutativity rules) are respected, the order of the operators (subject to the rules) can be exchanged. Perhaps the best way to convey the idea is to consider a few more examples.

1. $[\hat{D}_x, \hat{x}] = \hat{I}$, as demonstrated earlier. Thus, $\hat{D}_x\hat{x} - \hat{x}\hat{D}_x = \hat{I}$. Consequently, $\hat{D}_x\hat{x} = \hat{x}\hat{D}_x + \hat{I}$, and rather trivially, $\hat{x}\hat{D}_x = \hat{D}_x\hat{x} - \hat{I}$.

2. What about terms such as $\hat{D}_x\hat{x}^2$? To this end,

    $[\hat{D}_x, \hat{x}^2] = [\hat{D}_x, \hat{x}\hat{x}] = \hat{x}[\hat{D}_x, \hat{x}] + [\hat{D}_x, \hat{x}]\hat{x} = \hat{x}\hat{I} + \hat{I}\hat{x} = 2\hat{x}$. Equating the first and the last terms, $\hat{D}_x\hat{x}^2 - \hat{x}^2\hat{D}_x = 2\hat{x}$. Consequently, $\hat{D}_x\hat{x}^2 = \hat{x}^2\hat{D}_x + 2\hat{x}$, and, $\hat{x}^2\hat{D}_x = \hat{D}_x\hat{x}^2 - 2\hat{x}$.

3. What about terms such as $\hat{D}_x^2\hat{x}$? To this end,

    $[\hat{D}_x^2, \hat{x}] = [\hat{D}_x\hat{D}_x, \hat{x}] = \hat{D}_x[\hat{D}_x, \hat{x}] + [\hat{D}_x, \hat{x}]\hat{D}_x = \hat{D}_x\hat{I} + \hat{I}\hat{D}_x = 2\hat{D}_x$. Equating the first and the last terms, $\hat{D}_x^2\hat{x} - \hat{x}\hat{D}_x^2 = 2\hat{D}_x$. Consequently, $\hat{D}_x^2\hat{x} = \hat{x}\hat{D}_x^2 + 2\hat{D}_x$, and $\hat{x}\hat{D}_x^2 = \hat{D}_x^2\hat{x} - 2\hat{D}_x$.

4. What about terms such as $\hat{D}_x^2\hat{x}^3$? Proceed as follows:

$$\left[\hat{D}_x^2, \hat{x}^3\right] = \left[\hat{D}_x\hat{D}_x, \hat{x}^3\right] \tag{3.286a}$$

$$= \hat{D}_x\underbrace{\left[\hat{D}_x, \hat{x}^3\right]}_{3\hat{x}^2} + \underbrace{\left[\hat{D}_x, \hat{x}^3\right]}_{3\hat{x}^2}\hat{D}_x \tag{3.286b}$$

$$= 3\underbrace{\hat{D}_x\hat{x}^2}_{\hat{x}^2\hat{D}_x+2\hat{x}} + 3\hat{x}^2\hat{D}_x \tag{3.286c}$$

$$= 6\hat{x}^2\hat{D}_x + 6\hat{x}. \tag{3.286d}$$

Equating the first term (after expanding the commutator) and the last term,

$$\hat{D}_x^2\hat{x}^3 - \hat{x}^3\hat{D}_x^2 = 6\hat{x}^2\hat{D}_x + 6\hat{x}. \tag{3.287}$$

Consequently,

$$\hat{D}_x^2\hat{x}^3 = \hat{x}^3\hat{D}_x^2 + 6\hat{x}^2\hat{D}_x + 6\hat{x}. \tag{3.288}$$

Note that the $\hat{x}$-operators precede the $\hat{D}_x$-operators, at the R.H.S. In the last term $6\hat{x} = 6\hat{x}\hat{D}_x^0$, the operator $\hat{D}_x$ appears in disguise.

Thus the commutator relationship $[\hat{D}_x, \hat{x}] = \hat{I}$ and the general rules,

$$[AC, B] = A[C, B] + [A, B]C \tag{3.289a}$$
$$[A, CB] = C[A, B] + [A, C]B, \tag{3.289b}$$

and the uncompromising discipline of respecting the order of operators, are all what is needed to operate with powers of $\hat{D}_x$ and $\hat{x}$ as if they were algebraic terms. Furthermore, employing Formal Taylor Series Expansions of the functions $f(\cdot)$ and $g(\cdot)$ in $\left[f(\hat{D}_x), g(\hat{x})\right]$ allows "converting" terms of the form $f(\hat{D}_x)g(\hat{x})$ to terms of the form $g(\hat{x})f(\hat{D}_x)$, and *vice versa*.   □

3.16   Concluding remarks

The discussion in this chapter was based on three essential ideas: (i) Formal Taylor Series Expansion, (ii) canonical commutator relationship, and (iii) proof by mathematical induction. Many far reaching results could be inferred from the property of the canonical commutator relationship. Thereby, the internal structures of the operators, e.g., **A** and **B** satisfying the canonical commutator relationship $[\mathbf{A}, \mathbf{B}] = \mathbf{I}$, did not constitute the major objects of investigation. The following chapter is devoted to this task, with a strong desire to generalize the standard relationships. This ambitious goal besides being relevant on its own sake is also critically important in clarifying the prevailing content. The reader might have realized the impetus of zooming in and zooming out strategies for gaining insight. Even more importantly, the reader might have experienced that the metaphor of opening the box, peering into the box, scrutinizing and evaluating the content of the box is a powerful instrument in getting inspiration for how to augment or extend the standard contents.

The references [1] and [2] present a good starting point to further the ideas developed in this chapter, or view them from alternative vantage points. The reader will recognize that the pseudo axiomatic approach developed and pursued in this chapter differs considerably from the existing expositions. It is hoped that the style of presentation in this chapter, and for this matter, in the entire book, provides an overarching bird's view by introducing a meta language, a language which allows to talk about the prevailing language itself. This way, the existing relationships can be understood more clearly, and, ideally, novel relationships can be established. References [3] and [4] (written in German) contain a large number of elementary and interesting problems along with their solutions.

References

[1]   Barry G.A., *Algebraic Approach to Simple Quantum Systems*, Springer Verlag, 1994.
[2]   Willi-Hans S. and Yorik H., *Problems & Solutions in Quantum Computing & Quantum Information*, World Scientific, 2004.
[3]   Dietrich G., *Übungsaufgaben zur Quantumtheorie*, Verlag Karl Thiemig München, 1975 (in German).
[4]   Hans W.W., *Praktische Quantummechanik: Eine Vorlesung*, Rombach Hochschul Paperback, 1972 (in German).

Chapter 4
Generalized creation and annihilation operators

4.1 A brief guide through the chapter

Standard introduction of the differential operator d/dx requires the "vague" idea of evaluating a function at a given point. The discussion in this chapter starts with rendering this crucial notion precise. The concepts of the Dirac delta function $\delta(\cdot)$, and in particular, its γ-parametrized representation $\delta_\gamma(\cdot)$, loom large in interpreting what the evaluation of a function at a certain point means. Varying γ, the γ-parametrized representations $\delta_\gamma(\cdot)$ constitute a continuous sequence of well-behaved arbitrarily-differentiable functions. The presented analysis promises to provide a fresh perspective of the ideas underlying the first-, and consequently, the higher-order differential operators. The n-order differential operators d^n/dx^n are defined in terms of integrals involving n-order derivatives of $\delta_\gamma(\cdot)$. The developed formulations are algorithmic, and provide easy-to-implement recipes for numerical calculations. Subsequently, new symbols have been introduced to simplify and unify the calculation of derivatives of the products of functions, e.g., $f(x)g(x)$ and $f(x)g(x)h(x)$. As elsewhere in this book, the introduced procedures are based on inductive reasoning: the ideas propel their own further developments. The discussion continues with thoroughly examining several important commutator relationships involving the differential operator $\hat{D}$ (standing for d/dx) and the position operator $\hat{x}$, in one-, two-, and three-spatial dimensions. The presentation in this chapter culminates in the introduction of an original versatile scheme for the construction of a class of novel generalized annihilation operators $(\hat{b})$ and creation operators $(\hat{b}^\dagger)$. The generalized formula includes the canonical commutation relationship $[\hat{D}, \hat{x}] = \hat{\mathbb{I}}$, and the widely used fundamental commutation relationship $[\hat{b}, \hat{b}^\dagger] = \hat{\mathbb{I}}$, as simplest possible realizations. Consequently, two further major generalizations are presented. These findings are summarized and presented as a theorem alongside its proof. The entirety of the development in this chapter has its genesis in the position operator $\hat{x}$, the differential operator $\hat{D}$, and the canonical commutation relationship, $[\hat{A}, \hat{B}] = \hat{A}\hat{B} - \hat{B}\hat{A} = \hat{\mathbb{I}}$. The fact that the Formal Taylor Series Expansion plays an important role in these developments should not be surprising by now. The Formal Taylor Series Expansion breaks arbitrary functions down into ingredients which can readily be interpreted, augmented, and incorporated into commutators. One last insight should be brought to the attention of the reader. The notions of the differentiation and the commutation both require building the difference of two terms. The existence of the minus sign leads to the "astonishing proliferation" of results and relationships. The reader would benefit from the discussion in this chapter greatly by keeping in their minds the following fact: the presence of the minus sign in the definitions of the differentiation and the commutation brings along inherent self-regularization and self-renormalization effects. The large number of solved problems is aimed to make this idea clear. Even though not covered in this chapter for completeness it should be mentioned that the Lagrangian (as the difference between the kinetic energy and the potential energy) also plays a significant role in quantum physics.

4.2 Differential operators

The differential (derivative) operators play an all-embracing fundamental role in engineering and science. It is justified to scrutinize their definitions, properties, and representations. The functions, the derivatives of which are investigated, are assumed to possess the required smoothness and differentiability requirements. They are also expected to fall off to zero at $\pm\infty$ rapidly enough such that the integrals which arise in the calculations are integrable. In fact, the functions merely serve as Formal Aids. In order to investigate the structure and properties of an operator $\mathcal{L}$, the operator is applied to a test function $f(x)$. It is implicitly assumed that $f(x)$ is in the definition range of $\mathcal{L}$ and thus that $f(x)$ possess all the properties required for carrying out the intended processes. Having completed the calculations and manipulations, the function $f(x)$ is subsequently dismissed.

The first derivative: The first derivative of the function $f(x)$, denoted by $\frac{d}{dx}f(x)$, $f'(x)$, or $f^{(1)}(x)$, is defined in terms of an $\epsilon_1 \rightarrow 0^+$ limit process,

$$\frac{d}{dx}f(x) = \lim_{\epsilon_1 \rightarrow 0^+} \frac{1}{\epsilon_1}[f(x + \epsilon_1) - f(x)]. \tag{4.1}$$

Note that the coefficients of the terms within the square brackets, $\{+1, -1\}$, add up to zero. The universality and significance of this observation becomes clear in the course of discussion in this chapter. The structure of the expression at the R.H.S. of (4.1) prompts the first two of the many remarks which will follow.

Remark: The test function as a formal aid.

The sifting property of the Dirac delta function,

$$\int_{-\infty}^{\infty} d\xi\, \delta(\xi - x)f(\xi) = f(x), \tag{4.2}$$

can be employed to express the test function $f(x)$ evaluated at $x + \epsilon_1$ and x. Substituting the corresponding integral expressions for $f(x + \epsilon_1)$ and $f(x)$ in (4.1),

$$\frac{d}{dx}f(x) = \lim_{\epsilon_1 \rightarrow 0^+} \frac{1}{\epsilon_1}\{ \int_{-\infty}^{\infty} d\xi\, \delta\,[\xi - (x + \epsilon_1)]f(\xi) - \int_{-\infty}^{\infty} d\xi\, \delta\,(\xi - x)f(\xi)\}. \tag{4.3}$$

Factoring out $\int_{-\infty}^{\infty} d\xi$ to the left and $f(\xi)$ to the right,

$$\frac{d}{dx}f(x) = \lim_{\epsilon_1 \rightarrow 0^+} \frac{1}{\epsilon_1} \int_{-\infty}^{\infty} d\xi\, \{\delta\,[\xi - (x + \epsilon_1)] - \delta\,(\xi - x)\,\}f(\xi). \tag{4.4}$$

Since $f(\cdot)$ is arbitrary, in virtue of being a test function, its role as a formal aid is accomplished and thus can be dismissed,

$$\frac{d}{dx} \equiv \lim_{\epsilon_1 \rightarrow 0^+} \frac{1}{\epsilon_1} \int_{-\infty}^{\infty} d\xi\, \{\delta\,[\xi - (x + \epsilon_1)] - \delta\,(\xi - x)\,\}. \tag{4.5}$$

This integral expression involving the Dirac delta function is an alternative representation of $\frac{d}{dx}$ which explicates the sampling processes involved.

$\square$

Remark: Verbalization and articulation as tools for rendering contents obvious.

Verbalization and articulation of observed relationships and established facts are effective techniques for investigating the anatomy of algorithms, identifying the limits of their applicability, extending the scope of their validity, and ultimately, automating their implementation in digital computers. Verbalization and articulation also serve as tools for identifying possible shortfalls, gaps, and inconsistencies in formulations. Verbalization and articulation are effective vehicles for zooming into the internal structure of reasoning itself, scrutinizing the developed concepts, and thus enabling deep understanding. Verbalization and articulation open the Pandora's box and prompt one to grant a deep look into the machinery of what is going on in the box, and consequentially, hone the creativity of inquisitive minds.

Equation (4.1) provides an elementary example. The expression at the R.H.S. is the defining process of $(d/dx)f(x)$, and comprises several steps: (i) consider x and the neighboring point $x + \epsilon_1$ for a vanishingly small positive ϵ_1; (ii) consider the object (here, the function $f(x)$) onto which the operator d/dx is applied; (iii) evaluate $f(x)$ at $x + \epsilon_1$ and x (with an emphasis on the yet vague notion of "evaluation"); (iv) build the difference $f(x + \epsilon_1) - f(x)$; (v) consider the difference $(x + \epsilon_1) - x = \epsilon_1$; (vi) divide $f(x + \epsilon_1) - f(x)$ by ϵ_1; (vii) determine the limit of the resulting ratio with $\epsilon_1 > 0$ approaching zero.

The idea is to internalize the habit of communicating with an imaginary Robot which merely can carry out clear sound valid unambiguous instructions. Verbalization and articulation encourage one to empty one's mind. What is written down, is all what it is: partially writing down the ideas and partially holding some aspects of the formulation in the mind is a common mistake and a source of missing critically important details, and consequently, misguided conclusions. Learning to convey the contents of equations and formulae fully is revealing. To elaborate the latter point, consider the R.H.S. in (4.1). Upon assumption, $\epsilon_1 > 0$. Thus for every $x \in \mathbb{R}$, $x + \epsilon_1 > x$. The difference $f(x + \epsilon_1) - f(x)$ compares the value of $f(\cdot)$ at an advanced point $x + \epsilon_1$ to the value of $f(\cdot)$ at a preceding point. Assume $f(x + \epsilon_1) > f(x)$, and thus $f(x + \epsilon_1) - f(x) > 0$. Since $\epsilon_1 > 0$, $[f(x + \epsilon_1) - f(x)]/\epsilon_1 > 0$ for any ϵ_1. Thus, implying $(d/dx)f(x) > 0$. Conversely, in virtue of $\epsilon_1 > 0$, the condition $(d/dx)f(x) > 0$ implies $f(x + \epsilon_1) > f(x)$. Thus, the function is ascending with increasing x.

Verbalization and articulation maybe viewed as holy grails of understanding; they are diagonally opposite to shallow reading, and force one to see with the mind's eye rather than simply seeing and memorizing. The following equation and many other examples in this text will demonstrate and hopefully convince the reader of the powers of verbalization and articulation in designing algorithms.

□

Remark: Spreading and shifting the Dirac delta function.

Consider the result just obtained,

$$\frac{d}{dx} \equiv \lim_{\epsilon_1 \to 0^+} \frac{1}{\epsilon_1} \int_{-\infty}^{\infty} d\xi \left\{ \delta \left[\xi - (x + \epsilon_1) \right] - \delta \left(\xi - x \right) \right\}. \tag{4.6}$$

This formula utilizes the sifting property of the Dirac delta function. As such, it is merely a convenient bookkeeping vehicle. It renders the following idea a rigorous formal meaning: "Evaluate the function $f(x)$ at the neighboring points $\xi - x$ and $\xi - x - \epsilon_1$ with $\xi - x > \xi - x - \epsilon_1$, build the difference between the two function values, and divide the resulting difference by ϵ_1." The full power of this formula becomes manifest when the two Dirac delta functions in (4.6) are γ-parametrized, i.e., when $\delta_\gamma(\cdot)$ is used, with $\delta_\gamma(\cdot)$ approaching $\delta(\cdot)$ for $\gamma \to 0^+$,

$$\left(\frac{d}{dx} \right)_\gamma \equiv \lim_{\epsilon_1 \to 0^+} \frac{1}{\epsilon_1} \int_{-\infty}^{\infty} d\xi \left\{ \delta_\gamma \left[\xi - (x + \epsilon_1) \right] - \delta_\gamma \left(\xi - x \right) \right\}. \tag{4.7}$$

As elaborated in the previous chapter, the γ-parametrized representations $\delta_\gamma(\cdot)$ are arbitrarily smooth (differentiable) functions for any $\gamma > 0$. The γ-parametrized representations $\delta_\gamma(\cdot)$ are also integrable in ordinary Riemannian sense (in closed-form or numerically), and offer great utility in the regularization of ill-conditioned problems. It should also be pointed out that the γ-parametrization with γ written as a subindex, plays a substantially different role than the ϵ_1-translation with ϵ_1 appearing in the argument of one of the Dirac delta functions. Generally, the parameter appearing as the subindex regulates the "spreading," the "width" of $\delta_\gamma(x - \epsilon_1)$, while ϵ_1 arising in the argument is a measure of the "translation," "shift" of $\delta_\gamma(x)$, along the x-axis. In summary, any parametrized realization $\delta_\gamma(\cdot)$ can safely be viewed as an ordinary function, whereas the Dirac delta functions $\delta(\cdot)$ must be seen merely as a powerful symbol. With this understanding the next remark introduces a further interesting symbolic notation.

□

Remark: Symbolic representation of the differential d/dx.

In view of the property $\epsilon_1 > 0$, the inequality $\xi - (x + \epsilon_1) = \xi - x - \epsilon_1 < \xi - x$ holds true. Consequently, the order of the δ_γ-functions in (4.7) must be changed, if one wishes to employ the derivative formula for the γ-parametrized function $\delta_\gamma(\cdot)$,

$$\left(\frac{d}{dx}\right)_\gamma = -\lim_{\epsilon_1 \to 0^+} \frac{1}{\epsilon_1} \int_{-\infty}^{\infty} d\xi \left\{ \delta_\gamma\left(\xi - x\right) - \delta_\gamma\left[\xi - (x + \epsilon_1)\right] \right\}. \tag{4.8}$$

Due to the well-behavedness of $\delta_\gamma(\cdot)$, the order of integration and the limit process can be exchanged,

$$\left(\frac{d}{dx}\right)_\gamma = -\int_{-\infty}^{\infty} d\xi \underbrace{\lim_{\epsilon_1 \to 0^+} \frac{1}{\epsilon_1} \left\{ \delta_\gamma\left(\xi - x\right) - \delta_\gamma\left[\xi - (x + \epsilon_1)\right] \right\}}. \tag{4.9}$$

The under-braced term being an expression for the derivative $\delta_\gamma^{(1)}\left(\xi - x\right) \left(= \frac{d}{d\xi}\delta_\gamma\left(\xi - x\right)\right)$,

$$\left(\frac{d}{dx}\right)_\gamma = -\int_{-\infty}^{\infty} d\xi \delta_\gamma^{(1)}\left(\xi - x\right). \tag{4.10}$$

The representation at the R.H.S. permits obtaining a γ-regularized expression for the derivative of the function $f(x)$ with respect of x, i.e., $\frac{d}{dx}f(x)$,

$$\left(\frac{d}{dx}\right)_\gamma f(x) = -\int_{-\infty}^{\infty} d\xi \delta_\gamma^{(1)}\left(\xi - x\right) f(\xi). \tag{4.11}$$

In the limit $\gamma \to 0^+$,

$$\frac{d}{dx}f(x) = -\int_{-\infty}^{\infty} d\xi \delta^{(1)}\left(\xi - x\right) f(\xi). \tag{4.12}$$

$\square$

Problem: Show that the integral at the R.H.S. of (4.12) is an expression for $\frac{d}{dx}f(x)$.

Solution: Considering $\delta^{(1)}\left(\xi - x\right) = \frac{d}{d\xi}\delta\left(\xi - x\right)$, and appealing to integration by parts,

$$\frac{d}{dx}f(x) = -\left[\int_{-\infty}^{\infty} d\xi \frac{d}{d\xi}\delta\left(\xi - x\right) \underbrace{f(\xi)}\right] \tag{4.13a}$$

$$= -\left[\delta\left(\xi - x\right)f(\xi)\Big|_{-\infty}^{\infty} - \int_{-\infty}^{\infty} d\xi \delta\left(\xi - x\right) \frac{d}{d\xi}f(\xi)\right]. \tag{4.13b}$$

Since the test function $f(x)$ must vanish at $\pm\infty$, the first term at the R.H.S. does not contribute to the result. Thus,

$$\frac{d}{dx}f(x) = \int_{-\infty}^{\infty} d\xi \delta\left(\xi - x\right) \frac{d}{d\xi}f(\xi) \tag{4.14a}$$

$$= \frac{d}{dx}f(x). \tag{4.14b}$$

∎

The second derivative: The second derivative is the derivative of the first derivative,

$$\frac{d^2}{dx^2}f(x) \overset{(1)}{=} \frac{d}{dx}\left(\frac{d}{dx}f(x)\right) \tag{4.15a}$$

$$\overset{(2)}{=} \frac{d}{dx}\left(\lim_{\epsilon_1 \to 0}\frac{1}{\epsilon_1}[f(x+\epsilon_1)-f(x)]\right) \tag{4.15b}$$

$$\overset{(3)}{=} \lim_{\epsilon_2 \to 0}\frac{1}{\epsilon_2}\left\{\left(\lim_{\epsilon_1 \to 0}\frac{1}{\epsilon_1}[f(x+\epsilon_2+\epsilon_1)-f(x+\epsilon_2)]\right)\right.$$
$$\left. -\left(\lim_{\epsilon_1 \to 0}\frac{1}{\epsilon_1}[f(x+\epsilon_1)-f(x)]\right)\right\} \tag{4.15c}$$

$$\overset{(4)}{=} \lim_{\epsilon_2 \to 0}\lim_{\epsilon_1 \to 0}\frac{1}{\epsilon_2\epsilon_1}[f(x+\epsilon_2+\epsilon_1)-f(x+\epsilon_2)-f(x+\epsilon_1)+f(x)]. \tag{4.15d}$$

The transition (1) is the implementation of the definition of the second derivative.

The transition (2) is the implementation of $\frac{d}{dx}f(x)$. It requires evaluating the "term" in front of $\frac{d}{dx}$, i.e., $f(x)$, at the two neighboring points $x+\epsilon_1$ and x, resulting in $f(x+\epsilon_1)$ and $f(x)$.

The transition (3) is the implementation of $\frac{d}{dx}\left(\lim_{\epsilon_1 \to 0}\frac{1}{\epsilon_1}[f(x+\epsilon_1)-f(x)]\right)$. It requires evaluating the "term" in front of $\frac{d}{dx}$; i.e., $\left(\lim_{\epsilon_1 \to 0}\frac{1}{\epsilon_1}[f(x+\epsilon_1)-f(x)]\right)$ at the two neighboring points $x+\epsilon_2$ and x, resulting in $\left(\lim_{\epsilon_1 \to 0}\frac{1}{\epsilon_1}[f(x+\epsilon_2+\epsilon_1)-f(x+\epsilon_2)]\right)$ and $\left(\lim_{\epsilon_1 \to 0}\frac{1}{\epsilon_1}[f(x+\epsilon_1)-f(x)]\right)$.

The transition (4) is the manifestation of the linearity of the limit process.

Note that the coefficients of the terms within the square brackets in (4.15d), $\{+1, -1, -1, +1\}$, add up to zero.

The last equation is symmetric with respect to ϵ_1 and ϵ_2: the first term $f(x+\epsilon_2+\epsilon_1)$ involves $\epsilon_2 + \epsilon_1$, and thus ϵ_2 and ϵ_1 on equal footing. The second term $f(x+\epsilon_2)$ and the third term $f(x+\epsilon_1)$ involve ϵ_2 and ϵ_1, respectively. The fourth term $f(x)$ involves neither ϵ_2 nor ϵ_1. Furthermore, $\frac{1}{\epsilon_2\epsilon_1} = \frac{1}{\epsilon_1\epsilon_2}$. Thus, the order of the limit processes is immaterial.

Employing the $\epsilon_1 \leftrightarrow \epsilon_2$ symmetry, carrying out the limit processes with respect to ϵ_1 first, and then with respect to ϵ_2,

$$\frac{d^2}{dx^2}f(x) = \lim_{\epsilon_2 \to 0}\lim_{\epsilon_1 \to 0}\frac{1}{\epsilon_2\epsilon_1}[f(x+\epsilon_2+\epsilon_1)-f(x+\epsilon_2)-f(x+\epsilon_1)+f(x)] \tag{4.16a}$$

$$= \lim_{\epsilon_2 \to 0}\lim_{\epsilon_1 \to 0}\frac{1}{\epsilon_2\epsilon_1}\{[f(x+\epsilon_2+\epsilon_1)-f(x+\epsilon_2)]-[f(x+\epsilon_1)-f(x)]\} \tag{4.16b}$$

$$= \lim_{\epsilon_2 \to 0}\frac{1}{\epsilon_2}\left\{\underbrace{\lim_{\epsilon_1 \to 0}\frac{1}{\epsilon_1}[f(x+\epsilon_2+\epsilon_1)-f(x+\epsilon_2)]}_{f^{(1)}(x+\epsilon_2)}-\underbrace{\lim_{\epsilon_1 \to 0}\frac{1}{\epsilon_1}[f(x+\epsilon_1)-f(x)]}_{f^{(1)}(x)}\right\} \tag{4.16c}$$

$$= \lim_{\epsilon_2 \to 0}\frac{1}{\epsilon_2}[f^{(1)}(x+\epsilon_2)-f^{(1)}(x)] \tag{4.16d}$$

$$= f^{(2)}(x). \tag{4.16e}$$

Exploiting the $\epsilon_1 \leftrightarrow \epsilon_2$ symmetry, carrying out the limit processes with respect to ϵ_2 first, and then with respect to ϵ_1,

$$\frac{d^2}{dx^2}f(x) = \lim_{\epsilon_1 \to 0}\lim_{\epsilon_2 \to 0}\frac{1}{\epsilon_1\epsilon_2}[f(x+\epsilon_1+\epsilon_2) - f(x+\epsilon_1) - f(x+\epsilon_2) + f(x)] \tag{4.17a}$$

$$= \lim_{\epsilon_1 \to 0}\lim_{\epsilon_2 \to 0}\frac{1}{\epsilon_1\epsilon_2}\{[f(x+\epsilon_1+\epsilon_2) - f(x+\epsilon_1)] - [f(x+\epsilon_2) - f(x)]\} \tag{4.17b}$$

$$= \lim_{\epsilon_1 \to 0}\frac{1}{\epsilon_1}\left\{\underbrace{\lim_{\epsilon_2 \to 0}\frac{1}{\epsilon_2}[f(x+\epsilon_1+\epsilon_2) - f(x+\epsilon_1)]}_{f^{(1)}(x+\epsilon_1)} - \underbrace{\lim_{\eta \to 0}\frac{1}{\epsilon_2}[f(x+\epsilon_2) - f(x)]}_{f^{(1)}(x)}\right\} \tag{4.17c}$$

$$= \lim_{\epsilon_1 \to 0}\frac{1}{\epsilon_1}\left[f^{(1)}(x+\epsilon_1) - f^{(1)}(x)\right] \tag{4.17d}$$

$$= f^{(2)}(x). \tag{4.17e}$$

A further observation is that nothing prevents one from setting $\epsilon_2 = \epsilon_1$ in (4.15d),

$$\frac{d^2}{dx^2}f(x) = \lim_{\epsilon_1 \to 0}\frac{1}{\epsilon_1^2}[f(x+2\epsilon_1) - 2f(x+\epsilon_1) + f(x)]. \tag{4.18}$$

Note that the coefficients of the terms within the square brackets, $\{+1, -2, +1\}$, add up to zero.

Remark: Expressing d^2/dx^2 in terms of the Dirac delta functions.

Consider (4.17a), which is reproduced here, and expressed in terms of the Dirac delta functions,

$$\frac{d^2}{dx^2}f(x) = \lim_{\epsilon_1 \to 0^+}\lim_{\epsilon_2 \to 0^+}\frac{1}{\epsilon_1\epsilon_2}[f(x+\epsilon_1+\epsilon_2) - f(x+\epsilon_1) - f(x+\epsilon_2) + f(x)] \tag{4.19a}$$

$$= \lim_{\epsilon_1 \to 0^+}\lim_{\epsilon_2 \to 0^+}\frac{1}{\epsilon_1\epsilon_2}\left\{\int_{-\infty}^{\infty} d\xi\,\delta[\xi - (x+\epsilon_1+\epsilon_2)]f(\xi)\right.$$

$$- \int_{-\infty}^{\infty} d\xi\,\delta[\xi - (x+\epsilon_1)]f(\xi)$$

$$- \int_{-\infty}^{\infty} d\xi\,\delta[\xi - (x+\epsilon_2)]f(\xi)$$

$$\left.+ \int_{-\infty}^{\infty} d\xi\,\delta(\xi - x)f(\xi)\right\}. \tag{4.19b}$$

Factoring out $f(\xi)$, removing the round parentheses in the arguments of the Dirac delta functions, and reorganizing terms,

$$\frac{d^2}{dx^2}f(x) = \lim_{\epsilon_1 \to 0^+} \lim_{\epsilon_2 \to 0^+} \frac{1}{\epsilon_1 \epsilon_2} \left\{ -\int_{-\infty}^{\infty} d\xi\, \delta\,(\xi - x - \epsilon_1) + \int_{-\infty}^{\infty} d\xi\, \delta\,(\xi - x - \epsilon_1 - \epsilon_2) \right.$$
$$\left. + \int_{-\infty}^{\infty} d\xi\, \delta\,(\xi - x) - \int_{-\infty}^{\infty} d\xi\, \delta\,(\xi - x - \epsilon_2) \right\} f(\xi). \qquad (4.20)$$

Focus on the first and the second integrals at the R.H.S. Note that the argument $\xi - x - \epsilon_1$ is larger than the argument $\xi - x - \epsilon_1 - \epsilon_2$ by ϵ_2 (> 0). Next focus on the third and the fourth integrals at the R.H.S. Note that the argument $\xi - x$ is larger than the argument $\xi - x - \epsilon_2$ by ϵ_2. Considering these observations and bringing the limit $\epsilon_2 \to 0^+$ inside the integrals,

$$\frac{d^2}{dx^2}f(x) = \lim_{\epsilon_1 \to 0^+} \frac{1}{\epsilon_1} \left\{ -\int_{-\infty}^{\infty} d\xi \lim_{\epsilon_2 \to 0^+} \frac{1}{\epsilon_2} \underbrace{[\delta\,(\xi - x - \epsilon_1) - \delta\,(\xi - x - \epsilon_1 - \epsilon_2)]} \right.$$
$$\left. + \int_{-\infty}^{\infty} d\xi \lim_{\epsilon_2 \to 0^+} \frac{1}{\epsilon_2} \underbrace{[\delta\,(\xi - x) - \delta\,(\xi - x - \epsilon_2)]} \right\} f(\xi). \qquad (4.21)$$

Transferring $\lim_{\epsilon_2 \to 0^+}$ into the $\int_{-\infty}^{\infty} d\xi$ integrals can be made rigorously valid, by remembering that the Dirac delta functions stand symbolically for the limit of a sequence of smooth γ-parametrized functions.

The first under-braced term is the derivative of the Dirac delta function $\delta\,(\xi - x - \epsilon_1)$ with respect to ξ, while the second under-braced term is the derivative of the Dirac delta function $\delta\,(\xi - x)$ with respect to ξ,

$$\frac{d^2}{dx^2}f(x) = \lim_{\epsilon_1 \to 0^+} \frac{1}{\epsilon_1} \left\{ -\int_{-\infty}^{\infty} d\xi\, \delta^{(1)}\,(\xi - x - \epsilon_1) + \int_{-\infty}^{\infty} d\xi\, \delta^{(1)}\,(\xi - x) \right\} f(\xi). \qquad (4.22)$$

The argument $\xi - x$ is larger than the argument $\xi - x - \epsilon_1$ by ϵ_1 (> 0). Rearranging and transferring the limit $\epsilon_1 \to 0^+$ under the integral,

$$\frac{d^2}{dx^2}f(x) = \int_{-\infty}^{\infty} d\xi \lim_{\epsilon_1 \to 0^+} \frac{1}{\epsilon_1} \underbrace{\left[\delta^{(1)}\,(\xi - x) - \delta^{(1)}\,(\xi - x - \epsilon_1)\right]} f(\xi). \qquad (4.23)$$

The under-braced terms is equal to the derivative of $\delta^{(1)}\,(\xi - x)$, i.e., it is the second derivative of $\delta\,(\xi - x)$. Thus

$$\frac{d^2}{dx^2}f(x) = \int_{-\infty}^{\infty} d\xi\, \delta^{(2)}\,(\xi - x) f(\xi). \qquad (4.24)$$

With $f(x)$ being a test function, the following universal representation can be concluded,

$$\frac{d^2}{dx^2} \equiv \int_{-\infty}^{\infty} d\xi \delta^{(2)}(\xi - x). \tag{4.25}$$

With rigorous understanding that the γ-parametrized counterparts, developed further above, are equivalent,

$$\left(\frac{d^2}{dx^2}\right)_\gamma \equiv \int_{-\infty}^{\infty} d\xi \delta_\gamma^{(2)}(\xi - x). \tag{4.26}$$

It can be shown that (4.25) and (4.26), beyond their theoretical and aesthetical appeal, possess great utility in theory development as well as in designing customized algorithms for the numerical treatment of ill-posed problems.

□

Problem: Show that $\int_{-\infty}^{\infty} d\xi \delta^{(2)}(\xi - x)$ induces the second derivative with respect to x, i.e., $\frac{d^2}{dx^2}$.

Solution: The solution (the proof) consists of applying $\int_{-\infty}^{\infty} d\xi \delta^{(2)}(\xi - x)$ onto a test function $f(x)$, rolling over the two derivatives from $\delta^{(2)}(\xi - x)$ onto the test function, and utilizing the sifting property of the Dirac delta function. It should also borne in mind that the "boundary terms" are zero due to the localization property of the Dirac delta function: the Dirac delta functions and their derivatives to any order are zero at $\pm\infty$. Following this recipe, the individual steps must be self-explanatory.

$$\int_{-\infty}^{\infty} d\xi \delta^{(2)}(\xi - x) f(\xi) = \left[\delta^{(1)}(\xi - x) f(\xi)\right]\Big|_{-\infty}^{\infty} - \int_{-\infty}^{\infty} d\xi \delta^{(1)}(\xi - x) f^{(1)}(\xi) \tag{4.27a}$$

$$= -\int_{-\infty}^{\infty} d\xi \delta^{(1)}(\xi - x) f^{(1)}(\xi) \tag{4.27b}$$

$$= -\left\{\left[\delta^{(0)}(\xi - x) f^{(1)}(\xi)\right]\Big|_{-\infty}^{\infty} - \int_{-\infty}^{\infty} d\xi \delta^{(0)}(\xi - x) f^{(2)}(\xi)\right\} \tag{4.27c}$$

$$= \int_{-\infty}^{\infty} d\xi \delta(\xi - x) f^{(2)}(\xi) \tag{4.27d}$$

$$= \int_{-\infty}^{\infty} d\xi \delta(\xi - x) \left[\frac{d^2}{d\xi^2} f(\xi)\right] \tag{4.27e}$$

$$= \frac{d^2}{dx^2} f(x). \tag{4.27f}$$

■

Remark: Evaluating a function at a point, singling out a function in space, and the collapse of wavefunctions.

In the previous chapter, the construction of problem-specific Dirac delta functions occupied a prominent space, by emphasizing the γ-parametrization of the underlying sequence of functions. There, the resolution of identity dealt with the big picture associated with transforms and inverse transforms. In this chapter, the local behavior of functions is the focus, the differential operators. Delving much deeper on these insights might obscure the presentation. Thus, putting the above state of affairs on the record might suffice. The γ-parametrization can be considered as a powerful tool for the regularization and renormalization of illusive categories, such as divergences and infinities. To the curious reader: whether or not the γ-parametrization is the sought after formal mathematical vehicle to explain the collapse of the wave function remains in the realm of wide speculations. Picking up one eigenstate out of an infinity of possible eigenstates by an experiment can be considered as an act of evaluation not unlike the evaluation of a function at a certain point. The variational calculus, based on Lagrangian (the difference between the kinetic and potential energy), Feynman's path integral, leading to the optimum solution, i.e., picking up one function out of an infinity of possibilities, all allude to the process of evaluation. This book does not aim at any interpretation of quantum physics. It is just that digging deep, gaining insights, formulating hypotheses, and aiming to refute them are what inquisitive minds are supposed to do.

□

The third derivative: The third derivative can be obtained analogously.

$$\frac{d^3}{dx^3}f(x) \overset{(1)}{=} \frac{d}{dx}\left(\frac{d^2}{dx^2}f(x)\right) \tag{4.28a}$$

$$\overset{(2)}{=} \frac{d}{dx}\left(\lim_{\epsilon\to 0}\frac{1}{\epsilon^2}[f(x+2\epsilon)-2f(x+\epsilon)+f(x)]\right) \tag{4.28b}$$

$$\overset{(3)}{=} \lim_{\epsilon_3\to 0}\frac{1}{\epsilon_3}\left\{\lim_{\epsilon\to 0}\frac{1}{\epsilon^2}[f(x+\epsilon_3+2\epsilon)-2f(x+\epsilon_3+\epsilon)+f(x+\epsilon_3)]\right.$$
$$\left. -\lim_{\epsilon\to 0}\frac{1}{\epsilon^2}[f(x+2\epsilon)-2f(x+\epsilon)+f(x)]\right\} \tag{4.28c}$$

$$\overset{(4)}{=} \lim_{\epsilon_3\to 0}\frac{1}{\epsilon_3}\lim_{\epsilon\to 0}\frac{1}{\epsilon^2}[f(x+\epsilon_3+2\epsilon)-2f(x+\epsilon_3+\epsilon)+f(x+\epsilon_3)$$
$$-f(x+2\epsilon)+2f(x+\epsilon)-f(x)]. \tag{4.28d}$$

The transition (1) employs the definition of $\frac{d^3}{dx^3}$.

The transition (2) substitutes the expression for $\frac{d^2}{dx^2}f(x)$.

The transition (3) implements the definition of $\frac{d}{dx}$, by evaluating $\lim_{\epsilon\to 0}\frac{1}{\epsilon^2}[f(x+2\epsilon)-2f(x+\epsilon)+f(x)]$ at $x+\epsilon_3$ and x, taking the difference between them, dividing the resulting expression by ϵ_3, and letting ϵ_3 approach zero.

The transition (4) utilizes the linearity property of the limit process.

Setting $\epsilon_3 = \epsilon$,

$$\frac{d^3}{dx^3}f(x) = \lim_{\epsilon\to 0}\frac{1}{\epsilon^3}[f(x+3\epsilon)-2f(x+2\epsilon)+f(x+\epsilon)$$
$$-f(x+2\epsilon)+2f(x+\epsilon)-f(x)]. \tag{4.29}$$

Simplifying,

$$\frac{d^3}{dx^3}f(x) = \lim_{\epsilon\to 0}\frac{1}{\epsilon^3}[f(x+3\epsilon)-3f(x+2\epsilon)+3f(x+\epsilon)-f(x)]. \tag{4.30}$$

Note that the coefficients of the terms within the square brackets, $\{+1, -3, +3, -1\}$, add up to zero.

Problem: Consider the expression for $\frac{d^2}{dx^2}f(x)$ in (4.19a), which is reproduced here,

$$\frac{d^2}{dx^2}f(x) = \lim_{\epsilon_1\to 0^+}\lim_{\epsilon_2\to 0^+}\frac{1}{\epsilon_1\epsilon_2}[f(x+\epsilon_1+\epsilon_2)-f(x+\epsilon_1)-f(x+\epsilon_2)+f(x)]. \tag{4.31}$$

Here, $f(x)$ is a test function. Employing (4.31) deduce an expression for $\frac{d^3}{dx^3}$ in terms of the derivatives of the *Dirac delta function*.

Solution: The following steps are self-explanatory:

$$\frac{d^3}{dx^3}f(x) \overset{\text{def.}}{=} \frac{d}{dx}\left\{\frac{d^2}{dx^2}f(x)\right\} \tag{4.32a}$$

$$\overset{\text{def.}}{=} \lim_{\epsilon_3 \to 0^+} \frac{1}{\epsilon_3}\left\{\left(\frac{d^2}{dx^2}f\right)(x+\epsilon_3) - \left(\frac{d^2}{dx^2}f\right)(x)\right\} \tag{4.32b}$$

$$\overset{(4.31)}{=} \lim_{\epsilon_3 \to 0^+} \frac{1}{\epsilon_3}\left\{\lim_{\epsilon_1 \to 0^+}\lim_{\epsilon_2 \to 0^+} \frac{1}{\epsilon_1\epsilon_2}[f(x+\epsilon_1+\epsilon_2+\epsilon_3) - f(x+\epsilon_1+\epsilon_3) - f(x+\epsilon_2+\epsilon_3) + f(x+\epsilon_3)]\right.$$

$$\left. - \lim_{\epsilon_1 \to 0^+}\lim_{\epsilon_2 \to 0^+} \frac{1}{\epsilon_1\epsilon_2}[f(x+\epsilon_1+\epsilon_2) - f(x+\epsilon_1) - f(x+\epsilon_2) + f(x)]\right\}. \tag{4.32c}$$

Focus on (4.32b). The two terms in the curly brackets indicate the second derivative of $f(x)$ evaluated at $x+\epsilon_3$ and x, respectively.

Combining terms and simplifying,

$$\frac{d^3}{dx^3}f(x) = \lim_{\epsilon_1 \to 0^+}\lim_{\epsilon_2 \to 0^+}\lim_{\epsilon_3 \to 0^+} \frac{1}{\epsilon_1\epsilon_2\epsilon_3}$$
$$\times [f(x+\epsilon_1+\epsilon_2+\epsilon_3) - f(x+\epsilon_1+\epsilon_3) - f(x+\epsilon_2+\epsilon_3) + f(x+\epsilon_3)$$
$$- f(x+\epsilon_1+\epsilon_2) + f(x+\epsilon_1) + f(x+\epsilon_2) - f(x)]. \tag{4.33}$$

Reordering,

$$\frac{d^3}{dx^3}f(x) = \lim_{\epsilon_1 \to 0^+}\lim_{\epsilon_2 \to 0^+}\lim_{\epsilon_3 \to 0^+} \frac{1}{\epsilon_1\epsilon_2\epsilon_3}$$
$$\times [-f(x+\epsilon_1+\epsilon_2) + f(x+\epsilon_1+\epsilon_2+\epsilon_3)$$
$$+ f(x+\epsilon_1) - f(x+\epsilon_1+\epsilon_3)$$
$$+ f(x+\epsilon_2) - f(x+\epsilon_2+\epsilon_3)$$
$$- f(x) + f(x+\epsilon_3)]. \tag{4.34}$$

Remark: Symmetric function values.

The eight function values in (4.34) can be arranged in various forms to more easily exhibit the underlying symmetry of the ϵ-terms. One possibility is the following:

$$f(x+\epsilon_1+\epsilon_2+\epsilon_3)$$

$$-f(x+\epsilon_1+\epsilon_2) \qquad -f(x+\epsilon_1+\epsilon_3) \qquad -f(x+\epsilon_2+\epsilon_3)$$

$$f(x+\epsilon_1) \qquad\qquad f(x+\epsilon_2) \qquad\qquad f(x+\epsilon_3) \tag{4.35}$$

$$-f(x)$$

This arrangement reveals the symmetry with respect to any of the exchanges $\epsilon_1 \leftrightarrow \epsilon_2$, $\epsilon_1 \leftrightarrow \epsilon_3$, and $\epsilon_2 \leftrightarrow \epsilon_3$. Furthermore, it can be observed that the patterns of terms in (4.35) along with their parities (signs) allow writing down the formula in (4.34) simply-by-inspection. It is also obvious that these considerations can be extended to any number of $\epsilon_1, \epsilon_2, \epsilon_3, \ldots$, and consequently, to any order of derivatives desired.

□

Introducing the Dirac delta functions to express function evaluated at eight distinct points,

$$
\frac{d^3}{dx^3}f(x) = \lim_{\epsilon_1 \to 0^+} \lim_{\epsilon_2 \to 0^+} \lim_{\epsilon_3 \to 0^+} \frac{1}{\epsilon_1 \epsilon_2 \epsilon_3}
$$

$$
\times \Bigg[- \int_{-\infty}^{\infty} d\xi\, \delta\left[\xi - (x + \epsilon_1 + \epsilon_2)\right] f(\xi) + \int_{-\infty}^{\infty} d\xi\, \delta\left[\xi - (x + \epsilon_1 + \epsilon_2) - \epsilon_3\right] f(\xi)
$$

$$
+ \int_{-\infty}^{\infty} d\xi\, \delta\left[\xi - (x + \epsilon_1)\right] f(\xi) - \int_{-\infty}^{\infty} d\xi\, \delta\left[\xi - (x + \epsilon_1) - \epsilon_3\right] f(\xi)
$$

$$
+ \int_{-\infty}^{\infty} d\xi\, \delta\left[\xi - (x + \epsilon_2)\right] f(\xi) - \int_{-\infty}^{\infty} d\xi\, \delta\left[\xi - (x + \epsilon_2) - \epsilon_3\right] f(\xi)
$$

$$
- \int_{-\infty}^{\infty} d\xi\, \delta(\xi - x) f(\xi) + \int_{-\infty}^{\infty} d\xi\, \delta(\xi - x - \epsilon_3) f(\xi) \Bigg].
$$

(4.36)

$$
\frac{d^3}{dx^3}f(x) = \lim_{\epsilon_1 \to 0^+} \lim_{\epsilon_2 \to 0^+} \frac{1}{\epsilon_1 \epsilon_2} \int_{-\infty}^{\infty} d\xi \lim_{\epsilon_3 \to 0^+} \frac{1}{\epsilon_3}
$$

$$
\times \Big\{ -\delta\left[\xi - (x + \epsilon_1 + \epsilon_2)\right] + \delta\left[\xi - (x + \epsilon_1 + \epsilon_2) - \epsilon_3\right]
$$

$$
+ \delta\left[\xi - (x + \epsilon_1)\right] - \delta\left[\xi - (x + \epsilon_1) - \epsilon_3\right]
$$

$$
+ \delta\left[\xi - (x + \epsilon_2)\right] - \delta\left[\xi - (x + \epsilon_2) - \epsilon_3\right]
$$

$$
- \delta(\xi - x) + \delta(\xi - x - \epsilon_3) \Big\} f(\xi).
$$

(4.37)

$$
\frac{d^3}{dx^3}f(x) = \lim_{\epsilon_1 \to 0^+} \lim_{\epsilon_2 \to 0^+} \frac{1}{\epsilon_1 \epsilon_2} \int_{-\infty}^{\infty} d\xi
$$

$$
\times \Big\{ -\delta^{(1)}\left[\xi - (x + \epsilon_1 + \epsilon_2)\right]
$$

$$
+ \delta^{(1)}\left[\xi - (x + \epsilon_1)\right]
$$

$$
+ \delta^{(1)}\left[\xi - (x + \epsilon_2)\right]
$$

$$
- \delta^{(1)}(\xi - x) \Big\} f(\xi).
$$

(4.38)

$$
\frac{d^3}{dx^3}f(x) = \lim_{\epsilon_1 \to 0^+} \frac{1}{\epsilon_1} \int_{-\infty}^{\infty} d\xi \lim_{\epsilon_2 \to 0^+} \frac{1}{\epsilon_2}
$$

$$
\times \Big\{ \delta^{(1)}\left[\xi - (x + \epsilon_1)\right] - \delta^{(1)}\left[\xi - (x + \epsilon_1) - \epsilon_2\right]
$$

$$
- \delta^{(1)}(\xi - x) + \delta^{(1)}(\xi - x - \epsilon_2) \Big\} f(\xi).
$$

(4.39)

$$
\frac{d^3}{dx^3}f(x) = \lim_{\epsilon_1 \to 0^+} \frac{1}{\epsilon_1} \int_{-\infty}^{\infty} d\xi \Big\{ \delta^{(2)}\left[\xi - (x + \epsilon_1)\right] - \delta^{(2)}(\xi - x) \Big\} f(\xi).
$$

(4.40)

$$\frac{d^3}{dx^3}f(x) = \int_{-\infty}^{\infty} d\xi \lim_{\epsilon_1 \to 0^+} \frac{1}{\epsilon_1}\left[-\delta^{(2)}(\xi - x) + \delta^{(2)}(\xi - x - \epsilon_1)\right]f(\xi). \tag{4.41}$$

$$\frac{d^3}{dx^3}f(x) = \int_{-\infty}^{\infty} d\xi \left[-\delta^{(3)}(\xi - x)\right]f(\xi). \tag{4.42}$$

Since $f(x)$ is a test function, it can be omitted. Thus,

$$\frac{d^3}{dx^3} \equiv \int_{-\infty}^{\infty} d\xi \left[(-1)^3 \delta^{(3)}(\xi - x)\right]. \tag{4.43}$$

The following lemma is stated for completeness. Its derivation does not add anything substantial to the understanding of the subject matter.

Lemma: *For any $n \in \mathbb{N}$, the symbolic,*

$$\frac{d^n}{dx^n} \equiv (-1)^n \int_{-\infty}^{\infty} d\xi \, \delta^{(n)}(\xi - x) \tag{4.44}$$

and the regular,

$$\left(\frac{d^n}{dx^n}\right)_\gamma \equiv (-1)^n \int_{-\infty}^{\infty} d\xi \, \delta_\gamma^{(n)}(\xi - x), \tag{4.45}$$

relationships hold valid.

∎

Problem: Consider the following equalities:

$$\frac{d}{dx}f(x) = \lim_{\epsilon \to 0} \frac{1}{\epsilon}\left[f(x + \epsilon) - f(x)\right] \tag{4.46a}$$

$$\frac{d^2}{dx^2}f(x) = \lim_{\epsilon \to 0} \frac{1}{\epsilon^2}\left[f(x + 2\epsilon) - 2f(x + \epsilon) + f(x)\right] \tag{4.46b}$$

$$\frac{d^3}{dx^3}f(x) = \lim_{\epsilon \to 0} \frac{1}{\epsilon^3}\left[f(x + 3\epsilon) - 3f(x + 2\epsilon) + 3f(x + \epsilon) - f(x)\right] \tag{4.46c}$$

$$\frac{d^4}{dx^4}f(x) = \lim_{\epsilon \to 0} \frac{1}{\epsilon^4}\left[f(x + 4\epsilon) - 4f(x + 3\epsilon) + 6f(x + 2\epsilon) - 4f(x + \epsilon) + f(x)\right] \tag{4.46d}$$

$$\frac{d^5}{dx^5}f(x) = \lim_{\epsilon \to 0} \frac{1}{\epsilon^5}\left[f(x + 5\epsilon) - 5f(x + 4\epsilon) + 10f(x + 3\epsilon) - 10f(x + 2\epsilon) + 5f(x + \epsilon) - f(x)\right] \tag{4.46e}$$

The first three relationships were shown further above. It is immediate to establish the last two equalities, following the same process. In view of the expressions at the R.H.S. in (4.46), and reasoning inductively, a recipe for the calculation of derivatives of a function to any arbitrary order suggests itself. The problem here is the explication of the recipe.

Solution: Observing the above induction steps ($n = 1, \ldots, 5$), the induction hypothesis can be formulated as follows:

- The nth derivative involves the fraction $\frac{1}{\epsilon^n}$ in front of the $\lim\limits_{\epsilon \to 0}$ symbol.
- The nth derivative involves $n + 1$ function values: $f(x + n\epsilon), f(x + (n - 1)\epsilon), \ldots, f(x + \epsilon), f(x)$.
- The coefficient of $f(x + n\epsilon)$ is one.
- The absolute value of the coefficient of $f(x)$ is one.
- The coefficients of the function values in each derivative, alternate in sign.
- Writing the absolute values of the coefficients of $\frac{d}{dx} f(x), \ldots, \frac{d^5}{dx^5} f(x)$ as a pyramid,

$$
\begin{array}{ccccccc}
1 & & -1 & & & & \\
1 & & -2 & & 1 & & \\
1 & & -3 & & 3 & & -1 \\
1 & & -4 & & 6 & & -4 & & 1 \\
1 & & -5 & & 10 & & -10 & & 5 & & -1
\end{array}
\tag{4.47}
$$

explicates the fact that coefficients in each row add up to zero.
- Writing the absolute values of the coefficients of $\frac{d}{dx} f(x), \ldots, \frac{d^5}{dx^5} f(x)$ as a pyramid,

$$
\begin{array}{cccccc}
1 & 1 & & & & \\
1 & 2 & 1 & & & \\
1 & 3 & 3 & 1 & & \\
1 & 4 & 6 & 4 & 1 & \\
1 & 5 & 10 & 10 & 5 & 1
\end{array}
\tag{4.48}
$$

the Pascal's triangle emerges. Starting from the 2nd row the terms sandwiched between the "guarding" 1^{s} can be calculated from the addition of the two neighboring terms in the preceding row immediately above them. The Pascal's triangle arises in countless contexts in mathematical physics and algebra. The properties of the Pascal's triangle are widely known, and they will not be discussed any further here. It suffices to mention two examples. For calculating 4 in the 4th row, 1 and 3 in the 3rd row must be added. For calculating 6 in the 4th row, 3 and 3 in the 3rd row must be added. Let C_n^k denote the kth term in the nth row, with $n = 2, 3, \ldots$ and $k \in [2, n]$. Then

$$
C_n^k = C_{n-1}^{k-1} + C_{n-1}^k \quad (n = 2, 3, \ldots \text{ and } k \in [2, n]).
$$

∎

Remark: There are several other fine features of derivative operators which are worth mentioning. To this end it is instructive to list the Formal Taylor Series Expansions of $f(x + n\epsilon)$ for $n = 1, 2, 3, 4$, explicitly. The reader should be reminded that in this text $f^{(n)}(x)$ refers to the nth derivative ($n \in \mathbb{N}_0$), with $f^{(0)}(x)$ standing for $f(x)$. Due to its paramount significance in quantum physics, as a universal tool, it is justified to consider the Formal Taylor Series Expansion in a slightly different perspective, as a means for factorization. Consider the Formal Taylor Series Expansion of $f(x + \eta)$, i.e.,

$$f(x+\eta) = \underbrace{\frac{1}{0!}f^{(0)}(x)}_{g_0(x)}\overbrace{(\eta)^0}^{h_0(\eta)} + \underbrace{\frac{1}{1!}f^{(1)}(x)}_{g_1(x)}\overbrace{(\eta)^1}^{h_1(\eta)} + \underbrace{\frac{1}{2!}f^{(2)}(x)}_{g_2(x)}\overbrace{(\eta)^2}^{h_2(\eta)} + \underbrace{\frac{1}{3!}f^{(3)}(x)}_{g_3(x)}\overbrace{(\eta)^3}^{h_3(\eta)} + \cdots . \tag{4.49}$$

Thus,

$$f(x+\eta) = g_0(x)h_0(\eta) + g_1(x)h_1(\eta) + g_2(x)h_2(\eta) + g_3(x)h_3(\eta) + \cdots , \tag{4.50}$$

or, more compactly,

$$f(x+\eta) = \sum_{m=0}^{\infty} g_m(x)h_m(\eta). \tag{4.51}$$

Formal Taylor Series Expansion: given the function $f(\cdot)$, denote $g_m(\cdot) = \frac{1}{m!}f^{(n)}(\cdot)$, and $h_m(\cdot) = (\cdot)^m$. Assume $[x, y] = 0$. Then

$$f(x+y) = \sum_{m=0}^{\infty} g_m(x)h_m(y) = \sum_{m=0}^{\infty} g_m(y)h_m(x). \tag{4.52}$$

Or, more explicitly,

$$f(x+y) = \sum_{m=0}^{\infty} \frac{1}{m!}f^{(m)}(x)y^m = \sum_{m=0}^{\infty} \frac{1}{m!}f^{(m)}(y)x^m. \tag{4.53}$$

Before concluding the present remark, and as preparation for solving the next series of problems, it is appropriate to list the following Formal Taylor Series Expansions:

$$f(x+3\epsilon) = f(x) + f^{(1)}(x)(3\epsilon) + \frac{1}{2!}f^{(2)}(x)(3\epsilon)^2 + \frac{1}{3!}f^{(3)}(x)(3\epsilon)^3 + O(\epsilon^4) \tag{4.54a}$$

$$f(x+2\epsilon) = f(x) + f^{(1)}(x)(2\epsilon) + \frac{1}{2!}f^{(2)}(x)(2\epsilon)^2 + \frac{1}{3!}f^{(3)}(x)(2\epsilon)^3 + O(\epsilon^4) \tag{4.54b}$$

$$f(x+\epsilon) = f(x) + f^{(1)}(x)(\epsilon) + \frac{1}{2!}f^{(2)}(x)(\epsilon)^2 + \frac{1}{3!}f^{(3)}(x)(\epsilon)^3 + O(\epsilon^4) \tag{4.54c}$$

$$f(x) = f(x) \tag{4.54d}$$

In view of the fact that $f(x)$ at the R.H.S. stands for $f^{(0)}(x)$, the last equation can be interpreted as the definition for $f^{(0)}(x)$, rather than expressing the reflection property ($a = a$).

□

Problem: Interpret the terms in the following relationships,

$$\frac{d}{dx}f(x) \overset{\text{def.}}{=} \lim_{\epsilon\to 0} \frac{1}{\epsilon}[f(x+\epsilon) - f(x)] \tag{4.55a}$$

$$= \left(\frac{df}{dx}\right)(x). \tag{4.55b}$$

Solution: With reference to (4.55), several distinctions might be in order:

1. $\frac{d}{dx}f(x)$ alludes to applying the derivative operator $\frac{d}{dx}$ on the function $f(x)$.

2. The limit process in (4.55a) prescribes how the derivative operator must be applied: besides requiring the function values at two neighboring points $x + \epsilon$ and x, and building the ratio $\frac{1}{\epsilon}[f(x + \epsilon) - f(x)]$, the derivation process requires the notion of $\lim$. In essence it is stating that $\frac{1}{\epsilon}[f(x + \epsilon) - f(x)]$ represents some function depending on the independent variable x, i.e., $g(x; \epsilon)$, which in the limit $\epsilon \to 0$ results in $g(x)$.

3. $\left(\frac{df}{dx}\right)(x)$ is a function, and it refers to the aforementioned function $g(x)$.

4. This is the sense the equality $\frac{d}{dx}f(x) = \left(\frac{df}{dx}\right)(x)$ needs to be interpreted. Since there is no risk of misunderstanding, both $\frac{d}{dx}f(x)$ and $\left(\frac{df}{dx}\right)(x)$ will be used interchangeably.

5. Stopping at any instant while ϵ approaches zero, the fraction in (4.55a) represents an approximation for $\frac{d}{dx}f(x)$, i.e.,

$$\frac{d}{dx}f(x) \approx \frac{1}{\epsilon}[f(x + \epsilon) - f(x)]. \tag{4.56}$$

6. For estimating the order of the error in this approximation, subtract (4.54d) from (4.54c),

$$f(x + \epsilon) - f(x) = f^{(1)}(x)(\epsilon) + \frac{1}{2!}f^{(2)}(x)(\epsilon)^2 + \frac{1}{3!}f^{(3)}(x)(\epsilon)^3 + O(\epsilon^4) \tag{4.57}$$

Divide by ϵ,

$$\frac{1}{\epsilon}[f(x + \epsilon) - f(x)] = f^{(1)}(x) + \underbrace{\frac{1}{2!}f^{(2)}(x)(\epsilon) + \frac{1}{3!}f^{(3)}(x)(\epsilon)^2 + O(\epsilon^3)}_{\sim O(\epsilon)} \tag{4.58}$$

Rearrange,

$$f^{(1)}(x) = \frac{1}{\epsilon}[f(x + \epsilon) - f(x)] + O(\epsilon). \tag{4.59}$$

Thus,

$$f^{(1)}(x) \approx \frac{1}{\epsilon}[f(x + \epsilon) - f(x)], \tag{4.60}$$

with an error of the order of ϵ.

∎

Problem: Interpret the equation,

$$\frac{d^2}{dx^2}f(x) = \lim_{\epsilon \to 0}\frac{1}{\epsilon^2}[f(x + 2\epsilon) - 2f(x + \epsilon) + f(x)]. \tag{4.61}$$

Solution: Consider (4.54b)–(4.54d). When building $f(x + 2\epsilon) - 2f(x + \epsilon) + f(x)$, i.e., the term within the square brackets in (4.61), the contribution $f(x) - 2f(x) + f(x)$ cancels out. Furthermore, the two $f^{(1)}(x)$-terms add up to zero. Thus

$$f(x + 2\epsilon) - 2f(x + \epsilon) + f(x) = f^{(2)}(x)(\epsilon)^2 + f^{(3)}(x)(\epsilon)^3 + O(\epsilon^4). \tag{4.62}$$

Divide through by ϵ^2,

$$\frac{1}{\epsilon^2}[f(x + 2\epsilon) - 2f(x + \epsilon) + f(x)] = f^{(2)}(x) + f^{(3)}(x)\epsilon + O(\epsilon^2). \tag{4.63}$$

Rearrange,

$$f^{(2)}(x) = \frac{1}{\epsilon^2}[f(x + 2\epsilon) - 2f(x + \epsilon) + f(x)] + O(\epsilon). \tag{4.64}$$

This is the same as,

$$f^{(2)}(x) \approx \frac{f(x+2\epsilon) - 2f(x+\epsilon) + f(x)}{\epsilon^2}, \tag{4.65}$$

with an error of the order of ϵ, as in the previous case.

∎

Problem: Interpret the equation,

$$\frac{d^3}{dx^3} f(x) = \lim_{\epsilon \to 0} \frac{1}{\epsilon^3} \left[f(x+3\epsilon) - 3f(x+2\epsilon) + 3f(x+\epsilon) - f(x) \right]. \tag{4.66}$$

Solution: Consider (4.54a)–(4.54d). When building $f(x+3\epsilon) - 3f(x+2\epsilon) + 3f(x+\epsilon) - f(x)$, i.e., the term within the square brackets in (4.66), the $f(x)$-contributions cancel out. Furthermore, the three $f^{(1)}(x)$-contributions add up to zero. Additionally, the three $f^{(2)}(x)$-terms sum up to zero. Thus

$$f(x+3\epsilon) - 3f(x+2\epsilon) + 3f(x+\epsilon) - f(x) = f^{(3)}(x)(\epsilon)^3 + O(\epsilon^4). \tag{4.67}$$

Divide through by ϵ^3,

$$\frac{1}{\epsilon^3} \left[f(x+3\epsilon) - 3f(x+2\epsilon) + 3f(x+\epsilon) - f(x) \right] = f^{(3)}(x) + O(\epsilon). \tag{4.68}$$

Rearrange,

$$f^{(3)}(x) = \frac{1}{\epsilon^3} \left[f(x+3\epsilon) - 3f(x+2\epsilon) + 3f(x+\epsilon) - f(x) \right] + O(\epsilon). \tag{4.69}$$

This is the same as

$$f^{(3)}(x) \approx \frac{1}{\epsilon^3} \left[f(x+3\epsilon) - 3f(x+2\epsilon) + 3f(x+\epsilon) - f(x) \right], \tag{4.70}$$

with an error of the order of ϵ, as in the previous cases.

∎

The following examples demonstrate the application of the proposed formulae. Thereby, employing standard functions permits obtaining closed-form expressions and provides further insight.

Example: Determine the first derivative of $\sin(x)$ using the formulae established above.

The first derivative of $\sin(x)$:

$$\frac{d}{dx} \sin(x) = \lim_{\epsilon \to 0} \frac{1}{\epsilon} \left[\sin(x+\epsilon) - \sin(x) \right] \tag{4.71a}$$

$$= \lim_{\epsilon \to 0} \frac{1}{\epsilon} \left[\sin(x)\cos(\epsilon) + \cos(x)\sin(\epsilon) - \sin(x) \right] \tag{4.71b}$$

$$= \lim_{\epsilon \to 0} \frac{1}{\epsilon} \left\{ \sin(x)\left[\cos(\epsilon) - 1\right] + \cos(x)\sin(\epsilon) \right\} \tag{4.71c}$$

$$= \sin(x)\lim_{\epsilon \to 0} \frac{1}{\epsilon} \left[\cos(\epsilon) - 1 \right] + \cos(x)\lim_{\epsilon \to 0} \frac{1}{\epsilon} \left[\sin(\epsilon) \right]. \tag{4.71d}$$

With

$$\lim_{\epsilon \to 0} \frac{1}{\epsilon} [\cos(\epsilon) - 1] = \lim_{\epsilon \to 0} \frac{1}{\epsilon} \left[1 - \frac{1}{2!}\epsilon^2 + O(\epsilon^4) - 1 \right] \tag{4.72a}$$

$$= \lim_{\epsilon \to 0} \left[-\frac{1}{2}\epsilon + O(\epsilon^3) \right] \tag{4.72b}$$

$$= 0 \tag{4.72c}$$

and

$$\lim_{\epsilon \to 0} \frac{1}{\epsilon} [\sin(\epsilon)] = \lim_{\epsilon \to 0} \frac{1}{\epsilon} \left[\epsilon - \frac{1}{3!}\epsilon^3 + O(\epsilon^5) \right] \tag{4.73a}$$

$$= \lim_{\epsilon \to 0} \left[1 - \frac{1}{3!}\epsilon^2 + O(\epsilon^4) \right] \tag{4.73b}$$

$$= 1, \tag{4.73c}$$

Equation (4.71d) reads,

$$\frac{d}{dx} \sin(x) = \cos(x). \tag{4.74}$$

∎

Example: Determine the second derivative of $\sin(x)$ using the formulae established above.

The second derivative of $\sin(x)$:

$$\frac{d^2}{dx^2} \sin(x) = \lim_{\epsilon \to 0} \frac{1}{\epsilon^2} [\sin(x + 2\epsilon) - 2\sin(x + \epsilon) + \sin(x)] \tag{4.75a}$$

$$= \lim_{\epsilon \to 0} \frac{1}{\epsilon^2} \{ [\sin(x)\cos(2\epsilon) + \cos(x)\sin(2\epsilon)]$$
$$- 2[\sin(x)\cos(\epsilon) + \cos(x)\sin(\epsilon)]$$
$$+ \sin(x)\} \tag{4.75b}$$

$$= \lim_{\epsilon \to 0} \frac{1}{\epsilon^2} \{ \sin(x)[\cos(2\epsilon) - 2\cos(\epsilon) + 1] + \cos(x)[\sin(2\epsilon) - 2\sin(\epsilon)] \} \tag{4.75c}$$

$$= \sin(x)\lim_{\epsilon \to 0} \frac{1}{\epsilon^2} \underbrace{[\cos(2\epsilon) - 2\cos(\epsilon) + 1]} + \cos(x)\lim_{\epsilon \to 0} \frac{1}{\epsilon^2} \underbrace{[\sin(2\epsilon) - 2\sin(\epsilon)]}. \tag{4.75d}$$

This result together with,

$$\lim_{\epsilon \to 0} \frac{1}{\epsilon^2} [\cos(2\epsilon) - 2\cos(\epsilon) + 1] = \lim_{\epsilon \to 0} \frac{1}{\epsilon^2} \left\{ \left[1 - \frac{1}{2!}(2\epsilon)^2 + O(\epsilon^4) \right] - 2\left[1 - \frac{1}{2!}\epsilon^2 + O(\epsilon^4) \right] + 1 \right\} \tag{4.76a}$$

$$= \lim_{\epsilon \to 0} \frac{1}{\epsilon^2} [-\epsilon^2 + O(\epsilon^4)] \tag{4.76b}$$

$$= \lim_{\epsilon \to 0} [-1 + O(\epsilon^2)] \tag{4.76c}$$

$$= -1, \tag{4.76d}$$

and,

$$\lim_{\epsilon \to 0} \frac{1}{\epsilon^2} [\sin(2\epsilon) - 2\sin(\epsilon)] = \lim_{\epsilon \to 0} \frac{1}{\epsilon^2} \left\{ \left[(2\epsilon) - \frac{1}{3!}(2\epsilon)^3 + O(\epsilon^5) \right] - 2 \left[\epsilon - \frac{1}{3!}\epsilon^3 + O(\epsilon^5) \right] \right\} \quad (4.77a)$$

$$= \lim_{\epsilon \to 0} \frac{1}{\epsilon^2} \left[-\epsilon^3 + O(\epsilon^5) \right] \quad (4.77b)$$

$$= \lim_{\epsilon \to 0} \left[-\epsilon + O(\epsilon^3) \right] \quad (4.77c)$$

$$= 0, \quad (4.77d)$$

gives,

$$\frac{d^2}{dx^2} \sin(x) = -\sin(x). \quad (4.78)$$

■

Example: Determine the third derivative of $\sin(x)$ using the formulae established above.

The third derivative of $\sin(x)$:

$$\frac{d^3}{dx^3} \sin(x) = \lim_{\epsilon \to 0} \frac{1}{\epsilon^3} [\sin(x + 3\epsilon) - 3\sin(x + 2\epsilon) + 3\sin(x + \epsilon) - \sin(x)] \quad (4.79a)$$

$$= \lim_{\epsilon \to 0} \frac{1}{\epsilon^3} \{ [\sin(x)\cos(3\epsilon) + \cos(x)\sin(3\epsilon)]$$

$$- 3[\sin(x)\cos(2\epsilon) + \cos(x)\sin(2\epsilon)]$$

$$+ 3[\sin(x)\cos(\epsilon) + \cos(x)\sin(\epsilon)]$$

$$- \sin(x)\} \quad (4.79b)$$

$$= \lim_{\epsilon \to 0} \frac{1}{\epsilon^3} \{ \sin(x)[\cos(3\epsilon) - 3\cos(2\epsilon) + 3\cos(\epsilon) - 1]$$

$$+ \cos(x)[\sin(3\epsilon) - 3\sin(2\epsilon) + 3\sin(\epsilon)]\} \quad (4.79c)$$

$$= \sin(x)\lim_{\epsilon \to 0} \frac{1}{\epsilon^3} [\cos(3\epsilon) - 3\cos(2\epsilon) + 3\cos(\epsilon) - 1]$$

$$+ \cos(x)\lim_{\epsilon \to 0} \frac{1}{\epsilon^3} [\sin(3\epsilon) - 3\sin(2\epsilon) + 3\sin(\epsilon)]. \quad (4.79d)$$

This result together with,

$$\lim_{\epsilon \to 0} \frac{1}{\epsilon^3} [\cos(3\epsilon) - 3\cos(2\epsilon) + 3\cos(\epsilon) - 1] = \lim_{\epsilon \to 0} \frac{1}{\epsilon^3} \left\{ \left[1 - \frac{1}{2!}(3\epsilon)^2 + \frac{1}{4!}(3\epsilon)^4 + O(\epsilon^6) \right] \right. $$

$$- 3 \left[1 - \frac{1}{2!}(2\epsilon)^2 + \frac{1}{4!}(2\epsilon)^4 + O(\epsilon^6) \right]$$

$$\left. + 3 \left[1 - \frac{1}{2!}\epsilon^2 + \frac{1}{4!}\epsilon^4 + O(\epsilon^6) \right] - 1 \right\} \tag{4.80a}$$

$$= \lim_{\epsilon \to 0} \frac{1}{\epsilon^3} \left[\frac{3}{2}\epsilon^4 + O(\epsilon^6) \right] \tag{4.80b}$$

$$= \lim_{\epsilon \to 0} \left[\frac{3}{2}\epsilon + O(\epsilon^3) \right] \tag{4.80c}$$

$$= 0, \tag{4.80d}$$

and,

$$\lim_{\epsilon \to 0} \frac{1}{\epsilon^3} [\sin(3\epsilon) - 3\sin(2\epsilon) + 3\sin(\epsilon)] = \lim_{\epsilon \to 0} \frac{1}{\epsilon^3} \left\{ \left[(3\epsilon) - \frac{1}{3!}(3\epsilon)^3 + O(\epsilon^5) \right] \right.$$

$$- 3 \left[(2\epsilon) - \frac{1}{3!}(2\epsilon)^3 + O(\epsilon^5) \right]$$

$$\left. + 3 \left[\epsilon - \frac{1}{3!}\epsilon^3 + O(\epsilon^5) \right] \right\} \tag{4.81a}$$

$$= \lim_{\epsilon \to 0} \frac{1}{\epsilon^3} \left[-\epsilon^3 + O(\epsilon^5) \right] \tag{4.81b}$$

$$= \lim_{\epsilon \to 0} \left[-1 + O(\epsilon^2) \right] \tag{4.81c}$$

$$= -1, \tag{4.81d}$$

gives,

$$\frac{d^3}{dx^3}\sin(x) = -\cos(x). \tag{4.82}$$

∎

Example: Determine the first derivative of e^x using the formulae established above.

The first derivative of e^x:

$$\frac{d}{dx}e^x = \lim_{\epsilon \to 0} \frac{1}{\epsilon}\left(e^{x+\epsilon} - e^x \right) \tag{4.83a}$$

$$= e^x \lim_{\epsilon \to 0} \frac{1}{\epsilon}\left(e^\epsilon - 1 \right) \tag{4.83b}$$

$$= e^x \lim_{\epsilon \to 0} \frac{1}{\epsilon}\left\{ \left[1 + \epsilon + O(\epsilon^2) \right] - 1 \right\} \tag{4.83c}$$

$$= e^x \lim_{\epsilon \to 0} \frac{1}{\epsilon}\left[\epsilon + O(\epsilon^2) \right] \tag{4.83d}$$

$$= e^x \lim_{\epsilon \to 0} \left[1 + O(\epsilon) \right] \tag{4.83e}$$

$$= e^x. \tag{4.83f}$$

∎

Example: Determine the second derivative of e^x using the formulae established above.

The second derivative of e^x:

$$\frac{d^2}{dx^2}e^x = \lim_{\epsilon \to 0} \frac{1}{\epsilon^2}\left[e^{x+2\epsilon} - 2e^{x+\epsilon} + e^x\right] \qquad (4.84a)$$

$$= e^x \lim_{\epsilon \to 0} \frac{1}{\epsilon^2}\left[e^{2\epsilon} - 2e^{\epsilon} + 1\right] \qquad (4.84b)$$

$$= e^x \lim_{\epsilon \to 0} \frac{1}{\epsilon^2}\left\{\left[1 + (2\epsilon) + \frac{1}{2!}(2\epsilon)^2 + O(\epsilon^3)\right]\right.$$
$$\left. - 2\left[1 + \epsilon + \frac{1}{2!}\epsilon^2 + O(\epsilon^3)\right] + 1\right\} \qquad (4.84c)$$

$$= e^x \lim_{\epsilon \to 0} \frac{1}{\epsilon^2}\left(\epsilon^2 + O(\epsilon^3)\right) \qquad (4.84d)$$

$$= e^x \lim_{\epsilon \to 0}\left(1 + O(\epsilon)\right) \qquad (4.84e)$$

$$= e^x. \qquad (4.84f)$$

∎

Example: Determine the third derivative of e^x using the formulae established above.

The third derivative of e^x:

$$\frac{d^3}{dx^3}e^x = \lim_{\epsilon \to 0} \frac{1}{\epsilon^3}\left[e^{x+3\epsilon} - 3e^{x+2\epsilon} + 3e^{x+\epsilon} - e^x\right] \qquad (4.85a)$$

$$= e^x \lim_{\epsilon \to 0} \frac{1}{\epsilon^3}\left[e^{3\epsilon} - 3e^{2\epsilon} + 3e^{\epsilon} - 1\right] \qquad (4.85b)$$

$$= e^x \lim_{\epsilon \to 0} \frac{1}{\epsilon^3}\left\{\left[1 + (3\epsilon) + \frac{1}{2!}(3\epsilon)^2 + \frac{1}{3!}(3\epsilon)^3 + O(\epsilon^4)\right]\right.$$
$$- 3\left[1 + (2\epsilon) + \frac{1}{2!}(2\epsilon)^2 + \frac{1}{3!}(2\epsilon)^3 + O(\epsilon^4)\right]$$
$$+ 3\left[1 + \epsilon + \frac{1}{2!}\epsilon^2 + \frac{1}{3!}\epsilon^3 + O(\epsilon^4)\right]$$
$$\left. - 1\right\} \qquad (4.85c)$$

$$= e^x \lim_{\epsilon \to 0} \frac{1}{\epsilon^3}\left[\epsilon^3 + O(\epsilon^4)\right] \qquad (4.85d)$$

$$= e^x \lim_{\epsilon \to 0}\left[1 + O(\epsilon)\right] \qquad (4.85e)$$

$$= e^x. \qquad (4.85f)$$

∎

4.3   First- and higher order derivatives of the product of two functions

It is imperative to introduce "new" symbols to render manipulations simpler and more clear.

Problem: Calculate the derivative of $f(x)g(x)$ with respect to x.

Solution:

$$\frac{d}{dx}\left(f(x)g(x)\right) \overset{(1)}{=} \left(\frac{d}{dx}f(x)\right)g(x) + f(x)\left(\frac{d}{dx}g(x)\right) \tag{4.86a}$$

$$\overset{(2)}{=} \left(\frac{d^1}{dx^1}f(x)\right)\left(\frac{d^0}{dx^0}g(x)\right) + \left(\frac{d^0}{dx^0}f(x)\right)\left(\frac{d^1}{dx^1}g(x)\right) \tag{4.86b}$$

$$\overset{(3)}{=} f(x)\left(\frac{d^1}{dx^1}\right]\left[\frac{d^0}{dx^0}\right)g(x) + f(x)\left(\frac{d^0}{dx^0}\right]\left[\frac{d^1}{dx^1}\right)g(x) \tag{4.86c}$$

$$\overset{(4)}{=} f(x)\left\{\left(\frac{d^1}{dx^1}\right]\left[\frac{d^0}{dx^0}\right) + \left(\frac{d^0}{dx^0}\right]\left[\frac{d^1}{dx^1}\right)\right\}g(x) \tag{4.86d}$$

$$\overset{(5)}{=} f(x)\left\{\left(D^1\right]\left[D^0\right) + \left(D^0\right]\left[D^1\right)\right\}g(x) \tag{4.86e}$$

$$\overset{(6)}{=} f(x)\left\{D^1 D^0 + D^0 D^1\right\}g(x). \tag{4.86f}$$

In the above the transitions (1)–(6) have the following meaning:

(1) employs the chain rule of differentiation;
(2) introduces $\frac{d^1}{dx^1}h(x) = \frac{d}{dx}h(x)$ and $\frac{d^0}{dx^0}h(x) = h(x)$ for arbitrary "permissible" (test) functions $h(x)$;
(3) introduces the differential operators $\left(\frac{d^0}{dx^0}\right]$ and $\left(\frac{d^1}{dx^1}\right]$ acting on functions to their left, and the corresponding operators $\left[\frac{d^0}{dx^0}\right)$ and $\left[\frac{d^1}{dx^1}\right)$ acting on functions to their right, respectively;
(4) utilizes the linearity property of $\left(\frac{d^1}{dx^1}\right]\left[\frac{d^0}{dx^0}\right)$ and $\left(\frac{d^0}{dx^0}\right]\left[\frac{d^1}{dx^1}\right)$;
(5) introduces $\left(D^0\right]$ and $\left(D^1\right]$ and the corresponding $\left[D^0\right)$ and $\left[D^1\right)$;
(6) stipulates that there is no ambiguity in introducing $D^1 D^0$ and $D^0 D^1$. In both cases, the left operator is of the type $(\cdot]$, and the right one of the type $[\cdot)$.

The definitions in (4.86) suggest setting,

$$\frac{d}{dx} \equiv \left(\frac{d^1}{dx^1}\right]\left[\frac{d^0}{dx^0}\right) + \left(\frac{d^0}{dx^0}\right]\left[\frac{d^1}{dx^1}\right). \tag{4.87}$$

Or, more compactly,

$$\frac{d}{dx} \equiv \left(D^1\right]\left[D^0\right) + \left(D^0\right]\left[D^1\right), \tag{4.88}$$

or, even more compactly,

$$\frac{d}{dx} \equiv D^1 D^0 + D^0 D^1. \tag{4.89}$$

∎

Remark:

- Observe the terms $D^1 D^0$ and $D^0 D^1$ at the R.H.S. of (4.89). Adding the orders of the derivatives of each term gives unity: $1 + 0 = 0 + 1 = 1$.
- The number of the terms at the R.H.S., i.e., 2, signifies that the expression at the R.H.S. represents the first derivative.

- The number of the D's in each term at the R.H.S., i.e., 2, indicates that this representation has been prepared for $\frac{d}{dx}$ to operate on products of two functions in the form fg.

As will be demonstrated further below, these observations are generalizable.

$\square$

Exercise: To gain familiarity with the usage of (4.87), consider $f(x) = 1 \cdot f(x)$,

$$\frac{d}{dx}f(x) = \frac{d}{dx}(1 \cdot f(x)) \tag{4.90a}$$

$$= 1 \left\{ \left(\frac{d^1}{dx^1} \right) \left[\frac{d^0}{dx^0} \right] + \left(\frac{d^0}{dx^0} \right) \left[\frac{d^1}{dx^1} \right] \right\} f(x) \tag{4.90b}$$

$$= 1 \left(\frac{d^1}{dx^1} \right) \left[\frac{d^0}{dx^0} \right] f(x) + 1 \left(\frac{d^0}{dx^0} \right) \left[\frac{d^1}{dx^1} \right] f(x) \tag{4.90c}$$

$$= \underbrace{\left(\frac{d^1}{dx^1} 1 \right)}_{=0} \underbrace{\left(\frac{d^0}{dx^0}f(x) \right)}_{=f(x)} + \underbrace{\left(\frac{d^0}{dx^0} 1 \right)}_{=1} \underbrace{\left(\frac{d^1}{dx^1}f(x) \right)}_{=\frac{d}{dx}f(x)} \tag{4.90d}$$

$$= \frac{d}{dx}f(x). \tag{4.90e}$$

■

Exercise: Due to the symmetry in (4.87), the same result can be obtained by writing $f(x) = f(x) \cdot 1$,

$$\frac{d}{dx}f(x) = \frac{d}{dx}(f(x) \cdot 1) \tag{4.91a}$$

$$= f(x) \left\{ \left(\frac{d^1}{dx^1} \right) \left[\frac{d^0}{dx^0} \right] + \left(\frac{d^0}{dx^0} \right) \left[\frac{d^1}{dx^1} \right] \right\} 1 \tag{4.91b}$$

$$= f(x) \left(\frac{d^1}{dx^1} \right) \left[\frac{d^0}{dx^0} \right] 1 + f(x) \left(\frac{d^0}{dx^0} \right) \left[\frac{d^1}{dx^1} \right] 1 \tag{4.91c}$$

$$= \underbrace{\left(\frac{d^1}{dx^1}f(x) \right)}_{=\frac{d}{dx}f(x)} \underbrace{\left(\frac{d^0}{dx^0} 1 \right)}_{=1} + \underbrace{\left(\frac{d^0}{dx^0}f(x) \right)}_{=f(x)} \underbrace{\left(\frac{d^1}{dx^1} 1 \right)}_{=0} \tag{4.91d}$$

$$= \frac{d}{dx}f(x). \tag{4.91e}$$

■

Problem: Calculate $\frac{d^2}{dx^2}[f(x)g(x)]$.

Solution:

$$\frac{d^2}{dx^2}(fg) \;\stackrel{\text{def.}}{=}\; \frac{d}{dx}\left[\frac{d}{dx}(fg)\right] \tag{4.92a}$$

$$\stackrel{\text{chain rule}}{=} \frac{d}{dx}\left[\left(\frac{d}{dx}f\right)g + f\left(\frac{d}{dx}g\right)\right] \tag{4.92b}$$

$$\stackrel{\text{chain rule}}{=} \left(\frac{d^2}{dx^2}f\right)g + \underbrace{\left(\frac{d}{dx}f\right)\left(\frac{d}{dx}g\right) + \left(\frac{d}{dx}f\right)\left(\frac{d}{dx}g\right)} + f\left(\frac{d^2}{dx^2}g\right) \tag{4.92c}$$

$$= \left(\frac{d^2}{dx^2}f\right)\left(\frac{d^0}{dx^0}g\right) + 2\left(\frac{d^1}{dx^1}f\right)\left(\frac{d^1}{dx^1}g\right) + \left(\frac{d^0}{dx^0}f\right)\left(\frac{d^2}{dx^2}g\right) \tag{4.92d}$$

$$= f\left\{\left(\frac{d^2}{dx^2}\right)\left[\frac{d^0}{dx^0}\right] + 2\left(\frac{d^1}{dx^1}\right)\left[\frac{d^1}{dx^1}\right] + \left(\frac{d^0}{dx^0}\right)\left[\frac{d^2}{dx^2}\right]\right\}g \tag{4.92e}$$

$$= f\left\{\left(D^2\right)\left[D^0\right] + 2\left(D^1\right)\left[D^1\right] + \left(D^0\right)\left[D^2\right]\right\}g \tag{4.92f}$$

$$= f\left\{D^2D^0 + 2D^1D^1 + D^0D^2\right\}g. \tag{4.92g}$$

The conventions in (4.92) suggest setting,

$$\frac{d^2}{dx^2} \equiv D^2D^0 + 2D^1D^1 + D^0D^2. \tag{4.93}$$

■

Remark:

- Observe the terms D^2D^0, $2D^1D^1$, and D^0D^2 at the R.H.S. of (4.93). Adding the orders of the derivatives of each term gives two: $2+0 = 1+1 = 0+2 = 2$.
- The number of the terms at the R.H.S., i.e., 3, signifies that the expression at the R.H.S. represents the second derivative.
- The number of the D's in each term at the R.H.S., i.e., 2, indicates that this representation has been prepared for $\frac{d^2}{dx^2}$ to operate on products of two functions in the form fg.

□

Problem: Find the equivalent representation for $\frac{d^3}{dx^3}$.

Solution: Standard long derivation.

$$\frac{d^3}{dx^3}(fg) = \frac{d}{dx}\left[\frac{d^2}{dx^2}(fg)\right] \tag{4.94a}$$

$$= \frac{d}{dx}\left[\left(\frac{d^2}{dx^2}f\right)g + 2\left(\frac{d}{dx}f\right)\left(\frac{d}{dx}g\right) + f\left(\frac{d^2}{dx^2}g\right)\right] \tag{4.94b}$$

$$= \left(\frac{d^3}{dx^3}f\right)g + \underbrace{\left(\frac{d^2}{dx^2}f\right)\left(\frac{d}{dx}g\right)}$$

$$+ 2\underbrace{\left(\frac{d^2}{dx^2}f\right)\left(\frac{d}{dx}g\right)} + 2\underbrace{\left(\frac{d}{dx}f\right)\left(\frac{d^2}{dx^2}g\right)}$$

$$+ \underbrace{\left(\frac{d}{dx}f\right)\left(\frac{d^2}{dx^2}g\right)} + f\left(\frac{d^3}{dx^3}g\right) \tag{4.94c}$$

$$= \left(\frac{d^3}{dx^3}f\right)g + 3\left(\frac{d^2}{dx^2}f\right)\left(\frac{d}{dx}g\right) + 3\left(\frac{d}{dx}f\right)\left(\frac{d^2}{dx^2}g\right) + f\left(\frac{d^3}{dx^3}g\right). \tag{4.94d}$$

Employing the symbolic notation introduced above,

$$\frac{d^3}{dx^3} \equiv \left(\frac{d^3}{dx^3}\right)\left[\frac{d^0}{dx^0}\right] + 3\left(\frac{d^2}{dx^2}\right)\left[\frac{d^1}{dx^1}\right] + 3\left(\frac{d^1}{dx^1}\right)\left[\frac{d^2}{dx^2}\right] + \left(\frac{d^0}{dx^0}\right)\left[\frac{d^3}{dx^3}\right). \tag{4.95}$$

Written compactly,

$$\frac{d^3}{dx^3} \equiv \left(D^3\right)\left[D^0\right) + 3\left(D^2\right)\left[D^1\right) + 3\left(D^1\right)\left[D^2\right) + \left(D^0\right)\left[D^3\right). \tag{4.96}$$

Or, even more compactly,

$$\frac{d^3}{dx^3} \equiv D^3D^0 + 3D^2D^1 + 3D^1D^2 + D^0D^3. \tag{4.97}$$

∎

Remark:

- Observe the terms D^3D^0, $3D^2D^1$, D^1D^2, D^0D^3 at the R.H.S. of (4.97). Adding the orders of the derivatives of each term gives 3: $3 + 0 = 2 + 1 = 1 + 2 = 0 + 3 = 3$.
- The number of the terms at the R.H.S., i.e., 4, signifies that the expression at the R.H.S. represents the third derivative.
- The number of the D's in each term at the R.H.S., i.e., 2, indicates that this representation has been prepared for $\frac{d^3}{dx^3}$ to operate on products of two functions in the form fg.

□

Remark:

- In view of the expressions of $\frac{d^1}{dx^1}$, $\frac{d^2}{dx^2}$, $\frac{d^3}{dx^3}$, derived so far, it can be hypothesized that the coefficients in the formulae for $\frac{d^n}{dx^n}$ are given by the Pascal triangle.
- Or, even more conveniently and immediately,

$$\frac{d^3}{dx^3} \equiv \frac{3!}{3!0!}D^3D^0 + \frac{3!}{2!1!}D^2D^1 + \frac{3!}{1!2!}D^1D^2 + \frac{3!}{0!3!}D^0D^3. \tag{4.98}$$

- Before engaging in a proof by induction, it is instructive to examine the aforementioned claims in the case of $\frac{d^4}{dx^4}$. Using the values of the coefficients from the Pascal triangle,

$$\frac{d^4}{dx^4} \equiv D^4D^0 + 4D^3D^1 + 6D^2D^2 + 4D^1D^3 + D^0D^4. \tag{4.99}$$

- Expressing the coefficients in terms of factorials,

$$\frac{d^4}{dx^4} \equiv \frac{4!}{4!0!}D^4D^0 + \frac{4!}{3!1!}D^3D^1 + \frac{4!}{2!2!}D^2D^2 + \frac{4!}{1!3!}D^1D^3 + \frac{4!}{0!4!}D^0D^4. \tag{4.100}$$

This result motivates the formulation of the induction hypothesis, which is the content of the next problem.

□

Problem: Show that the following representation of $\frac{d^n}{dx^n}$ holds true,

$$\frac{d^n}{dx^n} \equiv \sum_{k=0}^{n} \frac{n!}{(n-k)!k!}D^{n-k}D^k. \tag{4.101}$$

Solution: Before getting engaged in the proof of (4.101) by mathematical induction, it is instructive to consider the following observations, which provide further insight into the structure of the representations established so far. Straightforward substitution of terms along with simple manipulations reveal important properties:

$$\frac{d^2}{dx^2} = \frac{d}{dx}\left(\frac{d}{dx}\right) \tag{4.102a}$$

$$\equiv D^1\left(D^1D^0 + D^0D^1\right)D^0 + D^0\left(D^1D^0 + D^0D^1\right)D^1 \tag{4.102b}$$

$$= \left(\underbrace{D^1D^1}_{=D^2}\underbrace{D^0D^0}_{=D^0} + \underbrace{D^1D^0}_{=D^1}\underbrace{D^1D^0}_{=D^1}\right) + \left(\underbrace{D^0D^1}_{=D^1}\underbrace{D^0D^1}_{=D^1} + \underbrace{D^0D^0}_{=D^0}\underbrace{D^1D^1}_{=D^2}\right) \tag{4.102c}$$

$$= D^2D^0 + D^1D^1 + D^1D^1 + D^0D^2 \tag{4.102d}$$

$$= D^2D^0 + 2D^1D^1 + D^0D^2. \tag{4.102e}$$

Proceeding similarly, by inserting $D^2D^0 + 2D^1D^1 + D^0D^2$ between any of the terms in $D^1D^0 + D^0D^1$,

$$\frac{d^3}{dx^3} = \frac{d}{dx}\left(\frac{d^2}{dx^2}\right) \tag{4.103a}$$

$$\equiv D^1\left(D^2D^0 + 2D^1D^1 + D^0D^2\right)D^0 + D^0\left(D^2D^0 + 2D^1D^1 + D^0D^2\right)D^1 \tag{4.103b}$$

$$= \left(\underbrace{D^1D^2}_{=D^3}\underbrace{D^0D^0}_{=D^0} + 2\underbrace{D^1D^1}_{=D^2}\underbrace{D^1D^0}_{=D^1} + \underbrace{D^1D^0}_{=D^1}\underbrace{D^2D^0}_{=D^2}\right)$$
$$+ \left(\underbrace{D^0D^2}_{=D^2}\underbrace{D^0D^1}_{=D^1} + 2\underbrace{D^0D^1}_{=D^1}\underbrace{D^1D^1}_{=D^2} + \underbrace{D^0D^0}_{=D^0}\underbrace{D^2D^1}_{=D^3}\right) \tag{4.103c}$$

$$= D^3D^0 + 2D^2D^1 + D^1D^2 + D^2D^1 + 2D^1D^2 + D^0D^3 \tag{4.103d}$$

$$= D^3D^0 + 3D^2D^1 + 3D^1D^2 + D^0D^3. \tag{4.103e}$$

Alternatively, due to $\frac{d}{dx}\left(\frac{d^2}{dx^2}\right) = \frac{d^2}{dx^2}\left(\frac{d}{dx}\right)$, inserting $\left(D^1D^0 + D^0D^1\right)$ between any of the terms in $D^2D^0 + 2D^1D^1 + D^0D^2$,

$$\frac{d^3}{dx^3} = \frac{d^2}{dx^2}\left(\frac{d}{dx}\right) \tag{4.104a}$$

$$\equiv D^2\left(D^1D^0 + D^0D^1\right)D^0 + 2D^1\left(D^1D^0 + D^0D^1\right)D^1 + D^0\left(D^1D^0 + D^0D^1\right)D^2 \tag{4.104b}$$

$$= \underbrace{D^2D^1}_{=D^3}\underbrace{D^0D^0}_{=D^0} + \underbrace{D^2D^0}_{=D^2}\underbrace{D^1D^0}_{=D^1}$$
$$+ 2\underbrace{D^1D^1}_{=D^2}\underbrace{D^0D^1}_{=D^1} + 2\underbrace{D^1D^0}_{=D^1}\underbrace{D^1D^1}_{=D^2}$$
$$+ \underbrace{D^0D^1}_{=D^1}\underbrace{D^0D^2}_{=D^2} + \underbrace{D^0D^0}_{=D^0}\underbrace{D^1D^2}_{=D^3} \tag{4.104c}$$

$$= D^3D^0 + D^2D^1 + 2D^2D^1 + 2D^1D^2 + D^1D^2 + D^0D^3 \tag{4.104d}$$

$$= D^3D^0 + 3D^2D^1 + 3D^1D^2 + D^0D^3. \tag{4.104e}$$

Based on the above observations and enhanced intuition the proof by mathematical induction of the induction hypothesis (4.101) starts with inserting $\left(D^1D^0 + D^0D^1\right)$ between D^{n-k} and D^k,

$$\frac{d^{n+1}}{dx^{n+1}} \stackrel{\text{def.}}{\equiv} \frac{d^n}{dx^n}\left(\frac{d}{dx}\right) \tag{4.105a}$$

$$\equiv \sum_{k=0}^{n} \frac{n!}{(n-k)!k!}D^{n-k}\left(D^1D^0 + D^0D^1\right)D^k \tag{4.105b}$$

$$= \sum_{k=0}^{n} \frac{n!}{(n-k)!k!}\left(\underbrace{D^{n-k}D^1D^0D^k}_{D^{(n+1)-k}\quad D^k} + \underbrace{D^{n-k}D^0D^1D^k}_{D^{n-k}\quad D^{k+1}}\right) \tag{4.105c}$$

$$= \sum_{k=0}^{n} \frac{n!}{(n-k)!k!}\left(D^{(n+1)-k}D^k + D^{n-k}D^{k+1}\right). \tag{4.105d}$$

Writing the 1st, 2nd, $(k-1)$st, kth, $(n-1)$th, and the nth terms explicitly,

$$\frac{d^{n+1}}{dx^{n+1}} = \frac{n!}{n!0!}\left(D^{(n+1)}D^0 + D^nD^1\right) \qquad (k=0)$$
$$+ \frac{n!}{(n-1)!1!}\left(D^{(n+1)-1}D^1 + D^{n-1}D^{1+1}\right) \qquad (k=1)\dots$$

$$\cdots \vdots$$

$$+ \frac{n!}{(n-(k-1))!(k-1)!} \left(D^{(n+1)-(k-1)}D^{k-1} + D^{n-(k-1)}D^{(k-1)+1} \right) \qquad (k-1)$$

$$+ \frac{n!}{(n-k)!k!} \left(D^{(n+1)-k}D^k + D^{n-k}D^{k+1} \right) \qquad (k)$$

$$\vdots$$

$$+ \frac{n!}{(n-(n-1))!(n-1)!} \left(D^{(n+1)-(n-1)}D^{n-1} + D^{n-(n-1)}D^{(n-1)+1} \right) \qquad (n-1)$$

$$+ \frac{n!}{(n-n)!n!} \left(D^{(n+1)-n}D^n + D^{n-n}D^{n+1} \right) \qquad (n). \tag{4.106}$$

Simplifying,

$$\frac{d^{n+1}}{dx^{n+1}} \equiv \left(\underbrace{D^{n+1}D^0}_{a} + \underbrace{D^n D^1}_{b} \right)$$

$$+ n \left(\underbrace{D^n D^1}_{b} + D^{n-1}D^2 \right)$$

$$\vdots$$

$$+ \frac{n!}{((n+1)-k)!(k-1)!} \left(D^{(n+1)-(k-1)}D^{k-1} + \underbrace{D^{(n+1)-k}D^k}_{c} \right)$$

$$+ \frac{n!}{(n-k)!k!} \left(\underbrace{D^{(n+1)-k}D^k}_{c} + D^{n-k}D^{k+1} \right)$$

$$\vdots$$

$$+ n \left(D^2 D^{n-1} + \underbrace{D^1 D^n}_{d} \right)$$

$$+ \left(\underbrace{D^1 D^n}_{d} + \underbrace{D^0 D^{n+1}}_{e} \right). \tag{4.107}$$

A pattern emerges. The first (a) and the last (e) terms do not have any counterparts. However, all other terms appear pairwise, as indicated in the cases b, c, and d. In view of this relationship,

$$\frac{d^{n+1}}{dx^{n+1}} \equiv D^{n+1}D^0$$

$$+ (n+1)D^n D^1$$

$$\vdots$$

$$+ \left[\frac{n!}{((n+1)-k)!(k-1)!} + \frac{n!}{(n-k)!k!} \right] D^{(n+1)-k}D^k \cdots$$

$$\ldots \; \vdots$$

$$+ (n + 1)D^1 D^n$$

$$+ D^0 D^{n+1}. \tag{4.108}$$

The terms in square brackets give,

$$\frac{n!}{((n+1)-k)!(k-1)!} + \frac{n!}{(n-k)!k!} = \frac{n!}{(n-k)!\,((n+1)-k)\,(k-1)!} + \frac{n!}{(n-k)!(k-1)!k} \tag{4.109a}$$

$$= \frac{n!}{(n-k)!(k-1)!} \left(\frac{1}{n+1-k} + \frac{1}{k} \right) \tag{4.109b}$$

$$= \frac{n!}{(n-k)!(k-1)!} \left[\frac{n+1}{(n+1-k)(k)} \right] \tag{4.109c}$$

$$= \frac{(n+1)!}{((n+1)-k)!k!}. \tag{4.109d}$$

$$\frac{d^{n+1}}{dx^{n+1}} \equiv D^{n+1} D^0 \qquad (k = 0)$$

$$+ (n + 1)D^n D^1 \qquad (k = 1)$$

$$\vdots$$

$$+ \frac{(n+1)!}{((n+1)-k)!k!} D^{(n+1)-k} D^k \qquad (k = 2, \cdots, n-1)$$

$$\vdots$$

$$+ (n + 1)D^1 D^n \qquad (k = n)$$

$$+ D^0 D^{n+1} \qquad (k = n + 1). \tag{4.110}$$

Thus,

$$\frac{d^{n+1}}{dx^{n+1}} \equiv \sum_{k=0}^{n+1} \frac{(n+1)!}{((n+1)-k)!k!} D^{(n+1)-k} D^k. \tag{4.111}$$

∎

Remark: From the above discussion a recurring pattern in the proof by mathematical induction emerges, which conveys three insights: (a) the method can be regarded as an effective tool for designing algorithms. (b) Carrying out several initial steps (typically : $n = 0, 1, 2, 3$) guides the intuition and provides sufficient insight into the way how the induction step should proceed. The initial steps are sometime tedious, and thus perceived as not necessary to be carried out in full detail. However, they are essential in the formulation of strategies. In the course of grasping the initial steps, the intricacies of the algorithm reveal themselves. (c) The process of carrying out the induction step discloses possible missing links necessary for fully grasping what is going on in the algorithm.

□

4.4   Further properties of derivative operators

Consider the derivative of the product of three functions, $f(x)g(x)h(x)$,

$$\frac{d}{dx}(fgh) = \left(\frac{d}{dx}f\right)gh + f\left(\frac{d}{dx}(gh)\right) \tag{4.112a}$$

$$= \left(\frac{d}{dx}f\right)gh + f\left\{\left(\frac{d}{dx}g\right)h + g\left(\frac{d}{dx}h\right)\right\} \tag{4.112b}$$

$$= \left(\frac{d}{dx}f\right)gh + f\left(\frac{d}{dx}g\right)h + fg\left(\frac{d}{dx}h\right) \tag{4.112c}$$

$$= \left(\hat{D}^1 f\right)gh + f\left(\hat{D}^1 g\right)h + fg\left(\hat{D}^1 h\right) \tag{4.112d}$$

$$= \left(\hat{D}^1 f\right)\left(\hat{D}^0 g\right)\left(\hat{D}^0 h\right) + \left(\hat{D}^0 f\right)\left(\hat{D}^1 g\right)\left(\hat{D}^0 h\right) + \left(\hat{D}^0 f\right)\left(\hat{D}^0 g\right)\left(\hat{D}^1 h\right) \tag{4.112e}$$

$$= \hat{D}^1 f \hat{D}^0 g \hat{D}^0 h + \hat{D}^0 f \hat{D}^1 g \hat{D}^0 h + \hat{D}^0 f \hat{D}^0 g \hat{D}^1 h. \tag{4.112f}$$

Considering the first and the last terms in (4.112), the representation,

$$\frac{d}{dx} \equiv \hat{D}^1 \hat{D}^0 \hat{D}^0 + \hat{D}^0 \hat{D}^1 \hat{D}^0 + \hat{D}^0 \hat{D}^0 \hat{D}^1, \tag{4.113}$$

suggests itself. To more fully appreciate the pattern that arises, consider the second derivative of the product of three functions $f(x)g(x)h(x)$, as follows:

$$\frac{d^2}{dx^2}(fgh) = \frac{d}{dx}\left(\frac{d}{dx}(fgh)\right) \tag{4.114a}$$

$$= \frac{d}{dx}\left\{\left(\frac{d}{dx}f\right)gh + f\left(\frac{d}{dx}g\right)h + fg\left(\frac{d}{dx}h\right)\right\} \tag{4.114b}$$

$$= \frac{d}{dx}\left\{\left(\frac{d}{dx}f\right)gh\right\} + \frac{d}{dx}\left\{f\left(\frac{d}{dx}g\right)h\right\} + \frac{d}{dx}\left\{fg\left(\frac{d}{dx}h\right)\right\} \tag{4.114c}$$

$$= \left\{\left(\frac{d^2}{dx^2}f\right)gh + \left(\frac{d}{dx}f\right)\left(\frac{d}{dx}g\right)h + \left(\frac{d}{dx}f\right)g\left(\frac{d}{dx}h\right)\right\}$$

$$+ \left\{\left(\frac{d}{dx}f\right)\left(\frac{d}{dx}g\right)h + f\left(\frac{d^2}{dx^2}g\right)h + f\left(\frac{d}{dx}g\right)\left(\frac{d}{dx}h\right)\right\}$$

$$+ \left\{\left(\frac{d}{dx}f\right)g\left(\frac{d}{dx}h\right) + f\left(\frac{d}{dx}g\right)\left(\frac{d}{dx}h\right) + fg\left(\frac{d^2}{dx^2}h\right)\right\} \tag{4.114d}$$

$$= \left\{ \left(\hat{D}^2 f \right) gh + \left(\hat{D}^1 f \right) \left(\hat{D}^1 g \right) h + \left(\hat{D}^1 f \right) g \left(\hat{D}^1 h \right) \right\}$$
$$+ \left\{ \left(\hat{D}^1 f \right) \left(\hat{D}^1 g \right) h + f \left(\hat{D}^2 g \right) h + f \left(\hat{D}^1 g \right) \left(\hat{D}^1 h \right) \right\}$$
$$+ \left\{ \left(\hat{D}^1 f \right) g \left(\hat{D}^1 h \right) + f \left(\hat{D}^1 g \right) \left(\hat{D}^1 h \right) + fg \left(\hat{D}^2 h \right) \right\}. \tag{4.114e}$$

In the last step, the differential operators were used in the obvious manner. Introducing $\hat{D}^0$,

$$\frac{d^2}{dx^2} (fgh) = \hat{D}^2 f \hat{D}^0 g \hat{D}^0 h + \hat{D}^1 f \hat{D}^1 g \hat{D}^0 h + \hat{D}^1 f \hat{D}^0 g \hat{D}^1 h$$
$$+ \hat{D}^1 f \hat{D}^1 g \hat{D}^0 h + \hat{D}^0 f \hat{D}^2 g \hat{D}^0 h + \hat{D}^0 f \hat{D}^1 g \hat{D}^1 h$$
$$+ \hat{D}^1 f \hat{D}^0 g \hat{D}^1 h + \hat{D}^0 f \hat{D}^1 g \hat{D}^1 h + \hat{D}^0 f \hat{D}^0 g \hat{D}^2 h. \tag{4.115}$$

Observe that the "diagonal terms" appear once, while the "off-diagonal" terms, being symmetric, appear twice. Thus

$$\frac{d^2}{dx^2} (fgh) = \hat{D}^2 f \hat{D}^0 g \hat{D}^0 h + \hat{D}^0 f \hat{D}^2 g \hat{D}^0 h + \hat{D}^0 f \hat{D}^0 g \hat{D}^2 h$$
$$+ 2\hat{D}^1 f \hat{D}^1 g \hat{D}^0 h + 2\hat{D}^1 f \hat{D}^0 g \hat{D}^1 h + 2\hat{D}^0 f \hat{D}^1 g \hat{D}^1 h. \tag{4.116}$$

The numeral 2 is dictated by the order of the derivative at the L.H.S.. The functions f, g, and h define three "placeholders," ("pigeonholes"). The super-indices at the R.H.S. are determined by distinct possibilities in which the numeral 2 can be distributed into three pigeonholes as whole numbers. It is immediate that 2 can be placed in three distinct locations, leading to $(2, 0, 0), (0, 2, 0,), (0, 0, 2)$. Furthermore, 2 can be expressed as the addition of 1 and 1. The two ones can be placed in three pigeonholes in three distinct ways; i.e., $(1, 1, 0), (1, 0, 1), (0, 1, 1)$. Since exchanging the ones in these arrangements leads to indistinguishable realization, they must be counted twice.

The symbolic notation,

$$\frac{d^2}{dx^2} \equiv \hat{D}^2 \hat{D}^0 \hat{D}^0 + \hat{D}^0 \hat{D}^2 \hat{D}^0 + \hat{D}^0 \hat{D}^0 \hat{D}^2$$
$$+ 2\hat{D}^1 \hat{D}^1 \hat{D}^0 + 2\hat{D}^1 \hat{D}^0 \hat{D}^1 + 2\hat{D}^0 \hat{D}^1 \hat{D}^1, \tag{4.117}$$

suggests itself. Considering the relationships,

$$\frac{2!}{2!0!0!} = \frac{2!}{0!2!0!} = \frac{2!}{0!0!2!} = 1 \tag{4.118a}$$

$$\frac{2!}{1!1!0!} = \frac{2!}{1!0!1!} = \frac{2!}{0!1!1!} = 2, \tag{4.118b}$$

the coefficients at the R.H.S. of (4.117) can be written in a unified and generalizable form,

$$\frac{d^2}{dx^2} \equiv \frac{2!}{2!0!0!} \hat{D}^2 \hat{D}^0 \hat{D}^0 + \frac{2!}{0!2!0!} \hat{D}^0 \hat{D}^2 \hat{D}^0 + \frac{2!}{0!0!2!} \hat{D}^0 \hat{D}^0 \hat{D}^2$$
$$+ \frac{2!}{1!1!0!} \hat{D}^1 \hat{D}^1 \hat{D}^0 + \frac{2!}{1!0!1!} \hat{D}^1 \hat{D}^0 \hat{D}^1 + \frac{2!}{0!1!1!} \hat{D}^0 \hat{D}^1 \hat{D}^1. \tag{4.119}$$

4.5 Commutator relationships

Problem: Given the position operator $\hat{x}$, the derivative operator $\hat{D}$, and the identity operator $\hat{\mathbb{I}}$, defined by,

$$\hat{x}f(x) = xf(x) \tag{4.120a}$$

$$\hat{D}f(x) = \frac{d}{dx}f(x) \tag{4.120b}$$

$$\hat{\mathbb{I}}f(x) = f(x), \tag{4.120c}$$

show that

$$[\hat{D}, \hat{x}] = \hat{\mathbb{I}}. \tag{4.121}$$

Solution: Long proof.

The following steps are exhaustive and self-explanatory.

$$[\hat{D}, \hat{x}]f(x) \stackrel{\text{def.}}{=} \left(\hat{D}\hat{x} - \hat{x}\hat{D}\right)f(x) \tag{4.122a}$$

$$\stackrel{\text{linearity}}{=} \hat{D}\left(\hat{x}f(x)\right) - \hat{x}\left(\hat{D}f(x)\right) \tag{4.122b}$$

$$\stackrel{\text{the action of } \hat{x}}{=} \hat{D}\left(xf(x)\right) - x\left(\hat{D}f(x)\right) \tag{4.122c}$$

$$\stackrel{\text{the action of } \hat{D}}{=} \underbrace{\left(\hat{D}x\right)}_{1}f(x) + x\left(\hat{D}f(x)\right) - x\left(\hat{D}f(x)\right) \tag{4.122d}$$

$$= f(x) \tag{4.122e}$$

$$= \hat{\mathbb{I}}f(x). \tag{4.122f}$$

Since $f(x)$ is arbitrary,

$$[\hat{D}, \hat{x}] = \hat{\mathbb{I}}. \tag{4.123}$$

■

Problem: Show that

$$[\hat{D}^2, \hat{x}] = 2\hat{D}. \tag{4.124}$$

Solution: Long proof.

$$[\hat{D}^2, \hat{x}]f(x) \stackrel{\text{def.}}{=} \left(\hat{D}^2\hat{x} - \hat{x}\hat{D}^2\right)f(x) \tag{4.125a}$$

$$\stackrel{\text{linearity}}{=} \hat{D}^2\left(\hat{x}f(x)\right) - \hat{x}\left(\hat{D}^2f(x)\right) \tag{4.125b}$$

$$\stackrel{\text{the action of } \hat{x}}{=} \hat{D}^2\left(xf(x)\right) - x\left(\hat{D}^2f(x)\right) \tag{4.125c}$$

$$\stackrel{\text{def. of } \hat{D}^2}{=} \hat{D}\left(\underbrace{\hat{D}\left(xf(x)\right)}\right) - x\left(\hat{D}^2f(x)\right) \tag{4.125d}$$

$$\overset{\text{the action of } \hat{D}}{=} \hat{D}\left(\overbrace{\left(\hat{D}x\right)f(x) + x\left(\hat{D}f(x)\right)}\right) - x\left(\hat{D}^2f(x)\right) \tag{4.125e}$$

$$= \hat{D}\left(f(x) + x\left(\hat{D}f(x)\right)\right) - x\left(\hat{D}^2f(x)\right) \tag{4.125f}$$

$$= \hat{D}f(x) + \hat{D}\left(x\left(\hat{D}f(x)\right)\right) - x\left(\hat{D}^2f(x)\right) \tag{4.125g}$$

$$= \hat{D}f(x) + \underbrace{\left(\hat{D}x\right)}_{1}\left(\hat{D}f(x)\right) + \underbrace{x\hat{D}\left(\hat{D}f(x)\right)}_{x(\hat{D}^2f(x))} - x\left(\hat{D}^2f(x)\right) \tag{4.125h}$$

$$= 2\hat{D}f(x). \tag{4.125i}$$

Since $f(x)$ is arbitrary,

$$[\hat{D}^2,\hat{x}] = 2\hat{D}. \tag{4.126}$$

Solution: Short proof.

$$[\hat{D}^2,\hat{x}] \overset{\text{def.}}{=} [\hat{D}\hat{D},\hat{x}] \tag{4.127a}$$

$$= \hat{D}\underbrace{[\hat{D},\hat{x}]}_{\hat{1}} + \underbrace{[\hat{D},\hat{x}]}_{\hat{1}}\hat{D} \tag{4.127b}$$

$$= 2\hat{D}. \tag{4.127c}$$

∎

Problem: Show that

$$[\hat{D}^3,\hat{x}] = 3\hat{D}^2. \tag{4.128}$$

Solution:

$$[\hat{D}^3,\hat{x}] \overset{\text{def.}}{=} [\hat{D}\hat{D}^2,\hat{x}] \tag{4.129a}$$

$$= \hat{D}\underbrace{[\hat{D}^2,\hat{x}]}_{2\hat{D}} + \underbrace{[\hat{D},\hat{x}]}_{\hat{1}}\hat{D}^2 \tag{4.129b}$$

$$= 3\hat{D}^2. \tag{4.129c}$$

∎

Problem: Show by mathematical induction that

$$[\hat{D}^n,\hat{x}] = n\hat{D}^{n-1}. \tag{4.130}$$

Solution:

$$[\hat{D}^{n+1},\hat{x}] \overset{\text{def.}}{=} [\hat{D}\hat{D}^n,\hat{x}] \tag{4.131a}$$

$$= \hat{D}\underbrace{[\hat{D}^n,\hat{x}]}_{\hat{D}^{n-1}} + \underbrace{[\hat{D},\hat{x}]}_{\hat{1}}\hat{D}^n \tag{4.131b}$$

$$= (n+1)\hat{D}^n. \tag{4.131c}$$

∎

Problem: Show that,

$$\left[f(\hat{D}), \hat{x}\right] = f^{(1)}(\hat{D}). \tag{4.132}$$

Remark: Two comments are in place.

- The symbol $f^{(1)}(\cdot)$ refers to the derivative of $f(\cdot)$ with respect to its argument, whatever the argument maybe. In the preset case the appearance of the derivative operator $\hat{D}$ as the argument in $f(\hat{D})$ and $f^{(1)}(\hat{D})$ might seem slightly counterintuitive, since the variable of the function is itself the differential operator. The symbol $f^{(1)}(\hat{D})$ must be interpreted as follows: given $f(\hat{D})$ replace $\hat{D}$ with an "ordinary" variable, say x, to obtain the "ordinary" function $f(x)$. Following the discussion at the beginning of the current chapter, the calculation of the derivative of $f(x)$ with respect to x, i.e., $f^{(1)}(x)$, is immediate. Having determined $f^{(1)}(x)$, the "ordinary" variable x can be replaced by the operator $\hat{D}$, thus leading to $f^{(1)}(\hat{D})$. Obviously choosing y instead of x would have implied $f(y)$, $f^{(1)}(y)$, and thus again $f^{(1)}(\hat{D})$. The fact that the choice of the argument is immaterial, is expressed in the representation $f(\cdot)$, where the $\cdot$ can stand for any mathematical entity, variable, differential operator $\hat{D}$, a general operator $\mathcal{L}$, or a matrix $\mathbf{A}$. Thus, given $f(x), f(\mathcal{L})$, or $f(\mathbf{A})$, the expression for $f^{(1)}(\cdot)$ can be obtained from $f(\cdot)$.
- Whenever statements concern general functions the Formal Taylor Series Expansion must be resort to, after the corresponding relationship involving general monomials have been established. This step was accomplished in the previous problem.

$\square$

Solution:

$$[f(\hat{D}), \hat{x}] \overset{\text{FTSE}}{=} \left[\sum_{n=0}^{\infty} \frac{1}{n!} f^{(n)}(0) \hat{D}^n, \hat{x}\right] \tag{4.133a}$$

$$\overset{\text{linearity}}{=} \sum_{n=0}^{\infty} \frac{1}{n!} f^{(n)}(0) \left[\hat{D}^n, \hat{x}\right] \tag{4.133b}$$

$$= \frac{1}{0!} f^{(0)}(0) \underbrace{\left[\hat{D}^0, \hat{x}\right]}_{\hat{0}} + \frac{1}{1!} f^{(1)}(0) \underbrace{\left[\hat{D}^1, \hat{x}\right]}_{\hat{\mathbb{1}} = \hat{D}^0} + \frac{1}{2!} f^{(2)}(0) \underbrace{\left[\hat{D}^2, \hat{x}\right]}_{2\hat{D}^1} + \cdots$$

$$+ \frac{1}{n!} f^{(n)}(0) \underbrace{\left[\hat{D}^n, \hat{x}\right]}_{n\hat{D}^{n-1}} + \frac{1}{(n+1)!} f^{(n+1)}(0) \underbrace{\left[\hat{D}^{n+1}, \hat{x}\right]}_{(n+1)\hat{D}^n} + \cdots \tag{4.133c}$$

$$= \underbrace{\frac{1}{1!}}_{\frac{1}{0!}} f^{(1)}(0) \hat{D}^0 + \underbrace{\frac{1}{}}_{\frac{1}{1!}} f^{(2)}(0) \hat{D}^1 + \cdots$$

$$+ \frac{1}{(n-1)!} f^{(n)}(0) \hat{D}^{n-1} + \frac{1}{(n)!} f^{(n+1)}(0) \hat{D}^n + \cdots \tag{4.133d}$$

$$\overset{f^{(k+1)} = (f^{(1)})^{(k)}}{=} \frac{1}{0!} \left(f^{(1)}(0)\right)^{(0)} \hat{D}^0 + \frac{1}{1!} \left(f^{(1)}(0)\right)^{(1)} \hat{D}^1 + \cdots$$

$$+ \frac{1}{(n-1)!} \left(f^{(1)}(0)\right)^{(n-1)} \hat{D}^{n-1} + \frac{1}{(n)!} \left(f^{(1)}(0)\right)^{(n)} \hat{D}^n \cdots \tag{4.133e}$$

$$= \sum_{n=0}^{\infty} \frac{1}{n!} \left(f^{(1)}(0) \right)^{(n)} \hat{D}^n \tag{4.133f}$$

$$\overset{\text{FTSE}}{=} f^{(1)}(\hat{D}). \tag{4.133g}$$

In the above the comments are meant to be understood as follows.

- The abbreviation FTSE stands for the Formal Taylor Series Expansion.
- The linearity property $[\alpha A + \beta B, C] = \alpha[A, C] + \beta[B, C]$ has been employed.
- $f^{(k+1)} = \left(f^{(1)} \right)^{(k)}$ means that the $(k+1)$-derivative of f can be regarded as the k-derivative of $f^{(1)}$. To more fully appreciate this idea, consider the following analysis:

$$f^{(m+1)}(0) \overset{\text{def.}}{=} \left[\frac{d^{m+1}}{dx^{m+1}} f(x) \right]_{x=0} \tag{4.134a}$$

$$= \left[\frac{d^m}{dx^m} \left(\frac{d}{dx} f(x) \right) \right]_{x=0}. \tag{4.134b}$$

Denote,

$$\frac{d}{dx} f(x) = g(x). \tag{4.135}$$

Then,

$$f^{(m+1)}(0) = \left[\frac{d^m}{dx^m} g(x) \right]_{x=0} \tag{4.136a}$$

$$= g^{(m)}(0) \tag{4.136b}$$

Thus,

$$\left[f\left(\hat{D} \right), \hat{x} \right] = \sum_{m=0}^{\infty} \frac{1}{m!} g^{(m)}(0) \hat{D}^m \tag{4.137a}$$

$$= g\left(\hat{D} \right). \tag{4.137b}$$

∎

The result obtained in the last problem is interesting if not astonishing. Thereby, it should be pointed out that the nature of the operators involved; i.e., $\hat{x}$ and $\hat{D}$ are the defining factor. Every two operators $\hat{A}$ and $\hat{B}$ that satisfy the commutation property $[\hat{A}, \hat{B}] = \hat{\mathbb{I}}$ lead to the same result. The proof of this fact constitutes the content of the following two problems.

Problem: Assuming the commutation relationship,

$$\left[\hat{A}, \hat{B} \right] = \hat{\mathbb{I}}, \tag{4.138}$$

show that,

$$\left[\hat{A}^n, \hat{B} \right] = n\hat{A}^{n-1}. \tag{4.139}$$

Solution: Proof by induction. Assuming the induction hypothesis, (4.139),

$$\left[\hat{A}^{n+1}, \hat{B}\right] = \left[\hat{A}\hat{A}^n, \hat{B}\right] \tag{4.140a}$$

$$= \hat{A}\underbrace{\left[\hat{A}^n, \hat{B}\right]}_{n\hat{A}^{n-1}} + \underbrace{\left[\hat{A}, \hat{B}\right]}_{\hat{\mathbb{I}}}\hat{A}^n \tag{4.140b}$$

$$= n\hat{A}^n + \hat{A}^n \tag{4.140c}$$

$$= (n+1)\hat{A}^n. \tag{4.140d}$$

∎

Consequently, it is the algebraic property of the commutation relationship $\left[\hat{A}, \hat{B}\right] = \hat{\mathbb{I}}$ which leads to the observed differentiation operation $\left[\hat{A}^{n+1}, \hat{B}\right] = (n+1)\hat{A}^n$, rather than the fact that $\hat{D}$ is the differential operator d/dx. The self-explanatory steps in the solution of the next problem should further elaborate the reasoning.

Problem: Given the commutation relationship $[\hat{A}, \hat{B}] = \hat{\mathbb{I}}$, and thus $[\hat{A}^n, \hat{B}] = n\hat{A}^{n-1}$ show that,

$$\left[f(\hat{A}), \hat{B}\right] = f^{(1)}(\hat{A}). \tag{4.141}$$

Solution: The proof is a quick repetition of the steps in the solution further above with slight modifications carried out *mutatis mutandis*.

$$[f(\hat{A}), \hat{B}] = \left[\sum_{n=0}^{\infty} \frac{1}{n!}f^{(n)}(0)\hat{A}^n, \hat{B}\right] \tag{4.142a}$$

$$= \sum_{n=0}^{\infty} \frac{1}{n!}f^{(n)}(0)\underbrace{\left[\hat{A}^n, \hat{B}\right]}_{} \tag{4.142b}$$

The $(n=0)$-term $\left[\hat{A}^0, \hat{B}\right] = \left[\hat{\mathbb{I}}, \hat{B}\right] = \hat{0}$ does not contribute. Thus, the summation starts with $n=1$,

$$[f(\hat{A}), \hat{B}] = \sum_{n=1}^{\infty} \frac{1}{n!}f^{(n)}(0)\underbrace{\left[\hat{A}^n, \hat{B}\right]}_{n\hat{A}^{n-1}} \tag{4.143a}$$

$$= \sum_{n=1}^{\infty} \frac{1}{n!}f^{(n)}(0)n\hat{A}^{n-1} \tag{4.143b}$$

$$= \sum_{n=1}^{\infty} \frac{1}{(n-1)!}f^{(n)}(0)\hat{A}^{n-1} \tag{4.143c}$$

$$= \sum_{n=1}^{\infty} \frac{1}{(n-1)!}\left(f^{(1)}\right)^{(n-1)}(0)\hat{A}^{n-1} \tag{4.143d}$$

$$\stackrel{m=n-1}{=} \sum_{m=0}^{\infty} \frac{1}{m!}\left(f^{(1)}\right)^{(m)}(0)\hat{A}^m \tag{4.143e}$$

$$= f^{(1)}(\hat{A}). \tag{4.143f}$$

∎

Problem: Show that $[\hat{D}, \hat{x}^n] = n\hat{x}^{n-1}$.

Solution: The proof by mathematical induction is immediate. The induction step,

$$[\hat{D}, \hat{x}] = \hat{\mathbb{I}}, \tag{4.144}$$

is the well-established formula. Assuming $[\hat{D}, \hat{x}^n] = n\hat{x}^{n-1}$, the induction step is,

$$[\hat{D}, \hat{x}^{n+1}] = [\hat{D}, \hat{x}\hat{x}^n] \tag{4.145a}$$

$$= \hat{x}\underbrace{[\hat{D}, \hat{x}^n]}_{n\hat{x}^{n-1}} + \underbrace{[\hat{D}, \hat{x}]}_{\hat{\mathbb{I}}}\hat{x}^n \tag{4.145b}$$

$$= n\hat{x}^n + \hat{x}^n \tag{4.145c}$$

$$= (n+)\hat{x}^n. \tag{4.145d}$$

∎

Problem: Given the commutation relationship $[\hat{D}, \hat{x}] = \hat{\mathbb{I}}$, and thus $[\hat{D}, \hat{x}^n] = n\hat{x}^{n-1}$ show that,

$$\left[\hat{D}, f(\hat{x})\right] = f^{(1)}(\hat{x}). \tag{4.146}$$

Solution:

$$\left[\hat{D}, f(\hat{x})\right] = \left[\hat{D}, \sum_{n=0}^{\infty} \frac{1}{n!} f^{(n)}(0)\hat{x}^n\right] \tag{4.147a}$$

$$= \sum_{n=0}^{\infty} \frac{1}{n!} f^{(n)}(0)\underbrace{\left[\hat{D}, \hat{x}^n\right]}. \tag{4.147b}$$

The $n = 0$ term $\left[\hat{D}, \hat{x}^0\right] = \left[\hat{D}, \hat{\mathbb{I}}\right] = \hat{0}$ does not contribute. Thus, the summation starts from $n = 1$,

$$\left[\hat{D}, f(\hat{x})\right] = \sum_{n=1}^{\infty} \frac{1}{n!} f^{(n)}(0)\underbrace{\left[\hat{D}, \hat{x}^n\right]} \tag{4.148a}$$

$$= \sum_{n=1}^{\infty} \frac{1}{n!} f^{(n)}(0)n\hat{x}^{n-1} \tag{4.148b}$$

$$= \sum_{n=1}^{\infty} \frac{1}{(n-1)!} \underbrace{f^{(n)}(0)}\hat{x}^{n-1} \tag{4.148c}$$

$$= \sum_{n=1}^{\infty} \frac{1}{(n-1)!} \left(f^{(1)}\right)^{(n-1)}(0)\hat{x}^{n-1} \tag{4.148d}$$

$$\stackrel{m=n-1}{=} \sum_{m=0}^{\infty} \frac{1}{m!} \left(f^{(1)}\right)^{(m)}(0)\hat{x}^m \tag{4.148e}$$

$$= f^{(1)}(\hat{x}). \tag{4.148f}$$

∎

Problem: Find a representation of $[\hat{D}^3, \hat{x}^4]$ by resolving with respect to $\hat{x}$.

Solution:

$$\left[\hat{D}^3, \hat{x}^4\right] = \left[\hat{D}^3, \hat{x}\hat{x}^3\right] \tag{4.149a}$$

$$= \hat{x}\underbrace{\left[\hat{D}^3, \hat{x}^3\right]} + \left[\hat{D}^3, \hat{x}\right]\hat{x}^3 \tag{4.149b}$$

$$= \hat{x}\underbrace{\left[\hat{D}^3, \hat{x}\hat{x}^2\right]} + \left[\hat{D}^3, \hat{x}\right]\hat{x}^3 \tag{4.149c}$$

$$= \hat{x}\left\{\hat{x}\left[\hat{D}^3, \hat{x}^2\right] + \left[\hat{D}^3, \hat{x}\right]\hat{x}^2\right\} + \left[\hat{D}^3, \hat{x}\right]\hat{x}^3 \tag{4.149d}$$

$$= \hat{x}^2\underbrace{\left[\hat{D}^3, \hat{x}^2\right]} + \hat{x}\left[\hat{D}^3, \hat{x}\right]\hat{x}^2 + \left[\hat{D}^3, \hat{x}\right]\hat{x}^3 \tag{4.149e}$$

$$= \hat{x}^2\underbrace{\left[\hat{D}^3, \hat{x}\hat{x}\right]} + \hat{x}\left[\hat{D}^3, \hat{x}\right]\hat{x}^2 + \left[\hat{D}^3, \hat{x}\right]\hat{x}^3 \tag{4.149f}$$

$$= \hat{x}^2\left\{\hat{x}\left[\hat{D}^3, \hat{x}\right] + \left[\hat{D}^3, \hat{x}\right]\hat{x}\right\} + \hat{x}\left[\hat{D}^3, \hat{x}\right]\hat{x}^2 + \left[\hat{D}^3, \hat{x}\right]\hat{x}^3 \tag{4.149g}$$

$$= \hat{x}^3\underbrace{\left[\hat{D}^3, \hat{x}\right]}_{3\hat{D}^2} + \hat{x}^2\underbrace{\left[\hat{D}^3, \hat{x}\right]}\hat{x} + \hat{x}\underbrace{\left[\hat{D}^3, \hat{x}\right]}\hat{x}^2 + \underbrace{\left[\hat{D}^3, \hat{x}\right]}\hat{x}^3 \tag{4.149h}$$

$$= \hat{x}^3\left(3\hat{D}^2\right)\hat{x} + \hat{x}^2\left(3\hat{D}^2\right)\hat{x} + \hat{x}\left(3\hat{D}^2\right)\hat{x}^2 + \left(3\hat{D}^2\right)\hat{x}^3 \tag{4.149i}$$

$$= \hat{x}^3\left(3\hat{D}^2\right)\hat{x}^0 + \hat{x}^2\left(3\hat{D}^2\right)\hat{x}^1 + \hat{x}^1\left(3\hat{D}^2\right)\hat{x}^2 + \hat{x}^0\left(3\hat{D}^2\right)\hat{x}^3. \tag{4.149j}$$

∎

Remark: Note that the successive "resolution" of $\hat{x}^4$ ultimately led to $\left[\hat{D}^3, \hat{x}\right]$ in (4.149h) with $\left[\hat{D}^3, \hat{x}\right] = 3\hat{D}^2$. The "cosmetic" inclusion of $\hat{x}^0 = \hat{\mathbb{I}}$ in the last step is merely to aid the formulation of the following observation. Considering the first and the last terms in (4.149), i.e.,

$$\left[\hat{D}^3, \hat{x}^4\right] = \hat{x}^3\left(3\hat{D}^2\right)\hat{x}^0 + \hat{x}^2\left(3\hat{D}^2\right)\hat{x}^1 + \hat{x}^1\left(3\hat{D}^2\right)\hat{x}^2 + \hat{x}^0\left(3\hat{D}^2\right)\hat{x}^3, \tag{4.150}$$

the following "algorithmic" hypothesis offers itself:

1. Given $[\hat{D}^3, \hat{x}^4]$, consider $\hat{x}^4$.
2. "Invest" one $\hat{x}$ (out of $\hat{x}^4$) and construct $[\hat{D}^3, \hat{x}] = 3\hat{D}^2$.
3. Consider the "remaining" $\hat{x}^3$ (out of the initial $\hat{x}^4$).
4. Exhaust multiplicative factorization of $\hat{x}^3$, and add the terms together, to obtain:

$$\hat{x}^3\hat{x}^0 + \hat{x}^2\hat{x}^1 + \hat{x}^1\hat{x}^2 + \hat{x}^0\hat{x}^3$$

5. Insert $3\hat{D}^2$, obtained in step (2), between any two terms in the factorized form in step (4):

$$\hat{x}^3\left(3\hat{D}^2\right)\hat{x}^0 + \hat{x}^2\left(3\hat{D}^2\right)\hat{x}^1 + \hat{x}^1\left(3\hat{D}^2\right)\hat{x}^2 + \hat{x}^0\left(3\hat{D}^2\right)\hat{x}^3$$

This is exactly the expression at the R.H.S. of (4.150).

□

Remark: There is nothing particular about the exponent 3 in $\left[\hat{D}^3,\hat{x}^4\right]$. Thus, replacing $\hat{D}^3$ with $\hat{D}^M$, it can be expected that,

$$\left[\hat{D}^M,\hat{x}^4\right] = \hat{x}^3\left(M\hat{D}^{M-1}\right)\hat{x}^0 + \hat{x}^2\left(M\hat{D}^{M-1}\right)\hat{x}^1 + \hat{x}^1\left(M\hat{D}^{M-1}\right)\hat{x}^2 + \hat{x}^0\left(M\hat{D}^{M-1}\right)\hat{x}^3, \tag{4.151}$$

which can be written compactly,

$$\left[\hat{D}^M,\hat{x}^4\right] = \sum_{n=0}^{4-1}\hat{x}^{(4-1)-n}\left(M\hat{D}^{M-1}\right)\hat{x}^n. \tag{4.152}$$

Furthermore, there is nothing particular about the exponent 4 of $\hat{x}^4$ in $\left[\hat{D}^3,\hat{x}^4\right]$ either. Thus, replacing $\hat{x}^4$ with $\hat{x}^N$, it can be expected that,

$$\left[\hat{D}^M,\hat{x}^N\right] = \sum_{n=0}^{N-1}\hat{x}^{(N-1)-n}\left(M\hat{D}^{M-1}\right)\hat{x}^n. \tag{4.153}$$

Finally, replacing $\hat{D}^M$ with the arbitrary $f(\hat{D})$, it can be hypothesized that,

$$\left[f\left(\hat{D}\right),\hat{x}^N\right] = \sum_{n=0}^{N-1}\hat{x}^{(N-1)-n}\left(f^{(1)}\left(\hat{D}\right)\right)\hat{x}^n. \tag{4.154}$$

$\square$

Problem: Show the validity of (4.154).

Solution: The proof by mathematical induction requires considering the initial step $N = 1$ (and thus $n = 0$),

$$\left[f\left(\hat{D}\right),\hat{x}\right] = f^{(1)}\left(\hat{D}\right). \tag{4.155}$$

This result was established earlier above. Assuming (4.154), for establishing the induction step proceed as follows:

$$\left[f\left(\hat{D}\right),\hat{x}^{N+1}\right] = \left[f\left(\hat{D}\right),\hat{x}\hat{x}^N\right] \tag{4.156a}$$

$$= \hat{x}\underbrace{\left[f\left(\hat{D}\right),\hat{x}^N\right]}_{\text{ind. hyp.}} + \underbrace{\left[f\left(\hat{D}\right),\hat{x}\right]}_{\text{proved}}\hat{x}^N \tag{4.156b}$$

$$= \hat{x}\left(\sum_{n=0}^{N-1}\hat{x}^{(N-1)-n}\left(f^{(1)}\left(\hat{D}\right)\right)\hat{x}^n\right) + f^{(1)}\left(\hat{D}\right)\hat{x}^N \tag{4.156c}$$

$$= \sum_{n=0}^{N-1}\hat{x}^{N-n}\left(f^{(1)}\left(\hat{D}\right)\right)\hat{x}^n + \hat{x}^0 f^{(1)}\left(\hat{D}\right)\hat{x}^N. \tag{4.156d}$$

Note that the first term at the R.H.S. of (4.156d) is obtained by writing $\hat{x}\hat{x}^{(N-1)-n} = \hat{x}^{N-n}$ and augmenting the second term $f^{(1)}\left(\hat{D}\right)\hat{x}^N$ to $\hat{x}^0 f^{(1)}\left(\hat{D}\right)\hat{x}^N$ since $\hat{x}^0 = \hat{\mathbb{I}}$. Furthermore, note that the last term in the sum in (4.156d); i.e., $n = N - 1$, is $\hat{x}^1\left(f^{(1)}\left(\hat{D}\right)\right)\hat{x}^{N-1}$. Merging the two terms,

$$\left[f\left(\hat{D}\right),x^{N+1}\right] = \sum_{n=0}^{N}\hat{x}^{N-n}\left(f^{(1)}\left(\hat{D}\right)\right)\hat{x}^n, \tag{4.157}$$

thus completing the proof.

■

Problem: Find a representation of $[\hat{D}^3, \hat{x}^4]$ by resolving with respect to $\hat{D}$.

Solution:

$$\left[\hat{D}^3, \hat{x}^4\right] = \left[\hat{D}\hat{D}^2, \hat{x}^4\right] \tag{4.158a}$$

$$= \hat{D}\underbrace{\left[\hat{D}^2, \hat{x}^4\right]} + \left[\hat{D}, \hat{x}^4\right]\hat{D}^2 \tag{4.158b}$$

$$= \hat{D}\underbrace{\left[\hat{D}\hat{D}, \hat{x}^4\right]} + \left[\hat{D}, \hat{x}^4\right]\hat{D}^2 \tag{4.158c}$$

$$= \hat{D}\left\{\hat{D}\left[\hat{D}, \hat{x}^4\right] + \left[\hat{D}, \hat{x}^4\right]\hat{D}\right\} + \left[\hat{D}, \hat{x}^4\right]\hat{D}^2 \tag{4.158d}$$

$$= \hat{D}^2\underbrace{\left[\hat{D}, \hat{x}^4\right]}_{4\hat{x}^3} + \hat{D}^1\underbrace{\left[\hat{D}, \hat{x}^4\right]}\hat{D}^1 + \underbrace{\left[\hat{D}, \hat{x}^4\right]}\hat{D}^2 \tag{4.158e}$$

$$= \hat{D}^2\left(4\hat{x}^3\right)\hat{D}^0 + \hat{D}^1\left(4\hat{x}^3\right)\hat{D}^1 + \hat{D}^0\left(4\hat{x}^3\right)\hat{D}^2. \tag{4.158f}$$

∎

Remark: Note that the successive resolution of $\hat{D}^3$ ultimately led to $\left[\hat{D}, \hat{x}^4\right]$ in (4.158e) with $\left[\hat{D}, \hat{x}^4\right] = 4\hat{x}^3$. The cosmetic inclusion of $\hat{D}^0 = \hat{\mathbb{I}}$ in the last step is merely to aid the formulation of the following observation. Considering the first and the last terms in (4.158), i.e.,

$$\left[\hat{D}^3, \hat{x}^4\right] = \hat{D}^2\left(4\hat{x}^3\right)\hat{D}^0 + \hat{D}^1\left(4\hat{x}^3\right)\hat{D}^1 + \hat{D}^0\left(4\hat{x}^3\right)\hat{D}^2, \tag{4.159}$$

the following pseudo algorithm offers itself:

1. Given $[\hat{D}^3, \hat{x}^4]$, consider $\hat{D}^3$.
2. "Invest" one $\hat{D}$ (out of $\hat{D}^3$) and construct $[\hat{D}, \hat{x}^4] = 4\hat{x}^3$.
3. Consider the "remaining" $\hat{D}^2$ (out of $\hat{D}^3$).
4. Exhaust multiplicative factorization of $\hat{D}^2$, and add the terms together, to obtain:

$$\hat{D}^2\hat{D}^0 + \hat{D}^1\hat{D}^1 + \hat{D}^0\hat{D}^2$$

5. Insert $4\hat{x}^3$, obtained in step (2), between any two terms in the factorized form in step (4):

$$\hat{D}^2\left(4\hat{x}^3\right)\hat{D}^0 + \hat{D}^1\left(4\hat{x}^3\right)\hat{D}^1 + \hat{D}^0\left(4\hat{x}^3\right)\hat{D}^2$$

This is exactly the expression at the R.H.S. of (4.159).

□

Remark: There is nothing particular about the exponent 4 in $\left[\hat{D}^3, \hat{x}^4\right]$. Thus, replacing $\hat{x}^4$ with $\hat{x}^N$, it can be expected that,

$$\left[\hat{D}^3, \hat{x}^N\right] = \hat{D}^2\left(N\hat{x}^{N-1}\right)\hat{D}^0 + \hat{D}^1\left(N\hat{x}^{N-1}\right)\hat{D}^1 + \hat{D}^0\left(N\hat{x}^{N-1}\right)\hat{D}^2, \tag{4.160}$$

which can be written compactly,

$$\left[\hat{D}^3, \hat{x}^N\right] = \sum_{m=0}^{3-1}\hat{D}^{(3-1)-m}\left(N\hat{x}^{N-1}\right)\hat{D}^m. \tag{4.161}$$

Furthermore, there is nothing particular about the exponent 3 of $\hat{D}^3$ in $\left[\hat{D}^3, \hat{x}^4\right]$ either. Thus, replacing $\hat{D}^3$ with $\hat{D}^M$, it can be expected that,

$$\left[\hat{D}^M, \hat{x}^N\right] = \sum_{m=0}^{M-1} \hat{D}^{(M-1)-m} \left(N\hat{x}^{N-1}\right) \hat{D}^m. \tag{4.162}$$

Finally, replacing $\hat{x}^N$ with the arbitrary $g(\hat{x})$, it can be hypothesized that,

$$\left[\hat{D}^M, g(\hat{x})\right] = \sum_{m=0}^{M-1} \hat{D}^{(M-1)-m} \left(g^{(1)}(\hat{x})\right) \hat{D}^m. \tag{4.163}$$

□

Problem: Show the validity of (4.163).

Solution: The proof by mathematical induction requires considering the initial step $M = 1$ (and thus $m = 0$),

$$\left[\hat{D}, g(\hat{x})\right] = g^{(1)}(\hat{x}). \tag{4.164}$$

This result was established earlier above. Assuming (4.163), for establishing the induction step proceeds as follows:

$$\left[\hat{D}^{M+1}, g(\hat{x})\right] = \left[\hat{D}\hat{D}^M, g(\hat{x})\right] \tag{4.165a}$$

$$= \hat{D}\underbrace{\left[\hat{D}^M, g(\hat{x})\right]}_{\text{ind. hyp.}} + \underbrace{\left[\hat{D}, g(\hat{x})\right]}_{\text{proved}}\hat{D}^M \tag{4.165b}$$

$$= \hat{D}\left(\sum_{m=0}^{M-1} \hat{D}^{(M-1)-m} \left(g^{(1)}(\hat{x})\right) \hat{D}^m\right) + g^{(1)}(\hat{x}) \hat{D}^M \tag{4.165c}$$

$$= \sum_{m=0}^{M-1} \hat{D}^{M-m} \left(g^{(1)}(\hat{x})\right) \hat{D}^m + \hat{D}^0 g^{(1)}(\hat{x}) \hat{D}^M. \tag{4.165d}$$

Note that the first term at the R.H.S. of (4.165d) is obtained by writing $\hat{D}\hat{D}^{(M-1)-m} = \hat{D}^{M-m}$ and augmenting the second term $g^{(1)}(\hat{x})\hat{D}^M$ to $\hat{D}^0 g^{(1)}(\hat{x})\hat{D}^M$ since $\hat{D}^0 = \hat{\mathbb{I}}$. Furthermore, note that the last term in the sum in (4.165d), i.e., $m = M - 1$, is $\hat{D}^1 \left(g^{(1)}(\hat{x})\right)\hat{D}^{M-1}$. Merging the two terms,

$$\left[\hat{D}^{M+1}, g(\hat{x})\right] = \sum_{m=0}^{M} \hat{D}^{M-m} \left(g^{(1)}(\hat{x})\right) \hat{D}^m, \tag{4.166}$$

thus completing the proof.

∎

Since the results established in the preceding discussion, merely assumed the canonical commutativity $\left[\hat{D}, \hat{x}\right] = \hat{\mathbb{I}}$, they are valid for any $\hat{A}$ and $\hat{B}$ with $\left[\hat{A}, \hat{B}\right] = \hat{\mathbb{I}}$. Consequently, the results can be cast in more general terms, as it is done in the following theorem.

Theorem: *For arbitrary* $M, N \in \mathbb{N}$, *and* $\left[\hat{A}, \hat{B}\right] = \hat{\mathbb{1}}$, *the following results are valid*:

$$\left[\hat{A}^M, \hat{B}^N\right] = \sum_{n=0}^{N-1} \hat{B}^{(N-1)-n} \left(M\hat{A}^{M-1}\right) \hat{B}^n \tag{4.167a}$$

$$\left[f\left(\hat{A}\right), \hat{B}^N\right] = \sum_{n=0}^{N-1} \hat{B}^{(N-1)-n} \left(f^{(1)}\left(\hat{A}\right)\right) \hat{B}^n \tag{4.167b}$$

$$\left[\hat{A}^M, \hat{B}^N\right] = \sum_{m=0}^{M-1} \hat{A}^{(M-1)-m} \left(N\hat{B}^{N-1}\right) \hat{A}^m \tag{4.167c}$$

$$\left[\hat{A}^M, g(\hat{B})\right] = \sum_{m=0}^{M-1} \hat{A}^{(M-1)-m} \left(g^{(1)}(\hat{B})\right) \hat{A}^m. \tag{4.167d}$$

∎

Interpretation and recipes: Equations (4.167) were established in the preceding series of problems and solutions. In the following, recipes for the construction of these relationships have been explicated.

Recipe for the construction of (4.167a):

- Focus on $\hat{B}^N$.
- Consider the factorization $\hat{B}^N = \hat{B}\hat{B}^{N-1}$ and proceed with $\hat{B}$ and $\hat{B}^{N-1}$ as prescribed.
- Focus on $\hat{A}^M$.
- "Invest" $\hat{B}$ to obtain $\left[\hat{A}^M, \hat{B}\right] = M\hat{A}^{M-1}$.
- Factorize $\hat{B}^{N-1}$ in N varieties of the form of $\hat{B}^{N-1-n}\hat{B}^n$ with $n \in \{0, 1, \ldots, N-1\}$.
- Sandwich $M\hat{A}^{M-1}$ between $\hat{B}^{N-1-n}$ and $\hat{B}^n$, i.e., $\hat{B}^{(N-1)-n}\left(M\hat{A}^{M-1}\right)\hat{B}^n$.
- Sum the resulting $\hat{B}^{(N-1)-n}\left(M\hat{A}^{M-1}\right)\hat{B}^n$ over all n with $n \in \{0, 1, \ldots, N-1\}$, leading to N terms at the R.H.S.

Recipe for the construction of (4.167b):
This relationship shows that the above recipe holds valid if the monomial $\hat{A}^M$ is replaced with any analytic function of $\hat{A}$, i.e., $f\left(\hat{A}\right)$.

- Focus on $\hat{B}^N$.
- Consider the factorization $\hat{B}^N = \hat{B}\hat{B}^{N-1}$ and proceed with $\hat{B}$ and $\hat{B}^{N-1}$ as prescribed.
- Focus on $f(\hat{A})$.
- "Invest" $\hat{B}$ to obtain $\left[f(\hat{A}), \hat{B}\right] = f^{(1)}(\hat{A})$.
- Factorize $\hat{B}^{N-1}$ in N varieties of the form of $\hat{B}^{N-1-n}\hat{B}^n$ with $n \in \{0, 1, \ldots, N-1\}$.
- Sandwich $f^{(1)}(\hat{A})$ between $\hat{B}^{N-1-n}$ and $\hat{B}^n$, i.e., $\hat{B}^{(N-1)-n}\left(f^{(1)}(\hat{A})\right)\hat{B}^n$.
- Sum the resulting $\hat{B}^{(N-1)-n}\left(f^{(1)}(\hat{A})\right)\hat{B}^n$ over all n with $n \in \{0, 1, \ldots, N-1\}$, leading to N terms at the R.H.S.

The relationship $[\hat{A}, \hat{B}] = \hat{\mathbb{1}}$, and thus $[\hat{B}, \hat{A}] = -\hat{\mathbb{1}}$, can be exploited to state the recipes for the remaining two relationships, by just reversing the order of the operators $\hat{A}$ and $\hat{B}$.

Recipe for the construction of (4.167c):

- Focus on $\hat{A}^M$.
- Consider the factorization $\hat{A}^M = \hat{A}\hat{A}^{M-1}$ and proceed with $\hat{A}$ and $\hat{A}^{M-1}$ as prescribed.

- Focus on $\hat{B}^N$.
- "Invest" $\hat{A}$ to obtain $\left[\hat{A}, \hat{B}^N\right] = N\hat{B}^{N-1}$.
- Factorize $\hat{A}^{M-1}$ in M varieties of the form of $\hat{A}^{M-1-m}\hat{A}^m$ with $m \in \{0, 1, \ldots, M-1\}$.
- Sandwich $N\hat{B}^{N-1}$ between $\hat{A}^{M-1-m}$ and $\hat{A}^m$, i.e., $\hat{A}^{(M-1)-m}\left(N\hat{B}^{N-1}\right)\hat{A}^m$.
- Sum the resulting $\hat{A}^{(M-1)-m}\left(N\hat{B}^{N-1}\right)\hat{A}^m$ over all m with $m \in \{0, 1, \ldots, M-1\}$, leading to M terms at the R.H.S.

The final result concerns the generalization of $\hat{B}^N$ to $g(\hat{B})$,

Recipe for the construction of (4.167d):

- Focus on $\hat{A}^M$.
- Consider the factorization $\hat{A}^M = \hat{A}\hat{A}^{M-1}$ and proceed with $\hat{A}$ and $\hat{A}^{M-1}$ as prescribed.
- Focus on $g(\hat{B})$.
- "Invest" $\hat{A}$ to obtain $\left[\hat{A}, g(\hat{B})\right] = g^{(1)}(\hat{B})$.
- Factorize $\hat{A}^{M-1}$ in M varieties of the form of $\hat{A}^{M-1-m}\hat{A}^m$ with $m \in \{0, 1, \ldots, M-1\}$.
- Sandwich $g^{(1)}(\hat{B})$ between $\hat{A}^{M-1-m}$ and $\hat{A}^m$, i.e., $\hat{A}^{(M-1)-m}\left(g^{(1)}(\hat{B})\right)\hat{A}^m$.
- Sum the resulting $\hat{A}^{(M-1)-m}\left(g^{(1)}(\hat{B})\right)\hat{A}^m$ over all m with $m \in \{0, 1, \ldots, M-1\}$, leading to M terms at the R.H.S.

Corollary: *In the special case that $M = N = L \in \mathbb{N}$, $f(\cdot) = g(\cdot) = h(\cdot)$, and $\left[\hat{A}, \hat{B}\right] = \hat{\mathbb{I}}$, the following results are valid*:

$$\left[\hat{A}^L, \hat{B}^L\right] = \sum_{l=0}^{L-1} \hat{B}^{(L-1)-l}\left(L\hat{A}^{L-1}\right)\hat{B}^l = \sum_{l=0}^{L-1} \hat{B}^l\left(L\hat{A}^{L-1}\right)\hat{B}^{(L-1)-l} \tag{4.168a}$$

$$\left[h\left(\hat{A}\right), \hat{B}^L\right] = \sum_{l=0}^{L-1} \hat{B}^{(L-1)-l}\left(h^{(1)}\left(\hat{A}\right)\right)\hat{B}^l = \sum_{l=0}^{L-1} \hat{B}^l\left(h^{(1)}\left(\hat{A}\right)\right)\hat{B}^{(L-1)-l} \tag{4.168b}$$

$$\left[\hat{A}^L, \hat{B}^L\right] = \sum_{l=0}^{L-1} \hat{A}^{(L-1)-l}\left(L\hat{B}^{L-1}\right)\hat{A}^l = \sum_{l=0}^{L-1} \hat{A}^l\left(L\hat{B}^{L-1}\right)\hat{A}^{(L-1)-l} \tag{4.168c}$$

$$\left[\hat{A}^L, h(\hat{B})\right] = \sum_{l=0}^{L-1} \hat{A}^{(L-1)-l}\left(h^{(1)}(\hat{B})\right)\hat{A}^l = \sum_{l=0}^{L-1} \hat{A}^l\left(h^{(1)}(\hat{B})\right)\hat{A}^{(L-1)-l}. \tag{4.168d}$$

The second sum in each line in (4.168) results from reordering of the terms in the preceding sum. These alternative representations allow the discovery of new relationships, as demonstrated next. ∎

Several inferences can be drawn from the relationships in (4.168). Consider the sums in (4.168a) and (4.168c), which are all valid representations for $\left[\hat{A}^L, \hat{B}^L\right]$. Equating them new interesting equalities emerge. In particular setting the first sum in (4.168c) equal to the second sum in (4.168a) implies,

$$\sum_{l=0}^{L-1} \hat{A}^{(L-1)-l}\left(L\hat{B}^{L-1}\right)\hat{A}^l = \sum_{l=0}^{L-1} \hat{B}^l\left(L\hat{A}^{L-1}\right)\hat{B}^{(L-1)-l}. \tag{4.169}$$

Canceling L from both sides, and substituting $\hat{B}^{L-1} = \hat{B}^{(L-1)-l}\hat{B}^l$ and $\hat{A}^{L-1} = \hat{A}^l\hat{A}^{(L-1)-l}$ for the terms in the round brackets at the L.H.S. and R.H.S., respectively,

$$\sum_{l=0}^{L-1} \hat{A}^{(L-1)-l}\left(\hat{B}^{(L-1)-l}\hat{B}^l\right)\hat{A}^l = \sum_{l=0}^{L-1} \hat{B}^l\left(\hat{A}^l\hat{A}^{(L-1)-l}\right)\hat{B}^{(L-1)-l}. \tag{4.170}$$

Utilizing the associativity property and regrouping terms,

$$\sum_{l=0}^{L-1} \left(\hat{A}^{(L-1)-l}\hat{B}^{(L-1)-l}\right)\left(\hat{B}^l\hat{A}^l\right) = \sum_{l=0}^{L-1} \left(\hat{B}^l\hat{A}^l\right)\left(\hat{A}^{(L-1)-l}\hat{B}^{(L-1)-l}\right). \tag{4.171}$$

Bringing the sum at the right-hand side to the left-hand side, and using the commutator definition,

$$\sum_{l=0}^{L-1} \left[\hat{A}^{(L-1)-l}\hat{B}^{(L-1)-l}, \hat{B}^l\hat{A}^l\right] = 0, \tag{4.172}$$

which is valid for $L \geq 1$. Substituting $K = L - 1$, and renaming the running index to k, lead to:

Corollary: *For any operators $\hat{A}$ and $\hat{B}$, satisfying the condition $[\hat{A}, \hat{B}] = \hat{\mathbb{I}}$, and $K \in \mathbb{N}_0$,*

$$\sum_{k=0}^{K} \left[\hat{A}^{K-k}\hat{B}^{K-k}, \hat{B}^k\hat{A}^k\right] = 0. \tag{4.173}$$

∎

Remark: It is revealing to study this equation for small values of K. Thereby, it turns out that even a stronger version of (4.173) is valid, which justifies the formulation of a separate corollary. In the following special cases $K = 0$, $K = 1$, $K = 2$, and $K = 3$ are analyzed.

- $K = 0$ ($k = 0$) leads to $[\hat{\mathbb{I}}, \hat{\mathbb{I}}] = 0$, which is trivially true.
- $K = 1$ ($k = 0, 1$),

$$\underbrace{\left[\hat{A}\hat{B}, \hat{\mathbb{I}}\right]}_{=\hat{0}} + \underbrace{\left[\hat{\mathbb{I}}, \hat{B}\hat{A}\right]}_{=\hat{0}} = 0. \tag{4.174}$$

As demonstratively shown, each term at the left-hand side is individually (and trivially) zero.

- $K = 2$ ($k = 0, 1, 2$),

$$\underbrace{\left[\hat{A}^2\hat{B}^2, \hat{\mathbb{I}}\right]}_{=\hat{0}} + \left[\hat{A}\hat{B}, \hat{B}\hat{A}\right] + \underbrace{\left[\hat{\mathbb{I}}, \hat{B}^2\hat{A}^2\right]}_{=\hat{0}} = 0. \tag{4.175}$$

The first and the last terms are individually (and trivially) zero, leading to,

$$\left[\hat{A}\hat{B}, \hat{B}\hat{A}\right] = 0. \tag{4.176}$$

The validity of (4.176) can be easily and directly established, by merely resorting to $[\hat{A}, \hat{B}] = \hat{\mathbb{I}}$,

$$\left[\hat{A}\hat{B}, \hat{B}\hat{A}\right] = \hat{A}\left[\hat{B}, \hat{B}\hat{A}\right] + \left[\hat{A}, \hat{B}\hat{A}\right]\hat{B} \tag{4.177a}$$

$$= \hat{A}\left\{\hat{B}\left[\hat{B}, \hat{A}\right] + \left[\hat{B}, \hat{B}\right]\hat{A}\right\} + \left\{\hat{B}\left[\hat{A}, \hat{A}\right] + \left[\hat{A}, \hat{B}\right]\hat{A}\right\}\hat{B} \tag{4.177b}$$

$$= \hat{A}\hat{B}\left[\hat{B}, \hat{A}\right] + \hat{A}\underbrace{\left[\hat{B}, \hat{B}\right]}_{\hat{0}}\hat{A} + \hat{B}\underbrace{\left[\hat{A}, \hat{A}\right]}_{\hat{0}}\hat{B} + \left[\hat{A}, \hat{B}\right]\hat{A}\hat{B} \tag{4.177c}$$

$$= \hat{A}\hat{B}\underbrace{\left[\hat{B}, \hat{A}\right]}_{-\hat{\mathbb{I}}} + \underbrace{\left[\hat{A}, \hat{B}\right]}_{\hat{\mathbb{I}}}\hat{A}\hat{B}. \tag{4.177d}$$

$$= -\hat{A}\hat{B} + \hat{A}\hat{B} = \hat{0}. \tag{4.177e}$$

The cases ($K = 0, 1, 2$) considered so far suggest that each contributing commutator in the respective sum vanishes individually. Additionally, the case $K = 2$, implying (4.176), heralds the emergence of a new class of relationships, which is exemplified next. Consider,

$$\left[\hat{A}\hat{B}, \hat{B}\hat{A}\right] \overset{\text{def.}}{=} \hat{A}\hat{B}\hat{B}\hat{A} - \hat{B}\hat{A}\hat{A}\hat{B} = \hat{A}\hat{B}^2\hat{A} - \hat{B}\hat{A}^2\hat{B} = 0. \tag{4.178}$$

Consequently, provided $[\hat{A}, \hat{B}] = \hat{\mathbb{I}}$,

$$\hat{A}\hat{B}^2\hat{A} = \hat{B}\hat{A}^2\hat{B}. \tag{4.179}$$

The investigation of the case corresponding to $K = 3$ enhances the intuition further and reveals additional structural details of the new class of relationships.

- $K = 3$ ($k = 0, 1, 2, 3$)

$$\underbrace{\left[\hat{A}^3\hat{B}^3, \hat{\mathbb{I}}\right]}_{=\hat{0}} + \left[\hat{A}^2\hat{B}^2, \hat{B}\hat{A}\right] + \left[\hat{A}\hat{B}, \hat{B}^2\hat{A}^2\right] + \underbrace{\left[\hat{\mathbb{I}}, \hat{B}^3\hat{A}^3\right]}_{=\hat{0}} = 0. \tag{4.180}$$

The first and the last terms are each individually (and trivially) zero, leading to,

$$\left[\hat{A}^2\hat{B}^2, \hat{B}\hat{A}\right] + \left[\hat{A}\hat{B}, \hat{B}^2\hat{A}^2\right] = 0. \tag{4.181}$$

Considering the first term, it is next shown that $\left[\hat{A}^2\hat{B}^2, \hat{B}\hat{A}\right] = \hat{0}$.

$$\left[\hat{A}^2\hat{B}^2, \hat{B}\hat{A}\right] = \hat{A}^2\underbrace{\left[\hat{B}^2, \hat{B}\hat{A}\right]} + \underbrace{\left[\hat{A}^2, \hat{B}\hat{A}\right]}\hat{B}^2 \tag{4.182a}$$

$$= \hat{A}^2\overbrace{\left\{\hat{B}\left[\hat{B}^2, \hat{A}\right] + \left[\hat{B}^2, \hat{B}\right]\hat{A}\right\}} + \overbrace{\left\{\hat{B}\left[\hat{A}^2, \hat{A}\right] + \left[\hat{A}^2, \hat{B}\right]\hat{A}\right\}}\hat{B}^2 \tag{4.182b}$$

$$= \hat{A}^2\hat{B}\left[\hat{B}^2, \hat{A}\right] + \hat{A}^2\underbrace{\left[\hat{B}^2, \hat{B}\right]}_{=\hat{0}}\hat{A} + \hat{B}\underbrace{\left[\hat{A}^2, \hat{A}\right]}_{=\hat{0}}\hat{B}^2 + \left[\hat{A}^2, \hat{B}\right]\hat{A}\hat{B}^2 \tag{4.182c}$$

$$= \hat{A}^2\hat{B}\underbrace{\left[\hat{B}^2, \hat{A}\right]}_{-2\hat{B}} + \underbrace{\left[\hat{A}^2, \hat{B}\right]}_{2\hat{A}}\hat{A}\hat{B}^2 \tag{4.182d}$$

$$= -2\hat{A}^2\hat{B}^2 + 2\hat{A}^2\hat{B}^2 = \hat{0}. \tag{4.182e}$$

For obtaining this result, $\left[\hat{B}^2, \hat{A}\right] = -2\hat{B}$ and $\left[\hat{A}^2, \hat{B}\right] = 2\hat{A}$ were utilized in (4.182d). In return, establishing $\left[\hat{B}^2, \hat{A}\right] = -2\hat{B}$ and $\left[\hat{A}^2, \hat{B}\right] = 2\hat{A}$, merely requires $\left[\hat{A}, \hat{B}\right] = \hat{\mathbb{I}}$. This completes the demonstration of,

$$\left[\hat{A}^2\hat{B}^2, \hat{B}\hat{A}\right] = \hat{0}. \tag{4.183}$$

Before generalizing this result, consider a consequence of (4.183),

$$\hat{0} \stackrel{(4.183)}{=} \left[\hat{A}^2\hat{B}^2, \hat{B}\hat{A}\right] \stackrel{\text{def.}}{=} \hat{A}^2\hat{B}^2\hat{B}\hat{A} - \hat{B}\hat{A}\hat{A}^2\hat{B}^2 \tag{4.184a}$$

$$= \hat{A}^2\hat{B}^3\hat{A} - \hat{B}\hat{A}^3\hat{B}^2. \tag{4.184b}$$

Therefore,

$$\hat{A}^2\hat{B}^3\hat{A} = \hat{B}\hat{A}^3\hat{B}^2. \tag{4.185}$$

$\square$

Remark: Given $\hat{A}^2\hat{B}^3\hat{A}$ at the L.H.S., the term $\hat{B}\hat{A}^3\hat{B}^2$ at the R.H.S. can be constructed by, first, replacing $\hat{A}$ with $\hat{B}$, and $\hat{B}$ with $\hat{A}$; and second, by mirror imaging the exponents: $231 \rightarrow 132$. Furthermore, it is worth noting that, on each side, the exponents of the guarding operators added together give the exponent of the guarded operator.

$\square$

The preceding remark suggests the formulation of,

Corollary: *Given $\hat{A}$ and $\hat{B}$ with $\left[\hat{A}, \hat{B}\right] = \hat{\mathbb{I}}$ and $M \in \mathbb{N}_0$,*

$$\hat{A}^M\hat{B}^{M+1}\hat{A} = \hat{B}\hat{A}^{M+1}\hat{B}^M. \tag{4.186}$$

Proof: The following steps are self-explanatory. The proof follows the steps in (4.182), that is essentially replacing 2 with M, and accounting for the consequences thereof.

$$\left[\hat{A}^M\hat{B}^M, \hat{B}\hat{A}\right] = \hat{A}^M\underbrace{\left[\hat{B}^M, \hat{B}\hat{A}\right]} + \underbrace{\left[\hat{A}^M, \hat{B}\hat{A}\right]}\hat{B}^M \tag{4.187a}$$

$$= \hat{A}^M\overbrace{\left\{\hat{B}\left[\hat{B}^M, \hat{A}\right] + \left[\hat{B}^M, \hat{B}\right]\hat{A}\right\}} + \overbrace{\left\{\hat{B}\left[\hat{A}^M, \hat{A}\right] + \left[\hat{A}^M, \hat{B}\right]\hat{A}\right\}}\hat{B}^M \tag{4.187b}$$

$$= \hat{A}^M \hat{B} \left[\hat{B}^M, \hat{A} \right] + \hat{A}^M \underbrace{\left[\hat{B}^M, \hat{B} \right]}_{=\hat{0}} \hat{A} + \hat{B} \underbrace{\left[\hat{A}^M, \hat{A} \right]}_{=\hat{0}} \hat{B}^M + \left[\hat{A}^M, \hat{B} \right] \hat{A} \hat{B}^M \tag{4.187c}$$

$$= \hat{A}^M \hat{B} \underbrace{\left[\hat{B}^M, \hat{A} \right]}_{-M\hat{B}^{M-1}} + \underbrace{\left[\hat{A}^M, \hat{B} \right]}_{M\hat{A}^{M-1}} \hat{A} \hat{B}^M \tag{4.187d}$$

$$= -M \hat{A}^M \hat{B}^M + M \hat{A}^M \hat{B}^M = \hat{0} . \tag{4.187e}$$

On the other hand,

$$\hat{0} \stackrel{(4.187)}{=} \left[\hat{A}^M \hat{B}^M, \hat{B} \hat{A} \right] \stackrel{\text{def.}}{=} \hat{A}^M \hat{B}^M \hat{B} \hat{A} - \hat{B} \hat{A} \hat{A}^M \hat{B}^M \tag{4.188a}$$

$$= \hat{A}^M \hat{B}^{M+1} \hat{A} - \hat{B} \hat{A}^{M+1} \hat{B}^M . \tag{4.188b}$$

Consequently,

$$\hat{A}^M \hat{B}^{M+1} \hat{A} = \hat{B} \hat{A}^{M+1} \hat{B}^M . \tag{4.189}$$

∎

Remark: In the following sections, more complex functions of the operators $\hat{A}$ and $\hat{B}$ will be constructed by merely requiring the canonical commutator identity $\left[\hat{A}, \hat{B} \right] = \hat{\mathbb{I}}$. The entire collection of the foregoing properties will still be valid for the constructed operators, since the only construction requirement is $\left[\hat{A}, \hat{B} \right] = \hat{\mathbb{I}}$, irrespective of the structural complexity of the functions involving $\hat{A}$ and $\hat{B}$. The far-reaching implications of $\left[\hat{A}, \hat{B} \right] = \hat{\mathbb{I}}$ are awe-inspiring. The fact that Heisenberg's uncertainty principle concerns operators $\hat{A}$ and $\hat{B}$ subject to $\left[\hat{A}, \hat{B} \right] = \hat{\mathbb{I}}$ adds to the astonishment and satisfaction.

□

In view of (4.181), since $\left[\hat{A}^2 \hat{B}^2, \hat{B} \hat{A} \right]$ turned out to be zero, $\left[\hat{A} \hat{B}, \hat{B}^2 \hat{A}^2 \right]$ must also vanish. It is of course possible to demonstrate $\left[\hat{A} \hat{B}, \hat{B}^2 \hat{A}^2 \right] = \hat{0}$ directly. The manipulations below, paralleling (4.182), show that the "implications" of $\left[\hat{A}, \hat{B} \right] = \hat{\mathbb{I}}$ alone suffice to achieve the goal.

$$\left[\hat{A} \hat{B}, \hat{B}^2 \hat{A}^2 \right] = \hat{B}^2 \underbrace{\left[\hat{A} \hat{B}, \hat{A}^2 \right]}_{} + \underbrace{\left[\hat{A} \hat{B}, \hat{B}^2 \right]}_{} \hat{A}^2 \tag{4.190a}$$

$$= \hat{B}^2 \overbrace{\left\{ \hat{A} \left[\hat{B}, \hat{A}^2 \right] + \left[\hat{A}, \hat{A}^2 \right] \hat{B} \right\}} + \overbrace{\left\{ \hat{A} \left[\hat{B}, \hat{B}^2 \right] + \left[\hat{A}, \hat{B}^2 \right] \hat{B} \right\}} \hat{A}^2 \tag{4.190b}$$

$$= \hat{B}^2 \hat{A} \left[\hat{B}, \hat{A}^2 \right] + \hat{B}^2 \underbrace{\left[\hat{A}, \hat{A}^2 \right]}_{=\hat{0}} \hat{B} + \hat{A} \underbrace{\left[\hat{B}, \hat{B}^2 \right]}_{=\hat{0}} \hat{A}^2 + \left[\hat{A}, \hat{B}^2 \right] \hat{B} \hat{A}^2 \tag{4.190c}$$

$$= \hat{B}^2 \hat{A} \underbrace{\left[\hat{B}, \hat{A}^2 \right]}_{-2\hat{A}} + \underbrace{\left[\hat{A}, \hat{B}^2 \right]}_{2\hat{B}} \hat{B} \hat{A}^2 \tag{4.190d}$$

$$= -2 \hat{B}^2 \hat{A}^2 + 2 \hat{B}^2 \hat{A}^2 = \hat{0} . \tag{4.190e}$$

Consider,

$$\hat{0} \overset{(4.190)}{=} \left[\hat{A}\hat{B}, \hat{B}^2\hat{A}^2\right] \overset{\text{def.}}{=} \hat{A}\hat{B}\hat{B}^2\hat{A}^2 - \hat{B}^2\hat{A}^2\hat{A}\hat{B} \tag{4.191a}$$

$$= \hat{A}\hat{B}^3\hat{A}^2 - \hat{B}^2\hat{A}^3\hat{B}. \tag{4.191b}$$

Therefore,

$$\hat{A}\hat{B}^3\hat{A}^2 = \hat{B}^2\hat{A}^3\hat{B}. \tag{4.192}$$

Remark: The construction law stated further above holds valid here too. Given $\hat{A}\hat{B}^3\hat{A}^2$ at the L.H.S. in (4.192), to obtain the term at the R.H.S. proceed as follows: first, $\hat{A} \leftrightarrow \hat{B}$ to obtain, $\hat{A}\hat{B}\hat{A} \rightarrow \hat{B}\hat{A}\hat{B}$. Second, determine the mirror image of the exponents, $132 \rightarrow 231$, to obtain $\hat{B}^2\hat{A}^3\hat{B}$ at the R.H.S.

□

Remark: Assuming $\left[\hat{A}, \hat{B}\right] = \hat{\mathbb{I}}$, it is straightforward to show that,

$$\left[\hat{A}\hat{B}, \hat{B}^N\hat{A}^N\right] = 0, \tag{4.193}$$

and thus,

$$\hat{A}\hat{B}^{N+1}\hat{A}^N = \hat{B}^N\hat{A}^{N+1}\hat{B}. \tag{4.194}$$

□

Remark: Due to the obvious similarity of the relationships, the next generalization considers $\left[\hat{A}^M\hat{B}^M, \hat{B}^2\hat{A}^2\right]$ and does not elaborate $\left[\hat{A}^2\hat{B}^2, \hat{B}^N\hat{A}^N\right]$, which can be shown similarly.

□

Corollary: For $\hat{A}$ and $\hat{B}$ satisfying $\left[\hat{A}, \hat{B}\right] = \hat{\mathbb{I}}$, and $M \in \mathbb{N}_0$,

$$\left[\hat{A}^M\hat{B}^M, \hat{B}^2\hat{A}^2\right] = \hat{0}, \tag{4.195}$$

and consequentially,

$$\hat{A}^M\hat{B}^{M+2}\hat{A}^2 = \hat{B}^2\hat{A}^{M+2}\hat{B}^M. \tag{4.196}$$

Proof: The proof follows the familiar pattern, in addition to utilizing the implication of $\left[\hat{A}^M\hat{B}^M, \hat{B}\hat{A}\right] = \hat{0}$, i.e., $\hat{A}^M\hat{B}^{M+1}\hat{A} = \hat{B}\hat{A}^{M+1}\hat{B}^M$:

$$\left[\hat{A}^M\hat{B}^M, \hat{B}^2\hat{A}^2\right] = \hat{A}^M\underbrace{\left[\hat{B}^M, \hat{B}^2\hat{A}^2\right]} + \underbrace{\left[\hat{A}^M, \hat{B}^2\hat{A}^2\right]}\hat{B}^M \tag{4.197a}$$

$$= \hat{A}^M\left\{\hat{B}^2\underbrace{\left[\hat{B}^M, \hat{A}^2\right]} + \left[\hat{B}^M, \hat{B}^2\right]\hat{A}^2\right\} \dots$$

$$\ldots + \left\{ \hat{B}^2 \left[\hat{A}^M, \hat{A}^2 \right] + \left[\hat{A}^M, \hat{B}^2 \right] \hat{A}^2 \right\} \hat{B}^M \tag{4.197b}$$

$$= \hat{A}^M \hat{B}^2 \left[\hat{B}^M, \hat{A}^2 \right] + \hat{A}^M \underbrace{\left[\hat{B}^M, \hat{B}^2 \right]}_{=\hat{0}} \hat{A}^2$$

$$+ \hat{B}^2 \underbrace{\left[\hat{A}^M, \hat{A}^2 \right]}_{=\hat{0}} \hat{B}^M + \left[\hat{A}^M, \hat{B}^2 \right] \hat{A}^2 \hat{B}^M \tag{4.197c}$$

$$= \hat{A}^M \hat{B}^2 \left[\hat{B}^M, \hat{A}^2 \right] + \left[\hat{A}^M, \hat{B}^2 \right] \hat{A}^2 \hat{B}^M. \tag{4.197d}$$

Substituting,

$$\left[\hat{B}^M, \hat{A}^2 \right] = \hat{A} \left(-M\hat{B}^{M-1} \right) + \left(-M\hat{B}^{M-1} \right) \hat{A} \tag{4.198a}$$

$$\left[\hat{A}^M, \hat{B}^2 \right] = \hat{B} \left(M\hat{A}^{M-1} \right) + \left(M\hat{A}^{M-1} \right) \hat{B} \tag{4.198b}$$

into (4.197d),

$$\left[\hat{A}^M \hat{B}^M, \hat{B}^2 \hat{A}^2 \right] = \hat{A}^M \hat{B}^2 \left\{ -M\hat{A}\hat{B}^{M-1} - M\hat{B}^{M-1}\hat{A} \right\}$$
$$+ \left\{ M\hat{B}\hat{A}^{M-1} + M\hat{A}^{M-1}\hat{B} \right\} \hat{A}^2 \hat{B}^M. \tag{4.199}$$

Simplifying,

$$\left[\hat{A}^M \hat{B}^M, \hat{B}^2 \hat{A}^2 \right] = M \{ -\hat{A}^M \hat{B}^2 \hat{A} \hat{B}^{M-1} \underbrace{-\hat{A}^M \hat{B}^{M+1}\hat{A} + \hat{B}\hat{A}^{M+1}\hat{B}^M}_{=\hat{0}} + \hat{A}^{M-1}\hat{B}\hat{A}^2\hat{B}^M \}. \tag{4.200}$$

The second and the third terms cancel out in virtue of the corollary established above, thus

$$\left[\hat{A}^M \hat{B}^M, \hat{B}^2 \hat{A}^2 \right] = M \{ -\hat{A}^M \hat{B}^2 \hat{A} \hat{B}^{M-1} + \hat{A}^{M-1}\hat{B}\hat{A}^2\hat{B}^M \}. \tag{4.201}$$

A closer look at the two terms at the R.H.S. suggests splitting $\hat{A}^M$ and $\hat{B}^M$, respectively, in the form $\hat{A}^M = \hat{A}^{M-1}\hat{A}$ and $\hat{B}^M = \hat{B}\hat{B}^{M-1}$,

$$\left[\hat{A}^M \hat{B}^M, \hat{B}^2 \hat{A}^2 \right] = M \{ -\hat{A}^{M-1}\hat{A}\hat{B}^2\hat{A}\hat{B}^{M-1} + \hat{A}^{M-1}\hat{B}\hat{A}^2\hat{B}\hat{B}^{M-1} \}. \tag{4.202}$$

Factoring out $\hat{A}^{M-1}$ and $\hat{B}^{M-1}$,

$$\left[\hat{A}^M \hat{B}^M, \hat{B}^2 \hat{A}^2 \right] = M\hat{A}^{M-1} \underbrace{\left(-\hat{A}\hat{B}^2\hat{A} + \hat{B}\hat{A}^2\hat{B} \right)}_{=\hat{0}} \hat{B}^{M-1} \tag{4.203a}$$

$$= \hat{0}. \tag{4.203b}$$

Once again, applying the corollary, it is seen that the terms in the brackets vanish. This completes the proof and, furthermore, implies,

$$\hat{0} \overset{(4.203)}{=} \left[\hat{A}^M \hat{B}^M, \hat{B}^2 \hat{A}^2 \right] \overset{\text{def.}}{=} \hat{A}^M \hat{B}^M \hat{B}^2 \hat{A}^2 - \hat{B}^2 \hat{A}^2 \hat{A}^M \hat{B}^M \tag{4.204a}$$

$$= \hat{A}^M \hat{B}^{M+2} \hat{A}^2 - \hat{B}^2 \hat{A}^{M+2} \hat{B}^M, \tag{4.204b}$$

$$\tag{4.204c}$$

and thus,

$$\hat{A}^M \hat{B}^{M+2} \hat{A}^2 = \hat{B}^2 \hat{A}^{M+2} \hat{B}^M. \tag{4.205}$$

■

Corollary: *For $\hat{A}$ and $\hat{B}$ satisfying $\left[\hat{A}, \hat{B}\right] = \hat{\mathbb{I}}$, and $M \in \mathbb{N}_0$,*

$$\left[\hat{A}^M \hat{B}^M, \hat{B}^3 \hat{A}^3\right] = \hat{0}, \tag{4.206}$$

and consequentially,

$$\hat{A}^M \hat{B}^{M+3} \hat{A}^3 = \hat{B}^3 \hat{A}^{M+3} \hat{B}^M. \tag{4.207}$$

Proof: The proof follows the familiar pattern, in addition to utilizing the implication of $\left[\hat{A}^M \hat{B}^M, \hat{B}^2 \hat{A}^2\right] = \hat{0}$, i.e., $\hat{A}^M \hat{B}^{M+2} \hat{A} = \hat{B} \hat{A}^{M+2} \hat{B}^M$:

$$\left[\hat{A}^M \hat{B}^M, \hat{B}^3 \hat{A}^3\right] = \hat{A}^M \underbrace{\left[\hat{B}^M, \hat{B}^3 \hat{A}^3\right]} + \underbrace{\left[\hat{A}^M, \hat{B}^3 \hat{A}^3\right]} \hat{B}^M \tag{4.208a}$$

$$= \hat{A}^M \left\{ \hat{B}^3 \left[\hat{B}^M, \hat{A}^3\right] + \left[\hat{B}^M, \hat{B}^3\right] \hat{A}^3 \right\}$$

$$+ \left\{ \hat{B}^3 \left[\hat{A}^M, \hat{A}^3\right] + \left[\hat{A}^M, \hat{B}^3\right] \hat{A}^3 \right\} \hat{B}^M \tag{4.208b}$$

$$= \hat{A}^M \hat{B}^3 \left[\hat{B}^M, \hat{A}^3\right] + \hat{A}^M \underbrace{\left[\hat{B}^M, \hat{B}^3\right]}_{=\hat{0}} \hat{A}^3$$

$$+ \hat{B}^3 \underbrace{\left[\hat{A}^M, \hat{A}^3\right]}_{=\hat{0}} \hat{B}^M + \left[\hat{A}^M, \hat{B}^3\right] \hat{A}^3 \hat{B}^M \tag{4.208c}$$

$$= \hat{A}^M \hat{B}^3 \left[\hat{B}^M, \hat{A}^3\right] + \left[\hat{A}^M, \hat{B}^3\right] \hat{A}^3 \hat{B}^M. \tag{4.208d}$$

Substituting,

$$\left[\hat{B}^M, \hat{A}^3\right] = \hat{A}^2 \left(-M \hat{B}^{M-1}\right) + \hat{A} \left(-M \hat{B}^{M-1}\right) \hat{A} + \left(-M \hat{B}^{M-1}\right) \hat{A}^2 \tag{4.209a}$$

$$\left[\hat{A}^M, \hat{B}^3\right] = \hat{B}^2 \left(M \hat{A}^{M-1}\right) + \hat{B} \left(M \hat{A}^{M-1}\right) \hat{B} + \left(M \hat{A}^{M-1}\right) \hat{B}^2 \tag{4.209b}$$

into (4.208d),

$$\left[\hat{A}^M \hat{B}^M, \hat{B}^3 \hat{A}^3\right] = \hat{A}^M \hat{B}^3 \left\{ -M \hat{A}^2 \hat{B}^{M-1} - M \hat{A} \hat{B}^{M-1} \hat{A} - M \hat{B}^{M-1} \hat{A}^2 \right\}$$

$$+ \left\{ M \hat{B}^2 \hat{A}^{M-1} + M \hat{B} \hat{A}^{M-1} \hat{B} + M \hat{A}^{M-1} \hat{B}^2 \right\} \hat{A}^3 \hat{B}^M. \tag{4.210}$$

Simplifying,

$$\left[\hat{A}^M \hat{B}^M, \hat{B}^3 \hat{A}^3\right] = M \left\{ -\hat{A}^M \hat{B}^3 \hat{A}^2 \hat{B}^{M-1} - \hat{A}^M \hat{B}^3 \hat{A} \hat{B}^{M-1} \hat{A} - \hat{A}^M \hat{B}^{M+2} \hat{A}^2 \right\}$$

$$+ M \left\{ \hat{B}^2 \hat{A}^{M+2} \hat{B}^M + \hat{B} \hat{A}^{M-1} \hat{B} \hat{A}^3 \hat{B}^M + \hat{A}^{M-1} \hat{B}^2 \hat{A}^3 \hat{B}^M \right\}. \tag{4.211}$$

The third and the forth terms at the R.H.S. cancel each other in virtue of $\hat{A}^M \hat{B}^{M+2} \hat{A}^2 = \hat{B}^2 \hat{A}^{M+2} \hat{B}^M$, which was a consequence of the preceding corollary, $\left[\hat{A}^M \hat{B}^M, \hat{B}^2 \hat{A}^2\right] = \hat{0}$.

Considering the first and the sixth terms at the R.H.S. and substituting $\hat{A}^M = \hat{A}^{M-1}\hat{A}$ and $\hat{B}^M = \hat{B}\hat{B}^{M-1}$,

$$-\hat{A}^M\hat{B}^3\hat{A}^2\hat{B}^{M-1} + \hat{A}^{M-1}\hat{B}^2\hat{A}^3\hat{B}^M = -\hat{A}^{M-1}\hat{A}\hat{B}^3\hat{A}^2\hat{B}^{M-1} + \hat{A}^{M-1}\hat{B}^2\hat{A}^3\hat{B}\hat{B}^{M-1} \tag{4.212a}$$

$$= \hat{A}^{M-1}\{-\hat{A}\hat{B}^3\hat{A}^2 + \hat{B}^2\hat{A}^3\hat{B}\}\hat{B}^{M-1}. \tag{4.212b}$$

The terms in the curly brackets at the R.H.S. add up to zero due to $\left[\hat{B}^2\hat{A}^2, \hat{A}\hat{B}\right] = \hat{0}$, which was established earlier in the discussion. Thus, the second and the third terms remain,

$$\left[\hat{A}^M\hat{B}^M, \hat{B}^3\hat{A}^3\right] = M\{-\hat{A}^M\underbrace{\hat{B}^3\hat{A}}\hat{B}^{M-1}\hat{A} + \hat{B}\hat{A}^{M-1}\underbrace{\hat{B}\hat{A}^3}\hat{B}^M\}. \tag{4.213}$$

Focus on the underbraced terms. The next move consists of changing the order of $\hat{A}$ and $\hat{B}$ in these terms:

$$\hat{B}^3\hat{A} \implies \begin{cases} \left[\hat{B}^3, \hat{A}\right] = -3\hat{B}^2 \\ \left[\hat{B}^3, \hat{A}\right] = \hat{B}^3\hat{A} - \hat{A}\hat{B}^3 \end{cases} \implies \hat{B}^3\hat{A} - \hat{A}\hat{B}^3 = -3\hat{B}^2 \implies \underbrace{\hat{B}^3\hat{A}} = \hat{A}\hat{B}^3 - 3\hat{B}^2 , \tag{4.214}$$

and,

$$\hat{B}\hat{A}^3 \implies \begin{cases} \left[\hat{B}, \hat{A}^3\right] = -3\hat{A}^2 \\ \left[\hat{B}, \hat{A}^3\right] = \hat{B}\hat{A}^3 - \hat{A}^3\hat{B} \end{cases} \implies \hat{B}\hat{A}^3 - \hat{A}^3\hat{B} = -3\hat{A}^2 \implies \underbrace{\hat{B}\hat{A}^3} = \hat{A}^3\hat{B} - 3\hat{A}^2 . \tag{4.215}$$

Substituting $\hat{B}^3\hat{A}$ and $\hat{B}\hat{A}^3$, from (4.214) and (4.215), respectively, into (4.213),

$$\left[\hat{A}^M\hat{B}^M, \hat{B}^3\hat{A}^3\right] = M\{-\hat{A}^M\left(\hat{A}\hat{B}^3 - 3\hat{B}^2\right)\hat{B}^{M-1}\hat{A} + \hat{B}\hat{A}^{M-1}\left(\hat{A}^3\hat{B} - 3\hat{A}^2\right)\hat{B}^M\}. \tag{4.216}$$

Simplifying,

$$\left[\hat{A}^M\hat{B}^M, \hat{B}^3\hat{A}^3\right] = M\{-\hat{A}^{M+1}\hat{B}^{M+2}\hat{A} + 3\hat{A}^M\hat{B}^{M+1}\hat{A} + \hat{B}\hat{A}^{M+2}\hat{B}^{M+1} - 3\hat{B}\hat{A}^{M+1}\hat{B}^M\}. \tag{4.217}$$

The first and the third terms taken together vanish,

$$\hat{0} = -\hat{A}^{M+1}\hat{B}^{M+2}\hat{A} + \hat{B}\hat{A}^{M+2}\hat{B}^{M+1} \impliedby \hat{A}^{M+1}\hat{B}^{M+2}\hat{A} \overset{\text{Corollary}}{=} \hat{B}\hat{A}^{M+2}\hat{B}^{M+1}. \tag{4.218}$$

The second and the forth terms, taken together, vanish,

$$\hat{0} = 3\hat{A}^M\hat{B}^{M+1}\hat{A} - 3\hat{B}\hat{A}^{M+1}\hat{B}^M \impliedby \hat{A}^M\hat{B}^{M+1}\hat{A} \overset{\text{Corollary}}{=} \hat{B}\hat{A}^{M+1}\hat{B}^M. \tag{4.219}$$

It follows that,

$$\left[\hat{A}^M\hat{B}^M, \hat{B}^3\hat{A}^3\right] = \hat{0}. \tag{4.220}$$

This completes the proof and, furthermore, implies,

$$\hat{0} \overset{(4.220)}{=} \left[\hat{A}^M\hat{B}^M, \hat{B}^3\hat{A}^3\right] \overset{\text{def.}}{=} \hat{A}^M\hat{B}^M\hat{B}^3\hat{A}^3 - \hat{B}^3\hat{A}^3\hat{A}^M\hat{B}^M \tag{4.221a}$$

$$= \hat{A}^M\hat{B}^{M+3}\hat{A}^3 - \hat{B}^3\hat{A}^{M+3}\hat{B}^M, \tag{4.221b}$$

and thus,

$$\hat{A}^M\hat{B}^{M+3}\hat{A}^3 = \hat{B}^3\hat{A}^{M+3}\hat{B}^M. \tag{4.222}$$

■

4.6 Position operator, derivative commutator relationships in 2-D and 3-D

Problem: Find a representation of $[\frac{\partial}{\partial x_1}, \hat{x}_2]$.

Solution: $[\frac{\partial}{\partial x_1}, \hat{x}_2]$ is an operator depending on x_1 and x_2. To determine a representation of $[\frac{\partial}{\partial x_1}, \hat{x}_2]$ and to study its properties it must be applied to a general (test) function $f(x_1, x_2)$ of the two variables x_1 and x_2:

$$\left[\frac{\partial}{\partial x_1}, \hat{x}_2\right] f(x_1, x_2) \quad = \quad \left(\frac{\partial}{\partial x_1}\hat{x}_2 - \hat{x}_2 \frac{\partial}{\partial x_1}\right) f(x_1, x_2) \tag{4.223a}$$

$$\stackrel{\text{action of } \hat{x}_2}{=} \frac{\partial}{\partial x_1}\underbrace{\left(\hat{x}_2 f(x_1, x_2)\right)}_{(x_2 f(x_1,x_2))} - \underbrace{\hat{x}_2 \frac{\partial}{\partial x_1}f(x_1, x_2)}_{x_2 \frac{\partial}{\partial x_1}f(x_1,x_2)} \tag{4.223b}$$

$$= \quad \underbrace{\frac{\partial}{\partial x_1}(x_2 f(x_1, x_2))}_{} - x_2 \frac{\partial}{\partial x_1}f(x_1, x_2) \tag{4.223c}$$

$$= \quad \overbrace{\underbrace{\left(\frac{\partial}{\partial x_1}x_2\right)}_{0}f(x_1, x_2) + x_2 \frac{\partial}{\partial x_1}f(x_1, x_2)} - x_2 \frac{\partial}{\partial x_1}f(x_1, x_2) \tag{4.223d}$$

$$= \quad 0. \tag{4.223e}$$

Since the choice of $f(x_1, x_2)$ is arbitrary,

$$\left[\frac{\partial}{\partial x_1}, \hat{x}_2\right] = \hat{0}. \tag{4.224}$$

∎

Remark: In view of (4.224) and the result $\left[\frac{d}{dx}, \hat{x}\right] = \hat{\mathbb{I}}$, obtained earlier,

$$\left[\frac{\partial}{\partial x_k}, \hat{x}_l\right] = \delta_{kl}\hat{\mathbb{I}}, \tag{4.225}$$

with δ_{kl} being the Kronecker delta symbol.
    Or, more compactly,

$$\left[\hat{D}_k, \hat{x}_l\right] = \delta_{kl}\hat{\mathbb{I}}. \tag{4.226}$$

□

Problem: This problem concerns the commutativity of mixed derivatives, $\left[\frac{\partial}{\partial x}, \frac{\partial}{\partial y}\right] = 0$.

Solution: To obtain an expression for the second derivative $\frac{d^2}{dx^2}$ of the function $f(x)$ of one variable, ϵ- and η limit processes were applied successively. A similar procedure can be applied to obtain mixed derivatives of functions depending on two or more independent variables. Thereby, several nuances manifest themselves.

Consider the derivative of $f(x, y)$ with respect to x, defined in terms of the limit process,

$$\frac{\partial}{\partial x} f(x, y) = \lim_{\epsilon \to 0} \frac{1}{\epsilon} [f(x + \epsilon, y) - f(x, y)] \qquad (4.227a)$$

$$= \left(\frac{\partial f}{\partial x} \right) (x, y) . \qquad (4.227b)$$

Remark: The notations $\frac{\partial}{\partial x} f(x, y)$ and $\left(\frac{df}{dx} \right) (x, y)$ signify two crucial ideas:

- $\frac{\partial}{\partial x} f(x, y)$ alludes to the fact that an action is going to take place, given the function $f(x, y)$. Thus the emphasis is on the differential operator $\frac{\partial}{\partial x}$.
- The R.H.S. in (4.227a) reveals the mechanism involved in the operation and suggests a recipe: evaluate the function $f(\cdot, \cdot)$ at the points $(x + \epsilon, y)$ and (x, y), build the difference $f(x + \epsilon, y) - f(x, y)$, divide the difference by ϵ, and determine the limit of the fraction when ϵ approaches zero. Consequently, the R.H.S. in (4.227a) is a process.
- $\left(\frac{df}{dx} \right) (x, y)$ is the outcome of the operation, which is a function.

□

To obtain the derivative of $\frac{\partial}{\partial x} f(x, y)$ with respect to y, the expression at the R.H.S. can be viewed as a function of y, evaluate this expression at $y + \eta$ and y, subtract the resulting expressions, divide the difference by η, and let $\eta \to 0$. In detail,

$$\frac{\partial}{\partial y} \left(\frac{\partial}{\partial x} f(x, y) \right) = \lim_{\epsilon \to 0} \frac{1}{\eta} \left\{ \left(\frac{\partial f}{\partial x} \right) (x, y) \Big|_{y \to y + \eta} - \left(\frac{\partial f}{\partial x} \right) (x, y) \Big|_{y \to y} \right\} \qquad (4.228a)$$

$$= \lim_{\epsilon \to 0} \frac{1}{\eta} \left\{ \underbrace{\left(\frac{\partial f}{\partial x} \right) (x, y + \eta)} - \underbrace{\left(\frac{\partial f}{\partial x} \right) (x, y)} \right\} \qquad (4.228b)$$

$$= \lim_{\eta \to 0} \frac{1}{\eta} \left\{ \lim_{\epsilon \to 0} \frac{1}{\epsilon} \underbrace{[f(x + \epsilon, y + \eta) - f(x, y + \eta)]} \right.$$

$$\left. - \lim_{\epsilon \to 0} \frac{1}{\epsilon} \underbrace{[f(x + \epsilon, y) - f(x, y)]} \right\} \qquad (4.228c)$$

$$= \lim_{\eta \to 0} \frac{1}{\eta} \left\{ \lim_{\epsilon \to 0} \frac{1}{\epsilon} [f(x + \epsilon, y + \eta) - f(x, y + \eta) - f(x + \epsilon, y) + f(x, y)] \right\} \qquad (4.228d)$$

Considering the four function values at the R.H.S., there is no distinguishing feature concerning the variables x and y. Also, in view of the "symmetric" pattern of the four function values, there is no compelling reason why the ϵ

limit process should precede that of the η limit process. Thus, changing the order of the limits, $\lim\limits_{\eta \to 0}\lim\limits_{\epsilon \to 0} = \lim\limits_{\epsilon \to 0}\lim\limits_{\eta \to 0}$, and also reordering the additive terms according to $-f(x, y + \eta) - f(x + \epsilon, y) = -f(x + \epsilon, y) - f(x, y + \eta)$, result in,

$$\frac{\partial}{\partial y}\left(\frac{\partial}{\partial x}f(x, y)\right) = \lim_{\epsilon \to 0}\frac{1}{\epsilon}\left\{\lim_{\eta \to 0}\frac{1}{\eta}[f(x + \epsilon, y + \eta) - f(x + \epsilon, y) - f(x, y + \eta) + f(x, y)]\right\} \tag{4.229a}$$

$$= \lim_{\epsilon \to 0}\frac{1}{\epsilon}\left\{\underbrace{\lim_{\eta \to 0}\frac{1}{\eta}[f(x + \epsilon, y + \eta) - f(x + \epsilon, y)]}_{}\right.$$

$$\left.\underbrace{- \lim_{\eta \to 0}\frac{1}{\eta}[f(x, y + \eta) - f(x, y)]}_{}\right\} \tag{4.229b}$$

$$= \lim_{\epsilon \to 0}\frac{1}{\epsilon}\left\{\underbrace{\left(\frac{\partial f}{\partial y}\right)(x + \epsilon, y)}_{} - \underbrace{\left(\frac{\partial f}{\partial y}\right)(x, y)}_{}\right\} \tag{4.229c}$$

$$= \frac{\partial}{\partial x}\left(\frac{\partial}{\partial y}f(x, y)\right). \tag{4.229d}$$

Since $f(x, y)$ is a test function,

$$\frac{\partial}{\partial x}\frac{\partial}{\partial y} = \frac{\partial}{\partial y}\frac{\partial}{\partial x} \implies \left[\frac{\partial}{\partial x}, \frac{\partial}{\partial y}\right] = \hat{0}. \tag{4.230}$$

Or, more generally,

$$\left[\frac{\partial}{\partial x_k}, \frac{\partial}{\partial x_l}\right] = \hat{0}, \quad k, l = 1, 2, 3. \tag{4.231}$$

Or, more generally and compactly,

$$\left[\hat{D}_k, \hat{D}_l\right] = \hat{0}, \quad k, l = 1, 2, 3. \tag{4.232}$$

∎

Remark: Summarizing the results,

$$[\hat{x}_k, \hat{x}_l] = \hat{0} \tag{4.233a}$$

$$\left[\hat{D}_k, \hat{D}_l\right] = \hat{0} \tag{4.233b}$$

$$\left[\hat{D}_k, \hat{x}_l\right] = \delta_{kl}\hat{\mathbb{I}}. \tag{4.233c}$$

□

Problem: Given,

$$\frac{\partial}{\partial y}\left(\frac{\partial}{\partial x}f(x, y)\right) = \lim_{\eta \to 0}\frac{1}{\eta}\left\{\lim_{\epsilon \to 0}\frac{1}{\epsilon}[f(x + \epsilon, y + \eta) - f(x, y + \eta) - f(x + \epsilon, y) + f(x, y)]\right\}, \tag{4.234}$$

calculate $\frac{\partial}{\partial x}\left(\frac{\partial}{\partial y}\left(\frac{\partial}{\partial x}f(x, y)\right)\right)$.

Solution:

$$\frac{\partial}{\partial x}\left(\frac{\partial}{\partial y}\left(\frac{\partial}{\partial x}f(x,y)\right)\right)$$

$$= \lim_{\epsilon_1 \to 0}\frac{1}{\epsilon_1}\left\{\lim_{\eta \to 0}\frac{1}{\eta}\left\{\lim_{\epsilon \to 0}\frac{1}{\epsilon}[f(x+\epsilon_1+\epsilon,y+\eta)-f(x+\epsilon_1,y+\eta)-f(x+\epsilon_1+\epsilon,y)+f(x+\epsilon_1,y)]\right\}\right.$$

$$\left.- \lim_{\eta \to 0}\frac{1}{\eta}\left\{\lim_{\epsilon \to 0}\frac{1}{\epsilon}[f(x+\epsilon,y+\eta)-f(x,y+\eta)-f(x+\epsilon,y)+f(x,y)]\right\}\right\}. \tag{4.235}$$

Compressing the notation slightly,

$$\frac{\partial}{\partial x}\left(\frac{\partial}{\partial y}\left(\frac{\partial}{\partial x}f(x,y)\right)\right) = \frac{\partial}{\partial x}\left(\frac{\partial}{\partial x}\left(\frac{\partial}{\partial y}f(x,y)\right)\right) \tag{4.236a}$$

$$= \frac{\partial^2}{\partial x^2}\left(\frac{\partial}{\partial y}f(x,y)\right), \tag{4.236b}$$

using $\epsilon_1 = \epsilon$, and simplifying,

$$\frac{\partial^2}{\partial x^2}\left(\frac{\partial}{\partial y}f(x,y)\right)$$

$$= \lim_{\epsilon \to 0}\frac{1}{\epsilon^2}\lim_{\eta \to 0}\frac{1}{\eta}\{[f(x+\epsilon+\epsilon,y+\eta)-f(x+\epsilon,y+\eta)-f(x+\epsilon+\epsilon,y)+f(x+\epsilon,y)]$$

$$- [f(x+\epsilon,y+\eta)-f(x,y+\eta)-f(x+\epsilon,y)+f(x,y)]\}. \tag{4.237a}$$

Simplifying and combining terms,

$$\frac{\partial^2}{\partial x^2}\left(\frac{\partial}{\partial y}f(x,y)\right) = \lim_{\epsilon \to 0}\frac{1}{\epsilon^2}\lim_{\eta \to 0}\frac{1}{\eta}\{f(x+2\epsilon,y+\eta)-2f(x+\epsilon,y+\eta)+f(x,y+\eta)$$

$$-f(x+2\epsilon,y)+2f(x+\epsilon,y)-f(x,y)\}. \tag{4.238a}$$

∎

Problem: Let $\hat{\nabla}$ denote the vector differential operator $\left(\frac{\partial}{\partial x},\frac{\partial}{\partial y},\frac{\partial}{\partial z}\right)^T$ and $\hat{\mathbf{r}}$ the vector position operator $(\hat{x},\hat{y},\hat{z})^T$. Let "·" denote the dot-product. Find an expression for $\hat{\nabla}\cdot\hat{\mathbf{r}}-\hat{\mathbf{r}}\cdot\hat{\nabla}$.

Solution: It is instructive to use indices. Thus, set

$$\hat{\nabla} = \left(\frac{\partial}{\partial x_1},\frac{\partial}{\partial x_2},\frac{\partial}{\partial x_3}\right)^T \tag{4.239a}$$

$$\hat{\mathbf{r}} = (\hat{x}_1,\hat{x}_2,\hat{x}_3)^T. \tag{4.239b}$$

Then,

$$\hat{\mathbf{\nabla}} \cdot \hat{\mathbf{r}} - \hat{\mathbf{r}} \cdot \hat{\mathbf{\nabla}} = \left(\frac{\partial}{\partial x_1} \hat{x}_1 + \frac{\partial}{\partial x_2} \hat{x}_2 + \frac{\partial}{\partial x_3} \hat{x}_3 \right) - \left(\hat{x}_1 \frac{\partial}{\partial x_1} + \hat{x}_2 \frac{\partial}{\partial x_2} + \hat{x}_3 \frac{\partial}{\partial x_3} \right) \tag{4.240a}$$

$$= \left(\frac{\partial}{\partial x_1} \hat{x}_1 - \hat{x}_1 \frac{\partial}{\partial x_1} \right) + \left(\frac{\partial}{\partial x_2} \hat{x}_2 - \hat{x}_2 \frac{\partial}{\partial x_2} \right) + \left(\frac{\partial}{\partial x_3} \hat{x}_3 - \hat{x}_3 \frac{\partial}{\partial x_3} \right) \tag{4.240b}$$

$$= \left[\frac{\partial}{\partial x_1}, \hat{x}_1 \right] + \left[\frac{\partial}{\partial x_2}, \hat{x}_2 \right] + \left[\frac{\partial}{\partial x_3}, \hat{x}_3 \right] \tag{4.240c}$$

$$= \sum_{i=1}^{3} \underbrace{\left[\frac{\partial}{\partial x_i}, \hat{x}_i \right]}_{\hat{\mathbb{1}}} \tag{4.240d}$$

$$= 3\hat{\mathbb{1}}. \tag{4.240e}$$

■

4.7 Commutation relations involving components of $\hat{\mathbf{\nabla}}$ and functions of $\hat{r}$

Problem: Find a representation for $\left[\frac{\partial}{\partial x}, f(\hat{r}) \right]$.

Before embarking on the solution of this problem, the following remark may be helpful.

Remark: Writing $\hat{r}$ rather than r alludes to the fact that $f(\hat{r})$ is a function of the vector position operator $\hat{r}$. Note that $r = \left(x^2 + y^2 \right)^{1/2}$ implies $\hat{r} = \left(\hat{x}^2 + \hat{y}^2 \right)^{1/2}$ and that $f(\hat{r})\psi(r) = f(r)\psi(r)$.

□

Solution: For obtaining equivalent representations for an operator and studying its properties, the operator needs to be applied to an arbitrary test function in its domain. Formal calculations, in virtue of being formal, assume that the employed function, e.g., $\psi(r)$, is in operator domain:

$$\left[\frac{\partial}{\partial x}, f(\hat{r}) \right] \psi(r) = \left[\frac{\partial}{\partial x} f(\hat{r}) - f(\hat{r}) \frac{\partial}{\partial x} \right] \psi(r) \tag{4.241a}$$

$$\overset{\text{action of } f(\hat{r})}{=} \frac{\partial}{\partial x} \left(\underbrace{f(\hat{r})\psi(r)}_{f(r)\psi(r)} \right) - \underbrace{f(\hat{r}) \frac{\partial}{\partial x} \psi(r)}_{f(r)\frac{\partial}{\partial x}\psi(r)} \tag{4.241b}$$

$$= \frac{\partial}{\partial x} (f(r)\psi(r)) - f(r) \frac{\partial}{\partial x} \psi(r) \tag{4.241c}$$

$$\overset{\text{product rule}}{=} \left(\frac{\partial}{\partial x} f(r) \right) \psi(r) + f(r) \left(\frac{\partial}{\partial x} \psi(r) \right) - f(r) \frac{\partial}{\partial x} \psi(r) \tag{4.241d}$$

$$= \left(\frac{\partial}{\partial x} f(r) \right) \psi(r) \tag{4.241e}$$

$$= \underbrace{\left(\frac{\partial}{\partial r} f(r) \right)}_{f^{(1)}(r)} \underbrace{\left(\frac{\partial}{\partial x} r \right)}_{\frac{x}{r}} \psi(r) \tag{4.241f}$$

$$= \quad f^{(1)}(r)\frac{x}{r}\psi(r) \tag{4.241g}$$

$$\overset{\text{operator rep.}}{=} f^{(1)}(\hat{r})\frac{\hat{x}}{\hat{r}}\psi(r). \tag{4.241h}$$

Since $\psi(r)$ is arbitrary,

$$\left[\frac{\partial}{\partial x}, f(\hat{r})\right] = f^{(1)}(\hat{r})\frac{\hat{x}}{\hat{r}}. \tag{4.242}$$

Note that in (4.241f) the order of the bracketed terms can be interchanged,

$$\left[\frac{\partial}{\partial x}, f(\hat{r})\right]\psi(r) = \underbrace{\left(\frac{\partial}{\partial x}r\right)}_{\frac{x}{r}}\underbrace{\left(\frac{\partial}{\partial r}f(r)\right)}_{f^{(1)}(r)}\psi(r) \tag{4.243a}$$

$$= \frac{x}{r}f^{(1)}(r)\psi(r). \tag{4.243b}$$

Thus,

$$\left[\frac{\partial}{\partial x}, f(\hat{r})\right] = \frac{\hat{x}}{\hat{r}}f^{(1)}(\hat{r}). \tag{4.244}$$

The commutativity of $\frac{\hat{x}}{\hat{r}}$ and $f^{(1)}(\hat{r})$ is a consequent of the property that the operators involved, i.e., $f^{(1)}(\hat{r})$ and $\frac{\hat{x}}{\hat{r}}$, are both functions of position operators and thus commutative.

∎

Remark: The outcome of the above analysis is a simple recipe,

$$\left[\frac{\partial}{\partial x}, f(\hat{r})\right] \implies \frac{\partial}{\partial x}f(r) = \frac{x}{r}f^{(1)}(r) \implies \frac{\hat{x}}{\hat{r}}f^{(1)}(\hat{r}). \tag{4.245}$$

This is another manifestation of the property that commutators "act" as if they were differential operators.

□

Problem: Find a representation for $\left[\frac{\partial}{\partial x_j}, \hat{r}^{-n}\right]$.

Solution: For solving this problem, the tedious path of the preceding problem can be bypassed and the identified recipe can be applied directly,

$$\left[\frac{\partial}{\partial x_j}, \hat{r}^{-n}\right] \implies \frac{\partial}{\partial x_j}\left(r^{-n}\right) \tag{4.246a}$$

$$= -nr^{-n-1}\frac{x_j}{r} \tag{4.246b}$$

$$= -nr^{-n-2}x_j \tag{4.246c}$$

$$= -nx_j r^{-n-2} \tag{4.246d}$$

$$\implies -n\hat{x}_j \hat{r}^{-n-2}. \tag{4.246e}$$

∎

Remark: The last transition results in the operator $-n\hat{x}_j\hat{r}^{-n-2}$, which is a function of the position operators $\hat{x}$, $\hat{y}$, and $\hat{z}$. The action of the operator $-n\hat{x}_j\hat{r}^{-n-2}$ on any "permissible" function $\psi(x,y,z)$ is merely the multiplication of the corresponding function $-nx_j r^{-n-2}$ by $\psi(x,y,z)$, i.e., $-nx_j r^{-n-2}\psi(x,y,z)$.

A question arises: In view of the aforementioned fact, is the last step, the transition from $-nx_j r^{-n-2}$ to $-n\hat{x}_j \hat{r}^{-n-2}$, necessary? The answer is yes. The originating commutator $\left[\frac{\partial}{\partial x_j}, \hat{r}^{-n}\right]$ is an operator which can be expressed as $-n\hat{x}_j \hat{r}^{-n-2}$,

$$\left[\frac{\partial}{\partial x_j}, \hat{r}^{-n}\right] = -n\hat{x}_j \hat{r}^{-n-2}. \tag{4.247}$$

However, the effect of applying $\left[\frac{\partial}{\partial x_j}, \hat{r}^{-n}\right]$ onto the function $\psi(x, y, z)$ can be represented by the function $-nx_j r^{-n-2}$,

$$\left[\frac{\partial}{\partial x_j}, \hat{r}^{-n}\right] \xrightarrow{\text{rep.}} -nx_j r^{-n-2}. \tag{4.248}$$

Thus while (4.247) is an equality, (4.248) is a representation.

$\square$

The following problem illustrates the power of the recipe further.

Problem: Find an equivalent expression for $\left[\frac{\partial}{\partial x_j}, \hat{x}_k \hat{r}^{-n}\right]$.

Solution:

$$\left[\frac{\partial}{\partial x_j}, \hat{x}_k \hat{r}^{-n}\right] \implies \frac{\partial}{\partial x_j}\left(x_k r^{-n}\right) \tag{4.249a}$$

$$= \underbrace{\left(\frac{\partial}{\partial x_j} x_k\right)}_{\delta_{jk}} r^{-n} + x_k \underbrace{\left(\frac{\partial}{\partial x_j} r^{-n}\right)}_{-nr^{-n-1}\frac{x_j}{r}} \tag{4.249b}$$

$$= \delta_{jk} r^{-n} - nr^{-n}\frac{x_j x_k}{r^2} \tag{4.249c}$$

$$= r^{-n}\left(\delta_{jk} - n\frac{x_j x_k}{r^2}\right) \tag{4.249d}$$

$$\implies \hat{r}^{-n}\left(\hat{\delta}_{jk} - n\frac{\hat{x}_j \hat{x}_k}{\hat{r}^2}\right). \tag{4.249e}$$

The position operators $\hat{x}_1$, $\hat{x}_2$, and $\hat{x}_3$, and functions thereof, in particular, $\hat{r}$, $\frac{1}{\hat{r}}$, $\hat{r}^{-n}$ and $\frac{1}{\hat{r}^2}$ are all commutative. The reason for this commutativity property is twofold. First, an arbitrary function of the position operators; i.e., $f(\hat{x}_1, \hat{x}_2, \hat{x}_3)$, can be expressed in terms of its Formal Taylor Series Expansion involving the product of monomials of the type, $\hat{x}_1^m \hat{x}_2^n \hat{x}_3^k$. Second, it holds that $[\hat{x}_i, \hat{x}_j] = \hat{0}$

∎

Problem: Find an equivalent expression for $\left[\hat{D}_j, \hat{x}_j \hat{r}^{-n}\right]$.

Solution: Since the commutator $\left[\hat{D}_j, \hat{x}_j \hat{r}^{-n}\right]$ involves the products of $\hat{D}_j$ and $\hat{x}_j \hat{r}^{-n}$ the Einstein summation formula is implied. Thus,

$$\left[\hat{D}_j, \hat{x}_j \hat{r}^{-n}\right] \overset{\text{def. of } \hat{D}_j}{=} \left[\frac{\partial}{\partial x_j}, \hat{x}_j \hat{r}^{-n}\right] \tag{4.250a}$$

$$\overset{\text{recipe}}{\Longrightarrow} \frac{\partial}{\partial x_j}\left(x_j r^{-n}\right) \tag{4.250b}$$

$$\overset{\text{Einstein Conv.}}{\Longrightarrow} \sum_{j=1}^{3} \frac{\partial}{\partial x_j}\left(x_j r^{-n}\right) \tag{4.250c}$$

$$= \sum_{j=1}^{3}\left\{\underbrace{\left(\frac{\partial}{\partial x_j} x_j\right)}_{1} r^{-n} + x_j \underbrace{\left(\frac{\partial}{\partial x_j} r^{-n}\right)}_{-n r^{-n-1}\frac{x_j}{r}}\right\} \tag{4.250d}$$

$$= \sum_{j=1}^{3}\left(r^{-n} - n\frac{x_j^2}{r^2} r^{-n}\right) \tag{4.250e}$$

$$= r^{-n} \sum_{j=1}^{3}\left(1 - n\frac{x_j^2}{r^2}\right) \tag{4.250f}$$

$$= r^{-n}\left\{\left(1 - n\frac{x_1^2}{r^2}\right) + \left(1 - n\frac{x_2^2}{r^2}\right) + \left(1 - n\frac{x_3^2}{r^2}\right)\right\} \tag{4.250g}$$

$$= r^{-n}\left(3 - n\underbrace{\frac{x_1^2 + x_2^2 + x_3^2}{r^2}}_{1}\right) \tag{4.250h}$$

$$= (3 - n) r^{-n} \tag{4.250i}$$

$$\Longrightarrow (3 - n) \hat{r}^{-n}. \tag{4.250j}$$

∎

Problem: Find an equivalent expression for $\left[\hat{\nabla}, f(\hat{r})\right]$.

Solution:

$$\left[\hat{\nabla}, f(\hat{r})\right] \overset{\text{recipe}}{\Longrightarrow} \nabla f(r) \tag{4.251a}$$

$$= \left(\frac{\partial}{\partial x_1}, \frac{\partial}{\partial x_2}, \frac{\partial}{\partial x_3}\right)^T f(r) \tag{4.251b}$$

$$= \left(\underbrace{\frac{\partial}{\partial x_1} f(r)}_{f^{(1)}(r)\frac{x_1}{r}}, \underbrace{\frac{\partial}{\partial x_2} f(r)}_{f^{(1)}(r)\frac{x_2}{r}}, \underbrace{\frac{\partial}{\partial x_3} f(r)}_{f^{(1)}(r)\frac{x_3}{r}}\right)^T \tag{4.251c}$$

$$= \frac{1}{r} f^{(1)}(r) \underbrace{(x_1, x_2, x_3)^T}_{\mathbf{r}} \tag{4.251d}$$

$$\implies \frac{1}{\hat{r}} f^{(1)}(\hat{r}) \hat{\mathbf{r}}. \tag{4.251e}$$

Note that since $f(\hat{r})$ is a function of position operators $\hat{x}_1$, $\hat{x}_2$, and $\hat{x}_3$, the derivative $f^{(1)}(\hat{r})$ also depends on the position operators. Thus the order of the operators $\frac{1}{\hat{r}}$, $f^{(1)}(\hat{r})$, and $\hat{\mathbf{r}}$ is immaterial. Furthermore, it should be noted that vectors are usually understood to be column vectors, e.g., $\hat{\nabla} = \left(\frac{\partial}{\partial x_1}, \frac{\partial}{\partial x_2}, \frac{\partial}{\partial x_3} \right)^T$, and $\mathbf{r} = (x_1, x_2, x_3)^T$, and consequently, $\hat{\mathbf{r}} = (\hat{x}_1, \hat{x}_2, \hat{x}_3)^T$. Thus, the transposition symbol in the above expressions.

∎

Problem: Apply the result in the previous problem to the special case of $\left[\hat{\nabla}, \hat{r}^{-n} \right]$.

Solution:

$$\left[\hat{\nabla}, f(\hat{r}) \right] \overset{\text{Eq. (4.251d)}}{\implies} \frac{1}{r} \left(\frac{d}{dr} r^{-n} \right) \mathbf{r} \tag{4.252a}$$

$$= \frac{1}{r} \left(-n r^{-n-1} \right) \mathbf{r} \tag{4.252b}$$

$$= -n r^{-n-2} \mathbf{r} \tag{4.252c}$$

$$\implies -n \hat{r}^{-n-2} \hat{\mathbf{r}}. \tag{4.252d}$$

∎

4.8   Commutators involving $\hat{\mathbf{r}} \cdot \hat{\nabla}$ and $\hat{\nabla} \cdot \hat{\nabla}$

In the following whenever the case may be, Einsteins implied summation rule, or any of the equivalent expressions will be employed, whichever serve to better understanding, e.g.,

$$\hat{\mathbf{r}} \cdot \hat{\nabla} \overset{\text{Einstein conv.}}{=} \hat{x}_j \hat{D}_j \overset{\text{standard}}{=} \sum_{j=1}^{3} \hat{x}_j \hat{D}_j \overset{\text{explicit}}{=} \sum_{j=1}^{3} \hat{x}_j \frac{\partial}{\partial x_j} \tag{4.253a}$$

$$\hat{\nabla}^2 = \hat{\nabla} \cdot \hat{\nabla} \overset{\text{Einstein conv.}}{=} \hat{D}_j \hat{D}_j \overset{\text{standard}}{=} \sum_{j=1}^{3} \hat{D}_j \hat{D}_j \overset{\text{explicit}}{=} \sum_{j=1}^{3} \frac{\partial}{\partial x_j} \frac{\partial}{\partial x_j} \overset{\text{equiv.}}{=} \sum_{j=1}^{3} \frac{\partial^2}{\partial x_j^2}. \tag{4.253b}$$

Problem: Find an equivalent expression for $\left[\hat{\nabla}^2, \hat{\mathbf{r}} \right]$.

Solution:

$$\left[\hat{\nabla}^2, \hat{\mathbf{r}}\right] = \left[\sum_{j=1}^{3} \hat{D}_j\hat{D}_j, \hat{\mathbf{r}}\right] \tag{4.254a}$$

$$\overset{\text{Einstein Conv.}}{=} \left[\hat{D}_j\hat{D}_j, \hat{\mathbf{r}}\right] \tag{4.254b}$$

$$= \hat{D}_j\left[\hat{D}_j, \hat{\mathbf{r}}\right] + \left[\hat{D}_j, \hat{\mathbf{r}}\right]\hat{D}_j \tag{4.254c}$$

$$\overset{\hat{\mathbf{r}}=(\hat{x}_1,\hat{x}_2,\hat{x}_3)^T}{=} \hat{D}_j\left[\hat{D}_j, \sum_{k=1}^{3}\hat{x}_k\mathbf{e}_k\right] + \left[\hat{D}_j, \sum_{k=1}^{3}\hat{x}_k\mathbf{e}_k\right]\hat{D}_j \tag{4.254d}$$

$$\overset{\text{Einstein Conv.}}{=} \hat{D}_j\underbrace{\left[\hat{D}_j, \hat{x}_k\mathbf{e}_k\right]} + \underbrace{\left[\hat{D}_j, \hat{x}_k\mathbf{e}_k\right]}\hat{D}_j \tag{4.254e}$$

$$= \hat{D}_j\left(\hat{x}_k\underbrace{\left[\hat{D}_j, \mathbf{e}_k\right]}_{\hat{0}} + \underbrace{\left[\hat{D}_j, \hat{x}_k\right]}_{\delta_{jk}}\mathbf{e}_k\right) + \left(\hat{x}_k\underbrace{\left[\hat{D}_j, \mathbf{e}_k\right]}_{\hat{0}} + \underbrace{\left[\hat{D}_j, \hat{x}_k\right]}_{\delta_{jk}}\mathbf{e}_k\right)\hat{D}_j \tag{4.254f}$$

$$= \hat{D}_j\hat{\delta}_{jk}\mathbf{e}_k + \hat{\delta}_{jk}\mathbf{e}_k\hat{D}_j \tag{4.254g}$$

$$\overset{\text{decompress } k}{=} \hat{D}_j\left(\sum_{k=1}^{3}\hat{\delta}_{jk}\mathbf{e}_k\right) + \left(\sum_{k=1}^{3}\hat{\delta}_{jk}\mathbf{e}_k\right)\hat{D}_j \tag{4.254h}$$

$$= \underbrace{\hat{D}_j\mathbf{e}_j}_{\mathbf{e}_j\hat{D}_j} + \mathbf{e}_j\hat{D}_j \tag{4.254i}$$

$$= 2\mathbf{e}_j\hat{D}_j \tag{4.254j}$$

$$\overset{\text{decompress } j}{=} 2\sum_{j=1}^{3}\mathbf{e}_j\hat{D}_j \tag{4.254k}$$

$$= 2\hat{\nabla}. \tag{4.254l}$$

This is a reminiscence of a result obtained earlier in one dimension. It concerns the role of $\hat{\mathbf{r}}$ in $\left[\hat{\nabla}^2, \hat{\mathbf{r}}\right]$, which is taking the derivative of $\hat{\nabla}^2$ with respect to $\hat{\nabla}$, and thus leading to $2\hat{\nabla}$.

∎

4.9 On the generalization of the canonical commutators

The canonical identity commutator $[\frac{d}{dx}, \hat{x}] = \hat{\mathbb{I}}$, conveniently represented as $[\hat{D}, \hat{x}] = \hat{\mathbb{I}}$, is the simplest possible (archetype) commutator which consists of the primary building block operators $\hat{x}$ and $\hat{D}$.

At this stage, it is reasonable to examine the structures of the next simplest, and eventually proceed to most general types of derived canonical identity commutators, composed of powers of $\hat{x}$ and $\hat{D}$. The following examples are meant to help the reader to develop an intuition about the structure of increasingly more complex identity commutators, and seamlessly guide them to the most general case. In the following, $\frac{d}{dx}$ and $\hat{D}$ will be used interchangeably to find a balance between clarity and efficiency.

Problem: Consider,

$$\hat{L}_1 = \alpha_{11}\frac{d}{dx} + \alpha_{12}\hat{x} \tag{4.255a}$$

$$\hat{L}_2 = \alpha_{21}\frac{d}{dx} + \alpha_{22}\hat{x}. \tag{4.255b}$$

Determine the coefficients α_{ij}, $i,j = 1,2$ for satisfying $\left[\hat{L}_1, \hat{L}_2\right] = \hat{\mathbb{I}}$.

Solution:

$$\left[\hat{L}_1, \hat{L}_2\right] = \left[\alpha_{11}\frac{d}{dx} + \alpha_{12}\hat{x}, \alpha_{21}\frac{d}{dx} + \alpha_{22}\hat{x}\right] \tag{4.256a}$$

$$= \left[\alpha_{11}\frac{d}{dx}, \alpha_{21}\frac{d}{dx}\right] + \left[\alpha_{11}\frac{d}{dx}, \alpha_{22}\hat{x}\right] + \left[\alpha_{12}\hat{x}, \alpha_{21}\frac{d}{dx}\right] + \left[\alpha_{12}\hat{x}, \alpha_{22}\hat{x}\right] \tag{4.256b}$$

$$= \alpha_{11}\alpha_{21}\underbrace{\left[\frac{d}{dx}, \frac{d}{dx}\right]}_{\hat{0}} + \alpha_{11}\alpha_{22}\left[\frac{d}{dx}, \hat{x}\right] + \alpha_{12}\alpha_{21}\underbrace{\left[\hat{x}, \frac{d}{dx}\right]}_{-\left[\frac{d}{dx}, \hat{x}\right]} + \alpha_{12}\alpha_{22}\underbrace{\left[\hat{x}, \hat{x}\right]}_{\hat{0}} \tag{4.256c}$$

$$= (\alpha_{11}\alpha_{22} - \alpha_{12}\alpha_{21})\underbrace{\left[\frac{d}{dx}, \hat{x}\right]}_{\hat{\mathbb{I}}} \tag{4.256d}$$

$$= \underbrace{(\alpha_{11}\alpha_{22} - \alpha_{12}\alpha_{21})}_{1}\hat{\mathbb{I}} \tag{4.256e}$$

$$= \hat{\mathbb{I}}. \tag{4.256f}$$

Thus, writing (4.255) in the matrix form,

$$\begin{bmatrix} \hat{L}_1 \\ \hat{L}_2 \end{bmatrix} = \begin{bmatrix} \alpha_{11} & \alpha_{12} \\ \alpha_{21} & \alpha_{22} \end{bmatrix} \begin{bmatrix} \frac{d}{dx} \\ \hat{x} \end{bmatrix}, \tag{4.257}$$

the condition in (4.256e) amounts to requiring the determinant of the coefficient matrix in (4.257) to be unity. By choosing

$$\begin{bmatrix} \alpha_{11} & \alpha_{12} \\ \alpha_{21} & \alpha_{22} \end{bmatrix} \implies \begin{bmatrix} \cos(\theta) & -\sin(\theta) \\ \sin(\theta) & \cos(\theta) \end{bmatrix} = \mathbf{R}(\theta), \tag{4.258}$$

it is ensured that for arbitrary θ, the determinant of the two-dimensional rotation matrix $\mathbf{R}$ is unity.

Thus,

$$\begin{bmatrix} \hat{L}_1(\theta) \\ \hat{L}_2(\theta) \end{bmatrix} = \begin{bmatrix} \cos(\theta) & -\sin(\theta) \\ \sin(\theta) & \cos(\theta) \end{bmatrix} \begin{bmatrix} \frac{d}{dx} \\ \hat{x} \end{bmatrix}. \tag{4.259}$$

For further discussion it is instructive to also consider (4.259) as two separate equations for the determination of $\hat{L}_1$ and $\hat{L}_2$, i.e.,

$$\hat{L}_1(\theta) = \cos(\theta)\frac{d}{dx} - \sin(\theta)\hat{x} \tag{4.260a}$$

$$\hat{L}_2(\theta) = \sin(\theta)\frac{d}{dx} + \cos(\theta)\hat{x}. \tag{4.260b}$$

■

Remark: An alternative compact representation of (4.259) offers itself, by recalling the relationship,

$$\begin{bmatrix} \cos(\theta) & -\sin(\theta) \\ \sin(\theta) & \cos(\theta) \end{bmatrix} = \cos(\theta)\underbrace{\begin{bmatrix} 1 & 0 \\ 0 & 1 \end{bmatrix}}_{\mathbf{I}} + \sin(\theta)\underbrace{\begin{bmatrix} 0 & -1 \\ 1 & 0 \end{bmatrix}}_{\mathbf{J}}$$

$$= \mathbf{I}\cos(\theta) + \mathbf{J}\sin(\theta)$$

$$= e^{\mathbf{J}\theta}. \tag{4.261}$$

Thus,

$$\begin{bmatrix} \hat{L}_1(\theta) \\ \hat{L}_2(\theta) \end{bmatrix} = e^{\mathbf{J}\theta}\begin{bmatrix} \frac{d}{dx} \\ \hat{x} \end{bmatrix}. \tag{4.262}$$

□

Remark: The special case $\theta = 0$. With $\cos(0) = 1$ and $\sin(0) = 0$, the operators $\hat{L}_1$ and $\hat{L}_2$ in (4.260) read,

$$\hat{L}_1(0) = \frac{d}{dx} \tag{4.263a}$$

$$\hat{L}_2(0) = \hat{x}, \tag{4.263b}$$

which are the archetypical operators $\frac{d}{dx}$ and $\hat{x}$, with

$$\left[\hat{L}_1(0), \hat{L}_2(0)\right] = \left[\frac{d}{dx}, \hat{x}\right] \tag{4.264a}$$

$$= \hat{\mathbb{I}}. \tag{4.264b}$$

□

Remark: The special case $\theta = \frac{\pi}{2}$. With $\cos(\frac{\pi}{2}) = 0$ and $\sin(\frac{\pi}{2}) = 1$, the operators $\hat{L}_1$ and $\hat{L}_2$ in (4.260) read,

$$\hat{L}_1(\frac{\pi}{2}) = -\hat{x} \tag{4.265a}$$

$$\hat{L}_2(\frac{\pi}{2}) = \frac{d}{dx}, \tag{4.265b}$$

with

$$\left[\hat{L}_1(\frac{\pi}{2}), \hat{L}_2(\frac{\pi}{2})\right] = \left[-\hat{x}, \frac{d}{dx}\right] \tag{4.266a}$$

$$= -\left[\hat{x}, \frac{d}{dx}\right] \tag{4.266b}$$

$$= -\left(-\left[\frac{d}{dx}, \hat{x}\right]\right) \tag{4.266c}$$

$$= \left[\frac{d}{dx}, \hat{x}\right] \tag{4.266d}$$

$$= \hat{\mathbb{I}}, \tag{4.266e}$$

□

Remark: The next special case is particularly interesting. Choosing $\theta = -\frac{\pi}{4}$, and thus $\cos\left(-\frac{\pi}{4}\right) = \cos\left(\frac{\pi}{4}\right) = \frac{\sqrt{2}}{2} = \frac{1}{\sqrt{2}}$, and $\sin\left(-\frac{\pi}{4}\right) = -\sin\left(\frac{\pi}{4}\right) = -\frac{\sqrt{2}}{2} = -\frac{1}{\sqrt{2}}$, the operators $\hat{L}_1$ and $\hat{L}_2$ in (4.260) read:

$$\hat{L}_1\left(-\frac{\pi}{4}\right) = \frac{1}{\sqrt{2}}\left(\frac{d}{dx} + \hat{x}\right) \qquad \Longleftrightarrow \qquad \hat{b} \tag{4.267a}$$

$$\hat{L}_2\left(-\frac{\pi}{4}\right) = \frac{1}{\sqrt{2}}\left(-\frac{d}{dx} + \hat{x}\right) \qquad \Longleftrightarrow \qquad \hat{b}^\dagger \tag{4.267b}$$

Incidentally, the operators $\hat{L}_1\left(-\frac{\pi}{4}\right)$ and $\hat{L}_2\left(-\frac{\pi}{4}\right)$ are, in literature, referred to as the annihilation operator $\hat{b}$ and the creation operator $\hat{b}^\dagger$, respectively.

Since by construction, $[\hat{L}_1(\theta), \hat{L}_2(\theta)] = \hat{\mathbb{I}}$, for arbitrary θ, and since $\hat{b} = \hat{L}_1(-\frac{\pi}{4})$ and $\hat{b}^\dagger = \hat{L}_2(-\frac{\pi}{4})$, it follows that,

$$[\hat{b}, \hat{b}^\dagger] = \hat{\mathbb{I}}. \tag{4.268}$$

The annihilation and creation operators possess distinguished properties which are detailed the Chapter 1 of Volume 2, including a direct proof of (4.268).

□

Remark: It is satisfying to observe that the archetypical operators $\frac{d}{dx} = \hat{L}_1(0)$ and $\hat{x} = \hat{L}_2(0)$ with $[\frac{d}{dx}, \hat{x}] = \hat{\mathbb{I}}$, and the critically investigated and widely used operators $\hat{b} = \hat{L}_1\left(-\frac{\pi}{4}\right)$, and $\hat{b}^\dagger = \hat{L}_2\left(-\frac{\pi}{4}\right)$, with $[\hat{b}, \hat{b}^\dagger] = \hat{\mathbb{I}}$, turn out to be just two special cases of the developed theory. It is even profoundly satisfying, as will be demonstrated on the following pages in this chapter, that the operators $\hat{L}_1$ and $\hat{L}_2$ can be further generalized to include arbitrary powers of $\hat{x}$, arbitrary orders of $\frac{d}{dx}$, and combinations thereof.

□

Remark: The final special case, considered in this series of examples, provides further insight into what happens if the angle θ varies, e.g., from $-\frac{\pi}{4}$ to $\frac{\pi}{4}$. Choosing $\theta = \frac{\pi}{4}$, and thus $\cos\left(\frac{\pi}{4}\right) = \sin\left(\frac{\pi}{4}\right) = \frac{\sqrt{2}}{2} = \frac{1}{\sqrt{2}}$, the operators $\hat{L}_1$ and $\hat{L}_2$ in (4.260) read:

$$\hat{L}_1\left(\frac{\pi}{4}\right) = \frac{1}{\sqrt{2}}\left(\frac{d}{dx} - \hat{x}\right) \qquad \overset{(4.267b)}{\Longleftrightarrow} \qquad -\hat{b}^\dagger \tag{4.269a}$$

$$\hat{L}_2\left(\frac{\pi}{4}\right) = \frac{1}{\sqrt{2}}\left(\frac{d}{dx} + \hat{x}\right) \qquad \overset{(4.267a)}{\Longleftrightarrow} \qquad \hat{b}. \tag{4.269b}$$

Then,

$$\left[\hat{L}_1\left(\frac{\pi}{4}\right),\hat{L}_2\left(\frac{\pi}{4}\right)\right] = \left[-\hat{b}^\dagger,\hat{b}\right] \tag{4.270a}$$

$$= -\left[\hat{b}^\dagger,\hat{b}\right] \tag{4.270b}$$

$$= -\left\{-\left[\hat{b},\hat{b}^\dagger\right]\right\} \tag{4.270c}$$

$$= \left[\hat{b},\hat{b}^\dagger\right] \tag{4.270d}$$

$$\overset{(4.268)}{=} \hat{\mathbb{I}}. \tag{4.270e}$$

□

Remark: Conventionally, the creation and annihilation operators are introduced in the context of solving the quantum harmonic oscillator problem. In contrast, the above mathematical derivation stands on its own right without any reference to quantum mechanics, or, for this matter, to any other discipline in physics. The construction of $\hat{L}_1$ and $\hat{L}_2$ is an example of the tenor of the present book. The existing and perceived complexities in the interpretation of quantum physics and the mathematical machinery should not be mixed up. A repository of relevant mathematical tools can be established first. They can be studied and mastered in virtue of their own merits. Subsequently, and whenever deemed necessary, they can be employed for describing and explaining the phenomena in quantum physics.

□

Remark: The theory of general relativity is another prominent example whereby the clear separation (demarcation) of mathematics and physics is critically important in grasping the underlying ideas. Many engineers and technologists perceive the theory of general relativity daunting and beyond the scope of their grasp. To satisfy their curiosity they turn to popular scientific expositions with mixed results. The level of sophistication of the required mathematical tools, e.g., the tensor calculus, the differential geometry, and the notions of covariant and contravariant coordinates, must not be confused with possibly counter-intuitive physical concepts and their interpretations. One is well advised to be aware of the mathematical machinery needed, acquire them, and finally master their usage. Engineers should engage in studying and examining the mathematical tools without obscuring the mathematics with the subtleties of physics nor shrouding the physics with the intricacies of mathematics. If the intricacies of mathematics and physics are mixed up, they create murkiness, mysteriousness, and nebulousness.

□

Problem: Consider,

$$\hat{L}_1 = \beta_{12}\frac{d^2}{dx^2} + \alpha_{11}\frac{d}{dx} + \alpha_{12}\hat{x} \tag{4.271a}$$

$$\hat{L}_2 = \beta_{22}\frac{d^2}{dx^2} + \alpha_{21}\frac{d}{dx} + \alpha_{22}\hat{x}. \tag{4.271b}$$

Or, more compactly,

$$\hat{L}_1 = \beta_{12}\hat{D}^2 + \alpha_{11}\hat{D} + \alpha_{12}\hat{x} \tag{4.272a}$$

$$\hat{L}_2 = \beta_{22}\hat{D}^2 + \alpha_{21}\hat{D} + \alpha_{22}\hat{x}. \tag{4.272b}$$

Determine α_{ij} $(i,j = 1, 2)$, β_{12}, and β_{22} to satisfy $\left[\hat{L}_1,\hat{L}_2\right] = \hat{\mathbb{I}}$.

Solution:

$$\left[\hat{L}_1, \hat{L}_2\right] = \left[\beta_{12}\hat{D}^2 + \alpha_{11}\hat{D} + \alpha_{12}\hat{x}, \ \beta_{22}\hat{D}^2 + \alpha_{21}\hat{D} + \alpha_{22}\hat{x}\right] \tag{4.273a}$$

$$= \left[\beta_{12}\hat{D}^2, \ \beta_{22}\hat{D}^2\right] + \left[\beta_{12}\hat{D}^2, \ \alpha_{21}\hat{D}\right] + \left[\beta_{12}\hat{D}^2, \ \alpha_{22}\hat{x}\right]$$

$$+ \left[\alpha_{11}\hat{D}, \ \beta_{22}\hat{D}^2\right] + \left[\alpha_{11}\hat{D}, \ \alpha_{21}\hat{D}\right] + \left[\alpha_{11}\hat{D}, \ \alpha_{22}\hat{x}\right]$$

$$+ \left[\alpha_{12}\hat{x}, \ \beta_{22}\hat{D}^2\right] + \left[\alpha_{12}\hat{x}, \ \alpha_{21}\hat{D}\right] + \left[\alpha_{12}\hat{x}, \ \alpha_{22}\hat{x}\right] \tag{4.273b}$$

$$= \beta_{12}\beta_{22}\underbrace{\left[\hat{D}^2, \hat{D}^2\right]}_{\hat{0}} + \beta_{12}\alpha_{21}\underbrace{\left[\hat{D}^2, \hat{D}\right]}_{\hat{0}} + \beta_{12}\alpha_{22}\left[\hat{D}^2, \hat{x}\right]$$

$$+ \alpha_{11}\beta_{22}\underbrace{\left[\hat{D}, \hat{D}^2\right]}_{\hat{0}} + \alpha_{11}\alpha_{21}\underbrace{\left[\hat{D}, \hat{D}\right]}_{\hat{0}} + \alpha_{11}\alpha_{22}\left[\hat{D}, \hat{x}\right]$$

$$+ \alpha_{12}\beta_{22}\underbrace{\left[\hat{x}, \hat{D}^2\right]}_{-[\hat{D}^2, \hat{x}]} + \alpha_{12}\alpha_{21}\underbrace{\left[\hat{x}, \hat{D}\right]}_{-[\hat{D}, \hat{x}]} + \alpha_{12}\alpha_{22}\underbrace{\left[\hat{x}, \hat{x}\right]}_{\hat{0}} \tag{4.273c}$$

$$= (\beta_{12}\alpha_{22} - \alpha_{12}\beta_{22})\underbrace{\left[\hat{D}^2, \hat{x}\right]}_{2\hat{D}} + (\alpha_{11}\alpha_{22} - \alpha_{12}\alpha_{21})\underbrace{\left[\hat{D}, \hat{x}\right]}_{\hat{1}}. \tag{4.273d}$$

Setting,

$$\alpha_{11} = \cos(\theta) \quad \alpha_{12} = -\sin(\theta)$$
$$\alpha_{21} = \sin(\theta) \quad \alpha_{22} = \cos(\theta) \tag{4.274}$$

ensures that,

$$\alpha_{11}\alpha_{22} - \alpha_{12}\alpha_{21} = 1. \tag{4.275}$$

Furthermore, requiring,

$$\beta_{12}\alpha_{22} - \alpha_{12}\beta_{22} = 0 \implies \beta_{12}\alpha_{22} = \alpha_{12}\beta_{22}. \tag{4.276}$$

Substituting the results $\alpha_{12} = -\sin(\theta)$ and $\alpha_{22} = \cos(\theta)$, from (4.258), in the latter equation,

$$\underbrace{\beta_{12}}_{-\sin(\theta)} \cos(\theta) = -\sin(\theta) \underbrace{\beta_{22}}_{\cos(\theta)}. \tag{4.277}$$

Note that the substitutions

$$\underbrace{\beta_{12}}_{-C_2\sin(\theta)} \cos(\theta) = -\sin(\theta) \underbrace{\beta_{22}}_{C_2\cos(\theta)}, \tag{4.278}$$

would even lead to a more general solution. As indicated in (4.278), the "obvious" choice of

$$\beta_{12} = -C_2\sin(\theta) \tag{4.279a}$$
$$\beta_{22} = C_2\cos(\theta), \tag{4.279b}$$

guarantees that the coefficient $\beta_{12}\alpha_{22} - \alpha_{12}\beta_{22}$ in (4.273d) vanishes,

$$\beta_{12}\alpha_{22} - \alpha_{12}\beta_{22} = 0. \tag{4.280}$$

Substituting the resultant α- and β-values into (4.271),

$$\hat{L}_1 = -C_2 \sin(\theta)\frac{d^2}{dx^2} + \cos(\theta)\frac{d}{dx} - \sin(\theta)\hat{x} \tag{4.281a}$$

$$\hat{L}_2 = C_2 \cos(\theta)\frac{d^2}{dx^2} + \sin(\theta)\frac{d}{dx} + \cos(\theta)\hat{x}. \tag{4.281b}$$

Organizing the terms in terms of the "markers" $\cos(\theta)$ and $\sin(\theta)$,

$$\hat{L}_1 = \cos(\theta)\left(\frac{d}{dx}\right) - \sin(\theta)\left(\hat{x} + C_2\frac{d^2}{dx^2}\right) \tag{4.282a}$$

$$\hat{L}_2 = \sin(\theta)\left(\frac{d}{dx}\right) + \cos(\theta)\left(\hat{x} + C_2\frac{d^2}{dx^2}\right), \tag{4.282b}$$

whereby $\hat{L}_1$ and $\hat{L}_2$ satisfy the desired commutativity property,

$$\left[\hat{L}_1, \hat{L}_2\right] = \hat{\mathbb{I}}, \tag{4.283}$$

by "design."

∎

The above procedure is quite general. The following problems are meant to enhance the intuition and demonstrate the claimed generality.

Problem: Consider,

$$\hat{L}_1 = \beta_{13}\frac{d^3}{dx^3} + \beta_{12}\frac{d^2}{dx^2} + \alpha_{11}\frac{d}{dx} + \alpha_{12}\hat{x} \tag{4.284a}$$

$$\hat{L}_2 = \beta_{23}\frac{d^3}{dx^3} + \beta_{22}\frac{d^2}{dx^2} + \alpha_{21}\frac{d}{dx} + \alpha_{22}\hat{x}. \tag{4.284b}$$

Determine the involved α- and β-parameters to satisfy $\left[\hat{L}_1, \hat{L}_2\right] = \hat{\mathbb{I}}$.

Solution: To safe space the abbreviations $\hat{D}^n = \frac{d^n}{dx^n}$ ($n \in \mathbb{N}$) will be utilized:

$$\left[\hat{L}_1, \hat{L}_2\right] = \left[\beta_{13}\hat{D}^3 + \beta_{12}\hat{D}^2 + \alpha_{11}\hat{D} + \alpha_{12}\hat{x}, \ \beta_{23}\hat{D}^3 + \beta_{22}\hat{D}^2 + \alpha_{21}\hat{D} + \alpha_{22}\hat{x}\right]. \tag{4.285}$$

Noting that $\left[\hat{D}^m, \hat{D}^n\right] = \hat{0}$ and $[\hat{x}, \hat{x}] = \hat{0}$,

$$\left[\hat{L}_1, \hat{L}_2\right] = \left[\beta_{13}\hat{D}^3, \alpha_{22}\hat{x}\right] + \left[\beta_{12}\hat{D}^2, \alpha_{22}\hat{x}\right] + \left[\alpha_{11}\hat{D}, \alpha_{22}\hat{x}\right]$$
$$+ \left[\alpha_{12}\hat{x}, \beta_{23}\hat{D}^3\right] + \left[\alpha_{12}\hat{x}, \beta_{22}\hat{D}^2\right] + \left[\alpha_{12}\hat{x}, \alpha_{21}\hat{D}\right]. \tag{4.286}$$

Factoring out the constants,

$$\left[\hat{L}_1, \hat{L}_2\right] = \beta_{13}\alpha_{22}\left[\hat{D}^3, \hat{x}\right] + \beta_{12}\alpha_{22}\left[\hat{D}^2, \hat{x}\right] + \alpha_{11}\alpha_{22}\left[\hat{D}, \hat{x}\right]$$
$$+ \alpha_{12}\beta_{23}\left[\hat{x}, \hat{D}^3\right] + \alpha_{12}\beta_{22}\left[\hat{x}, \hat{D}^2\right] + \alpha_{12}\alpha_{21}\left[\hat{x}, \hat{D}\right]. \tag{4.287}$$

Using the property $\left[\hat{x}, \hat{D}^n\right] = -\left[\hat{D}^n, \hat{x}\right]$ and organizing the terms,

$$\left[\hat{L}_1, \hat{L}_2\right] = (\beta_{13}\alpha_{22} - \alpha_{12}\beta_{23})\underbrace{\left[\hat{D}^3, \hat{x}\right]}_{3\hat{D}^2} + (\beta_{12}\alpha_{22} - \alpha_{12}\beta_{22})\underbrace{\left[\hat{D}^2, \hat{x}\right]}_{2\hat{D}} + (\alpha_{11}\alpha_{22} - \alpha_{12}\alpha_{21})\underbrace{\left[\hat{D}, \hat{x}\right]}_{\hat{\mathbb{I}}}. \tag{4.288}$$

Thus,

$$\left[\hat{L}_1, \hat{L}_2\right] = \underbrace{(\beta_{13}\alpha_{22} - \alpha_{12}\beta_{23})}_{0}3\hat{D}^2 + \underbrace{(\beta_{12}\alpha_{22} - \alpha_{12}\beta_{22})}_{0}2\hat{D} + \underbrace{(\alpha_{11}\alpha_{22} - \alpha_{12}\alpha_{21})}_{1}\hat{\mathbb{I}}. \tag{4.289}$$

A pattern emerges. The determination of the α-coefficients remains unaffected:

$$\alpha_{11} = \cos(\theta) \quad \alpha_{12} = -\sin(\theta)$$
$$\alpha_{21} = \sin(\theta) \quad \alpha_{22} = \cos(\theta) \tag{4.290}$$

The determination of the β-coefficients follows a simple recipe,

$$\underbrace{\beta_{13}}_{C_3\alpha_{12}}\alpha_{22} - \alpha_{12}\underbrace{\beta_{23}}_{C_3\alpha_{22}} = 0 \tag{4.291a}$$

$$\underbrace{\beta_{12}}_{C_2\alpha_{12}}\alpha_{22} - \alpha_{12}\underbrace{\beta_{22}}_{C_2\alpha_{22}} = 0. \tag{4.291b}$$

Note that $\beta_{13} (= C_3\alpha_{12})$ and $\beta_{23} (= C_3\alpha_{22})$ are the coefficients of $\hat{D}^3$ in the expressions for $\hat{L}_1$ and $\hat{L}_2$. Similarly, $\beta_{12} (= C_2\alpha_{12})$ and $\beta_{22} (= C_2\alpha_{22})$ are the coefficients of $\hat{D}^2$ in the expressions for $\hat{L}_1$ and $\hat{L}_2$. Thus the coefficients of $\hat{x}$ in the expressions of $\hat{L}_1$ and $\hat{L}_2$, i.e., α_{12} and α_{21} determine the coefficients of derivatives of higher order. Consequently,

$$\hat{L}_1 = \alpha_{12}C_3\hat{D}^3 + \alpha_{12}C_2\hat{D}^2 + \alpha_{11}\hat{D} + \alpha_{12}\hat{x} \tag{4.292a}$$
$$\hat{L}_2 = \alpha_{22}C_3\hat{D}^3 + \alpha_{22}C_2\hat{D}^2 + \alpha_{21}\hat{D} + \alpha_{22}\hat{x}. \tag{4.292b}$$

Rearranging,

$$\hat{L}_1 = \alpha_{11}\hat{D} + \alpha_{12}\left(\hat{x} + C_2\hat{D}^2 + C_3\hat{D}^3\right) \tag{4.293a}$$
$$\hat{L}_2 = \alpha_{21}\hat{D} + \alpha_{22}\left(\hat{x} + C_2\hat{D}^2 + C_3\hat{D}^3\right). \tag{4.293b}$$

More explicitly,

$$\hat{L}_1 = \cos(\theta)\frac{d}{dx} - \sin(\theta)\left(\hat{x} + C_2\frac{d^2}{dx^2} + C_3\frac{d^3}{dx^3}\right) \tag{4.294a}$$
$$\hat{L}_2 = \sin(\theta)\frac{d}{dx} + \cos(\theta)\left(\hat{x} + C_2\frac{d^2}{dx^2} + C_3\frac{d^3}{dx^3}\right). \tag{4.294b}$$

∎

The above considerations render $\hat{D}$ the coefficients $\cos(\theta)$, respectively, $\sin(\theta)$ while the higher orders of $\hat{D}$ can be chosen arbitrary, a fact which is manifested by the appearance of arbitrary coefficients C_2 and C_3. The following problem demonstrates that the preceding recipe allows for further generalization.

Problem: Consider,

$$\hat{L}_1 = \beta_{11}\frac{d}{dx} + \alpha_{11}\frac{d}{dx} + \alpha_{12}\hat{x} \qquad (4.295a)$$

$$\hat{L}_2 = \beta_{21}\frac{d}{dx} + \alpha_{21}\frac{d}{dx} + \alpha_{22}\hat{x}. \qquad (4.295b)$$

Determine the involved α- and β-parameters to satisfy $\left[\hat{L}_1, \hat{L}_2\right] = \hat{\mathbb{I}}$.

Solution:

$$\left[\hat{L}_1, \hat{L}_2\right] = \left[\beta_{11}\hat{D} + \alpha_{11}\hat{D} + \alpha_{12}\hat{x}, \ \beta_{21}\hat{D} + \alpha_{21}\hat{D} + \alpha_{22}\hat{x}\right]. \qquad (4.296)$$

Noting that $\left[\hat{D}, \hat{D}\right] = \hat{0}$ and $\left[\hat{x}, \hat{x}\right] = \hat{0}$,

$$\left[\hat{L}_1, \hat{L}_2\right] = \left[\beta_{11}\hat{D}, \alpha_{22}\hat{x}\right] + \left[\alpha_{11}\hat{D}, \alpha_{22}\hat{x}\right]$$
$$+ \left[\alpha_{12}\hat{x}, \beta_{21}\hat{D}\right] + \left[\alpha_{12}\hat{x}, \alpha_{21}\hat{D}\right]. \qquad (4.297)$$

Factoring out the constants,

$$\left[\hat{L}_1, \hat{L}_2\right] = \beta_{11}\alpha_{22}\left[\hat{D}, \hat{x}\right] + \alpha_{11}\alpha_{22}\left[\hat{D}, \hat{x}\right]$$
$$+ \alpha_{12}\beta_{21}\left[\hat{x}, \hat{D}\right] + \alpha_{12}\alpha_{21}\left[\hat{x}, \hat{D}\right]. \qquad (4.298)$$

Using the property $\left[\hat{x}, \hat{D}\right] = -\left[\hat{D}, \hat{x}\right]$ and organizing the terms,

$$\left[\hat{L}_1, \hat{L}_2\right] = (\beta_{11}\alpha_{22} - \alpha_{12}\beta_{21})\underbrace{\left[\hat{D}, \hat{x}\right]}_{\hat{\mathbb{I}}} + (\alpha_{11}\alpha_{22} - \alpha_{12}\alpha_{21})\underbrace{\left[\hat{D}, \hat{x}\right]}_{\hat{\mathbb{I}}}. \qquad (4.299)$$

Thus,

$$\left[\hat{L}_1, \hat{L}_2\right] = \underbrace{(\beta_{11}\alpha_{22} - \alpha_{12}\beta_{21})}_{0}\hat{\mathbb{I}} + \underbrace{(\alpha_{11}\alpha_{22} - \alpha_{12}\alpha_{21})}_{1}\hat{\mathbb{I}}. \qquad (4.300)$$

The determination of the α-coefficients remains unaffected:

$$\alpha_{11} = \cos(\theta) \qquad \alpha_{12} = -\sin(\theta)$$
$$\qquad (4.301)$$
$$\alpha_{21} = \sin(\theta) \qquad \alpha_{22} = \cos(\theta)$$

The determination of the β-coefficients follows the introduced simple recipe,

$$\underbrace{\beta_{11}}_{C_1\alpha_{12}}\alpha_{22} - \alpha_{12}\underbrace{\beta_{21}}_{C_1\alpha_{22}} = 0. \qquad (4.302)$$

Consequently,

$$\hat{L}_1 = C_1\alpha_{12}\hat{D} + \alpha_{11}\hat{D} + \alpha_{12}\hat{x} \qquad (4.303a)$$

$$\hat{L}_2 = C_1\alpha_{22}\hat{D} + \alpha_{21}\hat{D} + \alpha_{22}\hat{x}. \qquad (4.303b)$$

Rearranging,

$$\hat{L}_1 = \alpha_{11}\hat{D} + \alpha_{12}\left(\hat{x} + C_1\hat{D}\right) \qquad (4.304a)$$

$$\hat{L}_2 = \alpha_{21}\hat{D} + \alpha_{22}\left(\hat{x} + C_1\hat{D}\right). \qquad (4.304b)$$

More explicitly,

$$\hat{L}_1 = \cos(\theta)\frac{d}{dx} - \sin(\theta)\left(\hat{x} + C_1\frac{d}{dx}\right) \tag{4.305a}$$

$$\hat{L}_2 = \sin(\theta)\frac{d}{dx} + \cos(\theta)\left(\hat{x} + C_1\frac{d}{dx}\right). \tag{4.305b}$$

∎

Remark: Since adding a constant term, e.g., $C_0\hat{\mathbb{I}}$, to $\hat{L}_1$ and $\hat{L}_2$, does not affect the commutation relationship $\left[\hat{L}_1, \hat{L}_2\right] = \hat{\mathbb{I}}$, in virtue of the results obtained in the preceding problems,

$$\hat{L}_1 = \cos(\theta)\frac{d}{dx} - \sin(\theta)\left(\hat{x} + C_0\hat{\mathbb{I}} + C_1\frac{d}{dx} + C_2\frac{d^2}{dx^2} + C_3\frac{d^3}{dx^3}\right) \tag{4.306a}$$

$$\hat{L}_2 = \sin(\theta)\frac{d}{dx} + \cos(\theta)\left(\hat{x} + C_0\hat{\mathbb{I}} + C_1\frac{d}{dx} + C_2\frac{d^2}{dx^2} + C_3\frac{d^3}{dx^3}\right). \tag{4.306b}$$

Since C_0, C_1, C_2, and C_3 are arbitrary, from the above recipe can be conjectured that

$$\hat{L}_1 = \cos(\theta)\frac{d}{dx} - \sin(\theta)\left(\hat{x} + C_0\hat{\mathbb{I}} + C_1\frac{d}{dx} + C_2\frac{d^2}{dx^2} + C_3\frac{d^3}{dx^3} + \cdots + C_n\frac{d^n}{dx^n} + \cdots\right) \tag{4.307a}$$

$$\hat{L}_2 = \sin(\theta)\frac{d}{dx} + \cos(\theta)\left(\hat{x} + C_0\hat{\mathbb{I}} + C_1\frac{d}{dx} + C_2\frac{d^2}{dx^2} + C_3\frac{d^3}{dx^3} + \cdots + C_n\frac{d^n}{dx^n} + \cdots\right), \tag{4.307b}$$

having any finite or infinite number of terms, also satisfy the commutation property $\left[\hat{L}_1, \hat{L}_2\right] = \hat{\mathbb{I}}$. It is instructive to investigate the properties of

$$C_0\hat{\mathbb{I}} + C_1\frac{d}{dx} + C_2\frac{d^2}{dx^2} + C_3\frac{d^3}{dx^3} + \cdots + C_n\frac{d^n}{dx^n} + \cdots, \tag{4.308}$$

or, written more compactly, of

$$C_0\hat{\mathbb{I}} + C_1\hat{D} + C_2\hat{D}^2 + C_3\hat{D}^3 + \cdots + C_n\hat{D}^n + \cdots. \tag{4.309}$$

To this end, consider the Formal Taylor Series Expansion of an arbitrary univariate function $g(\xi)$,

$$g(\xi) = g(0) + g^{(1)}(0)\xi + \frac{1}{2!}g^{(2)}(0)\xi^2 + \frac{1}{3!}g^{(3)}(0)\xi^3 + \cdots + \frac{1}{n!}g^{(n)}(0)\xi^n + \cdots. \tag{4.310}$$

Since C_n in (4.309) are arbitrary, given any $g(\xi)$, they can be chosen to be

$$C_n = \frac{1}{n!}g^{(n)}(0). \tag{4.311}$$

Thus,

$$C_0\hat{\mathbb{I}} + C_1\hat{D} + C_2\hat{D}^2 + C_3\hat{D}^3 + \cdots + C_n\hat{D}^n + \cdots = g(\hat{D}). \tag{4.312}$$

□

The preceding remark leads to the following theorem.

Theorem: *Given* $g(\hat{D})$, *the operators*

$$\hat{L}_1 = \cos(\theta)\hat{D} - \sin(\theta)\left(\hat{x} + g(\hat{D})\right) \tag{4.313a}$$

$$\hat{L}_2 = \sin(\theta)\hat{D} + \cos(\theta)\left(\hat{x} + g(\hat{D})\right), \tag{4.313b}$$

satisfy the commutation relationship $\left[\hat{L}_1, \hat{L}_2\right] = \hat{\mathbb{I}}$.

Proof:

$$\left[\hat{L}_1.\hat{L}_2\right] = \left[\cos(\theta)\hat{D} - \sin(\theta)\left(\hat{x} + g(\hat{D})\right) , \ \sin(\theta)\hat{D} + \cos(\theta)\left(\hat{x} + g(\hat{D})\right)\right] \tag{4.314a}$$

$$= \cos(\theta)\sin(\theta)\underbrace{\left[\hat{D},\hat{D}\right]}_{\hat{0}} + \cos^2(\theta)\underbrace{\left[\hat{D},\hat{x} + g(\hat{D})\right]}_{[\hat{D},\hat{x}]}$$

$$- \sin^2(\theta)\underbrace{\left[\hat{x} + g(\hat{D}),\hat{D}\right]}_{[\hat{x},\hat{D}]} - \sin(\theta)\cos(\theta)\underbrace{\left[\hat{x} + g(\hat{D}),\hat{x} + g(\hat{D})\right]}_{\hat{0}} \tag{4.314b}$$

$$= \cos^2(\theta)\left[\hat{D},\hat{x}\right] - \sin^2(\theta)\underbrace{\left[\hat{x},\hat{D}\right]}_{-[\hat{D},\hat{x}]} \tag{4.314c}$$

$$= \left(\cos^2(\theta) + \sin^2(\theta)\right)\underbrace{\left[\hat{D},\hat{x}\right]}_{\hat{\mathbb{I}}} \tag{4.314d}$$

$$= \hat{\mathbb{I}}. \tag{4.314e}$$

∎

Analogously, consider the following type of problems.

Problem: Consider,

$$\hat{L}_1 = \alpha_{11}\frac{d}{dx} + \alpha_{12}\hat{x} + \gamma_{10}\hat{\mathbb{I}} + \gamma_{11}\hat{x} + \gamma_{12}\hat{x}^2 \tag{4.315a}$$

$$\hat{L}_2 = \alpha_{21}\frac{d}{dx} + \alpha_{22}\hat{x} + \gamma_{20}\hat{\mathbb{I}} + \gamma_{21}\hat{x} + \gamma_{22}\hat{x}^2. \tag{4.315b}$$

Determine the involved α- and γ-parameters to satisfy $\left[\hat{L}_1, \hat{L}_2\right] = \hat{\mathbb{I}}$.

Solution:

$$\left[\hat{L}_1,\hat{L}_2\right] = \left[\alpha_{11}\hat{D} + \alpha_{12}\hat{x} + \gamma_{10}\hat{\mathbb{I}} + \gamma_{11}\hat{x} + \gamma_{12}\hat{x}^2, \right.$$
$$\left. \alpha_{21}\hat{D} + \alpha_{22}\hat{x} + \gamma_{20}\hat{\mathbb{I}} + \gamma_{21}\hat{x} + \gamma_{22}\hat{x}^2\right] \tag{4.316a}$$

$$= \alpha_{11}\alpha_{21}\underbrace{\left[\hat{D},\hat{D}\right]} + \alpha_{11}\alpha_{22}\left[\hat{D},\hat{x}\right] + \alpha_{11}\gamma_{20}\underbrace{\left[\hat{D},\hat{\mathbb{I}}\right]} + \alpha_{11}\gamma_{21}\left[\hat{D},\hat{x}\right] + \alpha_{11}\gamma_{22}\left[\hat{D},\hat{x}^2\right]$$

$$+ \alpha_{12}\alpha_{21}\left[\hat{x},\hat{D}\right] + \alpha_{12}\alpha_{22}\underbrace{[\hat{x},\hat{x}]} + \alpha_{12}\gamma_{20}\underbrace{\left[\hat{x},\hat{\mathbb{I}}\right]} + \alpha_{12}\gamma_{21}\underbrace{[\hat{x},\hat{x}]} + \alpha_{12}\gamma_{22}[\hat{x},\hat{x}^2]$$

$$+ \gamma_{10}\alpha_{21}\underbrace{\left[\hat{\mathbb{I}},\hat{D}\right]} + \gamma_{10}\alpha_{22}\underbrace{\left[\hat{\mathbb{I}},\hat{x}\right]} + \gamma_{10}\gamma_{20}\underbrace{\left[\hat{\mathbb{I}},\hat{\mathbb{I}}\right]} + \gamma_{10}\gamma_{21}\underbrace{\left[\hat{\mathbb{I}},\hat{x}\right]} + \gamma_{10}\gamma_{22}\underbrace{\left[\hat{\mathbb{I}},\hat{x}^2\right]}$$

$$+ \gamma_{11}\alpha_{21}\left[\hat{x},\hat{D}\right] + \gamma_{11}\alpha_{22}\underbrace{[\hat{x},\hat{x}]} + \gamma_{11}\gamma_{20}\underbrace{\left[\hat{x},\hat{\mathbb{I}}\right]} + \gamma_{11}\gamma_{21}\underbrace{[\hat{x},\hat{x}]} + \gamma_{11}\gamma_{22}[\hat{x},\hat{x}^2]$$

$$+ \gamma_{12}\alpha_{21}\left[\hat{x}^2,\hat{D}\right] + \gamma_{12}\alpha_{22}[\hat{x}^2,\hat{x}] + \gamma_{12}\gamma_{20}\underbrace{\left[\hat{x}^2,\hat{\mathbb{I}}\right]} + \gamma_{12}\gamma_{21}[\hat{x}^2,\hat{x}] + \gamma_{12}\gamma_{22}[\hat{x}^2,\hat{x}^2]. \tag{4.316b}$$

The under-braced terms vanish, and the R.H.S. in (4.316b) simplifies greatly. Nonetheless, all terms will still be considered for possible exploitation of symmetry relationships. Considering, the equality $\left[\hat{A}, \hat{B}\right] = -\left[\hat{B}, \hat{A}\right]$, in the "lower triangle" entries,

$$
\begin{aligned}
\left[\hat{L}_1, \hat{L}_2\right] = {} & \alpha_{11}\alpha_{21}\underbrace{\left[\hat{D}, \hat{D}\right]} + \alpha_{11}\alpha_{22}\left[\hat{D}, \hat{x}\right] + \alpha_{11}\gamma_{20}\underbrace{\left[\hat{D}, \hat{\mathbb{I}}\right]} + \alpha_{11}\gamma_{21}\left[\hat{D}, \hat{x}\right] + \alpha_{11}\gamma_{22}\left[\hat{D}, \hat{x}^2\right] \\
& - \alpha_{12}\alpha_{21}\left[\hat{D}, \hat{x}\right] + \alpha_{12}\alpha_{22}\underbrace{\left[\hat{x}, \hat{x}\right]} + \alpha_{12}\gamma_{20}\underbrace{\left[\hat{x}, \hat{\mathbb{I}}\right]} + \alpha_{12}\gamma_{21}\underbrace{\left[\hat{x}, \hat{x}\right]} + \alpha_{12}\gamma_{22}\underbrace{\left[\hat{x}, \hat{x}^2\right]} \\
& - \gamma_{10}\alpha_{21}\underbrace{\left[\hat{D}, \hat{\mathbb{I}}\right]} - \gamma_{10}\alpha_{22}\underbrace{\left[\hat{x}, \hat{\mathbb{I}}\right]} + \gamma_{10}\gamma_{20}\underbrace{\left[\hat{\mathbb{I}}, \hat{\mathbb{I}}\right]} + \gamma_{10}\gamma_{21}\underbrace{\left[\hat{\mathbb{I}}, \hat{x}\right]} + \gamma_{10}\gamma_{22}\underbrace{\left[\hat{\mathbb{I}}, \hat{x}^2\right]} \\
& - \gamma_{11}\alpha_{21}\left[\hat{D}, \hat{x}\right] - \gamma_{11}\alpha_{22}\underbrace{\left[\hat{x}, \hat{x}\right]} - \gamma_{11}\gamma_{20}\underbrace{\left[\hat{\mathbb{I}}, \hat{x}\right]} + \gamma_{11}\gamma_{21}\underbrace{\left[\hat{x}, \hat{x}\right]} + \gamma_{11}\gamma_{22}\underbrace{\left[\hat{x}, \hat{x}^2\right]} \\
& - \gamma_{12}\alpha_{21}\left[\hat{D}, \hat{x}^2\right] - \gamma_{12}\alpha_{22}\underbrace{\left[\hat{x}, \hat{x}^2\right]} - \gamma_{12}\gamma_{20}\underbrace{\left[\hat{\mathbb{I}}, \hat{x}^2\right]} - \gamma_{12}\gamma_{21}\underbrace{\left[\hat{x}, \hat{x}^2\right]} + \gamma_{12}\gamma_{22}\underbrace{\left[\hat{x}^2, \hat{x}^2\right]}.
\end{aligned}
\tag{4.317}
$$

Considering the non-vanishing terms in the first row and in the first column and choosing,

$$\alpha_{11}\alpha_{22} - \alpha_{12}\alpha_{21} = 1 \tag{4.318a}$$

$$\alpha_{11}\gamma_{20} - \gamma_{10}\alpha_{21} = 0 \tag{4.318b}$$

$$\alpha_{11}\gamma_{21} - \gamma_{11}\alpha_{21} = 0 \tag{4.318c}$$

$$\alpha_{11}\gamma_{22} - \gamma_{12}\alpha_{21} = 0 \tag{4.318d}$$

Equation (4.318a) implies,

$$
\begin{pmatrix} \alpha_{11} & \alpha_{12} \\ \alpha_{21} & \alpha_{22} \end{pmatrix} = \begin{pmatrix} \cos(\theta) & -\sin(\theta) \\ \sin(\theta) & \cos(\theta) \end{pmatrix}.
\tag{4.319}
$$

The remaining three equations are of the same kind. Substituting for α_{11} and α_{12}, into (4.318b), e.g.,

$$\cos(\theta)\underbrace{\gamma_{20}}_{C_0\sin(\theta)} = \underbrace{\gamma_{10}}_{C_0\cos(\theta)}\sin(\theta). \tag{4.320}$$

As indicated, the solutions

$$\gamma_{10} = C_0\cos(\theta) \tag{4.321a}$$

$$\gamma_{20} = C_0\sin(\theta), \tag{4.321b}$$

are valid for arbitrary C_0 and any choice of θ.

Similarly, using (4.318c),

$$\alpha_{11}\gamma_{21} - \gamma_{11}\alpha_{21} = 0 \implies \cos(\theta)\gamma_{21} = \gamma_{11}\sin(\theta) \implies \begin{cases} \gamma_{11} = C_1\cos(\theta) \\ \gamma_{21} = C_1\sin(\theta) \end{cases}. \tag{4.322}$$

Finally, employing (4.318d),

$$\alpha_{11}\gamma_{22} - \gamma_{12}\alpha_{21} = 0 \implies \cos(\theta)\gamma_{22} = \gamma_{12}\sin(\theta) \implies \begin{cases} \gamma_{12} = C_2\cos(\theta) \\ \gamma_{22} = C_2\sin(\theta) \end{cases}. \tag{4.323}$$

Substituting the α- and γ-coefficients into (4.315),

$$\hat{L}_1 = \cos(\theta)\frac{d}{dx} - \sin(\theta)\hat{x} + C_0\cos(\theta)\hat{\mathbb{I}} + C_1\cos(\theta)\hat{x} + C_2\cos(\theta)\hat{x}^2 \tag{4.324a}$$

$$\hat{L}_2 = \sin(\theta)\frac{d}{dx} + \cos(\theta)\hat{x} + C_0\sin(\theta)\hat{\mathbb{I}} + C_1\sin(\theta)\hat{x} + C_2\sin(\theta)\hat{x}^2. \tag{4.324b}$$

Summarizing terms,

$$\hat{L}_1 = \cos(\theta)\left(\frac{d}{dx} + C_0\hat{\mathbb{I}} + C_1\hat{x} + C_2\hat{x}^2\right) - \sin(\theta)\hat{x} \tag{4.325a}$$

$$\hat{L}_2 = \sin(\theta)\left(\frac{d}{dx} + C_0\hat{\mathbb{I}} + C_1\hat{x} + C_2\hat{x}^2\right) + \cos(\theta)\hat{x}. \tag{4.325b}$$

An argument along the lines in the preceding problem enables the generalization of (4.325) to arbitrary finite or infinite number of terms,

$$\hat{L}_1 = \cos(\theta)\left(\frac{d}{dx} + C_0\hat{\mathbb{I}} + C_1\hat{x} + C_2\hat{x}^2 + \cdots + C_n\hat{x}^n + \cdots\right) - \sin(\theta)\hat{x} \tag{4.326a}$$

$$\hat{L}_2 = \sin(\theta)\left(\frac{d}{dx} + C_0\hat{\mathbb{I}} + C_1\hat{x} + C_2\hat{x}^2 + \cdots + C_n\hat{x}^n + \cdots\right) + \cos(\theta)\hat{x}. \tag{4.326b}$$

∎

This result inspires the formulation of the following theorem.

Theorem: *Given $f(\hat{x})$, the operators*

$$\hat{L}_1 = \cos(\theta)\left(\hat{D} + f(\hat{x})\right) - \sin(\theta)\left(\hat{x}\right) \tag{4.327a}$$

$$\hat{L}_2 = \sin(\theta)\left(\hat{D} + f(\hat{x})\right) + \cos(\theta)\left(\hat{x}\right), \tag{4.327b}$$

satisfy the commutation relationship $\left[\hat{L}_1, \hat{L}_2\right] = \hat{\mathbb{I}}$.

Proof:

$$\left[\hat{L}_1.\hat{L}_2\right] = \left[\cos(\theta)\left(\hat{D}+f(\hat{x})\right) - \sin(\theta)\left(\hat{x}\right) , \sin(\theta)\left(\hat{D}+f(\hat{x})\right) + \cos(\theta)\left(\hat{x}\right)\right] \tag{4.328a}$$

$$= \cos(\theta)\sin(\theta)\underbrace{\left[\hat{D}+f(\hat{x}),\hat{D}+f(\hat{x})\right]}_{\hat{0}} + \cos^2(\theta)\underbrace{\left[\hat{D}+f(\hat{x}),\hat{x}\right]}_{[\hat{D},\hat{x}]}$$

$$- \sin^2(\theta)\underbrace{\left[\hat{x},\hat{D}+f(\hat{x})\right]}_{[\hat{x},\hat{D}]} - \sin(\theta)\cos(\theta)\underbrace{\left[\hat{x},\hat{x}\right]}_{\hat{0}} \tag{4.328b}$$

$$= \cos^2(\theta)\left[\hat{D},\hat{x}\right] - \sin^2(\theta)\underbrace{\left[\hat{x},\hat{D}\right]}_{-[\hat{D},\hat{x}]} \tag{4.328c}$$

$$= \left(\cos^2(\theta) + \sin^2(\theta)\right)\underbrace{\left[\hat{D},\hat{x}\right]}_{\hat{\mathbb{I}}} \tag{4.328d}$$

$$= \hat{\mathbb{I}}. \tag{4.328e}$$

∎

The results obtained in the preceding problems, evaluated in the special case $\theta = \frac{\pi}{2}$, have been summarized in the following theorem.

Theorem: *Given arbitrary functions $f(\hat{x})$ and $g(\hat{D})$, the canonical commutation relationships,*

$$\left[\hat{L}_1^0, \hat{L}_2^0\right] = \hat{\mathbb{I}} \tag{4.329a}$$

$$\left[\hat{L}_1^a, \hat{L}_2^a\right] = \hat{\mathbb{I}} \tag{4.329b}$$

$$\left[\hat{L}_1^b, \hat{L}_2^b\right] = \hat{\mathbb{I}}, \tag{4.329c}$$

with

$$\hat{L}_1^0 = \frac{1}{\sqrt{2}}\left(\hat{D} - \hat{x}\right) \tag{4.330a}$$

$$\hat{L}_2^0 = \frac{1}{\sqrt{2}}\left(\hat{D} + \hat{x}\right), \tag{4.330b}$$

$$\hat{L}_1^a = \frac{1}{\sqrt{2}}\left(\hat{D} + f(\hat{x}) - \hat{x}\right) \tag{4.331a}$$

$$\hat{L}_2^a = \frac{1}{\sqrt{2}}\left(\hat{D} + f(\hat{x}) + \hat{x}\right), \tag{4.331b}$$

$$\hat{L}_1^b = \frac{1}{\sqrt{2}}\left(\hat{D} - \hat{x} - g(\hat{D})\right) \tag{4.332a}$$

$$\hat{L}_2^b = \frac{1}{\sqrt{2}}\left(\hat{D} + \hat{x} + g(\hat{D})\right). \tag{4.332b}$$

Proof: The proof of $\left[\hat{L}_1^0, \hat{L}_2^0\right] = \hat{\mathbb{I}}$:

$$\left[\hat{L}_1^0, \hat{L}_2^0\right] = \left[\frac{1}{\sqrt{2}}\left(\hat{D} - \hat{x}\right), \frac{1}{\sqrt{2}}\left(\hat{D} + \hat{x}\right)\right] \tag{4.333a}$$

$$= \frac{1}{2}\left[\hat{D} - \hat{x}, \hat{D} + \hat{x}\right] \tag{4.333b}$$

$$= \frac{1}{2}\{\underbrace{\left[\hat{D}, \hat{D}\right]}_{\hat{0}} + \underbrace{\left[\hat{D}, \hat{x}\right] - \left[\hat{x}, \hat{D}\right]}_{-[\hat{D},\hat{x}]} - \underbrace{\left[\hat{x}, \hat{x}\right]}_{\hat{0}}\} \tag{4.333c}$$

$$= \hat{\mathbb{I}}. \tag{4.333d}$$

The proof of $\left[\hat{L}_1^a, \hat{L}_2^a\right] = \hat{\mathbb{I}}$:

$$\left[\hat{L}_1^a, \hat{L}_2^a\right] = \left[\frac{1}{\sqrt{2}}\left(\hat{D}+f(\hat{x})-\hat{x}\right), \frac{1}{\sqrt{2}}\left(\hat{D}+f(\hat{x})+\hat{x}\right)\right] \tag{4.334a}$$

$$= \frac{1}{2}\left[\hat{D}+f(\hat{x})-\hat{x}, \hat{D}+f(\hat{x})+\hat{x}\right] \tag{4.334b}$$

$$= \frac{1}{2}\{\underbrace{\left[\hat{D},\hat{D}\right]}+\left[\hat{D},f(\hat{x})\right]+\left[\hat{D},\hat{x}\right]$$

$$+\left[f(\hat{x}),\hat{D}\right]+\underbrace{\left[f(\hat{x}),f(\hat{x})\right]}+\underbrace{\left[f(\hat{x}),\hat{x}\right]}$$

$$-\left[\hat{x},\hat{D}\right]-\underbrace{\left[\hat{x},f(\hat{x})\right]}-\underbrace{\left[\hat{x},\hat{x}\right]}\}. \tag{4.334c}$$

Noting that the under-braced terms vanish,

$$\left[\hat{L}_1^a, \hat{L}_2^a\right] = \frac{1}{2}\{\left[\hat{D},f(\hat{x})\right]+\left[\hat{D},\hat{x}\right]+\underbrace{\left[f(\hat{x}),\hat{D}\right]}_{-[\hat{D},f(\hat{x})]}-\underbrace{\left[\hat{x},\hat{D}\right]}_{-[\hat{D},\hat{x}]}\} \tag{4.335a}$$

$$= \hat{\mathbb{I}}. \tag{4.335b}$$

The proof of $\left[\hat{L}_1^b, \hat{L}_2^b\right] = \hat{\mathbb{I}}$:

$$\left[\hat{L}_1^b, \hat{L}_2^b\right] = \left[\frac{1}{\sqrt{2}}\left(\hat{D}-\hat{x}-g(\hat{D})\right), \frac{1}{\sqrt{2}}\left(\hat{D}+\hat{x}+g(\hat{D})\right)\right] \tag{4.336a}$$

$$= \frac{1}{2}\left[\hat{D}-\hat{x}-g(\hat{D}), \hat{D}+\hat{x}+g(\hat{D})\right] \tag{4.336b}$$

$$= \frac{1}{2}\{\underbrace{\left[\hat{D},\hat{D}\right]}+\left[\hat{D},\hat{x}\right]+\underbrace{\left[\hat{D},g(\hat{D})\right]}$$

$$-\left[\hat{x},\hat{D}\right]-\underbrace{\left[\hat{x},\hat{x}\right]}-\left[\hat{x},g(\hat{D})\right]$$

$$-\underbrace{\left[g(\hat{D}),\hat{D}\right]}-\left[g(\hat{D}),\hat{x}\right]-\underbrace{\left[g(\hat{D}),g(\hat{D})\right]}\}. \tag{4.336c}$$

Noting that the under-braced terms vanish,

$$\left[\hat{L}_1^b, \hat{L}_2^b\right] = \frac{1}{2}\{\underbrace{\left[\hat{D},\hat{x}\right]}-\left[\hat{x},\hat{D}\right]-\underbrace{\left[\hat{x},g(\hat{D})\right]}_{-[g(\hat{D}),\hat{x}]}-\left[g(\hat{D}),\hat{x}\right]\} \tag{4.337a}$$

$$= \hat{\mathbb{I}}. \tag{4.337b}$$

4.10 Concluding remarks

Expressing the differential operator $\hat{D}$ (standing for d/dx) in terms of integrals (involving representations of Dirac delta functions) offers unique possibilities to redefine how derivatives are conceived in numerical analyses. In the course of discussion, the concept of evaluating a function at a given point was made formally rigorous. After thoroughly discussing the relationships between the position operator $\hat{x}$ and the differential operator $\hat{D}$, the annihilation and creation operators and their generalizations were introduced in a variety of ways. As it is expected from any *bona fide* generalization, the introduced generalized annihilation and creation operators included the standard annihilation operator $\hat{b}$ and the standard creation operator $\hat{b}^{\dagger}$, as special cases. The Chapter 1 of Volume 2 reintroduces the standard annihilation operator $\hat{b}$ and the standard creation operator $\hat{b}^{\dagger}$, by considering the Schrödinger equation for solving the quantum harmonic oscillator problem. Thereby, perturbed and squeezed harmonic oscillators will also be introduced.

A fundamental question which is not treated in this book is the potential utilization of the introduced generalized annihilation and creation operators for designing optimally localized states (frames and dual frames). This problem offers itself as an excellent starting point for a rigorous investigation of the properties of the associated eigenfunctions of the generalized annihilation operator. Unpacking and solving this problem may offer a means for extending the notions of the standard coherent states and squeezed states from an engineering perspective. The amalgamation of ideas presented in this chapter with the comparatively modern tools in signal processing since 1985s (wavelets and dual wavelets, frames and dual frames, and allied topics) might build a fertile ground for constructing new generations of problem-tailored basis and dual basis functions. Deep insights and versatile tools developed in digital filter design combined with generalizations presented in this chapter promise to open up a new field of research and inquiry.

In the references [1–8], a sample of this author's works on the design of problem-specific Dirac delta functions have been provided. These works and the references therein describe methods for the construction of Dirac delta functions originating from the governing partial differential equations (PDEs) in mathematical physics. The techniques are built upon the so-called diagonalization and supplementation of PDEs in mathematical physics. The interested reader might find further details concerning the calculations in this chapter which had to be omitted. The comprehensive compilation of these results is the subject matter of an ongoing independent work.

References

[1] Baghai-Wadji A., 3D Electrostatic charge distribution on finitely-thick bus-bars in micro-acoustic devices: combined regularization in the Near- and far-field, *IEEE-TUFFC Transactions, Ultrasonics, Ferroelectrics and Frequency Control Transactions*, Special Issue, vol. 62, no. 6, pp. 1132–1144, 2015.

[2] Baghai-Wadji A., Dyadic universal functions and simultaneous near-field/far-field regularization of elasto-dynamic dyadic Green's functions for 3-D mass-loading analysis in microacoustic devices, *IEEE-TUFFC Transactions, Ultrasonics, Ferroelectrics and Frequency Control Transactions*, Special Issue, vol. 63, no. 10, pp. 1563–1575, 2016.

[3] Baghai-Wadji A., Boundary element method applied to micro-acoustic devices: zooming into the near-field, in: K. Nakamura (Ed.), *Ultrasonic Transducers: Materials Design for Sensors, Actuators and Medical Applications*, Woodhead Publishing, pp. 220–263, 2012.

[4] Baghai-Wadji A., Physics-based performance enhancement in computational electromagnetics: a review, in: *The 2015 IEEE Symposium Series on Computational Intelligence*, Cape Town, 2015, pp. 1190–1119.

[5] Baghai-Wadji A, D-Theorem (on regularization): Green's function-induced distributed elementary sources – first kind, in: *The 2014 IEEE International Symposium on Antennas and Propagation and USNC-URSI Radio Science Meeting*, Memphis, TN, USA, 6–12 July, 2014

[6] Baghai-Wadji A., S-Theorem (on regularization): Green's function-induced distributed elementary sources – second kind, in: *The 2014 IEEE International Symposium on Antennas and Propagation and USNC-URSI Radio Science Meeting*, Memphis, TN, USA, 6–12 July, 2014.

[7] Baghai-Wadji A., Self-consistent physics-based δ_η-regularized Green's function for 2D Poisson's equation in anisotropic dielectric media, in: *The Annual Review of Progress in Applied Computational Electromagnetics*, Florida, USA, March 2014.

[8] Baghai-Wadji A., Self-consistent physics-based δ_η-regularized Green's function for 3D Poisson's equation in anisotropic dielectric media, in: *The Annual Review of Progress in Applied Computational Electromagnetics*, Florida, USA, March 2014.

Index

www.ingramcontent.com/pod-product-compliance
Lightning Source LLC
Chambersburg PA
CBHW082010190326
41458CB00010B/3142

* 9 7 8 1 8 3 9 5 3 8 6 6 7 *